Samy A. Madbouly, Chaoqun Zhang (Eds.)
Biopolymers and Composites

Also of interest

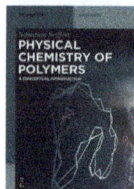

Physical Chemistry of Polymers
A Conceptual Introduction
Sebastian Seiffert, 2020
ISBN 978-3-11-067280-0, e-ISBN 978-3-11-067281-7

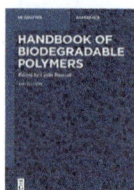

Handbook of Biodegradable Polymers
3rd Edition
Catia Bastioli (Ed.), 2020
ISBN 978-1-5015-1921-5, e-ISBN 978-1-5015-1196-7

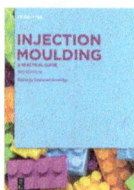

Injection Moulding
A Practical Guide
3rd Edition
Vanessa Goodship (Ed.), 2020
ISBN 978-3-11-065302-1, e-ISBN 978-3-11-065481-3

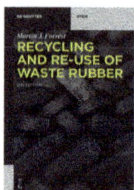

Recycling and Re-use of Waste Rubber
Martin J. Forrest, 2019
ISBN 978-3-11-064400-5, e-ISBN 978-3-11-064414-2

Physical Sciences Reviews
e-ISSN 2365-659X

Biopolymers and Composites

Processing and Characterization

Edited by
Samy A. Madbouly and Chaoqun Zhang

DE GRUYTER

Editor
Prof. Samy A. Madbouly
Pennsylvania State University
Behrend College
School of Engineering
4701 College Drive
Erie 16563
USA
sum1541@psu.edu

Prof. Chaoqun Zhang
South China Agricultural University
Key Laboratory for Biobased Materials
Energy of Ministry of Education
College of Materials and Energy
510642 Guangzhou
China
zhangcq@scau.edu.cn

ISBN 978-1-5015-2193-5
e-ISBN (PDF) 978-1-5015-2194-2
e-ISBN (EPUB) 978-1-5015-1539-2

Library of Congress Control Number: 2021940767

Bibliographic information published by the Deutsche Nationalbibliothek
The Deutsche Nationalbibliothek lists this publication in the Deutsche Nationalbibliografie; detailed bibliographic data are available on the Internet at http://dnb.dnb.de.

© 2021 Walter de Gruyter GmbH, Berlin/Boston
Cover image: XXLPhoto/iStock/Getty Images Plus
Typesetting: TNQ Technologies Pvt. Ltd.
Printing and binding: CPI books GmbH, Leck

www.degruyter.com

Contents

Guoqiang Zhu, Chengguo Liu and Chaoqun Zhang

Xing Zhou, Xin Zhang, Pu Mengyuan, Xinyu He, and Chaoqun Zhang

Samy Madbouly, Sean Edlis and Nicolas Ionadi

James Goodsel and Samy Madbouly

Samy A. Madbouly

List of contributing authors

Tanner Joseph Alauzen
Penn State Behrend
Plastics Engineering Technology
4701 College Drive, Erie, Pennsylvania
16563-4117
United States
tja5350@psu.edu

Emily Jean Archer
Penn State Behrend
Plastics Engineering Technology
4701 College Drive, Erie, Pennsylvania
16563-4117
United States
eja5296@psu.edu

Souvik Banerjee
Iowa State University
Materials Science and Engineering
Ames, Iowa
United States
sbanz@iastate.edu

Ty Burford
Penn State Erie The Behrend College School of
Engineering
Erie, Pennsylvania
United States
ty.burford98@gmail.com

John Coulter
Lehigh University
Mechanical Engineering and Mechanics
Bethlehem, Pennsylvania
United States
jc0i@lehigh.edu

Chuanshen Du
Iowa State University
Materials Science and Engineering
Ames, Iowa
United States
cdu11@iastate.edu

Sean P. Edlis
Penn State Behrend
Plastics Engineering Technology
4701 College Drive, Erie, Pennsylvania
16563-4117
United States
spe5141@psu.edu

Qi Fan
South China Agricultural University
Guangzhou, Guangdong
China
fanqiscience@126.com

James Goodsel
Penn State Behrend Erie, Pennsylvania
United States
jpg5738@psu.edu

Paul R. Gregory
Iowa State University
Materials Science and Engineering
Ames, Iowa
United States
paulgreg@iastate.edu

Yaya Hao
Xi'an University of Technology
Faculty of Printing, Packaging Engineering and
Digital Media Technology
Xi'an, Shaanxi
China
Hhaoyaya@163.com

Xinyu He
Xi'an University of Technology
Faculty of Printing, Packaging Engineering and
Digital Media Technology
Xi'an, Shaanxi
China
hxy5722@163.com

https://doi.org/10.1515/9781501521942-201

Nicolas Ionadi
Penn State Behrend
Plastics Engineering Technology
4701 College Drive, Erie, Pennsylvania
16563-4117
United States
nji5042@psu.edu

Sabrina Jedlicka
Lehigh University
Mechanical Engineering and Mechanics
Bethlehem, Pennsylvania
United States
ssj207@lehigh.edu

Medhat S. Farahat Khedr
Israd Hakim Jaafar
Utah Valley University
Mechanical Engineering
800 W University Parkway Orem, Orem, Utah
84058
United States
Israd.Jaafar@uvu.edu

Chengguo Liu
Institute of Chemical Industry of Forest Products
Nanjing, Jiangsu
China
liuchengguo@icifp.cn

Samy A. Madbouly
Penn State Behrend
Plastics Engineering Technology
4701 College Drive, Erie, Pennsylvania
16563-4117
United States
sum1541@psu.edu

Rongxian Ou
South China Agricultural University
Guangzhou, Guangdong
China
ourongxianvx@163.com

Mengyuan Pu
Xi'an University of Technology
Faculty of Printing, Packaging Engineering and
Digital Media Technology
Xi'an, Shaanxi
China
pumengyuan37@163.com

William Rieg
Penn State Erie The Behrend College School of
Engineering
Erie, Pennsylvania
United States
wsr5062@psu.edu

Shaelyn Ross
Penn State Behrend
Plastics Engineering Technology
4701 College Drive, Erie, Pennsylvania
16563-4117
United States
srr5433@psu.edu

Marissa Torretti
Penn State Behrend
Plastics Engineering Technology
Erie, Pennsylvania
United States
mlt37@psu.edu

Martin Thou
Iowa State University
Materials Science and Engineering
Ames, Iowa
United States
mthuo@iastate.edu

Marissa Torretti
Penn State Behrend
Plastics Engineering Technology
4701 College Drive, Erie, Pennsylvania
16563-4117
United States
mlt37@psu.edu

Qingwen Wang
South China Agricultural University
Guangzhou, Guangdong
China
weijun.yang@jiangnan.edu.cn

Weijun Yang
South China Agricultural University
Guangzhou, Guangdong
China
qwwang@scau.edu.cn

Chaoqun Zhang
South China Agricultural University
College of Materials and Energy
Guangzhou, Guangdong
China
zhangcq@scau.edu.cn

Xin Zhang
Xi'an University of Technology
Faculty of Printing, Packaging Engineering and
Digital Media Technology
Xi'an, Shaanxi
China
692794038@qq.com

Xing Zhou
Xi'an University of Technology
Faculty of Printing, Packaging Engineering and
Digital Media Technology
Xi'an, Shaanxi
China
zdxnlxaut@163.com

Guogiang Zhu
Institute of Chemical Industry of Forest Products
Nanjing, Jiangsu
China
18354256181@163.com

Paul Gregory, Souvik Banerjee, Chuanshen Du and Martin Thuo*

1 Introduction: biopolymers and biocomposites

Abstract: Biopolymers and biocomposites are an exciting class of ubiquitous materials. Interest in these materials has been driven in part by their biocompatibility/biodegradability, sustainability, potentially low-cost, renewability, being environmental benign, among other properties. These fascinating materials come in a range of forms from the DNA and RNA that is essential to life to the cellulose and collagen that mechanically reinforce tissues and as hybrid organic–inorganic composites like teeth. Herein, we summarize some aspects of the two classes of materials biopolymer and biocomposites, exploring specific examples while pointing to potential monomer sources, neoteric post-extraction modification and processing conditions. This lays the foundation to the following more specific chapters while illustrating the breadth of these material classes.

Keywords: biocomposites; biodegradation; biomimetic; biopolymers; cellulose; processing.

1.1 Introduction

Globalization, industrialization, a perilous planet, a rising population and concomitant advances in flexible/wearable devices are merely a few of the many reasons that has led to a heightened demand for plastics and/or flexible materials. With a chemical industry that is largely petroleum driven, synthetic plastics that are rapid to produce in large quantities have met the high demand for single use lightweight materials across disparate uses. This appetite for petroleum-based plastics, which are predominantly nonbiodegradable, has led to untenable environmental consequences. Most prominent is the effect of microplastics and their distribution across numerous ecosystems effecting both flora and fauna, and specifically effecting the largest and most diverse ecosystem, the ocean. To mitigate some of these challenges and to further the goal of green economies, which includes buffering against climate change, there is an urgent need for alternative materials or approaches to current petroleum based plastic uses. The principles of green chemistry, sustainability, eco-friendly engineering, and bio-alternatives are among the primary paths being explored. Biopolymers have emerged as a feasible alternative largely due to their ability to feed into existing renewable carbon or nitrogen cycles. As summarized herein, biopolymers can be sourced from numerous paths including lignin (chapter 2), sugars

*Corresponding author: **Martin Thuo**, Department of Materials Science and Engineering, Iowa State University, Ames, IA, USA; and Micro-Electronics Research Center, Ames, IA, USA; and Department of Electrical and Computer Engineering, Iowa State University, Ames, IA, USA, E-mail: mthuo@iastate.edu
Paul Gregory, Souvik Banerjee and Chuanshen Du, Department of Materials Science and Engineering, Iowa State University, Ames, IA, USA

This article has previously been published in the journal Physical Sciences Reviews. Please cite as: P. Gregory, S. Banerjee, C. Du and M. Thuo "Introduction: biopolymers and biocomposites" *Physical Sciences Reviews* [Online] 2021. DOI: 10.1515/psr-2020-0065 | https://doi.org/10.1515/9781501521942-001

(chapter 3 and 4) or plant oils (chapter 6). Unsurprisingly, the large number of sources give rise to an equally large number of applications (chapter 11 and 13). This wide range of applications is due to biopolymers range of diversity and complexity. For example, plant-oil-derived polymers are dominated by glycerides and analogous phospholipids (Figure 1.1) although a more elaborate, albeit low yielding or more volatile, alternatives exist in essential oils. Potential complexity in monomer pools that can be derived from the polyketide and/or shikimate biosynthesis pathways is immense, especially if hybrids from the glyceride and terpenoid families are developed. The presence of common moieties across these two classes of plant oils implies that analogous reactions can be adopted in a co-polymerization to increase the complexity of obtained polymers. Additionally, the C_5 periodicity in essential oils (terpenoids) implies that the structure of the resulting polymers can be readily evolved akin to a geometric progression, such progression that would not be possible in glyceride derived units.

Here in we provide a general overview of biopolymers, first looking at their basic classifications based on source, market shares, and potential uses, we then explore bio-composites giving examples from nature and conclude by highlighting their processing. Neoteric approaches that are under-development are also included to highlight the potential of these materials beyond current industry standards.

1.1.1 Classification of biopolymers

Biopolymers are a class of polymeric materials derived from flora and/or fauna and their synthetic analogs. Naturally, these materials are made by cells of the requisite

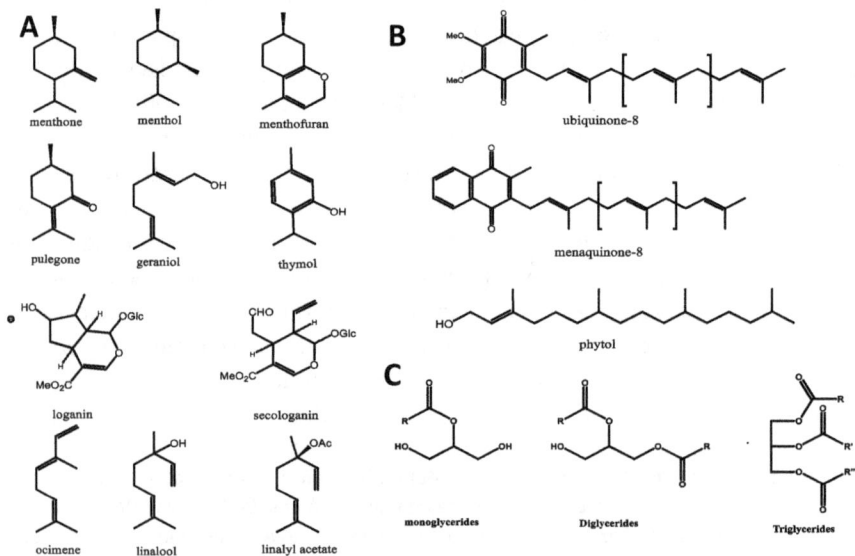

Figure 1.1: Chemical structure of bio-based molecules that can serve as potential monomers from a variety of biosynthetic pathways raging from less commonly used terpenoids and shikimates to widely adopted monomers lipids/glycerides.

source, but synthetic analogs can either be made from bioderived monomers or via a biomimetic approach. The dichotomy between natural or synthetic sources is the easiest classification of these materials (Figure 1.2) [1, 2]. Another binary classification, albeit sometimes not clear, is based on biodegradability of the polymer. Biopolymers can also be classified based on a parent monomer motif. This classification qualifies a core chemical functionality and groups structural analogs into one. For example;

- Polysaccharides: This group is one of the most abundant class of biopolymers, therefore they tend to be affordable. By volume, derivatives of the anomers α- and β-glucose, starch, and cellulose are the most abundant. Other analogs like chitin and chitosan only differ by substituents on the α-carbon to the anomeric center, a position that is known to alter the reactivity around the anomeric carbon (armed-disarmed glycosides). Polysaccharides have been widely used across disparate industries ranging from paper boards, print, textile, packaging, to use as food additives and thickening agents.

- Polypeptides: Given that life runs through amino acids, polymerization of these amino acids affords a new class of biopolymer, polypeptides [3]. Presence of a peptide bond renders these polymers readily degradable, for example, through hydrolysis to regenerate a carboxylate and amine. Availability of H-bond donors and acceptors in these polymers implies that they often present varied secondary and tertiary structures. These types of secondary bonds can also lead to functional microstructures and tunable properties.

- Polynucleotides: In 1869 [4], it was observed that a genetic code governs life. This code is carried in a deoxyribonucleic acid (DNA), a double-stranded biopolymer. Strong H-bonding across the two of these DNA strands induces helicity [5]. Based on the abundance of nucleic acids, other polymers have also been identified; however, based on the specific role they play in maintaining life, this class of polymers is produced in much smaller quantities compared to the polysaccharides.

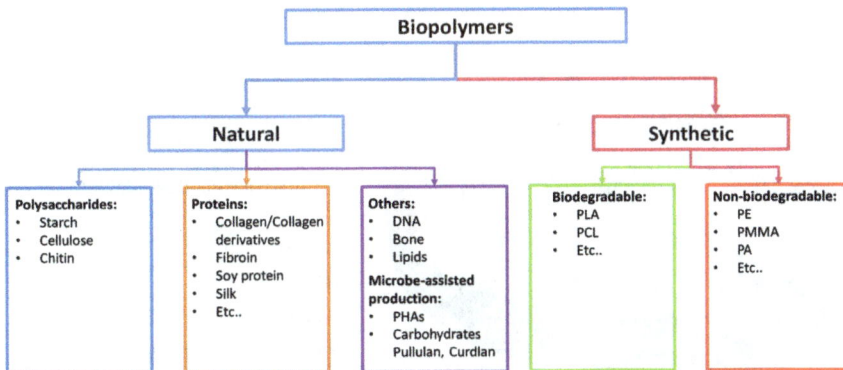

```
                          Biopolymers
                    ┌──────────┴──────────┐
               Natural                  Synthetic
        ┌─────────┴─────────┐        ┌──────┴───────┐
```

Polysaccharides:	Proteins:	Others:	Biodegradable:	Non-biodegradable:
• Starch	• Collagen/Collagen	• DNA	• PLA	• PE
• Cellulose	derivatives	• Bone	• PCL	• PMMA
• Chitin	• Fibroin	• Lipids	• Etc..	• PA
	• Soy protein	**Microbe-assisted**		• Etc..
	• Silk	**production:**		
	• Etc..	• PHAs		
		• Carbohydrates		
		Pullulan, Curdlan		

Figure 1.2: Classification of biopolymers based on material sources.

In the year 2018, a total of 2.11 million tons of biopolymers were produced globally (Figure 1.3) [1, 2]. Bio-based plastics, such as bio-based PE, PET, PA currently make up almost half of all biopolymers produced due to the wide application of plastics. The next dominant class is polysaccharides, which includes starch and cellulose derivatives. Among the 2.11 million tons of biopolymer produced in 2018, 56.8% of them are nonbiodegradable. Bio-based plastics such as PE, PET and PA are all nonbiodegradable, but their share in the market is expected to be decreasing, while biodegradable replacements, primarily PLA, will continue to grow.

1.1.2 Biocomposites

Nature provides a plethora of hybrid materials that despite being small (at the micrometer to nanometer scale) perform complex functions in plants and animals. These functions include load bearing by bones, actuation in muscles and tendons, protection in the form of skin or claws, among others. Such materials consist of small building blocks which are polymeric, ceramic, or a combination of the two [6, 7]. Examples include the polysaccharide cellulose which composes the majority of plant cell walls or the polypeptide collagen that acts as a structural element in tendons, ligaments or skin. These biopolymers are only capable of these vital functions as blends with other functional components [8, 9]. The components include natural ceramics and their composites such as bone, enamel or dentine which are composed of ceramic particles of hydroxyapatite, calcite or aragonite in a matrix of collagen. They further include natural polymers and their composites of previously introduced biopolymers such as polysaccharides and polypeptides, also including natural elastomers such as elastin, resilin, abductin or skin. Lastly, natural cellular materials such as wood or bamboo, that despite low density, bear unique properties due to the interplay between lattice structure, fibril structure and polymer networks such as cellulose and lignin [10].

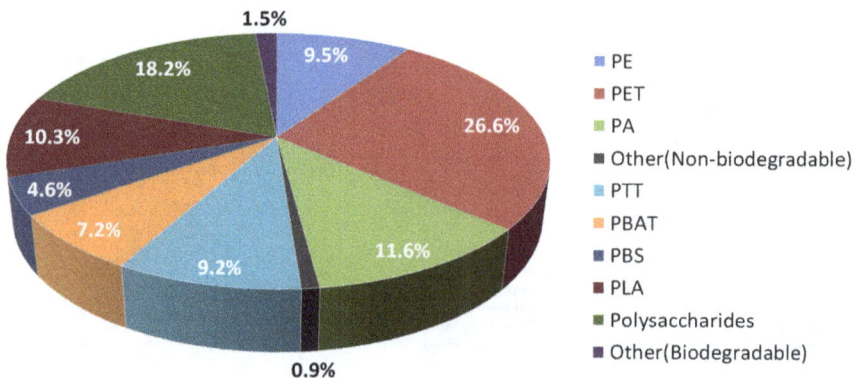

Figure 1.3: Distribution of biopolymers (2018) based on global production capacities [1, 2].

To illustrate the potential and versatility of biocomposites, we highlight three cases (bone, teeth and hair) focusing on the structure–property relations of these naturally occurring composites.

1.1.2.1 Bones

Bone is an interesting and vital biocomposite that is hierarchically assembled and comprised of organic and inorganic components. Bones are vital to vertebrate life for loading bearing and organ protection among other roles within various species. Bones are capable of these roles due their properties of being strong and tough and the ability to heal or repair after damage. Bones are composed of two regions, the cortical region which is hard and solid, and the trabecular region which is spongy or cellular. They are composed of collagen and mineral nanoparticles that along with a hierarchy of macro-to nano-scale give rise to bones unique properties [11]. These properties depend upon a cross-functional interplay at the various levels of the hierarchical structures [12, 13]. In a cross-section of a human femur, the compact and spongy regions of the bone can be seen (Figure 1.4). The compact region is composed of osteons that have a lamellar structure of 3–7 μm in thickness [14, 15]. These lamellas are formed of fiber bundles that are oriented along a particular axis.

These fibers, which are nanoscale in size, are composed of mineralized collagen fibrils roughly 100 nm in diameter and ~5–10 μm in length. Mineralization is the process of minerals growing in the spaces between the collagen fibers and for bone are composed of hydroxyapatite. The hydroxyapatite crystals are arranged parallel to one another and to the axis of the fibers [16]. Additionally the crystals are small, with thickness in the range of 1.5–4.5 nm [17, 18]. The deformation mechanism of bone is as of yet, not fully understood, but is a function of the collagen molecules. Within the collagen fibrils, there is a competition between molecular stretching and intermolecular sliding through breaking of H-bonds in the collagen molecule [19–21]. This sliding motion provides the fundamental for large plastic strains without brittle fracture (Figure 1.4).

1.1.2.2 Teeth

Teeth are another interesting biocomposite found in numerous vertebrates with an evolutionary history beginning 400 million years ago [24]. Like bones, teeth are a vital component to life performing functions such as eating and chewing, requiring them be both mechanically strong and chemically resistant. The fundamental structure of teeth consists of a calcified tissue called dentine, composed of collagen, proteins, and hydroxyapatite. Dentine surrounds the pulp, which is rich in fibroblast-like cells, blood vessels and nerves. The upper part of the dentine is usually covered by a layer of enamel, which is the hardest tissue of the body, and is collagen-free [25]. Enamel is hard, having a toughness value similar to glass, whereas dentine is tough but lack load

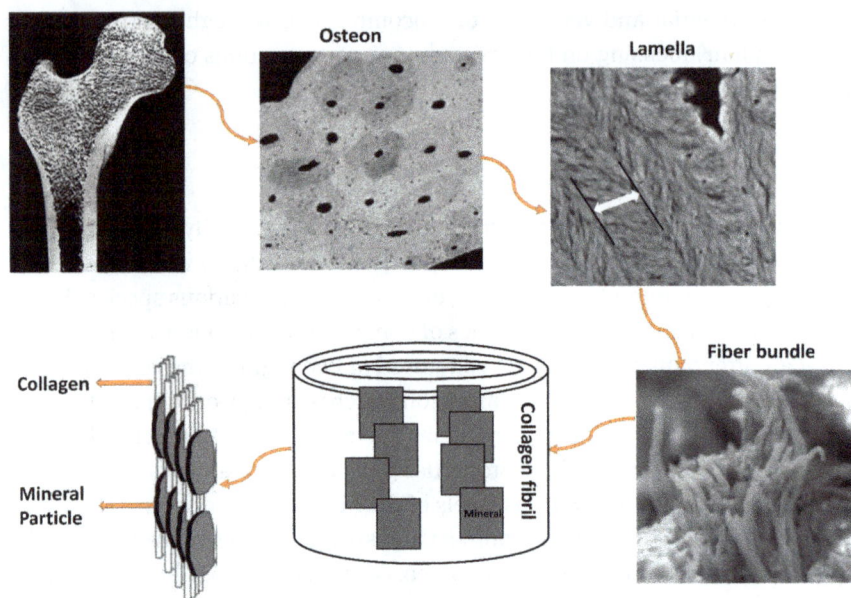

Figure 1.4: Hierarchical structure of a human femur. Interfaces of such hierarchical structures are shown which are at different length scales ranging from micro- to nano-meter scale. Bottom center sketch reproduced from Ref. 22 with permission. Copyright 2007, Elsevier. Remaining images reproduced from Ref. 23 with permission from the Royal Society of Chemistry.

bearing capacity [26, 27]. The unique structure of teeth makes them not only an interesting natural biocomposite, but as a candidate for biomimetic composites, discussed in the next section. The structure–property relations of teeth interest not just engineers but also evolutionary biologists. In this section, we discuss on the evolutionary history, structure and function of teeth among different vertebrates with specific examples in herbivorous dinosaurs *Triceratops* and humans, which are on the two opposite ends of the evolutionary spectrum.

The evolution of teeth began millions of years ago with fish, continuing to reptiles and amphibians, on to modern day vertebrates. The evolution of upper molar teeth began with reptilian cone teeth and advanced through the development of cusps, eventually reaching modern mammalian teeth (Figure 1.5A) [24]. Cusps can be made of individual bumps or elongated blade-like structures which are found in fish and reptiles. For mammals, the dentition system consists of incisors, canines, premolars and molars [28]. A cross-section of a tooth from *Triceratops horridus*, one of the dominant herbivores of the Mesozoic era can be seen in Figure 1.5B. The cross-section of the tooth displays six dominate regions. First, primitive reptilian tissues such as enamel, orthodentine, as well as derived coronal cementum (Figure 1.5B). Second, enamel on the outer edge, the hardest tissue that would be used to slice through plant matter (Figure 1.5C). Third, orthodentine on the outermost interior section (Figure 1.5D). Fourth, cementum-like tissue which helps

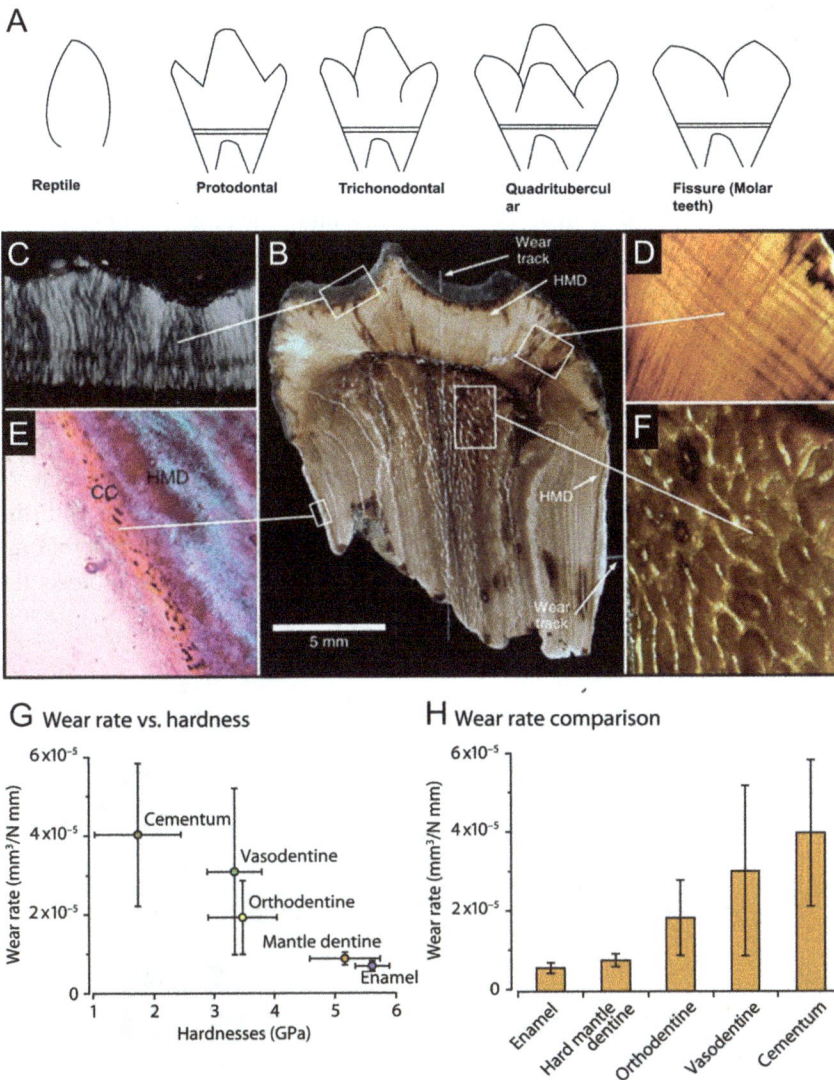

Figure 1.5: Evolutionary diversity and structure of teeth in animal kingdom. (A) Evolution of mammalian upper molar teeth across different stages starting from reptile to triangular to quadrate type teeth structure [24]. (B) Cross section of *T. horridus* teeth showing different layers of the teeth—enamel (C) orthodentine (D) coronal cementum (E) and vasodentine (F) (viewed with polarized microscopy with a wave plate filter), (G) hardness of different tissues versus wear rate ($n \geq 25$ per tissue) (H) Average triceratops dental tissue wear rates. Adapted from Ref. 29 with permission from AAAS.

the teeth sit in the jaw and acted as a load bearing material (Figure 1.5E). Fifth, material reported as vascodentine, i.e. dentine with vascularization (Figure 1.5F). Sixth, the core consisting of hard mantle dentine. The roles of vascodentine and orthodentine were to protect the dentine pulp against wear [29]. These various components have differing material properties and thus varying wear rates which can be observed in Figure 1.5G and H, with cementum having the highest rate and enamel with the lowest. The structure of teeth has evolved since the age of dinosaurs with mammalian teeth being mainly composed of enamel, dentin and cementum. In human teeth, dentin contains tubules that run perpendicularly and transport nutrients to the interior of teeth, while enamel consists of rods or prisms composed of hydroxyapatite crystallites. The structure of enamel is mainly of fiber-reinforced composite with defects known as tufts that are composed of closed cracks filled with organic matter and is one of the primary reasons of the toughness and the mechanical response of the enamel [30–32].

1.1.2.3 Hair

Hair is another biocomposite known for its organized microstructure consisting of various layers (Figure 1.6A) [33]. Hair is typically 50–100 mm in diameter with the outermost layer being formed of cuticle and the cortex composed of macrofibrils running parallel to one another and made of helical protein. The cuticle protects the

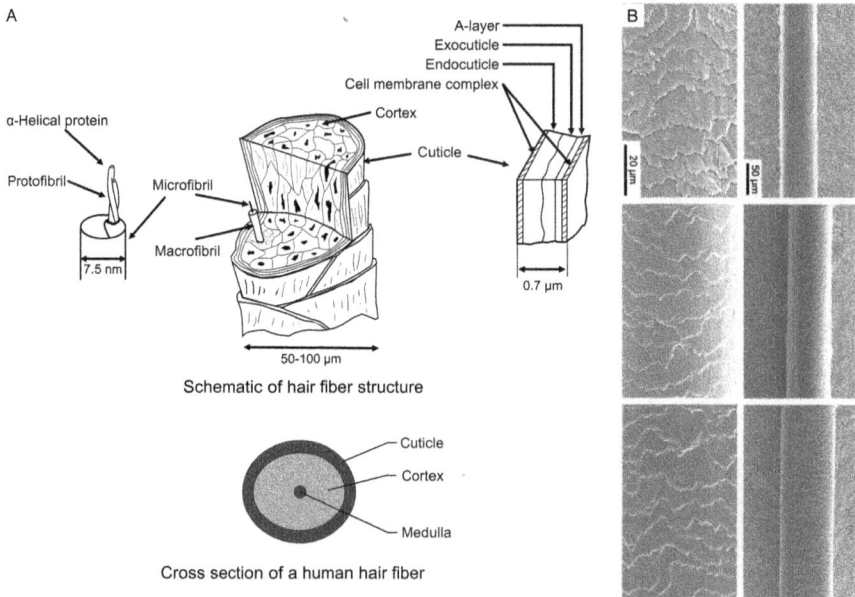

Figure 1.6: Schematic of human hair structure has been shown (A) cross-section and longitudinal illustration of human hair (B) SEM image of various hairs showing the cuticle layers. The panels are adapted from Ref. 34 with permission.

cortex and are flat overlapping scales that surround it (Figure 1.6B). Hair is mainly composed of keratin protein in the form of dead cells. It has high content of cytosine, an amino acid able to crosslink proteins using disulfide links. Hair additionally has numerous CO– and NH– groups that allow hydrogen bonds to form.

The cuticle in human hair is 0.5–1.0 μm thick and approximately 5–10 mm long and is composed of three major layers (Figure 1.6A). The first is the A-layer with a high cystine content, cystine being an amino acid. The second layer is the exocuticle, also rich in cystine, and third the endocuticle which is low in cystine content. The cortex is composed of cortical cells, which are 1–6 mm thick and 100 mm long running longi-tudinally along the hair fiber, and intercellular binding material [34]. The cortical cells are formed of macrofibrils consisting of filaments and matrix, and composed the major portion of the hair fiber composite. These filaments are made of twisted proteins with low cytosine content and the matrix which binds the cortical cells and cuticle together. The cuticle has higher values of hardness and elastic modulus than the cortex in the later direction, these properties were probed via SEM and nanoindentation [34]. The hardness is seen to decrease with indentation depth, primarily to the structure of the hair and the cytosine content.

Hair has shown potential for use in a range of applications as a hybrid composite ranging from agriculture to medicine. Fiber reinforced composites using human hair have already shown interesting and useful properties that could be improved even further [35].

1.1.3 Biomimetic composites

Nature has shown the ability to form numerous biocomposites with fascinating properties, a few of which were discussed in the previous section. These natural bio-composites are formed of only a few base elements such as C, N, H, O, Si, Ca, etc. but make use of hierarchical structure to produce their various properties [36]. It should be of no surprise that materials such as teeth or bone, with their strength and toughness or ability to heal, have inspired scientists to strive in mimicking them. The ability to mimic what nature has taken millions of years to perfect is no small challenge and controlling such processes at their micro- to nanoscale is difficult.

One successful technique is additive manufacturing (AM) and has been used to produce synthetic composites mimicking the complex hierarchical structures found in nature [37]. The AM approach relied on slip casting to fabricate composite mate-rials by dispersing 250-μm-thick and ~10-μm-diameter alumina platelets in a controlled way within metallic, polymeric and ceramic matrices, creating layers with a twisted-plywood structure [37]. The created composite resembled a tooth both in its structure and layered composition, albeit of different chemical composition (Figure 1.7A and B). As previously described, tooth enamel is a composite of parallel plates of hydroxyapatite and the collagen fibrils matrix. The synthetic composite was made of alumina platelets and silica nanoparticles as a hard cover to act as enamel. The authors also were successful in controlling the orientations of such layers as shown in Figure 1.7C.

However, the synthetic structures produced by Studart et al. (Figure 1.7D) and the twisted-plywood structure of the spider fang had significant differences (Figure 1.7C) [38]. The layers in the synthetic composites were thicker as compared to micrometer thickness in spider fangs. This thickness is the key factor in deciding crack propagation

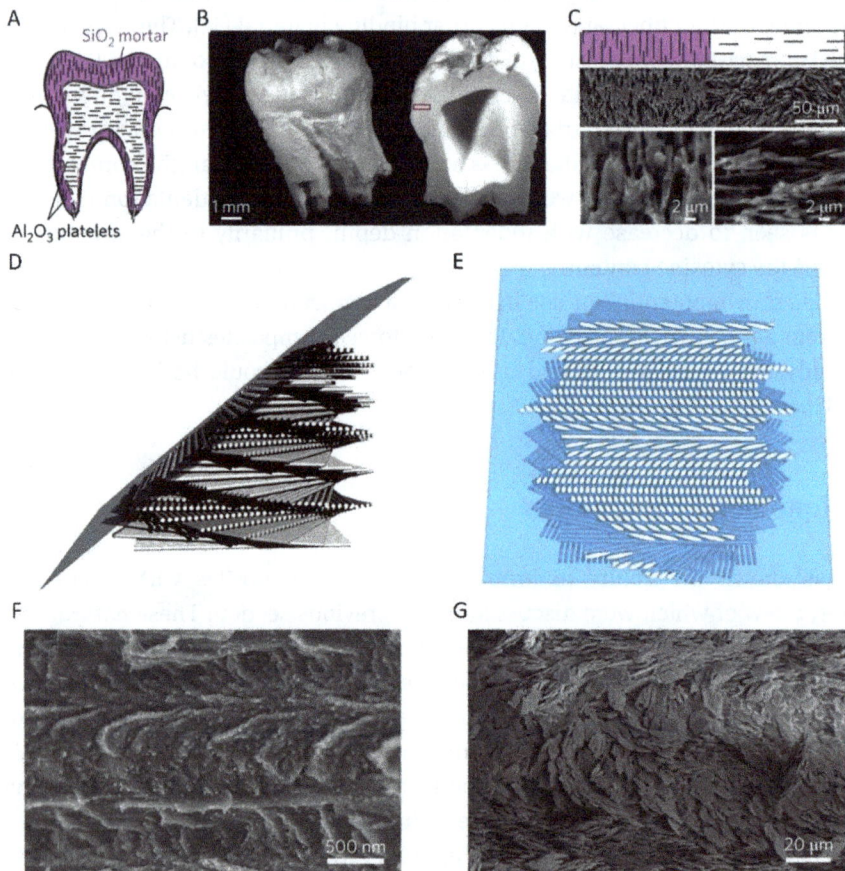

Figure 1.7: Tooth mimic. (A, B) Tooth-like bilayer composite made of alumina platelets and silica nanoparticles in a polymer matrix (C) scanning electron microscopy image of the orientations of the platelet. (D) Examples of heterogeneous composites-twisted-plywood-like structure of layered fibers, (E) characteristic arc-like patterns when cut in oblique cross-sections (F). Arc-like patterns in the scanning electron microscope image of a freeze-fracture surface of the fang of the wandering spider *Cupiennius salei* (courtesy of Yael Politi, Max Planck Institute of Colloids and Interfaces, Germany). (G) Twisted structures produced by magnetically assisted slip casting. The panels are adapted from Ref. 40 with permission from Nature Publishing Group.

in such layered structures because of the aspect ratio [39]. In natural materials, the orientation and position of such functional materials at nanoscale are controlled by cells but in synthetic biomimetic polymers such precision is hard to achieve as the manufacturing processes has to rely on self-organization which offers no control over local structure. Thus, current fabrication technologies do not offer such control over creating such hierarchical structures. Progress in this direction will certainly further improve the materials generated by additive manufacturing [40].

1.1.4 Processing of Bio-polymers

The processing and production of biopolymers can be extremely similar or even the same for those biopolymers that have the same chemical form or properties as those derived from petroleum. Biopolymers that vary from these traditional materials require additional work to ensure their ability to be processed. The processing of cellulose makes up a significant portion of biopolymer processing specifically in the cotton and paper industries. Production or extraction of cellulose and other biopolymers from previously under exploited sources such as biomass from farm waste show the potential for a new source of biopolymers from non-food sources.

1.1.4.1 Non-biodegradable bioplastics

Bioplastics are an ever-growing sector of the plastics industry and are composed of two main categories: biodegradable and nonbiodegradable. Bioplastics are defined as a biodegradable plastic or plastic derived from natural sources [41]. Biodegradable bioplastics include polylactic acids (PLA) and polyhydroxyalkanoates (PHA) and are discussed in the following section. Nonbiodegradable plastics make up a majority of the total bioplastics and is predominately polyethylene terephthalate (PET). Nonbio-degradable bioplastics are those plastics that are chemically identical to their petroleum-based counterparts but are instead derived from biomass precursors. These bioplastics are sometimes referred to as drop-in solutions due to having similar or the same properties as their synthetically derived counters [42]. There exists a second group of non-biodegradable bioplastics, that which is derived from a natural source with no synthetic analog and yet is still unable to biodegrade. These polythioesters have the potential to replace petroleum derived plastics that require high environmental resistance [43].

Two of the commonly used plastics, polyethylene (PE) and polypropylene (PP), can be manufactured completely from a biobase. For PE, ethanol is first produced via fermentation which is then converted to ethylene via dehydration. From there polymerization can be undergone typically, and the process can be seen in Figure 1.8. While classified as nonbiodegradable, work has shown the potential for biodegradation [44]. PP can be derived from a bio-source in numerous processes. These

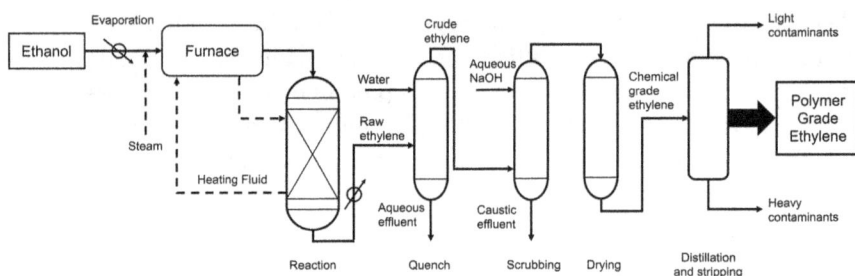

Figure 1.8: Process for the production of polymer grade ethylene for bio-based ethanol. Reproduced with permission from Ref. [44]. Copyright 2009, Taylor and Francis, Ltd.

include multistep processes and one similar to PE production where isopropanol can be derived from fermentation and dried to propylene which can then be polymerized just like synthetic PP [45].

Other widely used plastics such PET, polyamides (PA), or polyurethanes (PUR) can be produced at least partially from bio-sources. PET can be formed using ethanol as its base which is further made into monoethylene glycol. PET can then be formed with the addition of terephthalic acid, which is derived from petroleum [46]. Work to produce terephthalic acid from biomass at commercial scales is ongoing. PA can likewise be formed partially based on biomass and can form drop-in versions or even novel forms. PUR can be formed from biobases of vegetable oils that are further treated via oxidation, esterification or others. Biobased PUR has the potential to outpace synthetic sources due its improved versatility [47].

1.1.4.2 Biodegradable bioplastics

Biodegradable polymers of polylactic acid (PLA) and polyhydroxyalkanoate (PHA) are the most common. Their application would likely be far wider if not for the difficulties in processing them. Unlike drop-in polymers like the bioplastics discussed above, the properties of PLA and PHA make them unable to be dropped into an ongoing production process. Typical processing techniques such as extrusion, blow molding, injection molding or thermoforming are all possible with PLA and PHA but require modification of the parameters used and can come with challenges. Extrusion forming has difficulties with PLA due hydrolytic degradation, in injection molding PLA becomes brittle due to rapid aging. Blow molding is the technique used to form the numerous single use bottles currently used. PLA shows good candidacy for this use due the ability to rapidly degrade in the environment. As mentioned, the parameters required are varied and thus need to be changed and properly tuned to the requirements of PLA [48].

1.1.4.3 Cellulose

The most abundant biopolymer, cellulose is processed in two main ways, primarily in the cotton and paper industry. Cellulose can be difficult to dissolve in common solvents including water which can limit its potential [49]. Derivatives can be produced via derivatization in which materials such as methylcellulose or carboxymethyl cellulose, among others, can be produced [50, 51].

1.1.4.3.1 Cotton

Cotton is composed primarily of cellulose. Cotton is produced all over the world and is a vital component of the world's economy. Cotton is used primarily for textiles and around 25 million tons are produced annually. This is expected to increase due to ever expanding economies and growing populations [52].

The processing of cotton begins at the harvesting stage where mechanical harvesters pull the cotton bundles off of the plant. They are then bailed before being transported for ginning. Ginning is the process in which the seeds are separated from the fibers along with any other nonfiber objects. This cleans the cotton and preps it to be used in textile processing. Before being formed into a final product, the cotton must first undergo yarn formation. The cotton is first pulled from bales and mixed to a homogenous state. It is then cleaned, aligned and drawn together. Lastly it undergoes a twisting, which is done through various techniques that effect quality and strength. The spun yarn can then be used to make textiles that are broken into two main categories: woven and knitted. Woven fabrics are interlaced fibers at right angles while knitted fabrics have the yarn looped around one another just as in hand knitting. Nonwoven cotton is also used in various products where yarn is not required [53–55].

After fabric formation, the fabric moves on to textile mills that produce finished commercial products such as garments. Formation of a finished product requires the cotton fabric to undergo additional process steps. These steps include desizing (removal of sizing ingredient such as starch), mercerization (treatment with NaOH to improve luster and strength), bleaching (to produce white yarn), neutralization (addition of acid to neutralize the bases in previous steps), dyeing (impart color), and finished (softening or waterproofing, etc.) [56]. Each of these steps produce significant quantities of varying types of waste. Research proposes a vigorous recycling system is feasible for cotton textile production and will likely become required as environmental concerns grow [56].

1.1.4.3.2 Paper

Paper is comprised of a majority cellulose and is one of the largest biopolymer uses in the world, the production of which can be tracked back thousands of years. While the paper used then is vastly different than today's standards, the steps to produce it are still similar. These steps include: pulping of fibers, refining, dilution and suspension in slurry, formation (typically pressing), and then drying [57]. Prior to pulping, raw material must be prepared. This is most commonly wood which is debarked and broken down into chips that are uniform (or as close to uniform as possible) in size. Once chipped, the raw wood is transported to the next step: pulping. Pulping is the separation of the cellulose fibers from lignin and can done via a number of processes.

Pulping can be broken down into chemical and mechanical processes [57]. Mechanical, as the name suggests, is the breakdown of the wood chips via mechanical means. This method is often combined with thermal energy to loosen and weak the chips. Chemical processing is dominated by the Kraft process which uses a mix of sodium hydroxide and sodium sulphide to digest the wood (Figure 1.9). After being cooked in the solution, the chips are moved to new tank at a sufficient speed where they breakdown on impact. The sulphite process is less common but can create pulp that is easier to bleach. It uses a mix of sulfurous acid and bisulphite ions. There is additionally semi-chemical and secondary fiber pulping. Semichemical is a mixture of both chemical and mechanical processes and often includes the mixing of recycled fibers to improve processing. Secondary fiber pulping is the recycling of paper, where waste paper products are rewetted and broken down.

After pulping, the pulp is washed to remove any chemicals left over. This is often a balance between the level of cleanliness and the amount of water used. Using large quantities of water can lead to highly clean pulp but at the cost of significantly more waste produced. Once washed, the pulp is screened which cleans and removes any particles that could affect the paper formation such as uncooked pieces of wood. Pulp can then be bleached to create a brighter product for printing or tissue papers. Bleaching is also capable of improving the purity of the pulp. Bleaching is highly dependent on the lignin value of the pulp, where lower values allow easier bleaching.

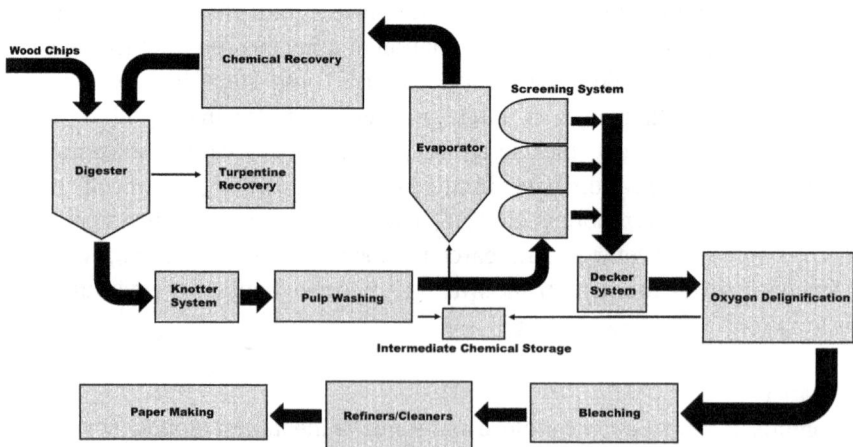

Figure 1.9: Outline of the paper making process using the Kraft chemical pulping technique. Reproduced with permission from Ref. [57]. Copyright 2015, Springer International Publishing.

The lignin value of pulp can be dependent on the source of the pulp and in the process used for pulping. For chemical and semichemical pulping, the reagents used are recovered and either recycled for continued use or in the form of containment solids, burned for energy production. Once the cleaned and bleached pulp is ready, additional additives can be included to change the final properties of the pulp and thus the type of paper it is suitable for. These additives include other types of fibers or minerals such as calcium carbonate, among others. Finally, the pulp can then be formed, pressed and dried into a final product. These steps depend on the type of paper being produced [57].

The previously described steps of the paper making process have all seen interest in reducing the cost or energy associated with them. The use of enzymes or microbes as treatment or pretreatment prior to the above steps can reduce energy costs, lower the quantity of chemicals used, and even improve the final properties of the end product [58].

1.1.4.4 Biomass extraction

Waste from farming or forestry is a large potential source of ligno-cellulosic biomass that can be harvested for biopolymers. Current processing is difficult and usually requires toxic pretreatment steps. Ligno-celluosic biomass contains cellulose, lignin and hemicellulose. Pretreatment is done to remove the lignin and hemicellulose while keeping the cellulose intact [59]. This is similar to the pulping step in paper making. The chemicals used are toxic and typically require large volumes. With the majority of cellulose being from wood pulp for paper production, there exists a need for smaller fractions used in such things as pharmaceuticals, among others [60]. Cellulose is able to be harvested from waste but is not a form suitable for paper; it thus needs to be applied in a different area [61]. There is also the ability to pretreat with organic solvents such as ethanol, methanol, or acetone. These have the potential benefits of easier recovery, lower toxicity, and the ability to recover and thus use the lignin as well as cellulose [62]. There is also the ability to do solid-state fermentation in which fungi or bacteria are used to breakdown biomass. This technique has issues with the fact that it takes much longer and yield is impacted by the fact that the bacteria or fungi used attack the cellulose along with the lignin and hemicellulose [63].

1.1.5 Neoteric processing and applications

We have so far seen a number of techniques for the processing of biopolymers. Plenty of which are typical or only lightly modified techniques for traditional polymers. Some of the most unique discussed is paper and cotton or natural fiber processing, which can be considered well-known or traditional (humans have been making paper and textiles for thousands of years) [64]. Outside of these traditional or standard processing techniques exist neoteric or new to world techniques. These new processes are developed for bio-polymers most often for new applications. These techniques are developed to impart new functionality into biopolymers by overcoming limitations in their nonmodified versions, to make them applicable to older techniques, or to convert them to higher value materials, among others. This section discusses a brief set of examples of these neoteric processes. The examples present are far from exhaustive and only represent a small section of new and upcoming processing techniques, of which, more and more are discovered and refined as time progresses.

1.1.5.1 Amphiphobicity

Surface modification of biopolymers is an often applied technique to impart new functionality. One of the most common is creating hydrophobicity in cellulose [65]. This can be accomplished via appropriate surface texture and lowering of the surface energy. One of the most straightforward methods to impart both is vapor deposition of alkylsilanes. Alkylsilanes undergo polymerization with water as the co-monomer found along and bound to the surface of cellulose [66]. Consumption of this water leaves the surface open for the polysiloxane to bind to the cellulose (Figure 1.10a$_i$). The polysiloxane forms islands that are stochastically distributed on the surface giving rise to an amphiphobic surface (Figure 1.10a$_{ii}$). That is, the polymer islands, along with the texture they impart, yields a hydrophobic (or super-hydrophobic) surface that still has hydrophilic cellulose present. This dual nature of the surface allows high water resistance but are still capable of absorption over long time scales and even the ability to biodegrade in the environment [67]. This simple processing technique imparts a large increase to the value of a biopolymer such as cellulose while maintaining biodegradability, one of the most desirable aspects of biopolymers.

1.1.5.2 Inverse thermal degradation

The same modification used in the above section can be used to alternate the thermal degradation of the treated cellulose. The crosslinked polymer formed on the surface has a higher ignition point than the cellulose below it. This mixing of ignition points leads to the inverting of the thermal degradation process in which a treated cellulose fiber should burn from the inside outward (Figure 1.10b$_i$) [68]. This process yields carbonaceous material that retains the structure of the base material (Figure 1.10b$_{ii}$). This is a low-cost process for the conversion of cellulose to lower energy carbon and has potential in catalysis, energy storage and composite reinforcement.

1.1.5.3 3D printing

One technique becoming more and more commonplace is 3D printing. 3D printing of traditional polymers is frequently easier done due to having a large set of material properties already known and understood (melting and glass transition temperature along with flow being some of the most important, depending on technique). The

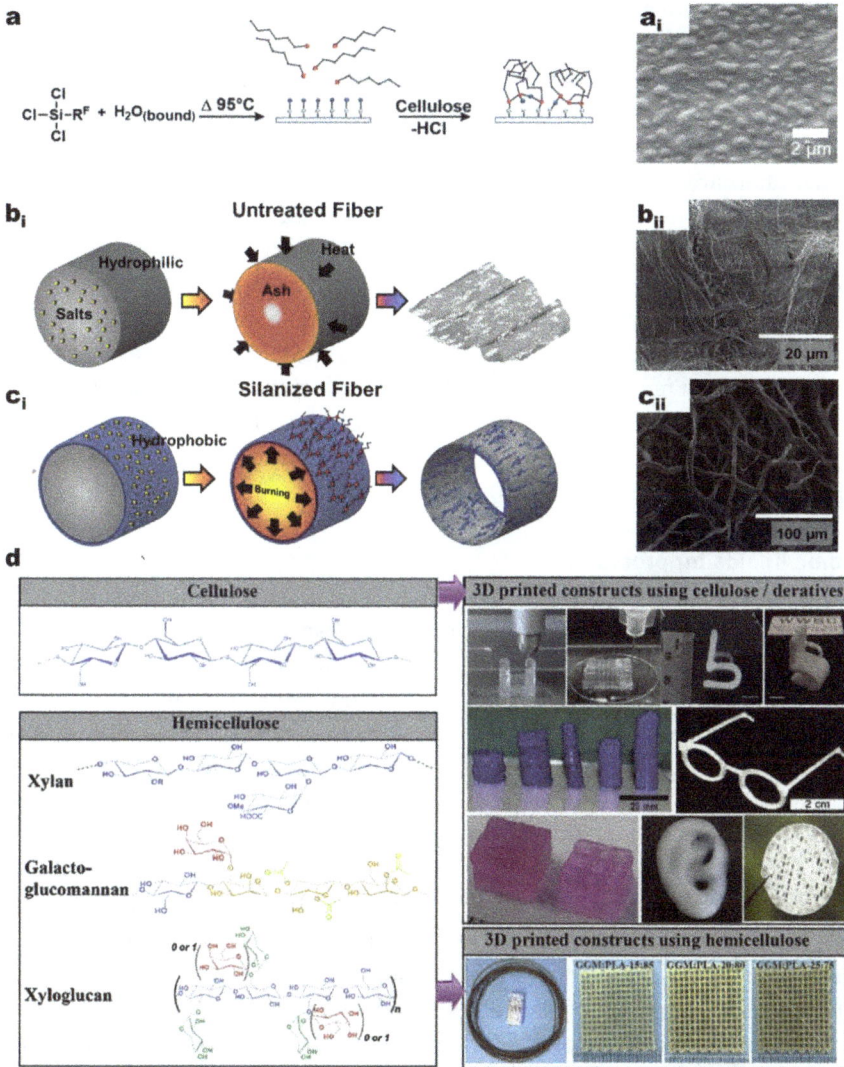

Figure 1.10: (a) Proposed mechanism of silane/cellulose reaction, (aᵢ) formation of polysiloxane silanes on paper surface [66], (b) thermal degradation of native cellulose fiber, (bᵢ) SEM of native paper surface, (c) inverted thermal degradation of silane treated cellulose, (cᵢ) retention of fiber network due to ⁱTD process [68], (d) examples of biopolymers and shapes constructed via 3D printing [69]. Copyright 2019, Elsevier. Reproduced with permission from Ref. 69.

potential use of 3D printing for biopolymers can be found in the medical field along with food and textile industries. Each of these have their own requirements based on the type of printing and the final use of the product. The printing of native cellulose is most commonly done with cellulose nanofibrils (CNFs) and can be seen in biomedical applications or as additive in food products [69, 70]. One major difficulty in the use of native cellulose in the form of CNFs in 3D printing is reaching sufficient dispersion. This is often overcome via modification of the cellulose in one form or another. One such alternative is cellulose acetate. Cellulose acetate dispersed in acetate solution allows easy printing of pure cellulose acetate parts after evaporation of the acetate solution [71]. Another alternative is in an alginate-CNF composite ink [72]. The combination of these biopolymers led to excellent printing properties and further, when combined with human cartilage cells, showed high cell viability. The excellent printing properties of the ink were further displayed in the printing of anatomically shaped human ear and sheep meniscus. 3D printing is also possible with chitosan, the second most abundant biopolymer. 3D printing of chitosan-based biopolymers lacks wider applicability due to a number of issues that must be overcome. These include difficulty incorporating cells due to the acidic nature of the ink, mechanical weakness and inability to degrade in the required environments [69]. Examples of 3D printed materials using cellulose and hemicellulose can be seen in Figure 1.10d.

1.1.5.4 Ionic liquids for processing

The desire to use biopolymers in ever more widespread applications requires processing of said biopolymers to be easy and to be environmentally friendly. Any benefit of a biodegradable or ecologically sustainable end product is reduced or lost completely if the processing emits or creates significant waste. While biopolymers have a number of beneficial properties, these come with the issue of low solubility or only reaching useful degrees of solubility in toxic or otherwise dangerous solvents [73]. One class of material being explored to overcome this is ionic liquids. Ionic liquids are salts or salt-like materials that are liquid at or around room temperature. Ionic liquids are not new to biopolymer processing (a patent on which was filed in 1934) but have seen wide-scale resurgence as biopolymers have grown in scope and application [74].

Ionic liquids are tunable solvents, the anion and cation components can be changed and thus the intermolecular forces tuned [75]. This tunability means that a solvent can be tailor made for a particular biopolymer. They also have the additional benefit of being recyclable, further increasing their value as a green alternative. They have shown to be remarkably good at dissolving cellulose or cellulose derivatives believed to be due to their ability to accept hydrogen bonds and interrupt the network of cellulose. This solubility has been shown up to 25 wt% and with the ability to reconstitute the cellulose from solution in forms such as powder, films, or fibers [74]. This ease of processing has been displayed for various biopolymers into films, fibers, aero- and hydrogels, along with composites in multiple forms. Specific examples include biomedical application of hybrid gels of starch and cellulose for drug delivery or films made of lignin and gelatin that displayed antibacterial, flexible, and biodegradable properties [76, 77]. Ionic liquids have also shown the ability to harvest biopolymers from previously untapped or prohibitively difficult sources such as waste from palm oil harvesting [78]. Additional sources include chitin from shrimp shells and keratin from camel hair [79]. Ionic liquids do have drawbacks like the potential for high viscosities depending on intermolecular forces. This high viscosity can prove difficult to use with certain biopolymers. They also lack in-depth study of properties such as toxicity to organisms and the environment, parameters for scaling and other material properties that are widely available for more typical or standard solvents [73]. These issues and ionic liquids as a whole are the focus of ongoing research.

1.1.5.5 Microemulsions

Microemulsions are a type of material that consists of a dispersed and continuous phase that employ surfactants to maintain a system of dispersed droplets and are found at the nanoscale [80, 81]. This presents as an excellent method to process nanoparticles of biopolymers. One such application of this process is for functional food. Functional food is the encapsulation and controlled delivery of nutraceuticals, typically in the form of proteins or polysaccharides. Biopolymers fit the needs of food products due to their properties such as gelation and emulsification. They are needed to protect bioactive agents during processing or consumption. To this end, the bioactive particles can be encapsulated with biopolymers via microemulsions [80]. There are multiple techniques possible that each have their own benefits and challenges. They are desirable though for their ability to create well-defined biopolymer nanoparticles. While varied, there are four main steps that are the same or similar. They are:
(1) Micro emulsification: a water in oil (or oil in water) microemulsion is created. This can be done using various known chemicals. A biopolymer solution is then introduced which diffuses into the microemulsions, creating biopolymer rich zones.
(2) Confined solidification: the biopolymers are solidified by using physical, chemical or enzyme based methods to induce gelation, crosslinking or precipitation.
(3) Isolation: the biopolymer nanoparticles, now solid, can be removed from the solution and purified.
(4) Drying: biopolymer nanoparticles, once cleaned and purified from solution, are dried and ready for use.

Single microemulsion is the most common approach using a one pot, single step process. In this case a biopolymer solution is added to a water in oil solution and through temperature change or addition of antisolvent, the biopolymer is isolated. As mentioned in the solidification step, crosslinking can also be used. This can be accomplished by adding a crosslinking agent after the biopolymer or including it within the biopolymer solution. This process has been done with proteins such as gelatin and whey and with polysaccharides like cellulose and chitin. Hybrid versions are also capable of creating core–shell particles with two different (or same) biopolymers. Additionally, there is a two-microemulsion method where two solutions of microemulsions are brought together. Through Brownian motion, the emulsions can collide with one another and mix. However, there still lacks an understanding on what effects the size of the particles such as biopolymer concentration, surfactant concentration, and water content.

1.1.5.6 Biofuel production

Biofuels are those fuels for various modes of transport that are derived from a biomass source. Biofuel production has seen expansion in recent years not only for environmental reasons but also economic and strategic goals such as limiting reliance on foreign sources. Biofuel comes in many different types such as ethanol, methanol, biodiesel, or biogas [82]. Bio-alcohols are commonly derived from fermentation of crops such as corn or potatoes which can be seen in Figure 1.11a. Fermentation yields a solution of around 12% ethanol that is then purified via distillation. Reaching 100% is difficult due to the formation of an azeotrope with water and thus requires additional processing. Ethanol can be mixed with gasoline or diesel in the 5–15% range that yields improved burning properties to gasoline and a reduction of smoke in diesel. Methanol can also be produced from biomass albeit with extra required steps such as gasification. Methanol production from biomass is less efficient and thus less common. Vegetable oils are capable of being used as fuel in modified diesel engines but can also be converted to biodiesel via transesterification. This process lowers the viscosity and allows it to be used in standard diesel engines. Biodiesel can be considered better than its petroleum-based counter due to lower flash point and the ability to biodegrade. The cost is often prohibitive for large-scale adoption due to the oils that would be ideal being high-value food products. The potential to use waste products from restaurants or animal processing is present but requires additional processing and comes with additional challenges. Bio-oil can be formed from high-temperature pyrolysis of biomass and can be converted with relatively high rates (~70%). Bio-oil is a strong candidate for the replacement of petroleum-based oils due to this. Gaseous biofuels can also be produced such as biogas or biohydrogen. Biogas is a mixture of methane and CO_2 that can be harvested from digesters. Any biomass, including waste from animals or sewers, can be used as feedstock combined with microorganisms that will convert the biomass to biogas, which can be used as a cheap fuel. Hydrogen may prove to be the key to environmentally friendly energy production in the future and with large-scale harvesting not yet being viable, biohydrogen may be the solution. Formation of hydrogen from biomass has multiple available pathways and are more efficient than thermo- or electro-chemical techniques. Numerous bacteria and other microorganisms are capable of hydrogen production but require specific biomass processing to be effective. An overview of value-added products from biomass can be seen in Figure 1.11b.

First generation biofuels are those previously discussed, the most common being bioethanol and biodiesel, that are produced from crops. Second generation biofuels are

a

```
                        ┌──────────────┐
                        │   BIOFUELS   │
                        └──────────────┘
                         /            \
                                    ┌──────────────┐
                                    │  BIOETHANOL  │
                                    └──────────────┘
           ┌──────────────┐        /    |      |      \
           │  BIODIESEL   │   ┌───────┐┌──────┐┌────────────┐┌──────────┐
           └──────────────┘   │ Wheat ││Maize ││ Sugar Beet ││ Potatoes │
          /    |    |    \     └───────┘└──────┘└────────────┘└──────────┘
```

| Rapeseed | Soybean | Palm | Sunflower |

b

┌──────────┐
│ BIOMASS │
└──────────┘

┌─────────────────────────────────────┐
│ GASIFICATION WITH PARTIAL OXIDATION │
└─────────────────────────────────────┘

┌───────────────┐
│ GAS CLEANING │
└───────────────┘

┌──────────────────────┐
│ GAS CONDITIONING   │
│ -Reforming   │
│ -Water-Gas Shift   │
│ -CO_2 Removal   │
│ -Recycle   │
└──────────────────────┘

┌──────────────────────────┐
│ FISHER-TROPSCH SYNTHESIS │
└──────────────────────────┘

┌────────────────────┐
│ PRODUCT UPGRADING │
└────────────────────┘

GREEN DIESEL	LIGHT PRODUCTS	HEAVY PRODUCTS	POWER
	-Gasoline	-Light wax	-Electricity
	-Kerosene	-Heavy wax	-Heat
	-LPG		
	-Methane		
	-Ethane		

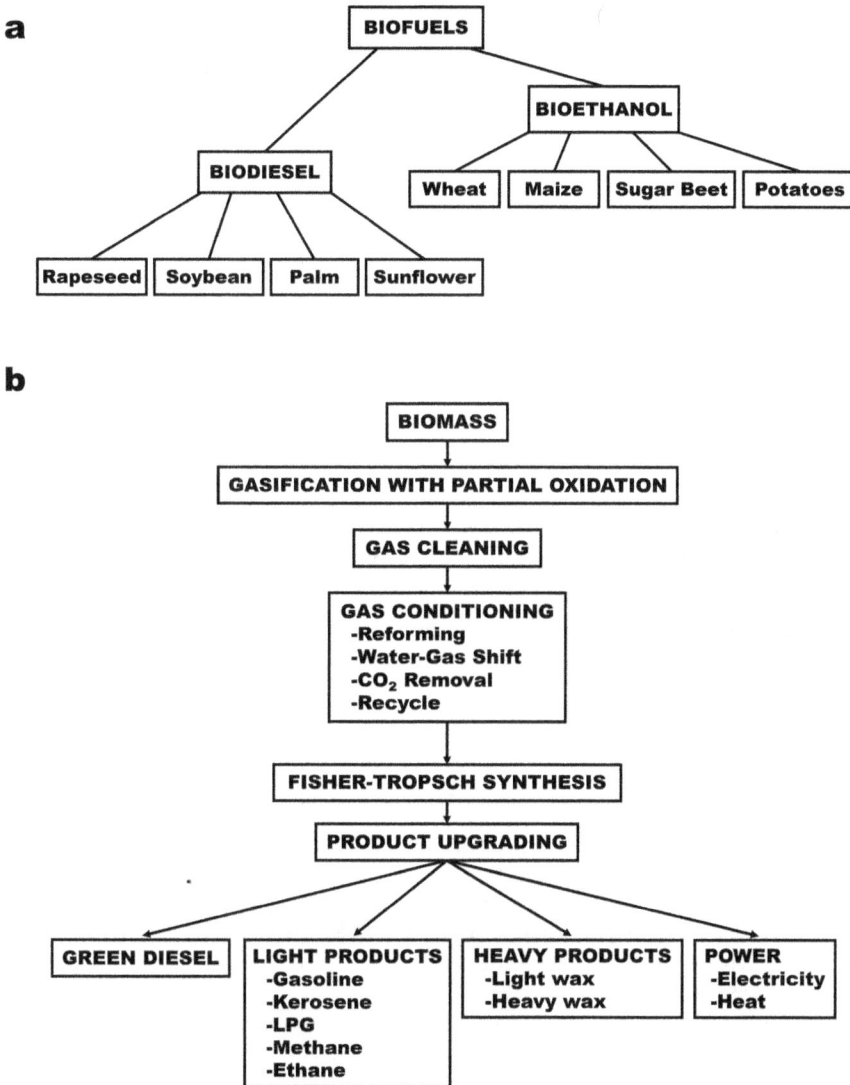

Figure 1.11: (a) Types of biofuels and what bio-based sources are used to produce them, (b) process overview of various end products that can made from biomass sources. Copyright 2008, Elsevier. Reproduced with permission from Ref. [82].

those that are formed from ligno-cellulosic feedstock or non-food biomass. Lingo-cellulosic biomass can include waste products from crops such as corn stocks and from dedicated sources such grasses or forests [83]. The desire to use non-food biomass comes from the competition of land use. The biomass is converted via thermo- or biochemical processes just as in first generation. Neither have shown superiority over the other and continue to have ongoing development to overcome technical and logistical issues.

1.1.6 Conclusion

We review a breadth of information on biopolymers and biocomposites to illustrate the vast space that is covered by these materials. The information provided herein, is only illustrative, not exhaustive, specifically to illustrate that although biopolymers and biocomposites have been known for eons, there is still a significant amount of work that needs to be explored if we are to adapt these materials beyond what nature intended them for. The following chapters highlight some of the opportunities and challenges in realizing this goal.

Author contributions: All the authors have accepted responsibility for the entire content of this submitted manuscript and approved submission.
Research funding: None declared.
Conflict of interest statement: The authors declare no conflicts of interest regarding this article.

References

1. Platform B. European bioplastics. Huerth: Institute for Bioplastics, Nova-Institute; 2015.
2. Bioplastics E. Bioplastics market data 2018. Berlin: European Bioplastics; 2018.
3. Yadav P, Yadav H, Shah VG, Shah G, Dhaka G. Biomedical biopolymers, their origin and evolution in biomedical sciences: a systematic review. J Clin Diagn Res 2015;9:ZE21.
4. Dahm R. Discovering DNA: Friedrich Miescher and the early years of nucleic acid research. Hum Genet 2008;122:565–81.
5. Consortium IHGS. Initial sequencing and analysis of the human genome. Nature 2001;409:860–921.
6. Currey JD. Hierarchies in biomineral structures. Science 2005;309:253.
7. Fratzl P, Weinkamer R. Nature's hierarchical materials. Prog Mater Sci 2007;52:1263–334.
8. Lakes R. Materials with structural hierarchy. Nature 1993;361:511–5.
9. Ortiz C, Boyce MC. Materials science. Bioinspired structural materials. Science 2008;319:1053–4.
10. Wegst UGK, Ashby MF. The mechanical efficiency of natural materials. Philos Mag 2004;84: 2167–86.
11. Launey ME, Buehler MJ, Ritchie RO. On the mechanistic origins of toughness in bone. Annu Rev Mater Res 2010;40:25–53.
12. SW, Wagner HD. The material bone: structure-mechanical function relations. Annu Rev Mater Sci 1998;28:271–98.
13. Rho JY, Kuhn-Spearing L, Zioupos P. Mechanical properties and the hierarchical structure of bone. Med Eng Phys 1998;20:92–102.
14. Ascenzi A, Bonucci E, Generali P, Ripamonti A, Roveri N. Orientation of apatite in single osteon samples as studied by pole figures. Calcif Tissue Int 1979;29:101–5.

15. Rho JY, Zioupos P, Currey JD, Pharr GM. Variations in the individual thick lamellar properties within osteons by nanoindentation. Bone 1999;25:295–300.
16. Landis WJ, Hodgens KJ, Arena J, Song MJ, McEwen BF. Structural relations between collagen and mineral in bone as determined by high voltage electron microscopic tomography. Microsc Res Tech 1996;33:192–202.
17. Fratzl P, Fratzl-Zelman N, Klaushofer K, Vogl G, Koller K. Nucleation and growth of mineral crystals in bone studied by small-angle X-ray scattering. Calcif Tissue Int 1991;48:407–13.
18. Fratzl P, Groschner M, Vogl G, Plenk H Jr., Eschberger J, Fratzl-Zelman N, et al. Mineral crystals in calcified tissues: a comparative study by SAXS. J Bone Miner Res 1992;7:329–34.
19. Buehler MJ. Nanomechanics of collagen fibrils under varying cross-link densities: atomistic and continuum studies. J Mech Behav Biomed Mater 2008;1:59–67.
20. Buehler MJ, Wong SY. Entropic elasticity controls nanomechanics of single tropocollagen molecules. Biophys J 2007;93:37–43.
21. Sun YL, Luo ZP, Fertala A, An KN. Stretching type II collagen with optical tweezers. J Biomech 2004; 37:1665–9.
22. Jäger I, Fratzl P. Mineralized collagen fibrils: a mechanical model with a staggered arrangement of mineral particles. Biophys J 2000;79:1737–46.
23. Fratzl P, Gupta HS, Paschalis EP, Roschger P. Structure and mechanical quality of the collagen–mineral nano-composite in bone. J Mater Chem 2004;14:2115–23.
24. Weller JM. Evolution of mammalian teeth. J Paleontol 1968;42:268–90.
25. Koussoulakou DS, Margaritis LH, Koussoulakos SL. A curriculum vitae of teeth: evolution, generation, regeneration. Int J Biol Sci 2009;5:226–43.
26. He LH, Swain MV. Contact induced deformation of enamel. Appl Phys Lett 2007;90:171916.
27. Chai H, Lee JJW, Constantino PJ, Lucas PW, Lawn BR. Remarkable resilience of teeth. Proc Natl Acad Sci U S A 2009;106:7289.
28. Jernvall J, Thesleff I. Tooth shape formation and tooth renewal: evolving with the same signals. Development (Cambridge, England) 2012;139:3487–97.
29. Erickson GM, Sidebottom MA, Kay DI, Turner KT, Ip N, Norell MA, et al. Wear biomechanics in the slicing dentition of the giant horned dinosaur Triceratops. Sci Adv 2015;1:e1500055.
30. Lawn BR, Lee JJW, Chai H. Teeth: among nature's most durable biocomposites. Annu Rev Mater Res 2010;40:55–75.
31. Osborn JW. The 3-dimensional morphology of the tufts in human enamel. Acta Anat 1969;73:481–95.
32. Sognnaes RF. The organic elements of the enamel; the gross morphology and the histological relationship of the lamellae to the organic framework of the enamel. J Dent Res 1950;29:260–9.
33. Robbins CR. Chemical and physical behavior of human hair. New York: Springer; 2013.
34. Wei G, Bhushan B, Torgerson PM. Nanomechanical characterization of human hair using nanoindentation and SEM. Ultramicroscopy 2005;105:248–66.
35. Verma A, Singh V, Verma SK, Sharma A. Human hair: a biodegradable composite fiber A review. Int J Waste Resour 2016;6:1–4.
36. Fratzl P. Biomimetic materials research: what can we really learn from nature's structural materials? J R Soc Interface 2007;4:637–42.
37. Le Ferrand H, Bouville F, Niebel TP, Studart AR. Magnetically assisted slip casting of bioinspired heterogeneous composites. Nat Mater 2015;14:1172–9.
38. Politi Y, Priewasser M, Pippel E, Zaslansky P, Hartmann J, Siegel S, et al. How to design an injection needle using chitin-based composite material. Adv Funct Mater 2012;22:2519–28.
39. Kolednik O, Predan J, Fischer FD, Fratzl P. Improvements of strength and fracture resistance by spatial material property variations. Acta Mater 2014;68:279–94.
40. Dunlop JWC, Fratzl P. Making a tooth mimic. Nat Mater 2015;14:1082–3.

41. Jambunathan P, Zhang K. Engineered biosynthesis of biodegradable polymers; 2016 (1476-5535 Electronic).
42. Andreeßen CA-O, Steinbüchel A. Recent developments in non-biodegradable biopolymers: precursors, production processes, and future perspectives; 2019 (1432-0614 Electronic).
43. Lütke-Eversloh T, Bergander K, Luftmann H, Steinbüchel A. Identification of a new class of biopolymer: bacterial synthesis of a sulfur-containing polymer with thioester linkages. Microbiology (Reading, England) 2001;147:11–9.
44. Morschbacker A. Bio-ethanol based ethylene. Polym Rev 2009;49:79–84.
45. Kikuchi Y, Oshita Y, Mayumi K, Hirao M. Greenhouse gas emissions and socioeconomic effects of biomass-derived products based on structural path and life cycle analyses: a case study of polyethylene and polypropylene in Japan. J Clean Prod 2017;167:289–305.
46. Prieto A. To be, or not to be biodegradable… that is the question for the bio-based plastics. Microbial Biotechn 2016;9:652–7.
47. Harmsen PFH, Hackmann MM, Bos HL. Green building blocks for bio-based plastics. Biofuels Bioprod Biorefin 2014;8:306–24.
48. Lee CH, Sapuan SM, Ilyas RA, Lee SH, Khalina A. Chapter 5 - Development and processing of PLA, PHA, and other biopolymers; In: Al-Oqla FM, Sapuan SM, editors. Advanced processing, properties, and applications of starch and other bio-based polymers. Amsterdam, Netherlands: Elsevier; 2020:47–63 pp.
49. Wang S, Lu A, Zhang L. Recent advances in regenerated cellulose materials. Prog Polym Sci 2016; 53:169–206.
50. Niu T, Wang X, Wu C, Sun D, Zhang X, Chen Z, et al. Chemical modification of cotton fabrics by a bifunctional cationic polymer for salt-free reactive dyeing. ACS Omega 2020;5:15409–16.
51. Wu J, Zhang J, Zhang H, He J, Ren Q, Guo M. Homogeneous acetylation of cellulose in a new ionic liquid. Biomacromolecules 2004;5:266–8.
52. Khan MA, Wahid A, Ahmad M, Tahir MT, Ahmed M, Ahmad S, et al. World cotton production and consumption: an overview; In: Ahmad S, Hasanuzzaman M, editors. Cotton production and uses: agronomy, crop protection, and postharvest technologies. Singapore: Springer Singapore;2020: 1–7 pp.
53. Seagull R, Alspaugh P, Cotton I, Texas Tech U, International Textile C. Cotton fiber development and processing : an illustrated overview. Lubbock, Tex: International Textile Center, Texas Tech University; 2001.
54. Saxonhouse G, Wright G. Technological evolution in cotton spinning, 1878–1933. In: The Japanese economy in retrospect. Singapore: World Scientific Publishing Company; 2010:323–46 pp.
55. Shabbir M, Mohammad F. Natural textile fibers: polymeric base materials for textile industry; In: Ikram S, Ahmed S, editors. Natural polymers: derivatives, blends, and composites. New Delhi, India: Nova Science Publishers Inc.; 2017, vol II:89–102 pp.
56. Babu B, Parande A, Sangeetha R, Prem T, Babu K, Kumar P. Textile technology: cotton textile processing: waste generation and effluent treatment. J Cotton Sci 2007;11.
57. Bajpai P. Basic overview of pulp and paper manufacturing process. In: Bajpai P, editor. Green chemistry and sustainability in pulp and paper industry. Cham: Springer International Publishing; 2015:11–39.
58. Viikari L, Suurnäkki A, Grönqvist S, Raaska L, Ragauskas A. Forest products: biotechnology in pulp and paper processing. In: Schaechter M, editor. Encyclopedia of microbiology, 3rd ed.. Oxford: Academic Press; 2009:80–94.
59. Poletto M, Ornaghi HL, Zattera AJ. Native cellulose: structure, characterization and thermal properties. Materials 2014;7:6105–19.
60. Tuck CO, Pérez E, Horváth IT, Sheldon RA, Poliakoff M. Valorization of biomass: deriving more value from waste. Science 2012;337:695.

61. Ass BAP, Ciacco GT, Frollini E. Cellulose acetates from linters and sisal: correlation between synthesis conditions in DMAc/LiCl and product properties. Bioresour Technol 2006;97:1696–702.
62. Kim Y, Yu A, Han M, Choi G-W, Chung B. Ethanosolv pretreatment of barley straw with iron(III) chloride for enzymatic saccharification. J Chem Technol Biotechnol 2010;85:1494–8.
63. Pandey A, Soccol CR, Mitchell D. New developments in solid state fermentation: I-bioprocesses and products. Process Biochem 2000;35:1153–69.
64. Zhang J, Lu H, Sun G, Flad R, Wu N, Huan X, et al. Phytoliths reveal the earliest fine reedy textile in China at the Tianluoshan site. Sci Rep 2016;6:18664.
65. Wei DW, Wei H, Gauthier AC, Song J, Jin Y, Xiao H. Superhydrophobic modification of cellulose and cotton textiles: methodologies and applications. J Bioresour Bioprod 2020;5:1–15.
66. Oyola-Reynoso S, Tevis I, Chen J, Chang B, Çinar S, Bloch J-F, et al. Recruiting physisorbed water in surface polymerization for bio-inspired materials of tunable hydrophobicity. J Mater Chem 2016;4: 14729–38.
67. Oyola-Reynoso S, Kihereko D, Chang BS, Mwangi JN, Halbertsma-Black J, Bloch J-F, et al. Substituting plastic casings with hydrophobic (perfluorosilane treated) paper improves biodegradability of low-cost diagnostic devices. Ind Crop Prod 2016;94:294–8.
68. Gregory PR, Martin A, Chang BS, Oyola-Reynoso S, Bloch JF, Thuo MM. Inverting thermal degradation (((i)TD) of paper using chemi- and physi-sorbed modifiers for templated material synthesis. Front Chem 2018;6:338.
69. Liu J, Sun L, Xu W, Wang Q, Yu S, Sun J. Current advances and future perspectives of 3D printing natural-derived biopolymers. Carbohydr Polym 2019;207:297–316.
70. Lille M, Nurmela A, Nordlund E, Metsä-Kortelainen S, Sozer N. Applicability of protein and fiber-rich food materials in extrusion-based 3D printing. J Food Eng 2018;220:20–7.
71. Pattinson SW, Hart AJ. Additive manufacturing of cellulosic materials with robust mechanics and antimicrobial functionality. Adv Mater Technol 2017;2:1600084.
72. Markstedt K, Mantas A, Tournier I, Martínez Ávila H, Hägg D, Gatenholm P. 3D bioprinting human chondrocytes with nanocellulose-alginate bioink for cartilage tissue engineering applications. (1526-4602 Electronic).
73. Mahmood H, Moniruzzaman M. Recent advances of using ionic liquids for biopolymer extraction and processing. Biotechnol J 2019;14:1900072.
74. Swatloski RP, Spear SK, Holbrey JD, Rogers RD. Dissolution of cellose with ionic liquids. J Am Chem Soc 2002;124:4974–5.
75. Freemantle M. Designer solvents. Chem Eng News Arch 1998;76:32–7.
76. Mehta MJ, Kumar A. Ionic liquid stabilized gelatin–lignin films: a potential UV-shielding material with excellent mechanical and antimicrobial properties. Chem Eur J 2019;25:1269–74.
77. Xu J, Tan X, Chen L, Li X, Xie F. Starch/microcrystalline cellulose hybrid gels as gastric-floating drug delivery systems. Carbohydr Polym 2019;215:151–9.
78. Mahmood H, Moniruzzaman M, Yusup S, Welton T. Ionic liquids assisted processing of renewable resources for the fabrication of biodegradable composite materials. Green Chem 2017;19:2051–75.
79. Tolesa LD, Gupta BS, Lee M-J. Chitin and chitosan production from shrimp shells using ammonium-based ionic liquids. Int J Biol Macromol 2019;130:818–26.
80. Asgari S, Saberi AH, McClements DJ, Lin M. Microemulsions as nanoreactors for synthesis of biopolymer nanoparticles. Trends Food Sci Technol 2019;86:118–30.
81. Klier J, Tucker CJ, Kalantar TH, Green DP. Properties and applications of microemulsions. Adv Mater 2000;12:1751–7.
82. Demirbas A. Biofuels sources, biofuel policy, biofuel economy and global biofuel projections. Energy Convers Manag 2008;49:2106–16.
83. Sims R, Mabee W, Saddler J, Taylor M. An overview of second generation biofuel technologies. Bioresour Technol 2009;101:1570–80.

Fan Qi, Zhang Chaoqun, Yang Weijun, Wang Qingwen and
Ou Rongxian*

2 Lignin-based polymers

Abstract: On the basis of the world's continuing consumption of raw materials, there was an urgent need to seek sustainable resources. Lignin, the second naturally abundant biomass, accounts for 15–35% of the cell walls of terrestrial plants and is considered waste for low-cost applications such as thermal and electricity generation. The impressive characteristics of lignin, such as its high abundance, low density, biodegradability, antioxidation, antibacterial capability, and its CO_2 neutrality and enhancement, render it an ideal candidate for developing new polymer/composite materials. In past decades, considerable works have been conducted to effectively utilize waste lignin as a component in polymer matrices for the production of high-performance lignin-based polymers. This chapter is intended to provide an overview of the recent advances and challenges involving lignin-based polymers utilizing lignin macromonomer and its derived monolignols. These lignin-based polymers include phenol resins, polyurethane resins, polyester resins, epoxy resins, etc. The structural characteristics and functions of lignin-based polymers are discussed in each section. In addition, we also try to divide various lignin reinforced polymer composites into different polymer matrices, which can be separated into thermoplastics, rubber, and thermosets composites. This chapter is expected to increase the interest of researchers worldwide in lignin-based polymers and develop new ideas in this field.

Keywords: epoxy resins; lignin-based polymers; lignin reinforced polymer composites; phenol resins; polyester resins; polyurethane resins.

2.1 Introduction

Lignin is the only renewable source of aromatic chemistry. It consists of three basic structural units: *p*-coumaryl alcohol, coniferyl alcohol, and sinapyl alcohol [1, 2], as shown in Figure 2.1. These basic units are connected by ether and carbon-carbon bonds [3].

*Corresponding author: Rongxian Ou, Key Laboratory for Biobased Materials and Energy of Ministry of Education, College of Materials and Energy, South China Agricultural University, Guangzhou, 510642, P. R. China; and Guangdong Provincial Laboratory of Lingnan Modern Agricultural Science and Technology, Guangzhou, P. R. China, E-mail: rongxian_ou@scau.edu.cn
Fan Qi, Zhang Chaoqun and Wang Qingwen, Key Laboratory for Biobased Materials and Energy of Ministry of Education, College of Materials and Energy, South China Agricultural University, Guangzhou, 510642, P. R. China; and Guangdong Provincial Laboratory of Lingnan Modern Agricultural Science and Technology, Guangzhou, P. R. China
Yang Weijun, The Key Laboratory of Synthetic and Biological Colloids, Ministry of Education, Jiangnan University, 214122 Wuxi, P. R. China

This article has previously been published in the journal Physical Sciences Reviews. Please cite as: F. Qi, Z. Chaoqun, Y. Weijun, W. Qingwen, and O. Rongxian "Lignin-based polymers" *Physical Sciences Reviews* [Online] 2021, 7. DOI: 10.1515/psr-2020-0066 | https://doi.org/10.1515/9781501521942-002

Lignin is low-cost and has many fascinating properties such as biodegradability, antioxidant activity, high thermal stability, and favorable stiffness, or high elastic modulus [4–6]. These advantages drive the development of lignin as a versatile product. However, the low reactivity and brittleness of lignin and its incompatibility with other polymer systems make it virtually unsuccessful in the manufacture of lignin-based high-performance materials [7, 8]. It is hoped that these shortcomings can be overcome by the chemical or physical modification of lignin and the synthesis of lignin-derived polymers [7, 9].

Lignin vascular plants were formed by chemical controlled random phenol free-radical coupling polymerization. However, the lignification theory is still under discussion, because enzyme pathways, called "dirigent" proteins, have been proposed. The composition, molecular weight, and content of lignin vary between plant species. The abundance of lignin usually decreases in the order of softwood > hardwood > grass. Schematic diagrams of hardwood and softwood lignin structures are shown in Figures 2.2 and 2.3, respectively. The structure is just a figure and does not indicate a specific order. The components derived from coniferyl, sinapyl, and *p*-coumaryl alcohol are shown in color, and some examples are given to illustrate the linkages among these components. The linkages include β-O-4, 4-O-5, β-5, β-1, 5–5, dibenzodioxocin, and β–β linkages, of which the β-O-4 linkage is dominant, consisting of >40% of the linkage structures.

Lignin is more sustainable than some other renewable polymer feedstocks, such as scarce mint or quercetin, and edible sugars and starch, as it is neither rare nor a source of food. So far, various studies have reported on different lignin valorization methods for modifying lignin and synthesizing lignin-based polymers [12, 13]. Some critical review articles have emphasized the problem of lignin being converted into functional materials. The mechanical and thermal characteristics of lignin-based copolymers, composites, and blends are discussed by Argyropoulos and coworkers [13]. The recent advances in lignin applications in developing green polymer composites and hydrogels are reviewed by Thakur and coworkers [14, 15]. The existing methods and strategies for preparing functional carbon materials form lignin by the thermochemical conversion are described by Liu and coworkers [16]. Kai and Zhao et al. summed up the

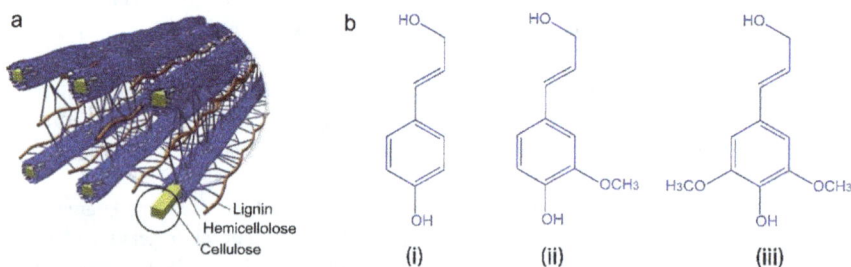

Figure 2.1: (a) The positional relationship between cellulose, lignin, and hemifibers in lignocellulose [10]. (b) Monolignol monomer, (i) *p*-coumaryl alcohol (4-hydroxyl phenyl, H), (ii) coniferyl alcohol (guaiacyl, G), (iii) sinapyl alcohol (syringyl, S).

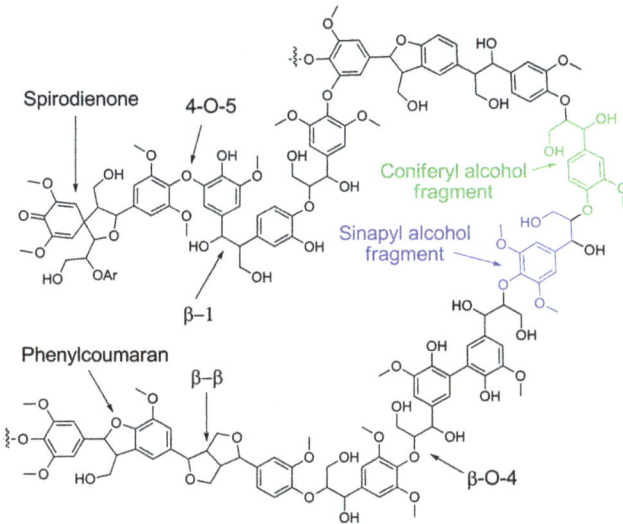

Figure 2.2: Schematic diagram of a hardwood lignin structure and some main linkages [11].

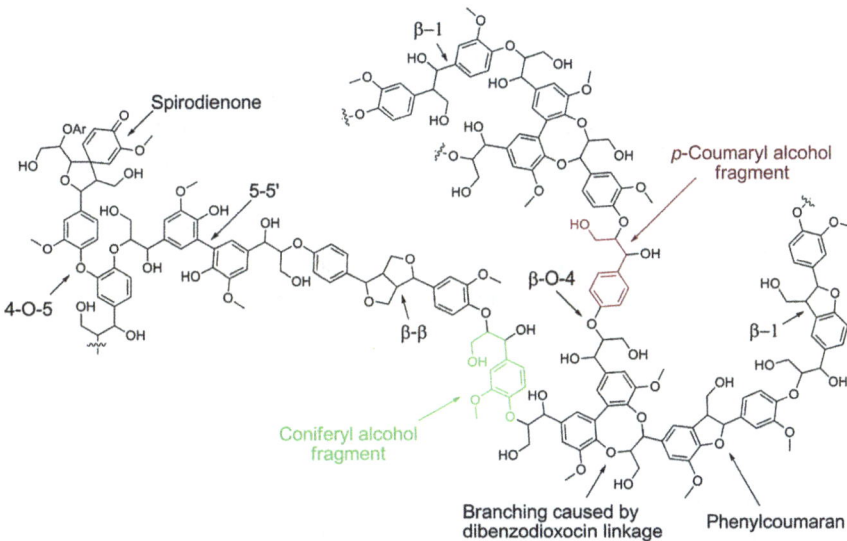

Figure 2.3: Schematic diagram of a softwood lignin structure and some main linkages [11].

recent advances in lignin from the perspective of different approaches to the synthesis of oligomeric copolymers, the resulting features, the various design methods to engineer lignin-based polymers and the potential applications of lignin-based functional materials [7, 17]. Vishtal et al. [18] summarized different methods of extracting lignin and pointed out the challenge of how to convert and utilize it in industrial applications. Lawoko et al. [19] reviewed the direct usage of lignin and its chemical modifications toward micro- and nanostructured materials. Sternberg and coworkers recently summarized the literature on lignin-derived polymers in the past 20 years, and focused on a comprehensive assessment of the greenness and sustainability of the latest technology for lignin-derived polymers [20].

2.2 Lignin recovery

At present, traditional pulp mills are the main suppliers of technical lignins. Most of the commercially available technical lignin types are fabricated from Kraft or sulfite pulping processes [21]. In addition to these two processes, there are other processes through which lignin can be recovered, such as soda pulping, steam explosion, acid decomposition, biological treatment, organic solvent recovery and others [22]. This section briefly discusses possible processes for recovering lignin.

2.2.1 Kraft pulping

Kraft pulping is the main source of industrial lignin [23]. In pulping, cellulose fibers are obtained from lignocellulosics by treating it with sodium sulfide and sodium hydroxide at high temperature and high pH [24]. Hemicellulose and some cellulose are degraded to isosaccharic acid. Thus, Kraft paper pulping liquid is composed of lignin (47%), hydroxy acids (28%, such as isosaccharic acid), inorganic substances and small amounts of other organic substances [25]. Lignin can be extracted from these waste pulping liquors as a precipitate by acidification. Almost all Kraft lignin is burned for energy recovery, but the United States, for example, produces about 35,000 tons per year for many chemical by-products [23].

2.2.2 Sulfite pulping

Sulfite pulping is a general term used to define many sulfate chemical wood pulp processes that are carried out at different levels of pH [26]. In the acid sulfite pulping process, wood is treated at low pH, high temperature and high pressure with sodium, ammonium, magnesium or calcium sulfite/bisulfite solution [27]. Through this process, lignin, hemicellulose and extracts are broken down and the fiber's cellulose is left behind. Hemicellulose is hydrolyzed critically to release water-soluble monosaccharides in spent sulfite liquor. Depending on the pulping process, lignosulfonate can be recovered either as sodium, calcium, magnesium, or ammonium salt [28]. Currently, only about 20% of all lignosulfonates produced are used as chemicals, such as economic binders for animal feed and road asphalt [29]. In addition, lignosulfonates can also be used as concrete or cement water-reducing agents and oil well drilling [29].

2.2.3 Soda pulping

Soda lignin is derived from the soda or soda-anthraquinone pulping process. Soda pulping is mainly used for annual crops such as straw, flax, bagasse and hardwood. Compared with the sulfate method, the main difference is the sulfur-free medium of the cooking liquor. Soda lignin does not contain sulfur, which means that soda lignin have a chemical composition that is closer to that of natural lignin than Kraft lignin and lignosulfonates.

2.2.4 Organosolv

Organosolv is a method of treating plant tissues with aqueous solutions of organic solvents (usually containing trace amounts of inorganic acids) [30]. These solvents include ethanol, methanol, butanol, phenol, acetic acid, ethyl acetate, and so on. So far, this method has not been commercially applied due to the amount of organic solvents consumed and the low quality of pulp fibers obtained [31]. The lignin molecular weight obtained by organic solvent extraction is comparatively low [32]. The data on physico-chemical characteristics of Kraft, lignosulfonate, soda, and organosolv lignin are shown in Table 2.1.

Table 2.1: Properties of technical lignins [7].

Technical lignins	Kraft	Lignosulfonate	Soda	Organosolv
Separation methods	Precipitation (pH change)	Ultrafiltration	Precipitation (pH change)	Precipitation (addition of non-solvent)
	Ultrafiltration		Ultrafiltration	Dissolved air flotation
Product status	Industrial	Industrial	Industrial	Laboratory/Pilot
Molecular weight ($\times 10^3$ g mol^{-1})	1.5–5 (<25)	1-50 (<150)	0.8–3 (<15)	0.5–5
Polydispersity	2.5–3.5	6–8	2.5–3.5	1.5–2.5
Sulfur (%)	1.0–3.0	3.5–8.0	0	0
Nitrogen (%)	0.05	0.02	0.2–1.0	0–0.3
Acid soluble lignin (%)	1–5	/	1–11	~2
Solubility	Alkali, some organic solvents (pyridine, DMF, and DMSO)	Water	Alkali	Wide range of organic solvents
T_g (°C)	140–150	130	140	90–110
T_d (°C)	340–370	250–260	360–370	390–400

T_g: Glass transition temperature; T_d: thermal decomposition temperature.

2.2.5 Acidolysis

During acid decomposition, plant material is saccharified by inorganic acids such as hydrochloric acid or sulfuric acid, and lignin is extracted and recovered as insoluble residues [33]. In the past, this process was used in Europe; In the Western Hemisphere, however, it was not considered as commercially important [34].

2.2.6 Enzymatic liberation

In the process of enzymatic hydrolysis, residues rich in lignin obtained by treating plant material with hemicellulase, cellulase, or pectinase [35]. This kind of lignin preparation is more similar to the "native" lignin because of the moderate processing environment employed [36].

2.3 Lignin derived polymers

2.3.1 Lignin as a macromonomer

The structure of lignin can be easily modified to develop novel materials, especially phenol and aliphatic hydroxyl groups [17]. Depending on the targeted application, lignin can be used with or without chemical modification. Without chemical conversion, lignin can be added directly to the polymer matrix, thereby reducing product costs and improving performance. For instance, unmodified lignin can be applied as an antioxidant, ultraviolet stabilizer, flame retardant, and additive to improve the rheological property of final products [37]. Although lignin has potential for direct industrial application, it can only be added in small quantities due to its weak mechanical properties and poor thermal stability. Lignin, on the other hand, can be chemically modified as the starting material for polymer synthesis or transformation into chemicals and fuels. There are four different ways to chemically modify lignin [38, 39]: (1) To depolymerize or fragment, using lignin as a carbon source or to break it down into small fragments containing aromatic rings; (2) Modifying lignin by synthesizing new chemical activity sites; (3) Chemical modification of hydroxyl groups in lignin structures; and (4) production of grafted copolymers. It is worth noting that the above chemical modification methods are highly dependent on the reactivity of lignin functional groups and the structural characteristics of lignin raw materials.

Lignin is represented in its aliphatic and phenolic hydroxyl groups at the C_α and C_γ positions on the side chains. Hydroxyl modification can lead to the formation of lignin derivatives. The effects of different functional groups on the modification of lignin have been explored (Figure 2.4a). Moreover, the modifications include the synthesis of new macromolecular monomers, which are more efficient and responsive by changing the

nature of chemical activity sites or increasing the activity of hydroxyl groups. Therefore, lignin's chemical reactivity is enhanced, the brittleness of lignin-based polymers
is decreased, and the solubility is improved, thus improving the processing of lignin. In
this way, a number of new chemical sites were introduced into the lignin structure
through several chemical modifications (Figure 2.4b), including amination, hydroxyalkylation, sulfomethylation, nitration, and sulfonation, etc.

Degradation of lignin to monolignols or phenolic monomers is an energy-
consuming process, which is one of the reasons why it is currently more valuable as a
fuel than a raw material [40]. Using lignin as a raw material for the synthesis of new
materials without any additional degradation, it is beneficial for the environment and
energy [6]. The structure of lignin varies greatly depending on its source and the process
of separation from the pulping [41]. Despite these differences, all the separated lignin
contained both aliphatic and phenolic hydroxyl groups [42]. These functions can be used
as connectors to create lignin-based macromonomers [43]. Currently, lignin as a nature
macromonomer can be used to synthesize polyurethane, polyester, epoxy, phenolic
resins and other polymers, which are briefly discussed in the next subsections [44].

2.3.1.1 Polyurethanes

Polyurethane is synthesized from diisocyanate and polyol by the poly-addition
reaction with terminal hydroxyl groups, and has polyurethane groups in the backbone

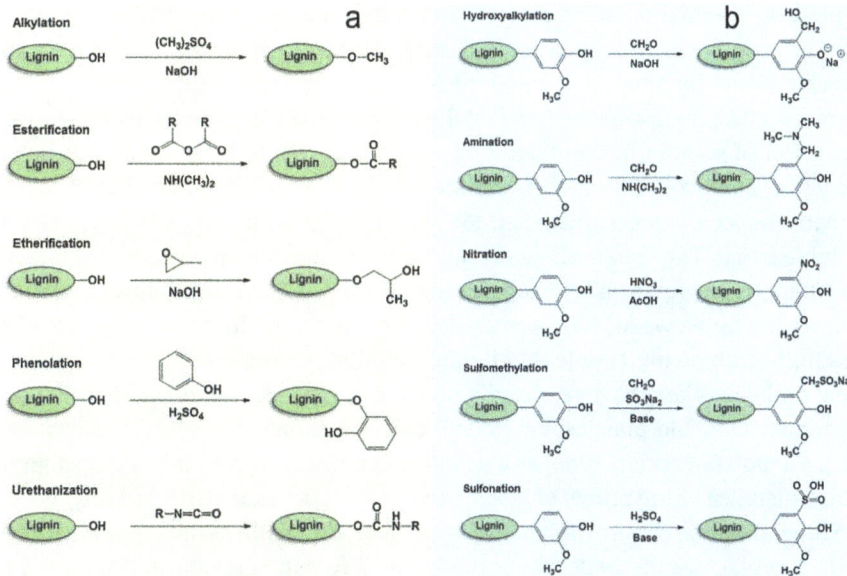

Figure 2.4: Schematic diagram of the chemical modifications of lignin. (a) Modification of the
hydroxyl groups; (b) Generation of new chemically active sites [7].

[45]. Lignin contains abundant aliphatic and phenolic hydroxyl functions, so it can easily replace the polyol component of polyurethane [46]. Moreover, biobased polyurethanes are more degradable than those originate from petroleum-based polyols [47, 48].

By controlling reaction temperature and reaction time, Duong and his colleagues produced a high-molecular-weight polyurethane [49]. Using Kraft lignin as raw material, molecular weights of the product up to 912,000 g/mol were obtained in only 3 h [49]. Hatakayama and coworkers combined these four types of technical lignin (ligonsulfonate, Kraft, solvolysis, and alcoholysis lignin) with polypropylene glycol (PPG) or polyethylene glycol (PEG), respectively, and with a mixture produced by reaction with diphenylmethane diisocyanate [50]. Whichever the type of lignin used, authors determined that lignin loading have a greater impact on the properties of synthetic materials than any other factor, such as the ratio of isocyanate/hydroxyl group or the molecular weight of the polyols.

Seeking after materials with enhanced stiffness, Xue and his colleagues prepared a range of lignin-based rigid polyurethane foams, adding up to 39.17 wt% lignin [51]. Compared with traditional polyurethane, these biofoams have increased thermal conductivity because of their larger cell volume and lower density. Unlike rigid polyurethane foams, flexible polyurethanes consist of a flexible and hard (cross-linked) blocks. Pohjanlehto and his colleagues synthesized a sugar-based xylaric acid, polyol, and reacted them with lignin and PEG to prepare a series of polyurethanes [52]. The increase in lignin content had almost no effect on Young's modulus or T_g, but the slight improvement in thermal stability showed that the lignin acted to stabilize the final product. Compared with the typical application of lignin blends, Yiamsawas and coworkers recently reported a method of synthesizing lignin-polyurethane nanocontainers with inverse microemulsion [53]. Before reacting with toluene diisocyanate, the suspension of sodium lignosulfonate in water is mixed with the surfactant polyethylene glycol polyricol. This method produced hollow and cross-linked lignin-based polyurethane nanocontainers (diameters of 311–390 nm) with the ability to contain water-soluble cargo. The results showed that these nanocontainers remained stable in aqueous solutions for weeks and can rapidly degrade within 24 h using natural enzymes.

However, as far as we know, there are few reports on lignin-based polyurethane polymers that combine high content of lignin and high performance.

Zhang and coworkers have reported a new poly(ε-caprolactone) modified lignin-based polyurethane bio-plastics exhibiting high performance [54]. As shown in Figure 2.5, the poly(ε-caprolactone) as a soft segment was incorporated into the lignin under the assistance of the bridge of hexamethylene diisocyanate (HDI). In this study, HDI with long aliphatic chains was used as an activator of lignin hydroxyl groups, and the poly(ε-caprolactone) is added as a biodegradable soft segment to improve the flexibility of the lignin-based polyurethane [54]. The effects of lignin content, –OH/–NCO molar ratio, and the molecular weight of poly(ε-caprolactone) on the properties of the obtained biobased polyurethane plastics were assessed [54]. The results showed that the biobased polyurethane film still has high strength, fracture elongation and tearing strength at 19.35 MPa, 188.36% and 38.94 kN/m with the content of lignin high as 37.3%, respectively. In addition, the sample has excellent thermostability and has a good solvent-resistance. Thus, lignin modification toward the urethane chemistry is an effective approach to develop lignin-based sustainable and high-performance materials [54].

In recent years, the free radical induced thiol-ene reaction has been widely concerned by materials science and synthetic chemistry because of its low sensitivity to oxygen, easy implementation, and high yield without complicated purification methods [55, 56]. As shown in Figure 2.6, Cao and coworkers have reported an novel lignin-based polyurethanes coatings, which were synthesized through polymerization of lignin-based polyol with HDI without catalysts, showed excellent corrosion resistance at high lignin content [57]. In this study, the reactivity and solubility of lignin-based polyol was increased, while the phenol hydroxyl groups of enzymic hydrolysis lignin were selectively transformed into aliphatic hydroxyls via alkylation and thiol-ene reaction [57]. Due to the good dispersion and crosslinking reaction of lignin-based polyol in polyurethane network, the obtained lignin-based polyurethanes coatings have excellent mechanical properties, high thermostability, and strong corrosion-resistance. These characteristics give lignin-based polyurethanes great potential for applications in areas requiring high mechanical strength, high thermostability, and high corrosion resistance, namely, coating, adhesives and electronics. However, further optimization of the lignin-based polyurethanes coatings formulation is needed to enhance the toughness of LPU coatings and elucidate the mechanism of corrosion-resistance [57].

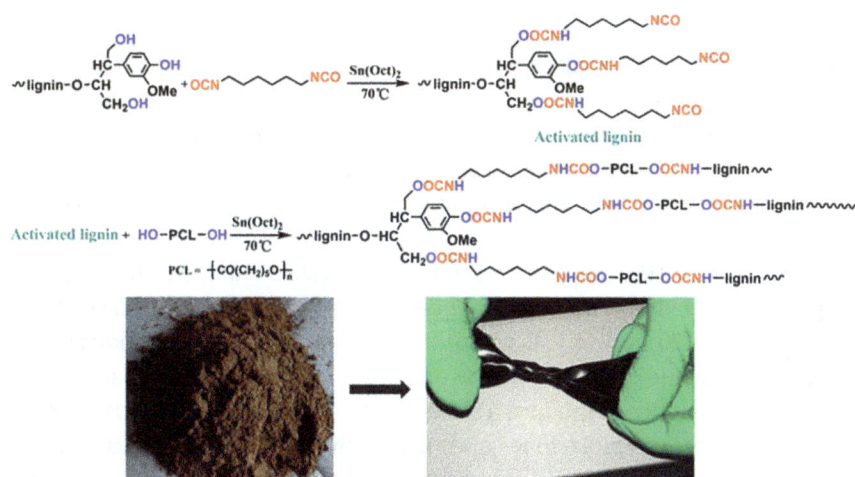

Figure 2.5: Synthesis route of the poly(ε-caprolactone) modified lignin-based polyurethane bio-plastics [54].

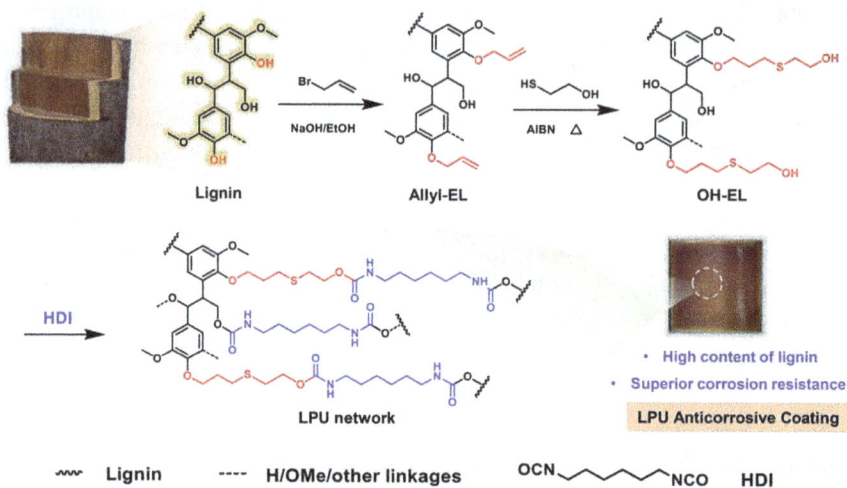

Figure 2.6: The synthesis route of lignin-based polyol and the schematic structure of anticorrosive coatings of thermosetting lignin-based polyurethanes [57].

Although above results support the concept of lignin incorporation, many properties require further adjustment before lignin-based polyurethanes e can be used as a commercial material. In addition, batch-to-batch variation in lignin will result in batch-to-batch variation in these systems.

2.3.1.2 Polyesters

Polyesters are polymers containing ester bonds that can be prepared in three different ways: esterification of dicarboxylic acid with diol or dihalides, self-polyesterification of hydroxycarboxyl acids, and the ring-opening polymerizations of lactones or cyclic esters [58]. The application of lignin as macromolecular monomer of polyester synthesis usually limits the polycondensation reaction, but the ring opening reaction of lactone is also reported.

Guo and coworkers [59] prepared a series of polyesters from dicarboxylic acid chlorides, polyethylene glycol, and lignin. These polymers have been exhibited to be able to melt between 120 and 140 °C without changing smell or color, indicating a potential commercial use [60]. Bonini and coworkers [61] reacted steam-explosion lignin with dodecandioyl dichloride to produce a series of low molecular weight (1915–6382 g/mol) polymers. Using ε-caprolactone and carboxylic acid-functionalized lignin, Matsushita and coworkers [62] used the condensation reaction of ε-caprolactone and carboxylic acid-functionalized lignin to prepare polyester. Compared with poly ε-caprolactone, the obtained polymer has higher melting point and higher loss and storage modulus. Sivasankarapillai and his colleagues [63] created a network of highly branched polymers by condensing lignin with a tri-branched carboxylic acid monomer, which was synthesized by the condensation of adipic acid with 1,1,1-triethanolamine.

Recent reports using another potentially biobased monomer, succinic anhydride (SAn), show the resulting materials has similar hardness and scratch resistance to petroleum-based polyester coatings [64]. Succinylated lignin dissolved in THF can be self-crosslinked by adding traditional crosslinking agents (such as diamines) (Figure 2.7) [64]. The resulting novel lignin-based thermosetting polyester coating systems are based on functionalization of the lignin-soluble fraction of Kraft lignin recovered by solvent extraction [64]. The synthetic polyesters were thoroughly characterized by chemical, physical and thermal properties to confirm the succinic anhydride was successfully covalent with lignin. These polyester coatings have film forming ability, thermostability, dynamic surface hardness, solvent resistance, and higher hydrophobicity than unmodified samples. The results of this study suggest that the development of lignin thermosetting polyester systems provide evidence of the potential bioderived coatings and adhesives materials [64].

The low solubility of lignin in common organic solvents limits the way of modification [65]. At present, a large number of researches show that lignin-based copolymers can play an active role in mechanical enhancement of block composites [66, 67]. However, few researches have been done on lignin-based porous materials. Lignin materials with porous structures may be more interesting than bulk materials because of their advantages of low density, high surface area, good absorbency and high permeability. Taking advantages, lignin-based porous composites have the

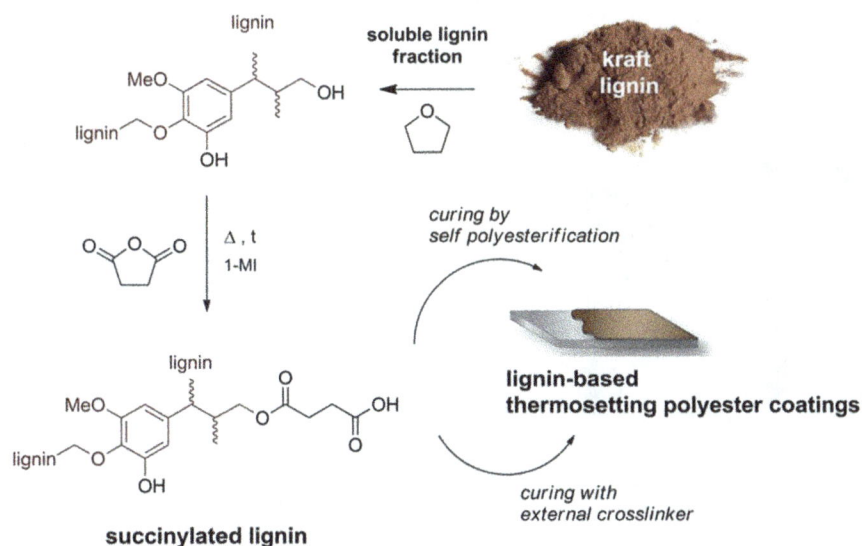

Figure 2.7: Synthetic scheme for the succinylated lignin for preparing crosslinked polyester coatings [64].

potential to be used in a variety of high-value applications, such as energy storage, catalysts, sensors and biomedical materials.

Recently, Kai and coworkers developed a class of novel lignin-poly (ε-caprolactone-co-lactide) copolymers (lignin-PCLLA) through ring-opening polymerization without solvent [68]. The obtained Lignin-PCLLA copolymers have adjustable molecular weights and glass transition temperatures [68]. It is blended with polyester by electrospinning to engineer the copolymer into ultrafine nanofibers (Figure 2.8). The size of the nanofibers is around 300–500 nm, and the tensile test showed that the mechanical properties of nanofibers can be improved by the mixture of lignin copolymer and polyester matrix. In addition, lignin nanofibers have good biocompatibility and antioxidant activity, demonstrating their great potential in healthcare applications [68].

2.3.1.3 Epoxide resins

Epoxy resin, one of the most commonly used thermoset resin materials in the fields of adhesives, coatings, composites, and electrical laminates, etc., is composed of monomers containing at least one epoxy group. Epoxy groups can be homopolymerized by cation or anion polymerization or copolymerized with comonomers, such as multifunctional amine, acids, anhydride, alcohols, and phenols [69]. The choice of epoxides and curing agents has a great influence on the mechanical, physical, and chemical properties of the products. Although lignin itself does not contain epoxy groups, the macromonomer phenolic can also be used as a curing agent/crosslinking agent for the synthesis of epoxide resins.

So far, the methods of adding bulk lignin to the thermosetting materials of epoxy resins can be summarized into three categories: the first type is the direct mixing of the

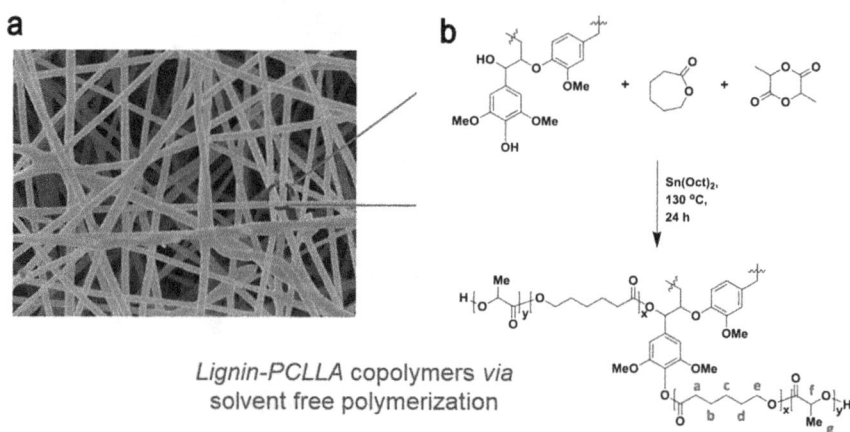

Lignin-PCLLA copolymers via solvent free polymerization

Figure 2.8: (a) SEM images of electrospun of lignin-PCLLA. Scale bars = 1 μm. (b) Synthetic scheme of ring-opening copolymerization of ε-caprolactone, L-lactide, and lignin [68].

lignin derivatives as fillers with the ordinary epoxide resin; the second type is modifying lignin by epoxidation; and the third type is modifying lignin or its derivatives to enhance reactivity, therewith epoxidation. It is important to note that epoxy prepolymers or glycol ethers should be liquid at room temperature or elevated temperature in order to fully contact and react with the curing agent to form a homogeneous cross-linking network. Nevertheless, most reports on the epoxidation of lignin is a non-molten solid, which cannot be directly cured by curing agents [8, 70, 71].

Recently, by mixing alkaline solution of Kraft lignin with cross-linking agent of polyethylene glycol diglycidyl ether, Nonaka and his colleagues have synthesized a water-soluble variation [72]. In this system, the transition temperature of glass can be adjusted conveniently by adding alternate crosslinkers. Engelmann and his colleagues combined a low molecular weight fraction of lignin with 1,3-glycerol diglycidyl ether to produce a battery of solvent-free resins containing up to 50 wt% lignin [73]. Comparing with those control group cured by pyrogallic acid, the higher lignin content of the lignin-based resins showed better thermostability.

Since most lignin is insoluble in organic solvents, such as enzymatic hydrolysis of lignin, and most lignin modification requires the use of organic solvents as reaction media, partial depolymerization can remarkably improve the solubility of lignin for further application [70]. Zhang and coworkers [70] have reported that Kraft lignin can be partially depolymerized through alkali catalysis in supercritical methanol to increase its solubility in organic solvents. By reacting with succinic anhydride, the partially depolymerized lignin is modified into lignin-based polycarboxylic acid (LPCA) [70]. LPCA can be applied to curing agent of epoxy resin. The results showed that LPCA could cure commercial epoxy resin at a similar temperature range. LPCA cured DER 353 resins showed a mild T_g and similar storage modulus to that of commercial anhydride cured products.

Glycidylation of lignin generally produce solid phase epoxy prepolymers, which will lead to the problem of poor compatibility with the curing agent. In order to solve this problem, Zhao and coworkers has reported a route to synthesize liquid lignin-containing epoxy prepolymers (Figure 2.9) [74]. In detail, lignin was phenolated by catechol (a renewable lignin derivatives), which has more phenolic hydroxyl groups than phenol [74]. The catecholized lignin was condensed with salicyl alcohol to form a novolac oligomer incorporating lignin, which was then glycidylated with epichloro-hydrin to prepare a fully liquid epoxy prepolymer [74]. These lignin-based epoxy prepolymers were cross-linked by diethylenetriamine to produce homogeneous lignin-based epoxy networks (catecholized lignin content reaching 40 wt%). Compared with the solid phase counterparts, the resulting liquid lignin-based epoxy prepolymers would present wider applications.

The statistical chemical structure and poly-dispersion of lignin macromolecules limit the properties of the final materials. Therefore, in material science, extraction of lignin from mild plant extracts (such as organic solvents, enzyme hydrolysis, and partial depolymerization) is preferred. However, these lignins are not available on a

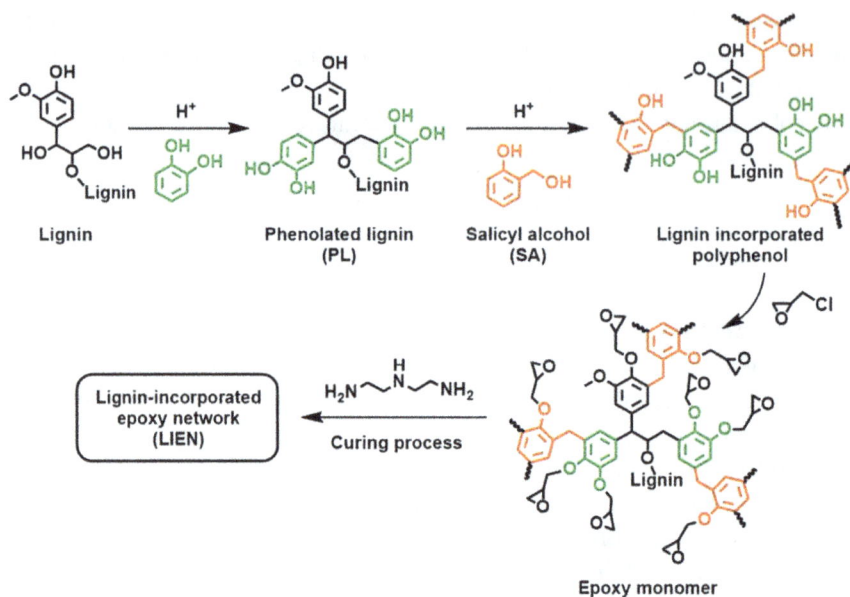

Figure 2.9: Synthetic scheme of catecholized lignin containing epoxy network [74].

large scale, currently, in contrast to the industrial Kraft lignin. Another way to overcome these problems is to use a purified fraction of industrial lignin.

Recently, Gioia and coworkers reported that the thermosetting epoxy resins was synthesized from high functionality and low molecular weight lignin Kraft, which was refined by solvent extraction [8]. The sequential extraction process used organic solvents such as methanol, ethanol, ethyl acetate, and acetone (Figure 2.10a). The extraction sequence was designed to recover the fractions with low dispersivity and gradually increasing molecular weight (Figure 2.10c). Then, through the reaction of epichlorohydrin with various lignin fractions, the oxirane groups were introduced by a glycidylation method (Figure 2.10b). The tensile strength of the resin varies from 1.2 to 5.0 MPa, increasing with the increase of molecular weight of the lignin fractions (Figure 2.10d). The proposed approach offers an unprecedented possibility of adjusting the network structure and performance of thermoset epoxy resins on the basis of real lignin fractions instead of lignin model compounds.

Generally, lignin imparts thermostability to the produced resins, but cannot be the sole crosslinking agent. Thus, another crosslinker must be added to obtain the desired performance. Feldman and coworkers studied a series of lignin-based epoxy resins with >40 wt% of lignin loading and other hardeners incorporated [75–77]. Addition of lignin can increase the adhesive strength of the tension, so it can reduce the cost of production of materials requiring robust adhesion [76]. In addition, the properties of the hardeners do not affect the physical performance of adhesives as much as the amount of lignin

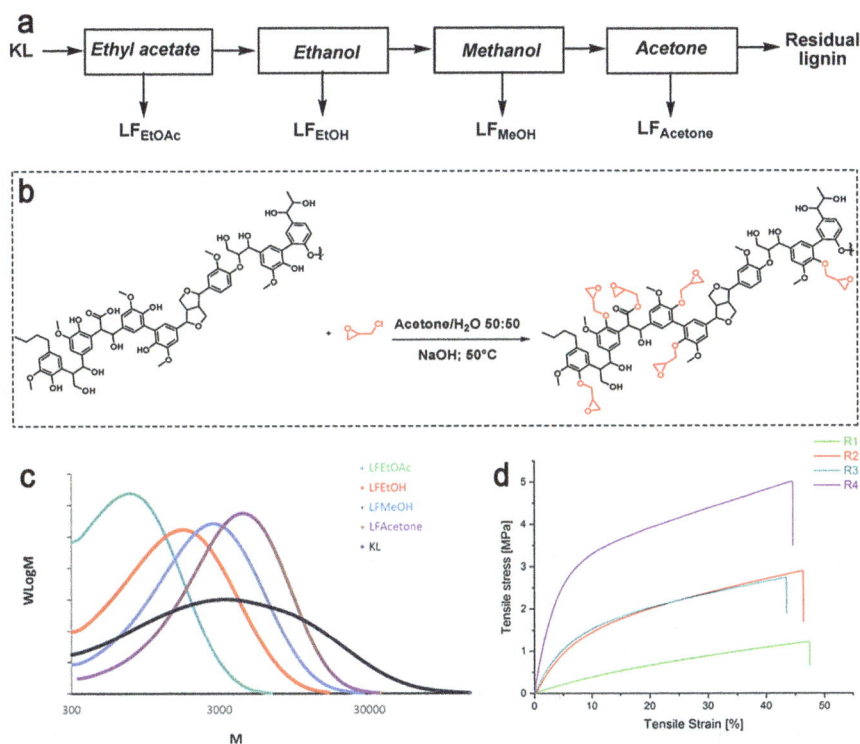

Figure 2.10: (a) Scheme of the sequential refining strategy; (b) Introduction of Oxirane Moieties on Lignin; (c) SEC analysis of the starting lignin compared with the obtained fractions; (d) The tensile stress–strain curves of the different lignin-based epoxy resins [8].

addition, which means that the least costly materials can be selected to reduce production costs [77].

2.3.1.4 Phenolic resins

Because of its unique thermal stability and flame retardancy, phenolic resin (a thermosetting rigid structure) has been widely used in the fields of petrochemical, construction, vehicles, ships, and aerospace industry [78]. The phenol/aldehyde ratio controls the cross-linking degree of the final product and determines whether the resin must be hardened by adding cross-linking agents [79]. For lignin-based phenolic resins, the phenolic chemical structure of lignin makes it a natural substitute for petroleum-based phenol in conventional synthesis schemes (Figure 2.11) [80]. As far as we know, there are no lignin-only phenolic resins, but there have been several attempts to replace partial phenol used with lignin to prepared lignin-based phenolic resins.

The most common type of phenolic resin is made from formaldehyde and phenol and is commonly used as adhesives. Hence, viscosity, cure rate, and the mechanical

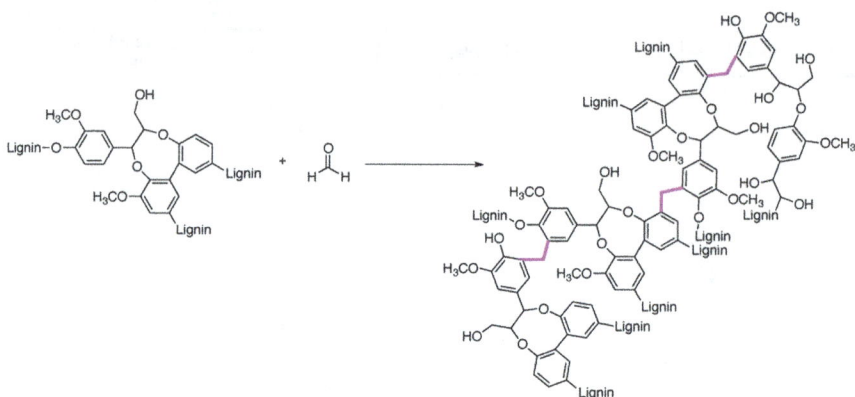

Figure 2.11: General scheme of phenolic resin produced by substituting phenol with lignin [4].

properties of the phenolic resin are important factors to determine their proper application. Alonso and coworkers explored the vitrification and gelation of lignin-based phenolic resins [81]. In these isometric experiments, they found that lignin-containing resins had lower activation energy than neat phenolic resin, resulting in lower curing degree at the point of gelation [81]. Vázques and his colleagues [82] added lignin to a phenolic resin applied as an plywood adhesive. Lignin was dissolved in phenol and reacts with formaldehyde to form an adhesive that combines about 20 wt% of the lignin incorporated into the final material. Although the adhesive took longer time to cure than the traditional phenolic resin, the strength of the plywood was the similar as that produced using the conventional method. Danielson and his colleagues also developed lignin-based phenolic resins as plywood adhesives, and further evaluate its potential commercial applications [83, 84]. The resins with different proportions of lignin were prepared. The results showed that boards adhered with lignin-based phenolic resin adding 20 and 60 wt% lignin had shear strength at the superior to that of standard phenol-formaldehyde resins. Due to the brittleness of lignin, the strength of adhesives was significantly reduced by adding 80 wt%. Additionally, Dennison et al. [83] found that 50 wt% lignin addition is the ideal biobased phenolic resin to maintain resin viscosity, adhesion, and storage stability.

In fact, lignin is usually less reactive than phenol because it reacts with formaldehyde in fewer locations, consequently, it is relatively difficult to obtain highly reactive resins acceptable for foaming. Thus, many methods, such as phenol [85], depolymerization [86], and liquefaction [87], have been used to modify lignin to improve its reactivity.

Recently, Gao and coworkers [88] presented an effective method to depolymerize, demethylate and phenolize lignin catalyzed by HBr in phenol solvents under microwave radiation with moderate temperature at 90 °C (Figure 2.12). The reaction mixture was directly used to developed phenolic foam without further separation and purification.

Compared with the control samples, the compressive strength and thermal insulation of processing lignin-based foams were increased by 150 and 26%, respectively [88]. The results of this study provide a high value-added application of lignin in the preparation of thermal insulation materials.

2.3.2 Lignin-derived phenolic compounds as monomers

Current studies have shown that there are three monolignols exists in plants: coniferyl alcohol, *p*-coumaryl alcohol, and sinapyl alcohol. Monolignols are functional phenolic substances that can be used in aromatic polymer systems such as lignin [4]. A variety of phenolic monomers similar in structure to monolignols, including vanillin, cresols, guaiacols, and other lignin derived phenolic compounds provided by lignin pyrolysis, supports the prospect of tunable properties by simply selecting monomers to meet specific requirements [89]. With this in mind, this section will include several lignin-derived phenolic compounds (vanilline, cinnamic acid, *p*-coumaric acid, and ferulic acid) as monomers in polymer synthesis.

2.3.2.1 Vanilline derived polymers

Vanillin is currently the most readily available pure monoaromatic phenol produced on an industrial scale with lignin, which means that it is renewable and does not compete with food sources [90]. Approximately 20,000 t of vanillin are yielded every

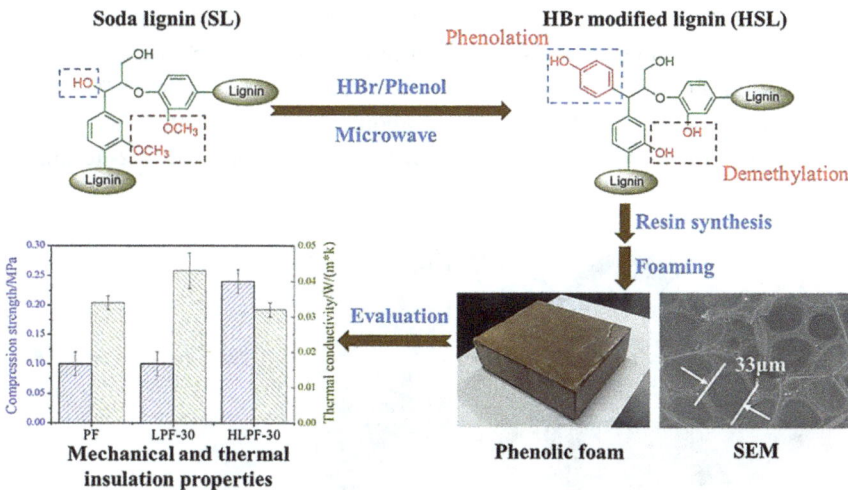

Figure 2.12: Schematic diagram of one-pot depolymerization, demethylation, and phenolation of lignin for phenolic foam preparation [88].

year, of which 15% comes from lignin [91]. Therefore, vanillin has the potential to become a key renewable aromatic building block.

Vanillin has many advantages. For example, it is a safe aromatic compound with two reactive functions that can be chemically modified. Therefore, vanillin can be regarded as a bifunctional compound, which is promising in the preparation of thermoplastic polymers.

In recent, Holmberg and coworkers [92] reported a general scheme for the synthesis of renewable homopolymers and block copolymers by the functionality of vanillin and reversible addition-fragmentation chain transfer polymerization.

The vanillin-based homopolymers showed glass conversion temperatures of 120 °C and degradation temperature over 300 °C, suggesting that these and similar polymers could be suitable substitutes for petroleum materials. In addition, using controlled techniques of polymerization, a vanillin-based homopolymer was chain-extended with lauryl methacrylate to prepare nanostructured block copolymers (Figure 2.13). The results showed that these elastomer copolymers can self-assemble in a poly(lauryl methacrylate) matrix with the form of a body-centered cubic array of vanillin-based nanospheres [92]. This work described a blueprint for controlling the polymerization of vanillin and its subsequent chain extension with a variety of comonomers, enabling the redesign and generation of novel modulated block copolymers.

Vanillin manufactures a variety of polymers, especially high-performance thermosetting materials such as epoxy resins [93] or polyphenoxylamines [94]. However, these studies mainly focus on the renewable nature of the resources and rarely address thermosets issues such as recycle trouble and flammability, etc.

Wang and coworkers prepared a novel Schiff base precursor from a rich and lignin derivative, vanillin, and produce malleable thermosetting materials combining high-performance, ultra-fast reprocessing, excellent monomer recovery, arbitrary permanent shape changes and excellent fire resistance (Figure 2.14) [95]. The Schiff base covalent adaptable networks showed high glass transition temperatures of about 178 °C, tensile strength of about 69 MPa, tensile modulus of about 1925 MPa. Meanwhile, the low activation energy of the bond exchange of 49–81 kJ mol^{-1} (reprocessed in 2–10 min at

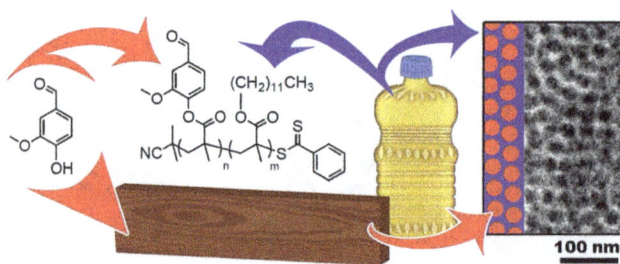

Figure 2.13: Schematic diagram of renewable homopolymers and block copolymers through the RAFT polymerization of vanillin [92].

180 °C) resulted in high malleability. These Schiff base covalent adaptive networks provide a good example for promoting the development of thermosetting polymer materials for nature biomass resources [95].

In addition, Zhang et al. [96] reported a novel biobased triepoxy (TEP) was synthesized based on vanillin and guaiacol as raw materials and cured with an anhydride as curing agent.

The cured TEP has the similar modulus, tensile strength, and thermal stability as bisphenol A epoxy resin. Moreover, under the catalytic action of Zn^{2+}, the cured TEP possessed dynamic trans-esterification reaction, which brought the sample stress relaxation and permitted reparability to the samples [96].

2.3.2.2 Cinnamic acid derived polymers

Resins made from compounds with more robust structures such as aromatic rings are more thermostable and preferable than more flexible compounds (like aliphatic chains) [4]. Lignin is used in many resin formulations, but its limited solubility is often a problem in the development and processing of new materials [97, 98]. Monolignols or lignin-derived phenolic compounds precursors contain the same desirable aromatic structure, and there is no problem of solubility of lignin macromolecule [4].

Kim and coworkers reacted cinnamic acid with a series of epoxy resins to produce a biobased cross-linked resin [99]. Then, the functional epoxy resin is cured with photo curing to make it have better thermostability and optical properties. Xin and coworkers

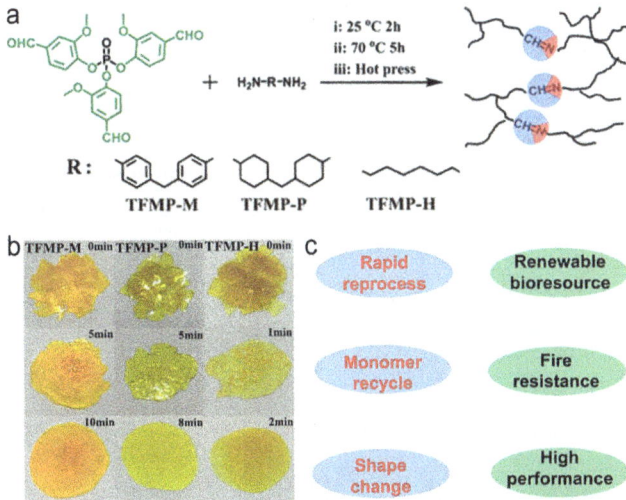

Figure 2.14: (a) Preparation scheme of the malleable Schiff base covalent adaptable networks; (b) Reprocess of Schiff base covalent adaptable networks at different times at 180 °C under 15 MPa pressure; (c) Outstanding qualities of the malleable thermosets based on lignin derivative vanillin [95].

[100] functionalized cinnamic acid by two synthetic steps (allylation then epoxidation) to form a double functional epoxide monomer. Then, this epoxy monomer was cross-linked with maleic anhydride derivatives as curing agents to produce a series of resins (Figure 2.15). The curing behaviors of the resins were alike to those of petroleum-based resins. Fache and coworkers [91] prepared epoxide resins from aromatic monomers mixture generated from the degradation process of lignin to vanillin. The mixtures (vanillin, *p*-hydroxybenzaldehyde, syringic acid, and syringaldehyde) suffered from Dakin oxidation to produced di-phenolic compounds. The obtained compound then reacts with epichlorohydrin to form the relevant epoxy compound [91].

2.3.2.3 *p*-coumaric acid derived polymers

Modified monolignols or precursors are more widely used in the literature because they produce monomer with good thermal stability.

Kaneko and his colleagues synthesized a biopolymer made by melt condensation of *p*-coumaric acid with sodium acetate and acetic acid. This is the first example of a liquid crystal polymer made from a single natural monomer, as shown in Figure 2.16 [101]. The polymer was prepared by photoinitiated cycloaddition in a liquid crystalline state and proved to have excellent cell compatibility [101]. Kaneko subsequently produced a biodegradable hyper-branched copolymer from the polymerization of 3,4-dihydrocinic cinnamate and *p*-coumaric acid [102]. The resulting material has a sufficiently high T_g to be processed, photoreactivity in liquid crystalline state, and readily degrades by hydrolysis. Matsusaki et al. copolymerized *p*-coumaric acid with lactic acid with and without solvent [103]. Although the content of *p*-coumaric acid in copolymers is much

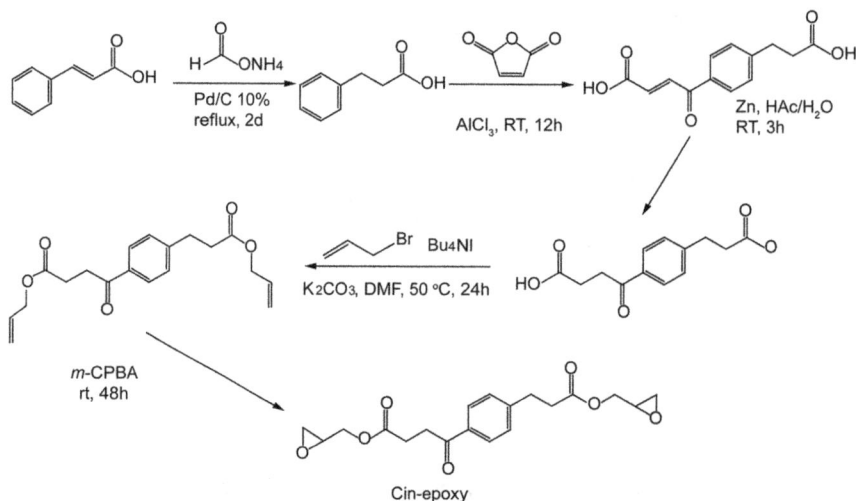

Figure 2.15: Synthesis scheme of cinnamic acid-derived epoxy [100].

lower than that of lactic acid, the obtained copolymers have good photoreactivity, high solubility, and biodegradability. Spiliopoulos and coworkers synthesized copolymer from *p*-hydroxy benzaldehyde and *p*-coumaric acid with thionyl chloride as catalyst under the condition of alkali [104]. The results showed that although copolymers are amorphous, once crosslinked, T_g of the resultant polyesters showed a significant increase.

The brittleness of the resulting *p*-coumaric acid derived polyester materials can be reduced by introducing other functionalities. Thi and his colleagues [105] reported report a way to synthesize a macromonomer by coupling 3,4-diacetoxycinnamoyl chloride with low molecular weight poly(L-lactic acid). The macro-monomer was polymerized with sodium acetate without solvents to form branched, polydispersion, and high-molecular-weight polyester [106]. The mechanical and thermal characteristics of polymers were superior than those of poly(L-lactic acid), and the degradation rate is accelerated with the addition of acetylated coumaric acid. Nagata and coworkers [107] prepared a macromonomer through capping both adipoyl dichloride with hydroxyl functionalized *p*-coumaric acid to generate a novel dichloride macro-monomer, as shown in Figure 2.17. The macro-monomer was further condensed with 1,6-hexanediol to prepare a high molecular weight copolyester resin.

2.3.2.4 Ferulic acid derived polymers

Ferulic acid can be achieved through hydrolysis of lignin based on alkalis or enzymes [108]. With an annual output of ~318 tons, it is one of the available phenolic molecular compound from lignin, second only to vanillin [109].

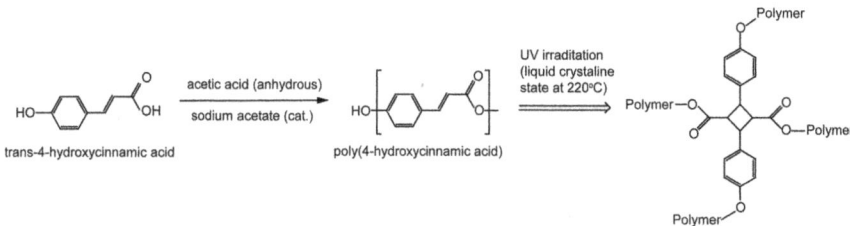

Figure 2.16: Projected cycloaddition of poly(*p*-coumaric acid) in a liquid crystalline state (220 °C) [101].

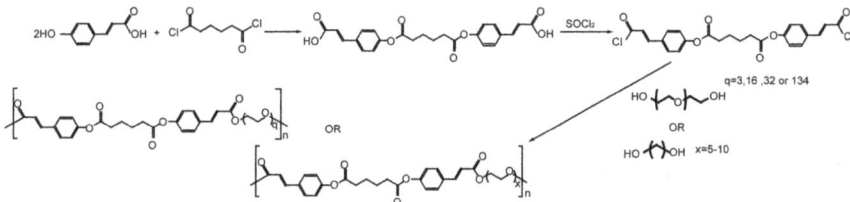

Figure 2.17: Scheme of *p*-coumaric acid dimer-based polyester from either aliphatic diols or polyethylene glycol [107].

Oulame and coworkers [110] prepared ferulic acid-based dimers via polycondensation of aliphatic diols with ferulic acid of different chain lengths or isosorbide. Then, the obtained diol was polymerized with 1,6-hexamethylene diisocyanate or 1,4-tolulene diisocyanate to produce a series of polyurethanes (Figure 2.18). These polyurethanes can be prepared with or without solvents, and showed high thermostability and controllable T_g. Although the molecular weight of these products is generally low, they support the concept of polyurethane material being synthesized from monolignol or its precursors. Chen and coworkers [111] recently synthesized polyurethane based on a dimeric cresol-based monomer. After dimerization, cresol reacts with epichlorohydrin to form a double epoxide, and then with CO_2 to produce a corresponding bis(cyclic carbonate) monomer. Then, the double functional monomer was reacted with hexamethylenediamine or isophoronediamine to generate polyurethane with excellent thermostability.

Polyphenoxyamine is a special phenolic resin, in which hexane heterocycles are synthesized via condensation of a phenol and an amine in the presence of formaldehyde [112]. The curing temperature and the physical properties of the polymer can be controlled according to the substitution of amine or phenol. Comí and his colleagues [113] synthesized a series of benzoxazines via the condensation of carboxylic acid (ferulic and coumaric acid) with formaldehyde and 1,3,5-triphenylhexahydro-1,3,5-triazine. These monomers require a curing temperature higher than their thermostable temperature, leading to decomposition instead of polymerization. Therefore, adding boron trifluoride as a ring opening catalyst can reduce curing temperature and form a variety of polymeric products, as shown in Figure 2.19. These products have proven to be thermostable and comparable to petroleum-based products [4].

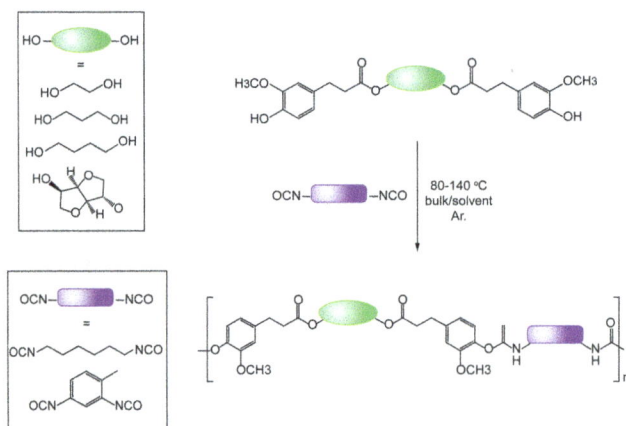

Figure 2.18: Scheme of the eight poly(ester-urethane) from ferulic acid-based diols [110].

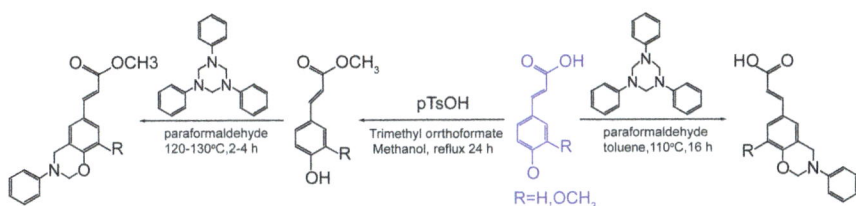

Figure 2.19: Synthetic scheme of *p*-coumaric acid and ferulic acid-based benzoxazines [113].

2.4 Lignin-based polymer composites

Polymer composites are materials achieved through strengthening polymer matrix with appropriate fibers/particle materials and can be defined in different ways [114–117]. The properties of the matrix vary from a synthetic to natural polymer depending on the application [118]. In most applications, matrices are obtained from synthetic polymers, more recently, biopolymers have been used as matrix materials [116]. Generally speaking, reinforcement materials are fibers or synthetic inlays, such as carbon, glass, and so on, or natural hemp, flax, coconut trees, bagasse, and so on. In recent years, the lignin-based polymer composites have drawn much attention in the research field at home and abroad [119, 120]. Lignin is being vigorously used as a low-cost, ecofriendly reinforcement material for the fabrication of high-performance composite materials [121, 122]. In the next section, the study of lignin-based polymer composites is described.

2.4.1 Lignin-based thermoplastic polymer composites

The ongoing lignin valorization studies, if successful, could make sustainable bio-manufacturing initiatives a reality. Thermoplastic plastics from lignin can be either lignin polymer alloys or functional (modified) lignin, or lignin copolymers.

Some researchers have developed lignin-enhanced thermoplastic composites. For example, a single screw extruder was used to prepare poly(ethylene terephthalate) (PET) composites reinforced with lignin [123]. The effects of lignin concentration on melting behavior and thermal stability were explored in detail by thermogravimetric analysis [123]. The results show that the content of lignin in the composites has great influence on the thermal stability. Experiments in the air show that adding lignin to the polymer composites is beneficial to forming protective surface shields, which has been found to reduce the diffusion of oxygen to the polymer body. Compared with pure PET, lignin in the composites also has a strong influence on the melt behavior of annealed specimens, which promotes the crystallization process [123]. Barzegar [124] have reported on the rheological behaviors and mechanical properties of lignin-reinforced polystyrene composites. Different polystyrene/lignin composites were developed by adding or not adding a linear triblock copolymer of styrene ethylbutylene styrene

(SEBS). The results indicated that the additive amount of lignin in the composites has significant impacts on the mechanical properties of the research system [124]. For instance, the bending and torsion modulus increase with the increase of the lignin load while its tensile properties decrease. The compatibilizer can remarkably improve the bending and tensile modulus of lignin/PS composites in this study. The increase in modulus is due to the improvement of interfacial adhesion between lignin and PS [124].

Both biomass and polymer industries are interested in developing low-cost lignin/ polymer composites with excellent mechanical properties. Nevertheless, because lignin tends to accumulate in polymer matrix, macroscopical phase separation at micron scale is usually obtained. The resulting lignin/polymer composites usually exhibit significantly less strength and toughness than the neat polymers [125]. In order to solve this problem, researchers have conducted works from different perspectives.

Among the synthetic biodegradable polymers, the poly(butylene adipate-co-terephthalate) (PBAT) is an aliphatic-aromatic polyester with outstanding flexibility, mechanical properties and biocompatibility, and is an ideal choice in the fields of packaging film, agriculture, and medical devices [126]. However, the high cost and low photostability of PBAT limit its widespread commercial use. Modified lignin derivatives can improve the UV-blocking, mechanical properties, and thermostability of the composites with PBAT as matrix [127]. Recently, Wang and coworkers have reported on a novel biodegradable lignin- (30–50 wt%)based composites with excellent properties, which are prepared by adding lignin into the PBAT matrix [127]. In order to improve the compatibility of lignin and PBAT, lignin was modified by a green esterification method under microwave assistance and without solvent [127] (Figure 2.20a and b). The characteristic of modified lignin has fewer inter-unit linkages, higher molecular weight, lower T_g and improved hydrophobicity. In addition, the mechanical properties of the lignin/ PBAT composites have been enhanced controllably by adding maleic anhydride (MAH) [127] (Figure 2.20c). The abtained lignin/PBAT composites have excellent UV resistance and fracture elongation, even at 40% of lignin loading, improved by 500% compared with that of unmodified specimens [127].

Generally, lignin does not form a compatible mixture with acrylonitrile butadiene styrene (ABS) thermoplastic matrix without being modified or with the assistance of the compatibility agent. Instead of using chemical modification with high cost and high energy consumption, Naskar and coworkers [126] have devised a simpler solution that improves the ductility and strength of the material through self-assembly of lignin with poly(ethylene oxide) (PEO) in ABS substrates induced by a hydrogen bonded interaction (Figure 2.21). In this study, a thermoplastic ABS polymer with ~30 wt% lignin was prepared through adding 10 wt% PEO, which has similar properties to pure resin. The ABS/lignin compositions containing PEO showed enhanced interface bonding of lignin with the thermoplastic ABS matrix. And even further, the formula was reinforced with short carbon fiber without pretreatment, exhibited outstanding mechanical properties and is suitable for automotive use [126].

a

b

R: –CH₃, –(CH₂)₆CH₃, –(CH₂)₁₀CH₃, or ∿∿∿═∿∿∿

c

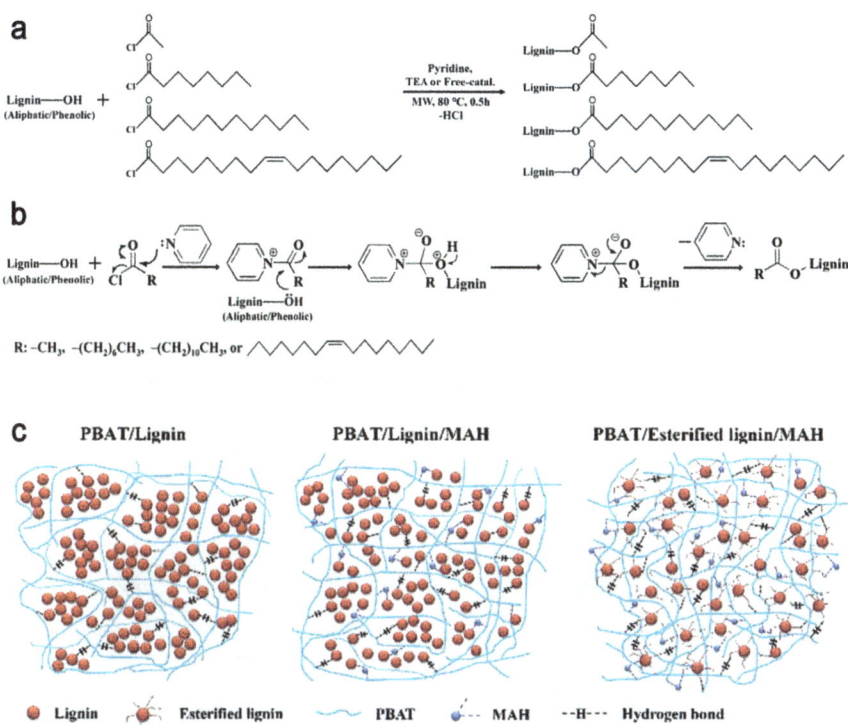

Figure 2.20: (a) schematic diagram of esterification of the phenolic/aliphatic hydroxyl groups of lignin with C_2–C_{18} acyl chlorides (TEA: triethylamine, MW: microwave); (b) the process of pyridine catalyzed lignin esterification with acyl chlorides; (c) schematic diagram of lignin/PBAT composites with esterified lignin (MAH as a compatibilizer) and unmodified lignin [127].

The issue of the easy agglomeration of lignin in a polymer matrix can also be improved recently by lignin nanoparticles (LNPs). Qiu et al. [128] have reported a simple method for preparing nanostructured biomimetic polymer materials by incorporating lignosulfonic acid (LA) into a biodegradable poly(vinyl alcohol) (PVA) matrix in the form of interspersed nanoparticles. Amphiphilic LA was prepared from sodium ligninsulfonate (LS) and then dispersed in water to form uniform nanoparticles. By mixing PVA with lignin solution, a biomimetic nanoparticle-separation structure and a strong intermolecular sacrificial hydrogen bond were obtained. As shown in Figure 2.22, this LNPs/PVA composite showed a maximum toughness of (172 ± 5) J/g, as well as a strong tensile strength of 98.2 MPa and a large fracture strain of 282%, which are much better than most engineering plastics. LA, which measures several hundred nanometers in size, was thought to be secondary spherical particles composed of loosely aggregated primary LA (several nanometers). The fragmentation of the secondary spherical particles and the scattering of primary nanoparticles in the PVA matrix lead to the dynamic fracture and reconstruction of the sacrificial

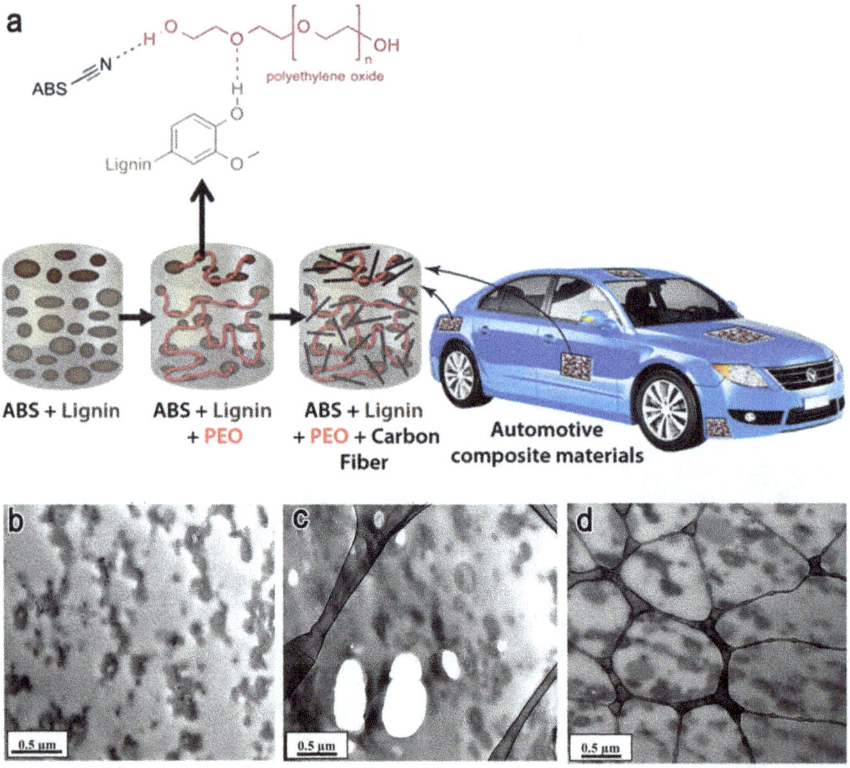

Figure 2.21: (a) Schematic diagram of lignin-based ABS polymer composites and short carbon fiber reinforced composites prepared from a blend improved with PEO; TEM micrographs of (b) neat ABS, (c) ABS/lignin (70/30), and (d) ABS/lignin/PEO (70/27/3) blends [126].

hydrogen bonds that were densely confined in the interphase between LA granules and PVA (Figure 2.23) [128]. This work is of great significance in the field of lignin-based composites and nanolignin applications [125].

2.4.2 Lignin-based rubber composites

A variety of polymer composites were developed with rubber as matrix and lignin as reinforcing agent or hard segment phase. Xiao and coworkers [129] have developed lignin enhanced styrene-butadiene rubber (SBR)/lignin-LDH (layered double hydroxide) composites using a melt blending method. In this research, layered dihydroxyl compounds (lignin-LDH) were synthesized by *in situ* synthesis method. The results showed that the mechanical properties of lignin-LDH/SBR were enhanced compared with the LDH/SBR composite samples [129]. Jiang and coworkers [130] have prepared nanocomposites by means of coprecipitation of lignin cationic polyelectrolyte

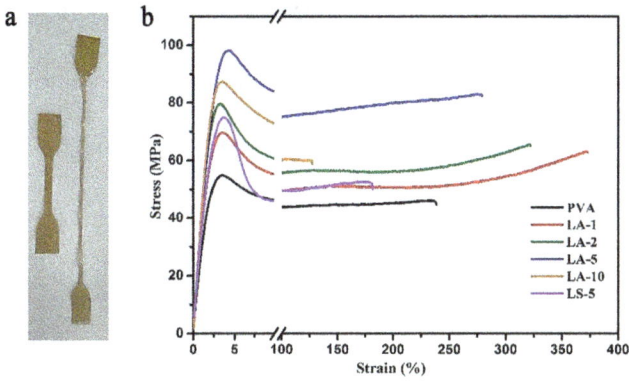

Figure 2.22: (a) Photo of the LA-5 film before and after extension. (b) Stress–strain curves of PVA/LA nanocomposites [128].

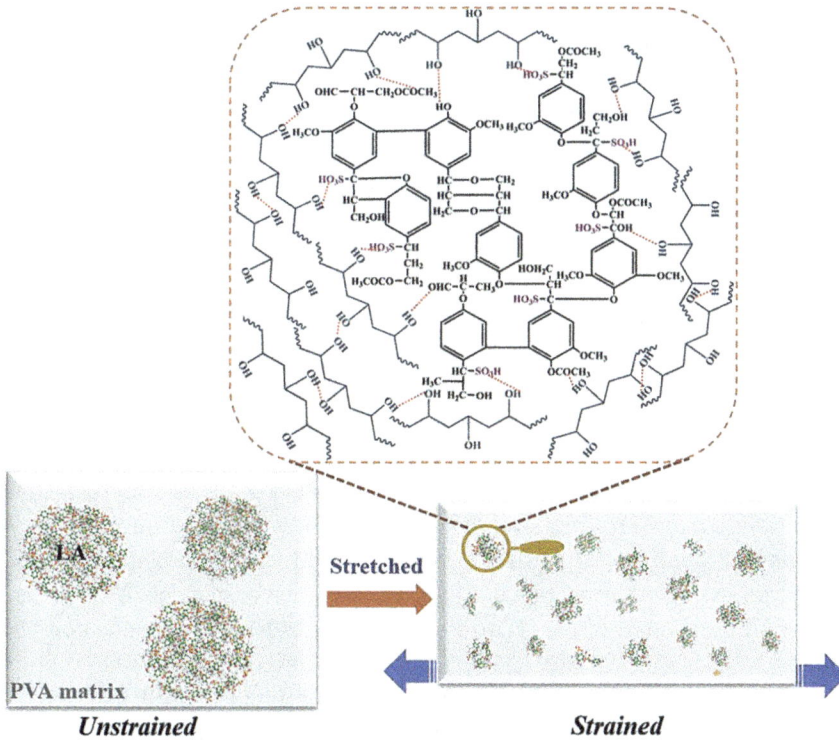

Figure 2.23: Enhancement mechanism of LA in PVA film [128].

complex and colloidal rubber latex. In this research, colloidal lignin-poly(diallyldimethylammonium chloride) (PDADMAC) complexes (LPCs) was prepared by self-assembly strategy. The homogenous distribution of LPCs in natural colloidal rubber latex was determined by SEM and dynamic mechanical analysis (DMA) analysis. The obtained nanocomposites showed improved thermal and mechanical properties [130].

In above cases, however, lignin was simply used as an reinforcing filler and no attempt is made to build advantageous interfacial interactions between the lignin and the rubber matrix.

Recently, Brook and coworkers [131] have developed a strategy to efficiently modify or disintegrate lignin with hydrosilanes of the Piers-Rubinsztajn (PR) reaction, which was catalyzed by $B(C_6F_5)_3$ (BCF, tris(pentafluorophenyl)borane). This process results in the conversion of phenol, aryl methoxy, and other ethers into silyl ethers or alkyl groups (Figure 2.24a and b). As a result, the lignin surface becomes silicified, making it more compatible with hydrophobic polymers. As a result, the lignin surfaces become silicified, rendering it more compatible with hydrophobic polymers [131]. This makes the raw softwood lignin as an enhanced filler or crosslinking additive for silicone rubber in a one-pot process. The resulting lignin-silicone elastomer has remarkable mechanical properties, even when filled with up to 40 wt% lignin. In addition, the research has presented a summary of the chemistry used in blowing foams and the ability to control the foam structures through changing treatment conditions and formulations (Figure 2.24c). This preparation strategy of lignin-based composites provides a new opportunities for better utilization of lignin involving silicone rubber [131, 132].

Recently, Naskar et al. [133] introduced a new method for synthesizing a novel class of high-performance renewable thermoplastic elastomers from acrylonitrile butadiene rubber (NBR) by introducing nanoscale dispersed lignin. The internal morphology and properties of this material can be regulated by a temperature-controlled mixture. The strategy is to replace the polystyrene fragments with lignin, similar to that found in soft matrices, as shown in Figure 2.25a and b. To further regulated lignin domain sizes, researchers selected NBR with 33, 41 and 51% acrylonitrile contents (namely NBR-41 and NBR-51, respectively) with Kraft softwood lignin (methanol extracted low-molecular-weight fractions) at ratio of 50/50 (w/w). Transmission electron microscopy (TEM) images of lignin/NBR-33 blends show the size of 0.2–2 μm domains (Figure 2.25c). The stress–strain curves of this mixture showed a typical behavior of reinforcement elastomers (Figure 2.25d). The tensile strength of the blend increases to above 30 MPa. This strain hardening is usually observed only in natural rubber because of the crystallization induced by strain (it's not common in nitrile rubber matrix). This new class of elastomers has great potential for sustainable products in the biomass industry [133].

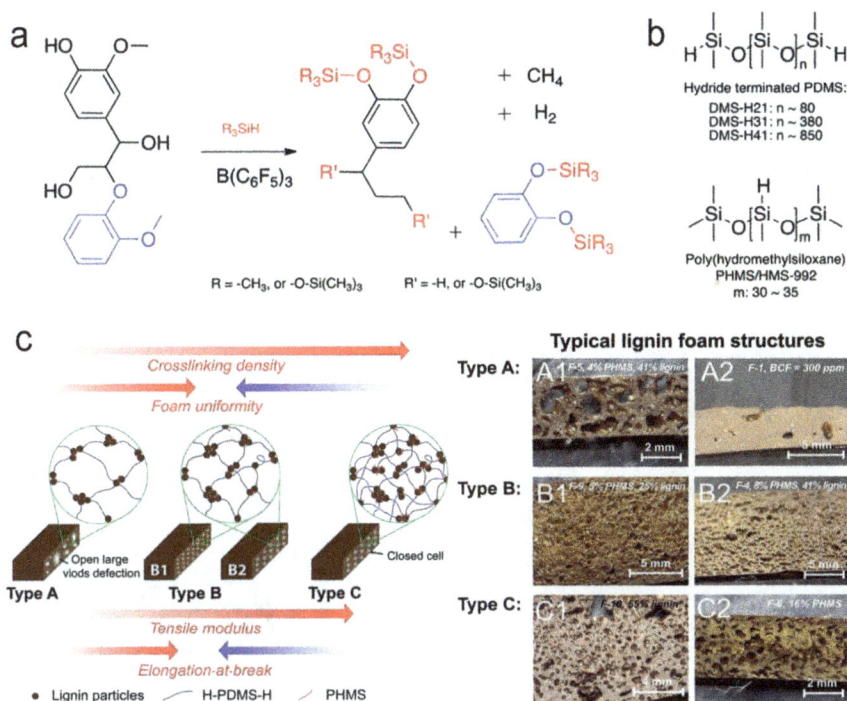

Figure 2.24: (a) Typical Piers–Rubinsztajn reactions at phenolic linkages of lignin; (b) The chemical structures of terminated PDMS and poly-(hydromethylsiloxane); (c) schematic diagram of lignin-silicone foam structure, and the mechanical properties they created: The tensile modulus increases with the increase of crosslinking density, and the elongation of fracture depends on the intersecting density and uniformity of foam [131].

2.4.3 Lignin-based thermosetting polymer composites

Developing thermosetting polymer composites using lignin as an additive is also reported. For example, Khalil and coworkers [134] have prepared epoxy matrix-based composites with empty fruit bunches as the reinforcement phase. In this study, lignin extracted from the black liquor was used as the curing agent in different ratios 15, 20, 25, and 30%, respectively. The results showed that the polymer composites with 25% lignin as the curing agent showed better mechanical properties than the composite materials prepared with commercial curing agent [134]. Doherty and coworkers [135] have developed lignin reinforced phenol-formaldehyde (PF) composites and applied them as coating with bagasse fibers as another reinforcement phase. The wetting properties showed that lignin and lignin-PF resin membranes are effective water-barrier coatings. Nevertheless, the contact angles of the lignin-PF resin membranes were found to is obviously smaller than that of commercial wax film. Lignin-based bio-oil was developed by Stanzione and his colleagues [136] for the preparation of biobased vinyl ester resin. In

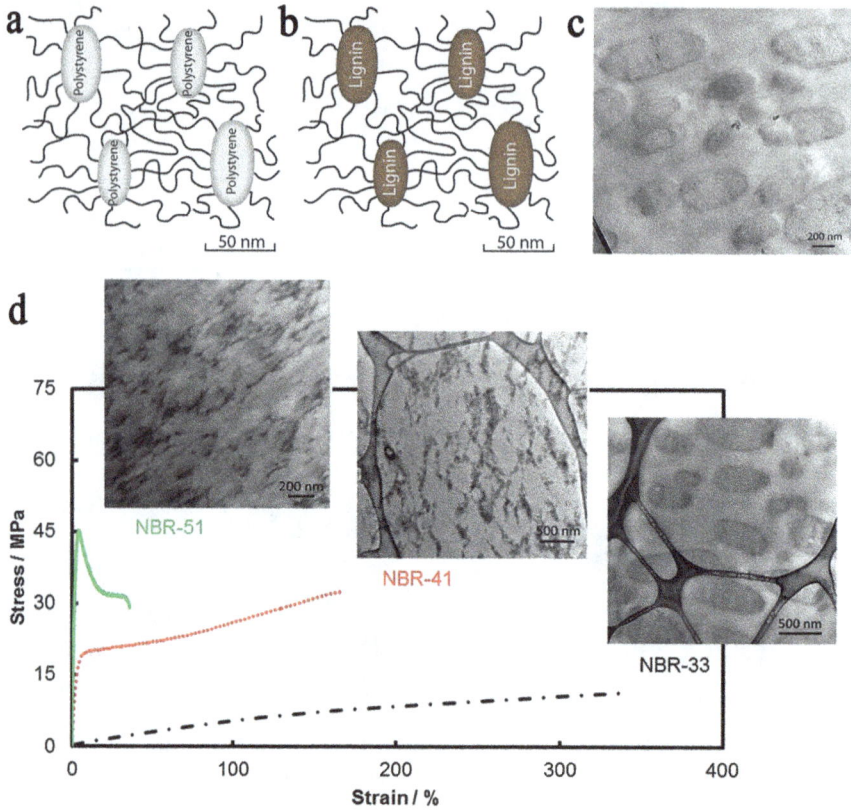

Figure 2.25: (a) Morphologic diagram of styrene-butadiene-styrene block copolymer. (b) Schematic diagram of lignin-based multiphase polymer in soft matrix. (c) TEM image of NBR-33/Kraft lignin composite (50/50). (d) Stress-strain curves of tensile mechanical strength of reactive composites of methanol extracted fraction of Kraft lignin and their corresponding TEM images [133].

this research, a lignin-based bio-oil mimic was prepared from methyl methacrylate and applied as a low viscosity vinyl ester resin. The thermal mechanical properties and thermostability of this thermoset were competitive with the commercially available petroleum-based and vinyl ester-based thermoset polymers [136].

2.5 Conclusions and outlooks

In the past 10 years, the field of lignin-based polymeric materials and composites has made great progress, and there is still room for innovation. In addition, it is important to recognize the technology of chemically modified lignin into products with integrated value without the use of expensive reagents or complex synthesis pathways. Indeed, because of their benefits, lignin-based polymers are a new class of ecofriendly and low-cost material. Despite the many advantages of lignin, the great differences and complexity of its structure have hampered the expansion from laboratory to industrial use. In order to overcome this limitation, a great deal of research should be undertaken to better understand the physics and chemistry of natural lignin. Advanced extraction methods or pretreatment processes should also be developed for lignin purification. Based on the current literature, compared with the traditional synthetic fiber reinforced polymer composites, the application of nano- or microsized lignin reinforced polymers still remain highly conceivable. As discussed in this chapter, in addition to the development of polymeric composites, the mechanical, physical and physicochemical properties of lignin and various polymer substrates can be utilized toward the commercialization of low-cost and ecofriendly lignin-based materials for a variety of applications.

Author contributions: All the authors have accepted responsibility for the entire content of this submitted manuscript and approved submission.
Research funding: None declared.
Conflict of interest statement: The authors declare no conflicts of interest regarding this article.

References

1. Ragauskas AJ, Beckham GT, Biddy MJ, Chandra R, Chen F, Davis MF, et al. Lignin valorization: improving lignin processing in the biorefinery. Science 2014;344:1246843.
2. Luterbacher JS, Azarpira A, Motagamwala AH, Lu F, Ralph J, Dumesic JA. Lignin monomer production integrated into the γ-valerolactone sugar platform. Energy Environ Sci 2015;8: 2657–63.
3. Shuai L, Amiri MT, Questell-Santiago YM, Heroguel F, Li YD, Kim H, et al. Formaldehyde stabilization facilitates lignin monomer production during biomass depolymerization. Science 2016;354:329–33.
4. Upton BM, Kasko AM. Strategies for the conversion of lignin to high-value polymeric materials: review and perspective. Chem Rev 2016;116:2275–306.
5. Li C, Zhao X, Wang A, Huber GW, Zhang T. Catalytic transformation of lignin for the production of chemicals and fuels. Chem Rev 2015;115:11559–624.
6. Sun Z, Fridrich B, de Santi A, Elangovan S, Barta K. Bright side of lignin depolymerization: toward new platform chemicals. Chem Rev 2018;118:614–78.
7. Kai D, Tan MJ, Chee PL, Chua YK, Yap YL, Loh XJ. Towards lignin-based functional materials in a sustainable world. Green Chem 2016;18:1175–200.
8. Gioia C, Lo Re G, Lawoko M, Berglund L. Tunable thermosetting epoxies based on fractionated and well-characterized lignins. J Am Chem Soc 2018;140:4054–61.
9. Liu H, Chung H. Lignin-based polymers via graft copolymerization. J Polym Sci Polym Chem 2017; 55:3515–28.
10. Doherty WO, Mousavioun P, Fellows CM. Value-adding to cellulosic ethanol: lignin polymers. Ind Crop Prod 2011;33:259–76.

11. Zakzeski J, Bruijnincx PCA, Jongerius AL, Weckhuysen BM. The catalytic valorization of lignin for the production of renewable chemicals. Chem Rev 2010;110:3552–99.
12. Gordobil O, Robles E, Egüés I, Labidi J. Lignin-ester derivatives as novel thermoplastic materials. RSC Adv 2016;6:86909–17.
13. Sen S, Patil S, Argyropoulos DS. Thermal properties of lignin in copolymers, blends, and composites: a review. Green Chem 2015;17:4862–87.
14. Thakur VK, Thakur MK, Raghavan P, Kessler MR. Progress in green polymer composites from lignin for multifunctional applications: a review. ACS Sustain Chem Eng 2014;2:1072–92.
15. Thakur VK, Thakur MK. Recent advances in green hydrogels from lignin: a review. Int J Biol Macromol 2015;72:834–47.
16. Liu W-J, Jiang H, Yu H-Q. Thermochemical conversion of lignin to functional materials: a review and future directions. Green Chem 2015;17:4888–907.
17. Zhao S, Abu-Omar MM. Materials based on technical bulk lignin. ACS Sustain Chem Eng 2021;9: 1477–93.
18. Vishtal AG, Kraslawski A. Challenges in industrial applications of technical lignins. BioResources 2011;6:3547–68.
19. Duval A, Lawoko M. A review on lignin-based polymeric, micro-and nano-structured materials. React Funct Polym 2014;85:78–96.
20. Sternberg J, Sequerth O, Pilla S. Green chemistry design in polymers derived from lignin: review and perspective. Prog Polym Sci 2021;113:101344.
21. Balakshin MY, Capanema EA, Sulaeva I, Schlee P, Huang Z, Feng M, et al. New opportunities in the valorization of technical lignins. ChemSusChem 2021;14:1016–36.
22. Calvo-Flores FG, Dobado JA. Lignin as renewable raw material. ChemSusChem 2010;3:1227–35.
23. Hu J, Zhang Q, Lee D-J. Kraft lignin biorefinery: a perspective. Bioresour Technol 2018;247:1181–3.
24. Gellerstedt G. Softwood Kraft lignin: raw material for the future. Ind Crop Prod 2015;77:845–54.
25. Crestini C, Lange H, Sette M, Argyropoulos DS. On the structure of softwood Kraft lignin. Green Chem 2017;19:4104–21.
26. Gratzl JS, Chen C-L. Chemistry of pulping: lignin reactions. Washington, DC, USA: ACS Publications; 2000.
27. Deshpande R, Giummarella N, Henriksson G, Germgård U, Sundvall L, Grundberg H, et al. The reactivity of lignin carbohydrate complex (LCC) during manufacture of dissolving sulfite pulp from softwood. Ind Crop Prod 2018;115:315–22.
28. Aro T, Fatehi P. Production and application of lignosulfonates and sulfonated lignin. ChemSusChem 2017;10:1861–77.
29. Fatehi P, Chen J. Extraction of technical lignins from pulping spent liquors, challenges and opportunities, Production of biofuels and chemicals from lignin. Heidelberg, Germany: Springer; 2016:35–54 pp.
30. El Hage R, Brosse N, Sannigrahi P, Ragauskas A. Effects of process severity on the chemical structure of Miscanthus ethanol organosolv lignin. Polym Degrad Stabil 2010;95:997–1003.
31. Konnerth H, Zhang J, Ma D, Prechtl MH, Yan N. Base promoted hydrogenolysis of lignin model compounds and organosolv lignin over metal catalysts in water. Chem Eng Sci 2015;123:155–63.
32. El Hage R, Brosse N, Chrusciel L, Sanchez C, Sannigrahi P, Ragauskas A. Characterization of milled wood lignin and ethanol organosolv lignin from miscanthus. Polym Degrad Stabil 2009;94: 1632–8.
33. Lou R, Wu S-B. Products properties from fast pyrolysis of enzymatic/mild acidolysis lignin. Appl Energy 2011;88:316–22.
34. Lei M, Wu S, Liang J, Liu C. Comprehensive understanding the chemical structure evolution and crucial intermediate radical in situ observation in enzymatic hydrolysis/mild acidolysis lignin pyrolysis. J Anal Appl Pyrol 2019;138:249–60.

35. Schubert WJ, Nord F. Investigations on lignin and lignification. II. The characterization of enzymatically liberated lignin. J Am Chem Soc 1950;72:3835–8.
36. Schubert W, Passannante A, Stevens GD, Bier M, Nord F. Investigations on lignin and lignification. XIII. Electrophoresis of native and enzymatically liberated Lignins 1. J Am Chem Soc 1953;75: 1869–73.
37. Jiang B, Yao Y, Liang Z, Gao J, Chen G, Xia Q, et al. Lignin-based direct ink printed structural scaffolds. Small 2020;16:e1907212.
38. Ren T, Qi W, Su R, He Z. Promising techniques for depolymerization of lignin into value-added chemicals. ChemCatChem 2019;11:639–54.
39. Laurichesse S, Avérous L. Chemical modification of lignins: towards biobased polymers. Prog Polym Sci 2014;39:1266–90.
40. del Río JC, Rencoret J, Gutiérrez A, Elder T, Kim H, Ralph J. Lignin monomers from beyond the canonical monolignol biosynthetic pathway: another brick in the wall. ACS Sustain Chem Eng 2020;8:4997–5012.
41. Thornburg NE, Pecha MB, Brandner DG, Reed ML, Vermaas JV, Michener WE, et al. Mesoscale reaction–diffusion phenomena governing lignin-first biomass fractionation. ChemSusChem 2020; 13:4495–509.
42. Liu X, Bouxin FP, Fan J, Budarin VL, Hu C, Clark JH. Recent advances in the catalytic depolymerization of lignin towards phenolic chemicals: a review. ChemSusChem 2020;13:4296.
43. Da Cunha C, Deffieux A, Fontanille M. Synthesis and polymerization of lignin-based macromonomers. III. Radical copolymerization of lignin-based macromonomers with methyl methacrylate. J Appl Polym Sci 1993;48:819–31.
44. Azhar SS, Ahn DJ. Purification of monomers leads to high-quality lignin macromonomers. Bristol, England: IOP Conference Series: Materials Science and Engineering, IOP Publishing; 2019:012021 p.
45. Petrović ZS. Polyurethanes from vegetable oils. Polym Rev 2008;48:109–55.
46. Sen S, Martin JD, Argyropoulos DS. Review of cellulose non-derivatizing solvent interactions with emphasis on activity in inorganic molten salt hydrates. ACS Sustain Chem Eng 2013;1:858–70.
47. Aniceto JP, Portugal I, Silva CM. Biomass-based polyols through oxypropylation reaction. ChemSusChem 2012;5:1358–68.
48. Lin S, Huang J, Chang PR, Wei S, Xu Y, Zhang Q. Structure and mechanical properties of new biomass-based nanocomposite: castor oil-based polyurethane reinforced with acetylated cellulose nanocrystal. Carbohydr Polym 2013;95:91–9.
49. Nam G-Y, Oh J-S, Park I-K, Luong ND, Yoon H-K, Lee S-H, et al. High molecular-weight thermoplastic polymerization of Kraft lignin macromers with diisocyanate. BioResources 2014;9:2359–71.
50. Hu TQ. Chemical modification, properties, and usage of lignin. New York, USA: Springer; 2002.
51. Xue B-L, Wen J-L, Sun R-C. Lignin-based rigid polyurethane foam reinforced with pulp fiber: synthesis and characterization. ACS Sustain Chem Eng 2014;2:1474–80.
52. Pohjanlehto H, Setälä HM, Kiely DE, McDonald AG. Lignin-xylaric acid-polyurethane-based polymer network systems: preparation and characterization. J Appl Polym Sci 2014;131:1–7.
53. Yiamsawas D, Baier G, Thines E, Landfester K, Wurm FR. Biodegradable lignin nanocontainers. RSC Adv 2014;4:11661–3.
54. Zhang Y, Liao J, Fang X, Bai F, Qiao K, Wang L. Renewable high-performance polyurethane bioplastics derived from lignin–poly(ε-caprolactone). ACS Sustain Chem Eng 2017;5:4276–84.
55. Mavila S, Sinha J, Hu Y, Podgórski M, Shah PK, Bowman CN. High refractive index photopolymers by Thiol–Yne "click" polymerization. ACS Appl Mater Interfaces 2021;13:15647–58.
56. Yilmaz G, Yagci Y. Light-induced step-growth polymerization. Prog Polym Sci 2020;100:101178.
57. Cao Y, Liu Z, Zheng B, Ou R, Fan Q, Li L, et al. Synthesis of lignin-based polyols via thiol-ene chemistry for high-performance polyurethane anticorrosive coating. Compos B Eng 2020;200: 108295.

58. Braun E, Levin BC. Polyesters: a review of the literature on products of combustion and toxicity. Fire Mater 1986;10:107–23.
59. Guo Z-X, Gandini A. Polyesters from lignin—2. The copolyesterification of Kraft lignin and polyethylene glycols with dicarboxylic acid chlorides. Eur Polym J 1991;27:1177–80.
60. Thanh Binh NT, Luong ND, Kim DO, Lee SH, Kim BJ, Lee YS, et al. Synthesis of lignin-based thermoplastic copolyester using Kraft lignin as a macromonomer. Compos Interface 2009;16: 923–35.
61. Bonini C, D'Auria M, Emanuele L, Ferri R, Pucciariello R, Sabia AR. Polyurethanes and polyesters from lignin. J Appl Polym Sci 2005;98:1451–6.
62. Malkappa K, Jana T. Simultaneous improvement of tensile strength and elongation: an unprecedented observation in the case of hydroxyl terminated polybutadiene polyurethanes. Ind Eng Chem Res 2013;52:12887–96.
63. Sivasankarapillai G, McDonald AG. Synthesis and properties of lignin-highly branched poly (ester-amine) polymeric systems, Biomass Bioenergy 2011;35:919–31.
64. Scarica C, Suriano R, Levi M, Turri S, Griffini G. Lignin functionalized with succinic anhydride as building block for biobased thermosetting polyester coatings. ACS Sustain Chem Eng 2018;6: 3392–401.
65. Fan Q, Liu T, Zhang C, Liu Z, Zheng W, Ou R, et al. Extraordinary solution-processability of lignin in phenol—maleic anhydride and dielectric films with controllable properties. J Mater Chem A 2019;7: 23162–72.
66. Jiang B, Chen C, Liang Z, He S, Kuang Y, Song J, et al. Lignin as a wood-inspired binder enabled strong, water stable, and biodegradable paper for plastic replacement. Adv Funct Mater 2019;30: 1906307.
67. Cao D, Zhang Q, Hafez AM, Jiao Y, Ma Y, Li H, et al. Lignin-derived holey, layered, and thermally conductive 3D scaffold for lithium dendrite suppression. Small Methods 2019;3:1800539.
68. Kai D, Zhang K, Jiang L, Wong HZ, Li Z, Zhang Z, et al. Sustainable and antioxidant lignin—polyester copolymers and nanofibers for potential healthcare applications. ACS Sustain Chem Eng 2017;5: 6016–25.
69. Auvergne R, Caillol S, David G, Boutevin B, Pascault J-P. Biobased thermosetting epoxy: present and future. Chem Rev 2014;114:1082–115.
70. Qin J, Woloctt M, Zhang J. Use of polycarboxylic acid derived from partially depolymerized lignin as a curing agent for epoxy application. ACS Sustain Chem Eng 2013;2:188–93.
71. Zhao S, Abu-Omar MM. Synthesis of renewable thermoset polymers through successive lignin modification using lignin-derived phenols. ACS Sustain Chem Eng 2017;5:5059–66.
72. Nonaka Y, Tomita B, Hatano Y. Synthesis of lignin/epoxy resins in aqueous systems and their properties. Holzforschung 1997;51:183–7.
73. Engelmann G, Ganster J. Bio-based epoxy resins with low molecular weight Kraft lignin and pyrogallol. Holzforschung 2014;68:435–46.
74. Zhao S, Huang X, Whelton AJ, Abu-Omar MM. Formaldehyde-free method for incorporating lignin into epoxy thermosets. ACS Sustain Chem Eng 2018;6:10628–36.
75. Feldman D, Banu D, Khoury M. Epoxy—lignin polyblends. III. Thermal properties and infrared analysis. J Appl Polym Sci 1989;37:877–87.
76. Feldman D, Banu D, Natansohn A, Wang J. Structure—properties relations of thermally cured epoxy—lignin polyblends. J Appl Polym Sci 1991;42:1537–50.
77. Wang J, Banu D, Feldman D. Epoxy—lignin polyblends: effects of various components on adhesive properties. J Adhes Sci Technol 1992;6:587–98.
78. Wang G, Liu X, Zhang J, Sui W, Jang J, Si C. One-pot lignin depolymerization and activation by solid acid catalytic phenolation for lightweight phenolic foam preparation. Ind Crop Prod 2018;124: 216–25.

79. Nair CR. Advances in addition-cure phenolic resins. Prog Polym Sci 2004;29:401–98.
80. Gardziella A, Pilato LA, Knop A. Phenolic resins: chemistry, applications, standardization, safety and ecology. Des Moines, Iowa, USA: Springer Science & Business Media; 2013.
81. Alonso M, Oliet M, Garcia J, Rodriguez F, Echeverría J. Gelation and isoconversional kinetic analysis of lignin–phenol–formaldehyde resol resins cure. Chem Eng J 2006;122:159–66.
82. Vazquez G, Antorrena G, González J, Mayor J. Lignin-phenol-formaldehyde adhesives for exterior grade plywoods. Bioresour Technol 1995;51:187–92.
83. Danielson B, Simonson R. Kraft lignin in phenol formaldehyde resin. Part 1. Partial replacement of phenol by Kraft lignin in phenol formaldehyde adhesives for plywood. J Adhes Sci Technol 1998; 12:923–39.
84. Danielson B, Simonson R. Kraft lignin in phenol formaldehyde resin. Part 2. Evaluation of an industrial trial. J Adhes Sci Technol 1998;12:941–6.
85. Podschun J, Saake B, Lehnen R. Reactivity enhancement of organosolv lignin by phenolation for improved bio-based thermosets. Eur Polym J 2015;67:1–11.
86. Cederholm L, Xu Y, Tagami A, Sevastyanova O, Odelius K, Hakkarainen M. Microwave processing of lignin in green solvents: a high-yield process to narrow-dispersity oligomers. Ind Crop Prod 2020;145:112152.
87. Zhu G, Jin D, Zhao L, Ouyang X, Chen C, Qiu X. Microwave-assisted selective cleavage of C C bond for lignin depolymerization. Fuel Process Technol 2017;161:155–61.
88. Gao C, Li M, Zhu C, Hu Y, Shen T, Li M, et al. One-pot depolymerization, demethylation and phenolation of lignin catalyzed by HBr under microwave irradiation for phenolic foam preparation. Compos B Eng 2021;205:108530.
89. Boerjan W, Ralph J, Baucher M. Lignin biosynthesis. Annu Rev Plant Biol 2003;54:519–46.
90. Fache M, Boutevin B, Caillol S. Vanillin production from lignin and its use as a renewable chemical. ACS Sustain Chem Eng 2015;4:35–46.
91. Fache M, Boutevin B, Caillol S. Epoxy thermosets from model mixtures of the lignin-to-vanillin process. Green Chem 2016;18:712–25.
92. Holmberg AL, Stanzione JF, Wool RP, Epps TH. A facile method for generating designer block copolymers from functionalized lignin model compounds. ACS Sustain Chem Eng 2014;2:569–73.
93. Wang S, Ma S, Xu C, Liu Y, Dai J, Wang Z, et al. Vanillin-derived high-performance flame retardant epoxy resins. Facile Synth Propert Macromol 2017;50:1892–901.
94. Van A, Chiou K, Ishida H. Use of renewable resource vanillin for the preparation of benzoxazine resin and reactive monomeric surfactant containing oxazine ring. Polymer 2014;55:1443–51.
95. Wang S, Ma S, Li Q, Yuan W, Wang B, Zhu J. Robust, fire-safe, monomer-recovery, highly malleable thermosets from renewable bioresources. Macromolecules 2018;51:8001–12.
96. Liu T, Hao C, Zhang S, Yang X, Wang L, Han J, et al. A self-healable high glass transition temperature bioepoxy material based on vitrimer chemistry. Macromolecules 2018;51:5577–85.
97. Mu L, Shi Y, Chen L, Ji T, Yuan R, Wang H, et al. [N-Methyl-2-pyrrolidone][C1-C4 carboxylic acid]: a novel solvent system with exceptional lignin solubility. Chem Commun (Camb) 2015;51:13554–7.
98. Fan Q, Liu T, Zhang C, Liu Z, Zheng W, Ou R, et al. Extraordinary solution-processability of lignin in phenol–maleic anhydride and dielectric films with controllable properties. J Mater Chem 2019;7: 23162–72.
99. Kim WG. Photocure properties of high-heat-resistant photoreactive polymers with cinnamate groups. J Appl Polym Sci 2008;107:3615–24.
100. Xin J, Zhang P, Huang K, Zhang J. Study of green epoxy resins derived from renewable cinnamic acid and dipentene: synthesis, curing and properties. RSC Adv 2014;4:8525–32.
101. Kaneko T, Matsusaki M, Hang TT, Akashi M. Thermotropic liquid-crystalline polymer derived from natural cinnamoyl biomonomers. Macromol Rapid Commun 2004;25:673–7.

102. Kaneko T, Thi TH, Shi DJ, Akashi M. Environmentally degradable, high-performance thermoplastics from phenolic phytomonomers. Nat Mater 2006;5:966–70.
103. Matsusaki M, Kishida A, Stainton N, Ansell CW, Akashi M. Synthesis and characterization of novel biodegradable polymers composed of hydroxycinnamic acid and d, l-lactic acid. J Appl Polym Sci 2001;82:2357–64.
104. Spiliopoulos IK, Mikroyannidis JA. Unsaturated polyamides and polyesters prepared from 1, 4-bis (2-carboxyvinyl) benzene and 4-hydroxycinnamic acid. J Polym Sci Polym Chem 1996;34: 2799–807.
105. Thi TH, Matsusaki M, Akashi M. Thermally stable and photoreactive polylactides by the terminal conjugation of bio-based caffeic acid. Chem Commun 2008:3918–20.
106. Thi TH, Matsusaki M, Akashi M. Development of photoreactive degradable branched polyesters with high thermal and mechanical properties. Biomacromolecules 2009;10:766–72.
107. Nagata M, Hizakae S. Synthesis and characterization of photocrosslinkable biodegradable polymers derived from 4-hydroxycinnamic acid. Macromol Biosci 2003;3:412–9.
108. Truong HT, Van Do M, Duc Huynh L, Thi Nguyen L, Tuan Do A, Thanh Xuan Le T, et al. Ultrasound-Assisted, base-catalyzed, homogeneous reaction for ferulic acid production from γ-oryzanol. J Chem 2018;2018:1–9.
109. Huang Y-C, Chen Y-F, Chen C-Y, Chen W-L, Ciou Y-P, Liu W-H, et al. Production of ferulic acid from lignocellulolytic agricultural biomass by *Thermobifida fusca* thermostable esterase produced in *Yarrowia lipolytica* transformant. Bioresour Technol 2011;102:8117–22.
110. Oulame MZ, Pion F, Allauddin S, Raju KV, Ducrot P-H, Allais F. Renewable alternating aliphatic-aromatic poly (ester-urethane) s prepared from ferulic acid and bio-based diols. Eur Polym J 2015; 63:186–93.
111. Chen Q, Gao K, Peng C, Xie H, Zhao ZK, Bao M. Preparation of lignin/glycerol-based bis (cyclic carbonate) for the synthesis of polyurethanes. Green Chem 2015;17:4546–51.
112. Ghosh N, Kiskan B, Yagci Y. Polybenzoxazines—new high performance thermosetting resins: synthesis and properties. Prog Polym Sci 2007;32:1344–91.
113. Comí M, Lligadas G, Ronda JC, Galià M, Cádiz V. Renewable benzoxazine monomers from "lignin-like" naturally occurring phenolic derivatives. J Polym Sci Polym Chem 2013;51:4894–903.
114. GREGoRová A, Kosikova B, Osvald A. The study of lignin influence on properties of polypropylene composites. Wood Res 2005;50:41–8.
115. Košíková B, Demianova V, Kačuráková M. Sulfur-free lignins as composites of polypropylene films. J Appl Polym Sci 1993;47:1065–73.
116. Yue X, Chen F, Zhou X. Synthesis of lignin-g-MMA and the utilization of the copolymer in PVC/ wood composites. J Macromol Sci B 2012;51:242–54.
117. Baumberger S, Lapierre C, Monties B. Utilization of pine Kraft lignin in starch composites: impact of structural heterogeneity. J Agric Food Chem 1998;46:2234–40.
118. Kadla J, Kubo S, Venditti R, Gilbert R, Compere A, Griffith W. Lignin-based carbon fibers for composite fiber applications. Carbon 2002;40:2913–20.
119. Megiatto JD, Jr, Silva CG, Rosa DS, Frollini E. Sisal chemically modified with lignins: correlation between fibers and phenolic composites properties. Polym Degrad Stabil 2008;93:1109–21.
120. Nenkova SK. Study of sorption properties of lignin-derivatized fibrous composites for the remediation of oil polluted receiving waters. BioResources 2007;2:408–18.
121. Morandim-Giannetti AA, Agnelli JAM, Lanças BZ, Magnabosco R, Casarin SA, Bettini SH. Lignin as additive in polypropylene/coir composites: thermal, mechanical and morphological properties. Carbohydr Polym 2012;87:2563–8.
122. Pupure L, Varna J, Joffe R, Pupurs A. An analysis of the nonlinear behavior of lignin-based flax composites. Mech Compos Mater 2013;49:139–54.

123. Canetti M, Bertini F. Influence of the lignin on thermal degradation and melting behaviour of poly (ethylene terephthalate) based composites. E-Polymers 2009;9:1–10.
124. Reza Barzegari M, Alemdar A, Zhang Y, Rodrigue D. Mechanical and rheological behavior of highly filled polystyrene with lignin. Polym Compos 2012;33:353–61.
125. Sun R. Across the board: runcang sun on lignin nanoparticles. ChemSusChem 2020;13:4768–70.
126. Akato K, Tran CD, Chen J, Naskar AK. Poly(ethylene oxide)-assisted macromolecular self-assembly of lignin in ABS matrix for sustainable composite applications. ACS Sustain Chem Eng 2015;3: 3070–6.
127. Wang H-M, Wang B, Yuan T-Q, Zheng L, Shi Q, Wang S-F, et al. Tunable, UV-shielding and biodegradable composites based on well-characterized lignins and poly(butylene adipate-co-terephthalate). Green Chem 2020;22:8623–32.
128. Zhang X, Liu W, Yang D, Qiu X. Biomimetic supertough and strong biodegradable. polymeric materials with improved thermal properties and excellent UV-blocking performance. Adv Funct Mater 2019;29:1806912.
129. Xiao S, Feng J, Zhu J, Wang X, Yi C, Su S. Preparation and characterization of lignin-layered double hydroxide/styrene-butadiene rubber composites. J Appl Polym Sci 2013;130:1308–12.
130. Jiang C, He H, Jiang H, Ma L, Jia D. Nano-lignin filled natural rubber composites: preparation and characterization. Express Polym Lett 2013;7:480–93.
131. Zhang J, Fleury E, Brook MA. Foamed lignin–silicone bio-composites by extrusion and then compression molding. Green Chem 2015;17:4647–56.
132. Zhang J, Chen Y, Sewell P, Brook MA. Utilization of softwood lignin as both crosslinker and reinforcing agent in silicone elastomers. Green Chem 2015;17:1811–9.
133. Tran CD, Chen J, Keum JK, Naskar AK. A new class of renewable thermoplastics with extraordinary performance from nanostructured lignin-elastomers. Adv Funct Mater 2016;26:2677–85.
134. Khalil HA, Marliana M, Alshammari T. Material properties of epoxy-reinforced biocomposites with lignin from empty fruit bunch as curing agent. BioResources 2011;6:5206–23.
135. Doherty W, Halley P, Edye L, Rogers D, Cardona F, Park Y, et al. Studies on polymers and composites from lignin and fiber derived from sugar cane. Polym Adv Technol 2007;18:673–8.
136. Stanzione JF, III, Giangiulio PA, Sadler JM, La Scala JJ, Wool RP. Lignin-based bio-oil mimic as biobased resin for composite applications. ACS Sustain Chem Eng 2013;1:419–26.

Xing Zhou*, Yaya Hao, Xin Zhang, Xinyu He and Chaoqun Zhang*

3 Cellulose-based polymers

Abstract: The presented chapter deals with structure, morphology, and properties aspects concerning cellulose-based polymers in both research and industrial production, such as cellulose fibers, cellulose membranes, cellulose nanocrystals, and bacterial cellulose, etc. The idea was to highlight the main cellulose-based polymers and cellulose derivatives, as well as the dissolution technologies in processing cellulose-based products. The structure and properties of cellulose are introduced briefly. The main attention has been paid to swelling and dissolution of cellulose in order to yield various kinds of cellulose derivatives through polymerization. The main mechanisms and methods are also presented. Finally, the environmental friendly and green cellulose-based polymers will be evaluated as one of the multifunctional and smart materials with significant progress.

Keywords: bulk structure; cellulose; nanocrystalline.

3.1 Introduction

Owing to the shortage of petroleum resources and the increasing demanding in environment-friendly polymers, cellulose-based polymers have attracted a great deal of attentions in developing various products, such as fibers, membranes, aerogel, hydrogel, bioplastic, etc. [1–3]. Actually, bioresources originated from agricultural-related plants and animals have been widely used for millennia [1]. In the process of human evolution, the history of using wood is very long. Human ancestors used wood to make various tools (hoes, hoe sticks, etc.), as well as build houses, bridges, carriages, furniture, daily necessities, paper, charcoal, etc. [2, 3]. In 105 AD, Ts'ai Lun, Chinese traditional religion inventor of paper, improved the technique of papermaking by using bark, fishing nets, and bamboo to press into paper, which is known as one of the four great inventions in China [4]. Both wood and paper are rich in cellulose, which drives a tremendous development of in biomass based industries at that time. Cellulose is commonly found in plants, such as cotton, wood, flax, grass, etc. It is the main component of plant cell walls. The cellulose content of dry wood is 40–60%, and that of

*Corresponding authors: Xing Zhou, Faculty of Printing, Packaging Engineering and Digital Media Technology, Xi'an University of Technology, Xi'an 710048, P. R. China; and School of Materials Science and Engineering, Xi'an University of Technology, Xi'an 710048, P. R. China,
E-mail: zdxnlxaut@163.com; and **Chaoqun Zhang**, College of Materials and Energy, South China Agricultural University, Guangzhou 510642, P. R. China, E-mail: zhangcq@scau.edu.cn
Yaya Hao, Xin Zhang and Xinyu He, Faculty of Printing, Packaging Engineering and Digital Media Technology, Xi'an University of Technology, Xi'an 710048, P. R. China

This article has previously been published in the journal Physical Sciences Reviews. Please cite as: X. Zhou, Y. Hao, X. Zhang, X. He and C. Zhang "Cellulose-based polymers" *Physical Sciences Reviews* [Online] 2021, 6. DOI: 10.1515/psr-2020-0067 | https://doi.org/10.1515/9781501521942-003

cotton and flax exceeds 90% [5]. Through photosynthesis, plants can produce tens of billions of tons of cellulose each year, which is absolutely an inexhaustible renewable resource [6]. In addition, there is also cellulose in algae and fungi, such as valonia ventricosa, glaucocystis, gluconobacter, pseudomonas, and rhizobium. Meanwhile, celluloses derived from animals, called the tunicin, are also investigated and employed in both research and industry [7]. Cellulose, as a homogeneous compound, was first named after the treatment of wood with nitric acid and sodium hydroxide solution alternately by the French scientist Anselme Payen in 1838 [8, 9]. It was not until 1932 that the German scientist Hermann Staudunger confirmed the polymer form of cellulose [10]. Since then, cellulose is defined as a polysaccharide consisting of a linear chain of several hundred to many thousands of $\beta(1 \rightarrow 4)$ linked D-glucose units, whose chemical formula is $(C_6H_{10}O_5)_n$ (the n of degree of polymerization). As shown in Figure 3.1, the chemical structure of cellulose contains three active hydroxyls in 2, 3, and 6 positions of β-glucose, supplying derivable performance.

Since the foundation of cellulose, numerous researches and attentions have been devoted into the study of celluloses and cellulose-based materials [11]. The major achievements of the cellulose research process can be briefly described as follows.

(1) Under severe conditions, cellulose can be completely hydrolyzed with proper treatment by sulfuric acid, and the product is determined to be glucose. According to this, the empirical chemical formula of cellulose is determined as $(C_6H_{10}O_5)_n$.

(2) The elemental analysis results (C, H, O) of cotton, ramie, and regenerated cellulose fibers are consistent with the theoretical calculation of dextran (glucose residue). The element C accounts for 44.4%, H accounts for 6.2%, and O accounts for 49.4% in cellulose, indicating further that cellulose is composed of only glucose units.

(3) It is found that cellulose can generate tri-substituted esters or ethers, suggesting that there are three hydroxyl groups on each glucose unit. In addition, it can also be further attested by the acquirement of a single trimethyl-glucose by degradation of trimethyl-cellulose obtained by completely methylating cellulose.

(4) Cellobiose, cellotriose, cellotetraose, cellopentaose, and higher oligosaccharides are produced in the process of cellulose hydrolysis, indicating that cellulose has a long-chain structure.

(5) It is found that the cellulose is composed of β-glucose by analyzing the structure of β-glycosidase.

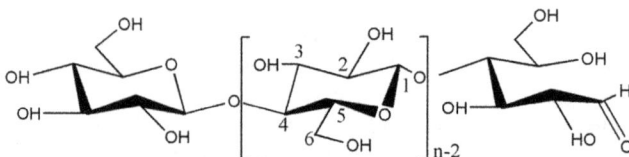

Figure 3.1: The chemical structure of cellulose.

(6) Glucose that is decomposed from cellulose possesses a six-membered ring structure (pyran ring).

(7) The disaccharide hydrolysate of completely methylated cellulose is 2,3,4,6-tetramethylglucose, indicating the presence of 1,4-glucoside bonds.

(8) The investigation of crystallization, morphology, and phase behavior of cellulose-based polymers promotes the blending between cellulose and synthetic polymers, preparing fiber composite materials with significant performance (Canadian scientist R. St. John Manley; German scientist Peter Zugenmaier research)

(9) Preparation and key technologies of nanomicrocrystalline cellulose and bacterial nanocellulose (sucrose synthase)

(10) Low-temperature dissolution mechanism of cellulose (Academician Zhang Lina)

As one of the most abundant naturally polymers, cellulose has attracted dramatically increasing attention in various industries due to the development of environmental protection and green economy [12]. Notably, cellulose has been also used in human and veterinary medicine delivery due to the biocompatibility, which supply a path to combine biopolymer with medicine [13]. Therefore, it is foreseeable that cellulose and cellulose-based polymers may play a much more important role in human's life.

3.2 The categories of cellulose polymers in different forms

3.2.1 Cellulose fibers

Apart from the natural cellulose, the regenerated cellulose fiber is commonly produced by wet spinning of a concentrated solution of cellulose or its derivatives originated from cotton linter pulp, wood pulp, bamboo pulp, etc. [14]. Owing to the advantages of unique luster, good moisture absorption, air permeability, and antistatic properties, the products based on cellulose fiber are favored by consumers, such as the functional paper [15]. Furthermore, the dramatically increasing depletion of petrochemical resources received extensive attention to the environmental pollution all over the world [16, 17]. The regenerated cellulose fibers, such as viscose fiber, copper ammonia fiber, and lyocell fibers, have been potentially considered to replace the synthetic polymer fibers [18, 19]. In addition, a variety of new cellulose fibers may also obtained by developing the variety of solvents for dissolving cellulose, including alkaline/urea aqueous solutions, ionic liquids, and cellulose carbamate systems.

Viscose fiber was first discovered by British chemist Cross in 1892. It is obtained by reacting cotton or wood pulp with NaOH, following by reacting with CS_2 to produce yellow viscose solution. Since a large amount of toxic CS_2 is used, this method is not suitable for the green and economic cellulose industry. Copper ammonia fiber is obtained by dissolving cotton and flax cellulose in a copper ammonia solution at room temperature, following by regeneration of cellulose from the copper ammonia solution using a dilute sulfuric acid aqueous solution. However, the copper, waste acid, and waste water generated in the production of cupro-ammonia fiber are harmful to the environment [20]. Lyocell fiber is directly obtained by dissolving wood pulp in N-methylmorpholine-N-oxide (NMMO) aqueous solution, following by dry-jet-wet spinning. Owing to the free of chemical reaction, high strength, high wet modulus, excellent dimensional stability, and easily blending with other fibers (linen, cashmere, wool, etc.), it is called "green fiber". Meanwhile, the effluent is nonhazardous which is produced in the Lyocell process [19]. However, two key problems are still to be solved: (1) NMMO is easy to decompose at high temperature. (2) The by-products generated during the dissolution process pollute lyocell fiber [21, 22].

The green solvent to dissolve and prepare cellulose fibers is essential. Cellulose–NaOH–urea aqueous solution spinning fiber was significantly designed and developed by Zhang's group [23–26] They found that cellulose can be dissolved in the mass fraction of 7% NaOH/12% urea aqueous solution rapidly within 2 min after precooling to –12 °C, as well as in the 4.6% LiOH/15% urea aqueous solution with better solubleness for cellulose [27]. Cellulose has unique thermal and cold-induced gelation behavior in NaOH/urea aqueous solution, which depends on the molecular weight, concentration, and solution temperature of cellulose [28]. This concentrated cellulose solution is suitable for spinning cellulose fibers and it remains stable at 0–5 °C for a long time. The cellulose fiber obtained by wet spinning shows good gloss, soft hand feeling. It presents morphology of round cross section with excellent mechanical property, which is similar to cupra and lyocell fibers [29].

Cellulose carbamate-NaOH solution spinning uses cellulose carbamate as the spinning solution, which is dissolved in aqueous NaOH [30]. It is also called Carbacell spinning process, which is similar to the viscose method. During the dissolving process of the cellulose carbamate in NaOH solution, carbamate is easily hydrolyzed in alkaline medium. Thus, it should be dissolved and stored at low temperature (below 0 °C). This cellulose derivative solution can be used to prepare Carbacell fiber by solidifying in sulfuric acid or sodium carbonate solution. Carbacell fiber presents an uniform structure with the morphology of egg-shaped cross-section, which is similar to Lyocell® fiber.

Cellulose-ionic liquid solution spinning employs ionic liquid as solvent to dissolve cellulose. It is similar to that of lyocell fiber and both of wet spinning and dry jet-wet spinning technology can be used to prepare this kind of fiber [26]. It has been reported that there are mainly five kinds of ionic liquid solutions for dissolving cellulose, such as 1-butyl-3-methylimidazole chloride salt ([Bmim]Cl), 1-allyl-3-methylimidazole chloride salt ([Amim]Cl), 1-methyl-3-ethylimidazole chloride (Emim]Cl), 1-butyl-3-methylimidazole acetate ([Bmim]Ac) and 1-ethyl-3-methylimidazole acetate ([Emim]Ac) [31–34]. The mechanical properties of the regenerated cellulose fibers obtained from the five ionic liquids are similar to that of the lyocell fibers [35].

3.2.2 Cellulose membrane

Cellulose membrane is an important part of membrane material with the advantages of good mechanical properties, hydrophilicity, less adsorption of proteins and blood cells, resistance to gamma rays, heat resistance, stability, biocompatibility, and safety [26]. The abundant of hydroxyl groups make it to be modified physically and chemically. Meanwhile, it can be completely decomposed into CO_2 and water under the action of microorganisms without causing environmental pollution after being discarded [36]. Because of its good gas permeability, cellulose membrane is usually used as food preservation films and packaging materials, such as cellophane and cuprophane. It can also be used as dialysis membranes to remove low molecular weight substances through solute diffusion with a pore size of less than 10 μm [37, 38]. Until now, the immersion-deposition processes are most widely used to prepare cellulose membrane [39]. This process was determined by three main components of macromolecules, solvent and nonsolvent, as well as the diffusion and convection behaviors between solvent and nonsolvent [40]. When the polymer solution is immersed in the coagulation bath, the solvent and nonsolvent mutually diffuse with the coagulant through the solution interface, resulting in liquid–liquid phase separation of the polymer solution. This is key to the preparation of cellulose membranes due to the strongly impacts of morphology on the transport properties. According to the previous works, the typical thermodynamic phase diagram of the ternary system immersed in the precipitation solidification process is qualitatively illustrated in Figure 3.2 [40]. As shown in region A in Figure 3.2, the solution becomes progressively more viscous with the increasing polymer concentration. In this system, the chain motion and diffusion slow down and the polymer plus any residual solvent effectively reverts to a glass. This results in a dense morphology with little or no porosity. The intersection of the binodal line and the spinodal line is the critical phase separation point. Phase separation occurs in the metastable region between the two lines at higher polymer concentration than the critical point. The spinodal line is the unstable phase separation zone, and the rest are homogeneous zones. In the membrane solidification process of the immersion precipitation method, the liquid–liquid phase separation occurs spontaneously when the nonsolvent content in the polymer solution increases to the region bounded by binodal line (B and D regions in Figure 3.2) with the broken of thermodynamic equilibrium of the system. Generally, there are two thermodynamic liquid–liquid phase separation processes, namely spinodal liquid–liquid separation (C region in Figure 3.2) and binodal liquid–liquid separation (B and D regions in Figure 3.2). For binodal liquid–liquid phase separation, generally, when the critical point of the system is at a relatively low polymer concentration and the composition of the system enters the metastable phase separation zone (B region) from the above critical point between the binodal line and the spinodal, nucleation and growth of nonsolvent droplets occurs. This results in the formation of liquid-filled pores in a continuous polymer matrix with polymer-depleted phase. When the system composition enters the metastable miscible phase separation zone (D region) below the critical point between the binodal line and the spinodal line, polymer-rich phase nucleation occurs with liquid–liquid phase separation. The polymer-rich phase droplets disperse in the continuous phase of polymer-depleted phase, which is composed of solvents, nonsolvents, and a small amount of polymer. These polymer-rich phase droplets continue to grow under the function of the concentration gradient until it solidifies into a film by phase transition. Most of the formed films present a granular morphology with a diameter between 25 and 200 nm. As for spinodal liquid–liquid phase separation, the system composition enters the unstable phase separation zone (C region) within the spinodal from the critical point. In this case, the polymer solution directly enters the nonsteady phase separation zone bounded by spinodal lines. Polymer-rich and polymer-depleted phases separate initially by the progressive growth of concentration waves' of constant wavelength but increasing amplitude, forming a dual continuous phase. This results in a mutually continuous, interwoven network of polymer domains and pores.

Several factors, such as coagulation bath composition, coagulation temperature, additives, and post-treatment conditions, have great influences on the structure and performance of cellulose membrane by using immersion precipitation method [41–44]. For example, the impact of coagulation baths on the morphology, structure, and performance of cellulose membranes was conducted by Zhang et al. in the green solvent NaOH/urea system [45, 46]. The coagulation bath includes H_2SO_4, CH_3COOH, H_2SO_4/Na_2SO_4, Na_2SO_4, $(NH_4)_2SO_4$, H_2O, C_2H_5OH, and $(CH_3)_2CO$. The results show that the surface pore size and cross-sectional pore size of the cellulose membrane vary between 265–730 and 173–266 nm, respectively. When the coagulation bath is acid, the prepared cellulose membrane presents relatively small pore size and narrow pore size distribution. Inversely, organic solvent leads to big pore size and broad pore size

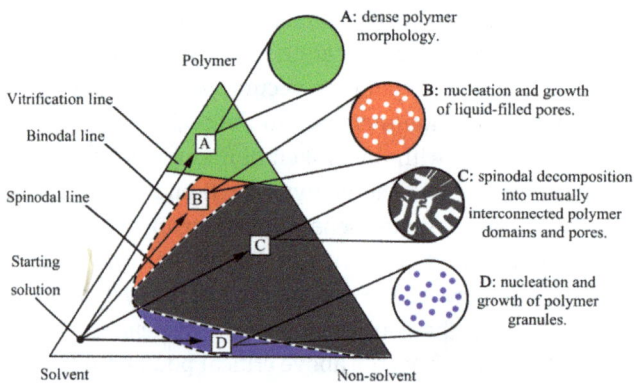

Figure 3.2: This process was impacted by the factors including components of coagulating bath, temperature, additives, orientation, and even drying conditions [40]. @2002 Elsevier.

distribution. Notably, the mechanical properties of the cellulose membrane regenerated by the acid and salt mixed coagulation bath are stronger than that of organic solvents. It is reported that cellulose membrane prepared by H_2SO_4/Na_2SO_4 has the highest light transmittance, the densest structure and the strongest mechanical properties. It has been the most common coagulation bath for the preparation of cellulose in NaOH/urea system. Furthermore, when the coagulation time is 5 min, the pore size of the cellulose membrane is the smallest with the densest structure, indicating that the cellulose regeneration process is also affected by time [47]. The lower is the coagulation bath temperature, the smaller is the pore size, and the denser is the structure of cellulose membrane, which also leads to higher mechanical properties and light transmittance [48]. Additives can change the chain conformation and dispersion state of polymers, thereby affecting the structure and performance of cellulose membrane [49]. Orientation process may facilitate the cellulose to form ordered supramolecular structures and enhance intermolecular interactions, which can improve the performance significantly, especially the optical properties [50, 51].

3.2.3 Cellulose gel

There are mainly two types of gel materials that are aerogel and hydrogel. Aerogel may be the lowest density solid with high permeability nanopore three-dimensional network structure, extremely high porosity, extremely low density, high specific surface area and other characteristics. Cellulose aerogels was firstly developed from the cellulose nitrate solution by Kistler in the year of 1931–1932 [52, 53]. Since then, Weatherwax et al. [54] prepared cellulose aerogel with the specific surface area of ca. 200 m^2/g via supercritical fluid drying process. Tan et al. [55] enhanced the specific surface area of cellulose aerogel to 389 m^2/g by employing cellulose acetate and toluene diisocyanates as raw materials via supercritical fluid drying process. The isocyanates were mainly used as cross-linking agent to improve the covalent bond interactions. Alternatively, the dissolution of cellulose in solvents, such as NaOH/urea solution, LiCl/DMAc, LiCl/DMSO, NMMO, and ionic liquid [56–58], can form cellulose aerogel by the displacement of water or ethyl alcohol via supercritical fluid drying process. Owing to the easily formation of strong hydrogen bonds with strong capillary force between cellulose and water, direct drying of the cellulose wet gel usually results in the complete collapse of the gel pores. Thus, the bulk structure of cellulose aerogels needs to be precisely controlled to minimize the effect of surface tension during the preparation process [59, 60].

    Cellulose hydrogels are directly converted from cellulose molecules or nanocellulose. There are mainly three types of cellulose hydrogels [61]. The first one is cellulose physical gel, which is formed by regeneration of cellulose solution. The cellulose is firstly dissolved in solvents to destroy the original crystalline structure. Meanwhile, the hydrogen bonds are reorganized during the dissolving process, inducing the formation of hydrogel network structure. The second type is to promote the reaction between a cross-linking agent and cellulose solution to form a chemical gel with a cross-linked structure. The cross-linking agent may completely destroy the original intra- and intermolecular hydrogen bonds of cellulose, releasing free hydroxyl groups with a high swelling rate. The third one is nanocellulose gel, which is prepared by directly arranging cellulose nanocrystals or fibers to form a fine network structure. The raw materials are mainly bacterial cellulose, cellulose nanocrystals, and cellulose nanofibers. Cellulose gels are green and possess significant performance with a great potential application in the areas of separation, adsorption, catalysis, optoelectronics, sensors, biomedicine, etc. [62].

3.2.4 Cellulose bioplastics

Generally, plastic refers to a material that deforms under hot pressing and causes structural changes in aggregate state. Cellulose cannot be melt-processed due to strong hydrogen bonds between molecules without softening temperature [63]. Therefore, it is not possible to directly use cellulose slurry to make plastics for a long time. Although cellulose derivatives obtained from a solvated state can be used as thermoplastics after the addition of additives, it is a chemical agent with high cost, such as cellulose ester [64]. Recently, cellulosic plastics are built from renewable biomass macromolecules through the process of hot pressing. In this process, the cellulose molecular chains are induced to orient in all directions on a dimension, resulting in deformation of the cellulose. Zhang et al. [26, 65] designed novel cellulose bioplastics (CBP) by using cellulose hydrogels through simple hot-pressing process. On the basis of the removal of cellulose molecules in the hydrogel state, the hot-pressing induced the transition of its aggregated structure, with a radial orientation of the cellulose aggregates occurred in the planar direction of the plate, leading to the plastic deformation, as shown in Figure 3.3. The CBP exhibited excellent properties compared with common commercial plastics, such as excellent mechanical properties, good thermal stabilities and low coefficients of thermal expansion. Furthermore, the whole life cycle assessment showed that the cellulose bioplastic is an environmentally friendly material.

Recently, Yu et al. [66] also reported a new type of cellulosic plastic that was developed using cellulosic materials on the basis of bionic structure design concept according to the so-called "directed deformation assembly" process, as shown in Figure 3.4. Cellulose nanofibers (CNF) and titanium dioxide-coated mica flakes (TiO_2-Mica) were combined to prepare a high-performance sustainable structural material with a bionic structure (storage modulus of ca. 20 GPa and thermal resistance of 25–200 °C). It presents better mechanical and thermal properties than petroleum-based plastics (shown in Figure 3.4b), which may be expected to be a significantly substitute for petroleum-based plastics. In addition, its thermal diffusion coefficient is also higher

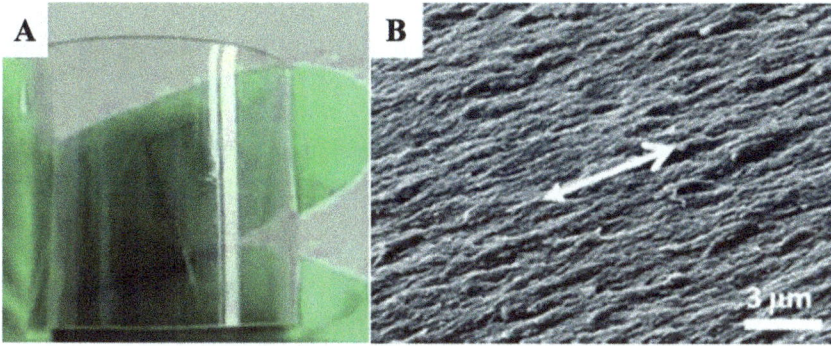

Figure 3.3: (A) Photogram of the cellulose bioplastic. (B) SEM image of the inner part of the cellulose bioplastic [65]. @2013 RSC.

than that of most engineering plastics, which is conducive to heat dissipation, thereby further ensuring the reliability of practical applications.

3.2.5 Nanocrystalline cellulose

Nanocrystalline cellulose refers to the high crystallinity nanometer rod-shaped or needle-shaped cellulose obtained through acid hydrolysis of large-size cellulose, as

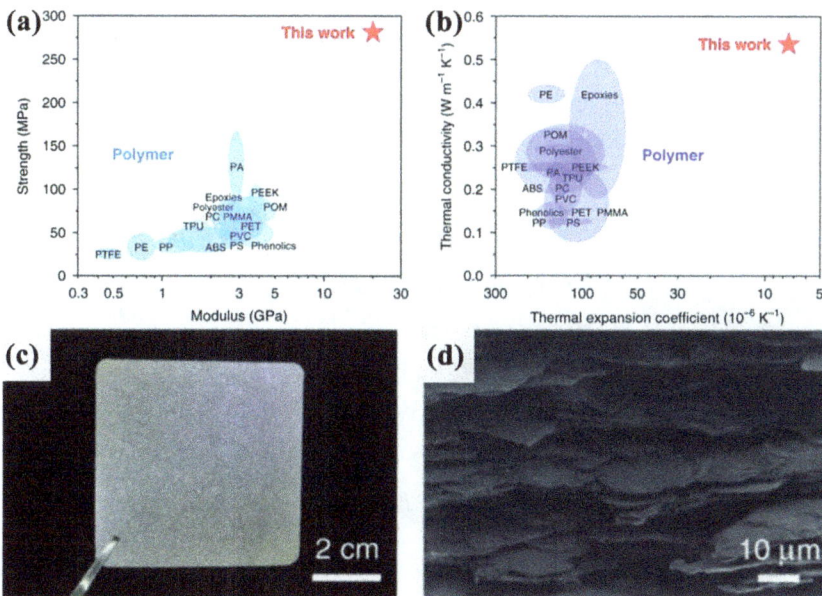

Figure 3.4: (a, b) Comparison of mechanical and thermal properties of all-natural bioinspired structural material with typical polymers. (c) All-natural bioinspired structural material. (d) Fracture surface of all-natural bioinspired structural material [66]. @2020 Springer.

shown in Figure 3.5 [17, 36, 67]. For preparation of cellulose nanocrystals, high-concentration inorganic acid is generally used to hydrolyze the amorphous and sub-crystalline domains in cellulose, retaining the crystalline domains [68]. Cellulose nanocrystals have attracted a huge amount of attention due to the excellent mechanical properties, reinforcing effect in composite materials, special rheology and liquid crystallinity [69]. During the preparation of nanocrystalline cellulose, the hydrolysis function can remove the defects (amorphous and subcrystalline regions) in cellulose and maintain a complete crystalline structure [70–72]. It suggests that this process mainly relies on the difference in hydrolysis kinetics between the amorphous and crystalline domains to prepare cellulose nanocrystals. The polysaccharide bond on the surface of the fibril is broken firstly, following by the amorphous region. And then, the crystalline region is exposure to the solution. After the hydrolysis, a large amount of water is added to dilute the acid solution to remove the residual acid and impurities via centrifugation and dialysis processes [50]. The main impact factors for the structure and performance of nanocrystalline cellulose are acid, material ratio, temperature, time, and so on during the hydrolysis process. For example, the properties of cellulose nanocrystals obtained by sulfuric acid and hydrochloric acid are different. Sulfuric acid may react with the hydroxyl groups on the surface of the nanocrystal to form charged sulfate groups, improving the dispersibility in water and reducing the thermal stability [73–75]. When the hydrochloric acid is employed, no charged groups on the surface of cellulose nanocrystal appear, limiting the dispersibility of the product in water. The cellulose nanocrystal is also easy to aggregate and flocculate. The particles size of nanocrystalline cellulose from these two acids is similar, even for the industrial production [76]. Actually, the cellulose nanocrystal from various resources may present different sizes and morphologies under microscopes. The length of nanocrystalline cellulose prepared from high crystalline tunicates and seaweed may be several microns, while that from low crystalline wood and hemp fibers

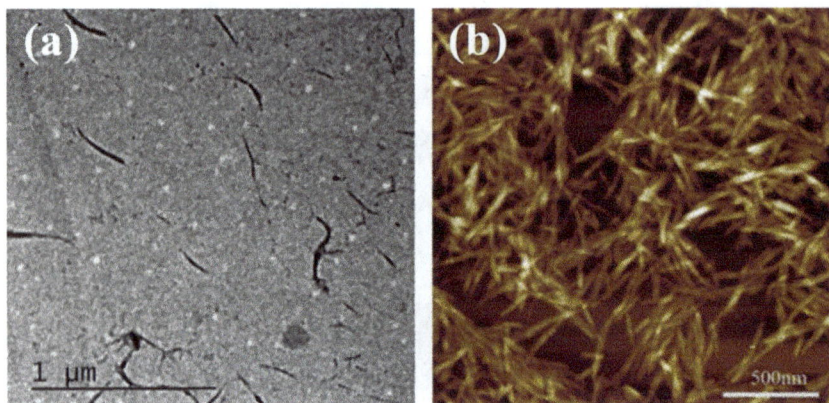

Figure 3.5: Microstructure of nanocrystalline cellulose (a) TEM; (b) AFM [17] @ 2019 Elsevier.

is in a few hundred nanometers [77]. Although these differences, the current industrially produced CNCs is positive, as well as the commercial-scale applications [76].

The size of cellulose nanocrystals is mainly determined by raw materials and hydrolysis conditions, as listed in Table 3.1 [72]. The width of nanocrystals is basically about 10 nm, and the length ranges from tens of nanometers to several microns. For example, the width of the nanocrystals extracted from wood, cystis, and tunicates are generally 3–5, 10–20, and 10–20 nm respectively with the length of 100–200 nm, more than 1000 and 500–2000 nm respectively [72, 78, 79]. Zhou et al. reported that the nanocrystals extracted from waste paper are 10–25 nm wide with the length of 50–770 nm [17, 50, 67]. It is obvious that the aspect ratio of nanocrystals varies greatly. The aspect ratio of nanocrystals made from cotton is 10–30, while that of tunicates is about 70 [80].

Table 3.1: The length and width of cellulose nanocrystals from various sources [17, 72].

Source	Length/nm	Width/nm	Technique
Bacterial	100–1000	5–10	TEM
	100–1000	30–50	TEM
Cotton	100–150	5–10	TEM
	70–170	~7	TEM
	200–300	8	TEM
	255	15	DLS
	250–210	5–11	AFM
Cotton linter	100–200	10–20	SEM-FEG
	25–320	6–70	TEM
	300–500	15–30	AFM
Microcrystalline cellulose	35–265	3–48	TEM
	250–270	23	TEM
	~500	10	AFM
Ramie	150–250	6–8	TEM
	50–150	5–10	TEM
Sisal	100–500	3–5	TEM
	150–280	3.5–6.5	TEM
Tunicate	/	8.8 × 18.2	SANS
	1160	16	DLS
	500–1000	10	TEM
	1000–3000	15–30	TEM
	100–1000	15	TEM
	1073	28	TEM
Valonia	>1000	10–20	TEM
Soft wood	100–200	3–4	TEM
	100–150	4–5	AFM
Hard wood	140–150	4–5	AFM
Waste paper	50–770	10–25	TEM + AFM

TEM, transmission electron microscopy; DLS, dynamic light scattering; AFM, atomic force microscopy; SEM–FEG, field emission scanning electron microscope; SANS, small angular neutron scattering.

Besides the characteristic morphology of the nanocrystalline cellulose, the excellent properties, including mechanical property, thermal resistance, and rheology, have also been analyzed and investigated [77]. The calculated theoretical value of the mechanical strength is between 100 and 160 GPa according to the structure, inter-molecular and intramolecular hydrogen bonds of cellulose type I crystals [81–83]. Diddens et al. reported the mechanical strength of an axial elastic modulus of 220 GPa and a radial direction of 15 GPa according to the relationship between the speed of sound and the diffusion of phonon in the crystal region under inelastic X-ray scattering [81]. It is reported that the thermal decomposed temperature of nanocrystalline cellulose is in the range of 200–300 °C, which is affected by the source of nanocrystals, heating rate, size, and morphology. Meanwhile, the thermal expansion index of nanocrystalline cellulose is about 1×10^{-7} K^{-1}, suggesting an excellent material with low thermal expansion coefficient [84]. Among the various attractive properties, one of the most interesting and important property of nanocrystalline cellulose should be the liquid crystal phenomenon [11]. Such behavior should be expected of any asymmetric rod-like or plate-like particle. Stiff rod-like particles (polymer micro-objects, viruses, rod-like alumina) are known to show liquid crystallinity, and the same to nano-crystalline cellulose. CNCs can be considered rigid-rods due to the stiffness and aspect ratio of micromorphology. Thus, one could expect nematic behavior where the rods align under certain conditions. Cellulose crystal has an axial spiral twisted structure similar to a screw (Figure 3.6a). This chiral nematic structure causes the nanocrystal to present a spiral arrangement in the vertical direction in suspension with a certain angle between every two planes of the crystals, showing a helical twist normal to the long axis of the rod (Figure 3.6b). In this model, each plane is rotated by a phase angle, which is dependent on concentration. Such alignment can result in optical bandgaps giving iridescent/pearlescent behavior due to the birefringence of individual domains typically producing fingerprint patterns (Figure 3.7) for chiral nematic as opposed to domain-like (or cross-like patterns) for simple nematic.

When referring to the rheological properties of cellulose nanocrystals, it mainly focuses on the gel behavior and liquid crystal or ordered behavior through viscometric

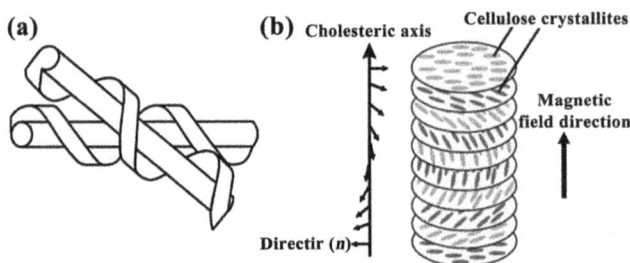

Figure 3.6: Schematic representations of the chiral nematic phase (a) and the tight packing achieved by chiral interaction of screwlike rods [85]. @2001 Wiley-VCH.

Figure 3.7: Chiral nematic texture of the anisotropic phase of a cellulose suspension [86]. @1996 American Chemical Society.

measurements or rheological characterization [11]. The liquid crystal transition under shear behavior is observed by studying the liquid crystal line of the suspension. As illustrated in Figure 3.8, the low-concentration cellulose sodium suspension exhibits shear-thinning behavior and is dependent on the concentration at low frequencies, but has a smaller relationship with the concentration at high frequencies [87]. However, at higher concentrations where the suspensions were lyotropic, the suspensions show anomalous behavior with viscosity showing shear-thinning behavior at low rates, a semi-plateau region where the shear-thinning is less pronounced as the rate is raised, and then a precipitous drop in viscosity at a critical rate. The rates at which such transitions in the flow behavior occur are concentration dependent. The explanation for such behavior is that at a critical shear rate, the nanocrystals align due to their rod-like nature, greatly easing their flow. Under enough shear the chirality of the suspensions breaks down in favor of a simple nematic structure. Additionally, the time

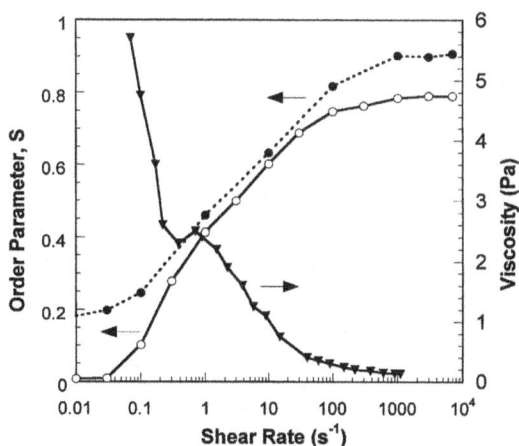

Figure 3.8: Order parameter and viscosity of CNC as a function of shear rate. Open circles are smaller aspect ratio than closed circles. Viscosity is for the smaller aspect ratio particles [87]. @1998 American Chemical Society.

constant of relaxation is highly dependent upon aspect ratio with higher aspect ratios staying aligned for longer times after shear (Figure 3.8). CNC suspensions in the dilute regime were shear thinning and this behavior increased as concentration was raised and showed concentration dependence at low rates and very little concentration dependence at high rates. In contrast to the sulfuric acid treated crystals which showed some shear thinning and no time-dependent behavior, HCl-derived crystals showed much higher shear thinning behavior, thixotropy at high concentrations and antithixotropy at dilute concentrations.

3.2.6 Bacterial cellulose

Bacterial cellulose (BC) is mainly synthesized by bacteria under different conditions, such as acetobacter, agrobacterium, rhizobium, sarcina, and so on [11]. In 1886, A.J. Brown synthesized bacterial cellulose by using acetobacter xylinum for the first time, whose microfiber structure was similar to the extracellular matrix. In the 1970s, the scientific research of bacterial cellulose appeared significantly [88] BC is fairly unique compared to the other cellulose microfibril sources in that there is an accessible strategy to alter the BC microfibril biosynthesis process. Meanwhile, the starting microfibril configuration within the pellicle (Figure 3.9) is usually retained in engineered BC materials when used [89]. The acetobacter xylinum, belonging to genus acetobacter glucono, is one of the most thoroughly studied bacteria with the strongest ability to produce cellulose [90]. There are four main processes for biosynthesis of bacterial cellulose, including polymerization, secretion, assembly, and crystallization. The cultivation of bacterial cellulose is mostly based on a combination of static culture, stirring culture, as well as continuous culture [91]. BC has great potential application in many fields, such as multipurpose tissue scaffolds, food additives, sound vibration membranes, artificial skin, artificial blood vessels, tissue engineering scaffolds, etc. [92]

The chemical structure of BC is similar to that of plant cellulose, while its physical form and molecular structure are quite different. The valuable character of BC is that it can be *in situ* controlled in shape, molecular structure, and compound by adjusting the bacteria, the shape of reactor, and the composition of the culture medium during the biosynthesis process. The cellulose content in BC is above 99% without impurities of lignin and hemicellulose [94, 95]. In the case of acetobacter xylinum, it secretes cellulose microfibrils through micro-holes on the cell wall, and forms ribbons with a width of 30–100 nm and thickness of 3–8 nm under the function of hydrogen bonding (Figure 3.9). Finally, a three-dimensional nanofiber network structure forms with the size of 1/100 comparing to plant cellulose [96]. Figure 3.9b–d depicts the three-dimensional network structure of BC and its significant difference from ordinary plant cellulose [93]. It is obvious that this nanofiber network structure makes a larger specific surface area for BC than plant cellulose. In addition, BC has the characteristics of high

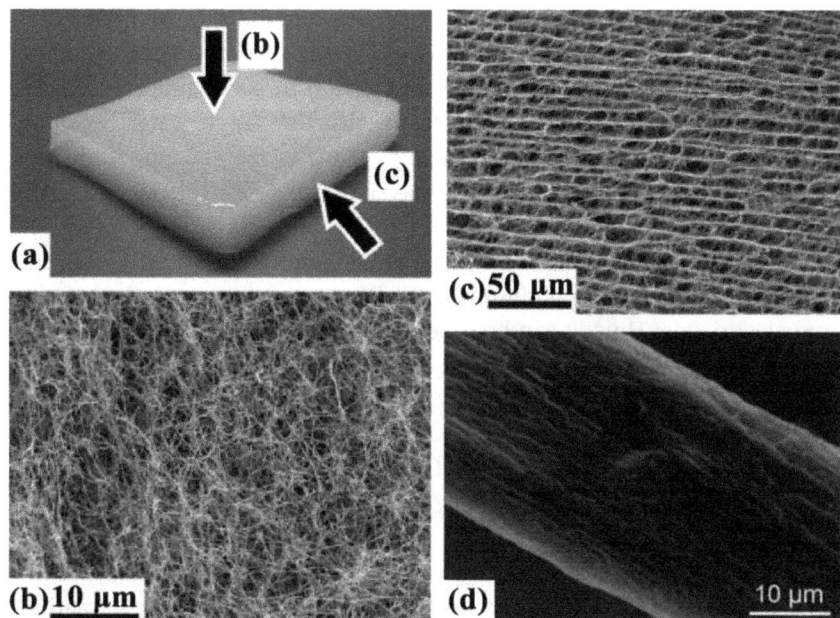

Figure 3.9: Images of BC pellicle. (a) Optical image of BC pellicles showing the directionality for images b and c. Field-emission SEM image showing (b) the low density in-plane BC network, and (c) the transverse structure [89]. @2008 WILEY-VCH. (d) Common pulp from plant cellulose [93]. @2005 Springer.

polymerization degree of 4000–10,000 and high crystallinity of 80–90%. Furthermore, BC hydrogel with a three-dimensional nanofiber network structure can be obtained by static culture of acetobacter xylinum, which has high strength, as well as high elastic modulus of ca. 15 GPa [11]. Astley et al. [97] had confirmed this by producing neat BC hydrogels with elastic modulus of ca. 14.0 MPa. It is reported that the containing of either xyloglucan or pectin may decrease the elastic modulus slightly in the culture solutions. BC shows great potential applications in our daily life. In terms of food, owing to the good water holding capacity, a unique gel-like translucent texture, BC is widely used in jelly, beverage, candy, canned food, food thickener, forming agent, dispersant, binding agent, etc. It can be used as a food supplement to give products a better taste, such as "Nata-coconut" product in some Southeast Asian countries [98]. BC is used to improve the dry and wet strength, durability, water absorption, and other properties of paper, which not only solves the problem of strength, but also eliminates the pretreatment and pulping process of plant fibers [99]. BC can also be used to prepare functional composites. Because of the large surface area, the interaction between BC and molecular compounds (water, etc.) or nanoparticles is improved. When BC is immersed in the carbon nanotube dispersion, it may present significant conductivity [100]. BC can be compounded with high-transparent polymers, such as epoxy resin or acrylic acid. A nanocomposite may present a transparency close to that of the polymer matrix with the great improvement in mechanical properties and thermal stability, holding valuable potential application in packaging and display areas [101]. Owing to the merits in nanonetwork structure, high water retention, plasticity, and biocompatibility of biosynthesis, BC opens up the important and strongly expanding fields of personal care, medicine, and life sciences [11]. The nanostructure and morphology of BC make it suitable as a scaffold for fixing and transferring cells. Artificial skin and auxiliary materials made of BC have high mechanical strength under humid conditions, contributing to the circulation of gas, moisture and electrolytes. Thus, it can be used as a substitute for skin in burns, skin ulcers, and skin transplant operations.

3.3 The structure and properties of cellulose

3.3.1 Chemical structure

Cellulose is a kind of carbohydrate which can be divided into two conformations (α- and β-) according to the location of the C-1 and C-2 hydroxy groups on the glucopyranose ring. Cellulose is a linear polymer composed of repeating cellobiose units connected by β-(1 $\rightarrow$ 4)-D-glycosidic bonds. The hydroxyl group is located at the positions of C-2, C-3, and C-6 respectively on each an hydroglucose unit, which possesses typical reaction properties of primary and secondary alcohols. As shown in Figure 3.10, the molecular structure of cellulose, cellohexaose, cellobiose, and glucose indicates that cellulose presents typical diol structures [102]. The an hydroglucose unit of the cellulose chain has a $^4C_1$ chair conformation. The free hydroxyl is located in the ring plane, and the hydrogen atom is located in the vertical position. The hydroxyl groups at the end of the cellulose chain exhibit different characteristics, such as the reductive C-1 terminal hydroxyl group and the oxidizing C-4 terminal hydroxyl group.

As a natural polymer, the cellulose molecular conformation is similar to that of synthetic polymers. It presents a variety of conformations in solution including random coils, spirals, worms, rods, and aggregates [103]. It is difficult for cellulose to form dispersion in molecular-level in most solutions, which mainly exhibits a rigid chain conformation and is easy to form aggregates [104]. Cellulose possesses fibrous structure with high crystallinity composed of repeating β-(1 $\rightarrow$ 4)-D-glycosidic units. It is stacked and arranged in different ways to build multiple crystal forms constructing homogeneous polycrystalline substance [11]. According to the difference of molecular chain polarity, cellulose polycrystals can be divided into two types: one is the parallel chain crystal (cellulose I_α, I_β, III_I, IV_I), and the other is the antiparallel chain crystal (cellulose II, IV_{II}). Cellulose III_{II} may have an antiparallel chain structure due to the rotation disorder of O6 hydroxyl group. The chain conformation is even close to 2_1 helix axis. There are some differences for the adjacent units or the O6 rotation position of the

Cellulose

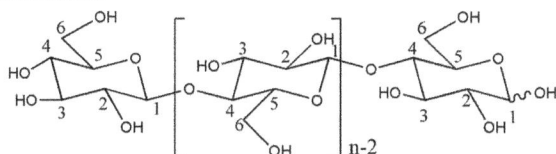

Cellohexose

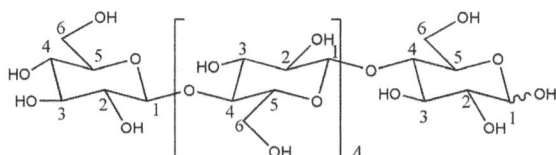

Cellobiose

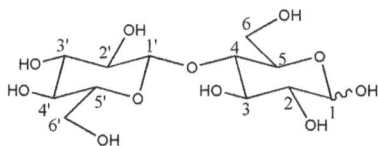

Glucose

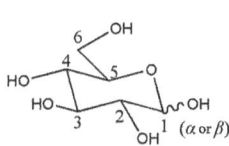

Figure 3.10: Chemical structures of the main cellulose molecules.

cellulose molecular in different crystal forms. However, the unit cells of all cellulose crystal forms may be double-stranded or single-stranded rather than a large-sized subunit cell structure of algae cellulose, even with a cellulose-conducting channel for the unique cellulose from *Rhodobacter sphaeroides* [11, 105, 106]. Currently, the crystalline variants of cellulose I_α, I_β, II, III$_I$, III$_{II}$, and IV$_{II}$ have been confirmed practically. The unit cell size of repeating unit is ca. 10.3 Å, which supports the fact of repeating cellobiose unit in cellulose. The quantization of cellulose crystals in different forms can be investigated according to the X-ray diffraction patterns of various cellulose unit cells, the infrared spectrum of hydrogen bond arrangement in the hydroxyl stretching vibration region, and the high-resolution solid CP/MAS ^{13}C-NMR elements for chain conformation and stacking [107]. Whether the cellulose molecular is parallel or anti-parallel, different cellulose crystal forms may undergo mutual transformation. The transformation of the chain conformation happens through the interaction with the solvent under certain conditions, suggesting the key role of solvent in cellulose area. Cellulose I is a naturally form of cellulose, which commonly contains cellulose I_α with a single-chain triclinic crystal structure and cellulose I_β with a two-chain monoclinic crystal structure. By heat treatment, Cellulose I_α can be transformed into cellulose I_β, which is more stable than cellulose I_α [108, 109]. Cellulose II is a crystalline modification of cellulose I that is regenerated in solution or after mercerization. Because of the formation of large amount of hydrogen bonds, cellulose II may be the most stable crystal with a denser unit cell structure than cellulose I. Cellulose III can be obtained by immersing cellulose in liquid ammonia or organic amines (methylamine, ethylamine, ethylenediamine, etc.). The cellulose III prepared with cellulose I and cellulose II respectively is defined as cellulose III$_I$ and cellulose III$_{II}$ respectively, generating cellulose IV$_I$ and IV$_{II}$ respectively after heat treatment. Notably, cellulose IV$_I$ and IV$_{II}$ can also be reversibly converted to cellulose I and II. Wada et al. [110] proposed that cellulose IV is mainly composed of cellulose I$_\beta$ converted from high crystalline cellulose I$_\beta$.

3.3.2 Aggregation structure

Cellulose is composed of linear molecules without branches. The cellulose presents rope-like bundles, each of which consists of 100–200 parallel cellulose macromolecular chains under the function of hydrogen bonds to form microfibrils with a diameter of 10–30 nm [72]. Several microfibers are gathered into a bundle to form a fibril. The regular sequences form tiny crystalline regions (about 85%) and irregular sequences form amorphous regions (about 15%) [111]. Natural plant fibers have a complex multilevel structure. A fiber is composed of several cellulose microfibers, and each microfiber is composed of several cellulose molecular chains. A large amount of intramolecular, intermolecular hydrogen bonds and van der Waals forces maintain the self-assembled macromolecular structure in fibril form. Strong acids, alkalis, enzymes, and even mechanical force can make the fibers fibrillate to split into microfibers, removing the amorphous domains of the microfibers to obtain cellulose whiskers and nanocrystals [112].

Since cellulose molecular contains a large amount of hydroxyl groups, forming intermolecular and intramolecular interactions among them. Cellulose macromolecules are firmly bonded together. In this way, the O6 (oxygen at position 6), O2′, O3, and O5′ are connected to make the entire polymer chain into a ribbon with high rigidity and high degree of regularity. After the formation of a lattice, interchain hydrogen bonds can also forms between the O6 and O3 of the adjacent cellulose molecule. The hydrogen bond acts as a physical cross-linking point in the molecule to form a three-dimensional network structure with high stability. Owing to the large number of hydrogen bonds in cellulose, it is hard to destroy this interaction, as well as the cellulose chains. In addition to intermolecular and intramolecular hydrogen bonds, this kind of interaction also appears between the cellulose and water molecules. They are mainly the hydroxyl group of cellulose, the oxygen bridge O4 of cellobiose and the oxygen O5 of the pyran ring of glucose residue.

Owing to the highly regular molecular, cellulose is easy to crystallize. The unique structure, consisting of 85% regular crystalline domains and 15% irregularly and loosely amorphous domains, presents voids in low density. A cellulose molecule may pass through several microcrystalline domains and amorphous domains. The cellulose molecules are connected to each other by the binding force between molecules in crystals, that is, the crystalline domains are also connected by cellulose molecules [107]. Notably, the crystalline domains are separated by the loosely and chaotically amorphous domains, forming the alternatively dense and loose phases to construct bulk structure of cellulose [113].

3.4 Swelling and dissolution of cellulose

3.4.1 Solvents and principles of cellulose dissolution

It is an important scientific issue for the solvent and dissolution of cellulose whether in theoretical research or in industry. Before the 1950s, copper ammonia solution was the mainly used solvent to dissolve cellulose. Copper ethylene diamine complex aqueous solution and tetraethylamine hydroxide aqueous solution were occasionally used for cellulose structural analysis. In the 1960s and 1970s, a large number of complex aqueous solutions based on transition metals were used to dissolve cellulose. Among them, sodium tartrate ferrite solution and cadmium ethylenediamine solution attracted huge attention to analyze cellulose structure. In the following decades, organic solvents such as dimethylacetamide (DMAc)/LiCl, N-methylmorpholine-N-oxide (NMMO) and ionic liquids were developed to replace transition metals in Japan, United States and Germany. In 2003, Lina Zhang [26] developed a new solvent system of alkali/urea aqueous solution that quickly dissolves cellulose at low temperature, which opened another new path in cellulose technology.

According to the organic chemistry, cellulose solvents can be divided into derivatized solvents and nonderivatized solvents with different dissolution rate and effect. Nonderivatized solvent refers to dissolving cellulose by intermolecular forces, which can generally be regarded as an acid-base interaction between the cellulose and the solvent [34]. Notably, there is strong interaction between the aqueous transition metal complex and cellulose without covalent bond. In this case, it is also regarded as a nonderivatized solvent. Derivatized solvents refer to dissolving cellulose by covalently bonding with cellulose to form ethers, esters, and acetals. In this solvent, cellulose derivatives can be decomposed into regenerated cellulose by changing the composition or pH of the solution. Both the two types of solvents include aqueous and nonaqueous solutions, as depicted in Figure 3.11 [114]. The following sections will focus on the main solvents for dissolution of cellulose according to the classification in Figure 3.11.

3.4.2 Nonderivatizing solvents

There are single and multicomponent systems for nonderivatizing solvents. Although various kinds of these solvents have been developed, only a few have shown a potential

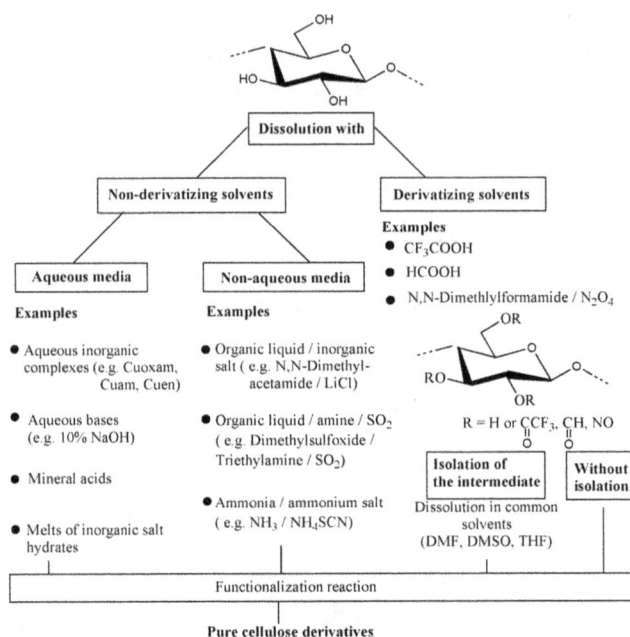

Figure 3.11: The classification of solvents for cellulose [114]. @2001 Elsevier.

for a controlled and homogeneous functionalization of polysaccharides, especially for cellulose. Limitations are mainly high toxicity, high reactivity of the solvents leading to undesired side reactions, and the loss of solubility during reactions yielding inhomogeneous conditions by formation of gels and pastes which can be hardly mixed and even by formation of deswollen particles of low reactivity which set down in the reaction medium [114]. The main idea of this part is to discuss the main cellulose solvents and their functions.

3.4.2.1 Inorganic compound/organic amine mixture containing nitrogen or sulfur

The combination of organic solvents and some simple inorganic compounds containing nitrogen or sulfur, such as SO_2, NH_3, N_2H_4, and certain organic amine mixtures, are the two major origins of cellulose nonderivatized solvent systems. The first type of cellulose nonderivatized organic solvent system is composed of SO_2 ($SOCl_2$)/aliphatic amine/polar organic solvent. The aliphatic amines are mainly ethylene diamine and the polar organic solvents, including dimethyl formamide (DMF), methyl sulfoxide (DMSO), dimethyl acetamide (DMAc), and formamide, etc. [26]. Their interaction with cellulose is assumed as electron donor–acceptor interaction, as well as possibly covalent bond interactions [115]. Owing to the strong corrosiveness and high pollution of SO_2, it is not suitable in the as known green chemistry of cellulose. The second type of cellulose solvent is composed of an amino-containing active compound/polar organic solvent. The addition of ethanolamine or appropriate inorganic salt (such as NaBr) can promote the dissolution of cellulose. Because of the complex chemical composition of this system, the interaction between them and cellulose is still unclear.

3.4.2.2 Amine oxide system

One of the most successful nonderivatization solvents of cellulose is *n*-methyl-morpholine-N-oxide (NMMO). NMMO is a kind of aliphatic cyclic tertiary amine oxide, which is obtained by the reaction of diethylene glycol and ammonia to form mor-pholine, following by methylation and oxidization of H_2O_2. In 1939, Charles Grae-nacher, et al. firstly discovered that tertiary amine oxides such as trimethylamine oxide, triethylamine oxide, and dimethylcyclohexylamine oxide can dissolve cellu-lose. It was found that NMMO and NMMO·H_2O showed good solubility for cellulose, which has used in large-scale production of cellulose fibers in industry [116].

3.4.2.3 *N,N*-dimethylacetamide/lithium chloride (DMAc/LiC)

Cellulose can be dissolved in lithium salts, including LiCl or LiBr, and some polar aprotic liquids, such as DMAc/LiCl, *N*-methylpyrrolidone (NMP)/LiCl, and *N,N*-dime-thylacetamide (DMAc)/LiBr, etc. Since water should be avoided in the dissolution process, it is necessary to dry the solvent of cellulose, followed by the substitution of the original solvent of cellulose. As listed in Table 3.2, the characterizations of several typical LiCl or LiBr-containing polar aprotic solvents are described, as well as the cellulose activation methods [26]. The system DMAc/LiCl shows an enormous potential for the analysis of cellulose and for the preparation of a wide variety of derivatives [114]. Owing to the advantages of colorless and dissolution succeeds without or at least with negligible degradation even in case of high molecular weight polysaccharides, this system is usefulness in analyzing cellulose structure, such as cotton linters or bacterial cellulose.

In order to dissolve cellulose completely, it is necessary to activate cellulose by solvent replacement with water, methanol and DMAc in advance. Although there is no change for the crystalline structure of cellulose in solvent activation process, the ag-gregation state of cellulose microfibers varies significantly. In this way, solvent mol-ecules may be greatly accessible to cellulose. After the preactivation, cellulose with even high molecular weight can be dissolved in solvent without molecular chain degradation. One of the effective processes to dissolve cellulose is to add a quantitative amount of DMAc to the dried cellulose following by the heating in a nitrogen atmo-sphere, which can avoid the oxidative degradation of cellulose due to the increase in temperature. And then the LiCl is added with continuously stirring to ensure the complete dissolution of cellulose. The DMAc/LiCl solvent is not only suitable for the analysis of cellulose, but also is a good medium for the homogeneous derivatization.

Table 3.2: Lithium salt/polar aprotic solvent of cellulose [26].

Composition	Activation method	Typical solution composition[a] (parts by weight)
DMAc/LiCl	All known methods	A wide range
NMP/LiCl	All known methods	6/8.5/85.5
DMF/LiCl	Solvent replacement after swelling in liquid ammonia	3/10/87
DMEU/LiCl	Heating in solvent; solvent replacement after swelling in water	2/5.5/92.5 or 5/5/90
DMPU/LiCl	Solvent replacement after swelling in water	3.5/5/91.5
DMAc/LiBr	Heating in solvent	3/20/77
NMP/LiBr	Heating in solvent	1/18/81
HMPT/LiCl	Heating in solvent	5/11/84
DMSO/LiCl	Solvent replacement	3/8/89

[a]Solvent composition order: cellulose/salt/solvent. Abbreviation: DMAc, N,N-dimethylacetamide; MMP, N-methylpyrrolidine; DMEU, dimethylethylene urea; DMPU, dimethylpropylene urea; HMPT, hexamethylphosphoric acid triamide.

3.4.2.4 Ionic liquid

Ionic liquid is an organic liquid composed of anions and cations with strong polarity, nonvolatile, nonflammable, good solubility for most inorganic and organic compounds, and stability to most reagents. In 1934, Charles Graenacher et al. found that cellulose could be dissolved in N-ethylpyridine chloride, which also was a homogeneous derivatization reaction medium for cellulose. In 2002, Rogers et al. [117] found that cellulose can be dissolved in dialkylimidazole halide ionic liquids, and 1-butyl-3-methylimidazolium chloride ([Bmim]Cl) showed excellent solubility to cellulose. Since then, Zhang et al. reported that 1-allyl-3-methylimidazolium chloride ([Amim]Cl) could also dissolve cellulose, which was able to be used for homogeneous acetylation of cellulose [118, 119]. Until now, ionic liquids that dissolve cellulose are mainly the cations including imidazolium salts, pyridinium salts, ammonium salts, and quaternary phosphonium derivatives, as well as the most anions including X^-, SCN^-, BF_4^-, PF_6^-, etc. Among them, [Bmim]Cl, [Amim]Cl and 1-ethy-3-methylimidazolium acetate ([Emim]Ac) are currently the most used ionic liquids. It is reported that [Emim]Ac may be the most effective in dissolving cellulose, and [Amim]Cl presents the strongest ability to dissolve wood [120]. The role of the anions and cations in ionic liquids has not been fully understood in the dissolution of cellulose. It is generally believed associated ion pairs dissociate to form free anions and cations in ionic liquids under heating. The free anions may form hydrogen bonds with the hydrogen atoms on the hydroxyl groups in cellulose molecular, which weakens the hydrogen bonding between the cellulose molecules or within the molecule. Meanwhile, the cations interact with the oxygen atoms on the hydroxyl groups to further reduce the intramolecular cellulose. Under the comprehensive interaction of anions, cations, and hydroxyl groups of cellulose, the dissolution of cellulose happens significantly. At present, the basic theoretical and technical problems such as large-scale synthesis of ionic liquids, purification, and recovery technology, biocompatibility, toxicity, stability evaluation, etc. are still needed to be solved urgently.

3.4.2.5 Transition metal/amine (or ammonia) complex aqueous solution

Transition metal complex aqueous solutions belong to aqueous nonderivatized solvents. They are mainly used for analyzing cellulose structure, including copper ammonia (Cuoxam), copper ethylenediamine (Cuen), cadmium ethylenediamine (Cadoxen), and sodium ferrate tartrate (FeTNa) solutions by mean of ^{1}H-NMR analysis after chain degradation and HPLC analysis after complete depolymerization [114]. The compound system can completely dissolve a high polymerization degree cellulose. The transition metal complex and the dissolution of cellulose are controlled by both the pH-dependent coordination balance and the bonding of cellulose hydroxyl group. Generally, Normann compounds, which can deprotonate the cellulose hydroxyl group and prevent the cross-linking of cellulose molecules to form diglycol structure, are used as solvent for cellulose [121, 122]. In the process of cellulose swelling and dissolution, the intermolecular and intramolecular hydrogen bonds are broken. The complexed deprotonated OH of the adjacent an hydroglucose generate stronger hydrogen bond in the form of $OH\cdots O^-$, which leads to a significant increase in the rigidity of the cellulose chain in solution.

3.4.2.6 Tetraalkylammonium hydroxide aqueous solution

Tetraalkylammonium hydroxide (R_4NOH) aqueous solution belongs to aqueous nonderivatized solvents. The interaction between R_4NOH and cellulose is similar to alkali metal hydroxide. The swelling degree of cellulose increases with the concentration of the R_4NOH aqueous solution as a swelling agent. Moreover, due to the volume effect of the substituents, the increase of R_4NOH substituents molar volume leads to an increase swelling capacity at a certain solution concentration. Some R_4NOH with sufficiently substituents at certain concentration may also dissolve cellulose significantly, such as trimethylbenzylammonium hydroxide (Triton B), dimethyldibenzylam-monium hydroxide (Triton F), trimethylbenzyl trimethylbenzylammonium hydroxide (TBAH), tetrabutylphospho-nium hydroxide (TBPH) and tetraethylammonium hydroxide (TEAH) [123]. However, the dissolution effect of cellulose will decrease as the substituent molar volume increases when the concentration of the R_4NOH exceeds the maximum value in aqueous solution. The dissolution of cellulose in R_4NOH mainly arises from the destruction of cellulose hydrogen bonds and the hindrance of molecular self-association. In R_4NOH aqueous solution, a complex, which is formed among the cellulose hydroxyl group, water molecules, and the polar ends of the R_4NOH dipole ion, quite sensibly breaks the hydrogen bond of the cellulose crystal domains. The nonpolar substituents of solvent penetrate into the interior of the cellulose to separate the cellulose molecular chains, dispersing, and completely dissolving the cellulose. Cellulose fibers with good performance can be obtained from the cellulose-R_4NOH solution using wet spinning process. Some solvents, including tetraalkylammonium hydroxide, trimethylbenzylammonium hydroxide, and dimethyldibenzylammonium hydroxide, play a role in the etherification of cellulose, especially for the homogeneous alkylation reaction. However, there are some problems for this solvent system, such as high solvent cost and hard to recycle [124].

3.4.2.7 Alkali metal hydroxide aqueous solution

The alkali metal hydroxide aqueous solutions are the key solvents belonging to aqueous nonderivatized system, which are commonly used as swelling agents or solvents for cellulose. The interaction between these solvents and cellulose is greatly important for both processing and derivatization of cellulose [125]. It is suggested that the dissolution degree for cellulose is influenced by the type and concentration of alkali metal hydroxide with the incresing sequence of LiOH/urea aqueous solution > NaOH/urea aqueous solution ≫ KOH/urea aqueous solution. Among these three alkali metal hydroxides, NaOH is employed mostly. The dissolution of cellulose in NaOH solution is exothermic, indicating that the solubility of cellulose increases by decreasing the temperature. In the process of cellulose mercerization, NaOH aqueous solution with the concentration of 8–10% (mass fraction) presents a strong swelling effect on natural cellulose at low temperature. However, only a small part (possibly low molecular weight part) of cellulose dissolved in NaOH aqueous solution. During this swelling process, the interaction between alkali metal hydroxide and cellulose induces the variations, such as the destruction of H-bonding, decrease of supramolecular order, changes in chain conformation, and solvent hydration layer structure, partial anionization of hydroxyl groups of cellulose [126].

Zhang et al. investigated the dissolution of cellulose in alkali/urea (or thiourea) aqueous solutions. They found that cellulose can be completely dissolved in a concentration (mass fraction) of 6% NaOH/4% urea aqueous solution and 6% NaOH/5% thiourea aqueous solution using the freeze-thaw cycle method [127, 128]. However, this method is not suitable for large-scale application. After that, it was found that cellulose could be dissolved directly and quickly in NaOH/urea aqueous solution by changing the NaOH concentration, urea concentration, and solvent precooling temperature [27]. The obtained cellulose solution is stable with the fastest dissolution speed in the nonderivatized aqueous system. It is reported that the optimal conditions to completely dissolve cellulose are with the concentration of NaOH and urea of 6–10%, 2–20% (mass fraction) respectively and the solvent precooling temperature of −20 to −5 °C. The mechanism of the cellulose rapid dissolution in NaOH/urea aqueous solution at low temperature has been investigated. The urea-NaOH-cellulose assemble into a sheath-like structure under the function of hydrogen bonding induced by low temperature between solvent molecules (NaOH hydrate, urea hydrate, water) and cellulose macromolecules. Moreover, the extended wormlike cellulose chains easily aggregated in parallel to form nanofibers in the aqueous solution, providing an important guidance for constructing novel nanomaterials [26]. The dissolution mechanism and chain conformation of cellulose in LiOH/urea aqueous solution are similar to those of NaOH, but it presents a stronger solubility than those of NaOH and even can dissolve cellulose with a polymerization degree of 2500 [129, 130].

3.4.3 Derivatization solvent

Derivatization solvents refer to the solvents that dissolve cellulose by forming covalent bonds with cellulose molecules to produce ethers, esters, and acetals. The character is that the cellulose derivatives generated in derivatization solvents can be decomposed into regenerated cellulose by changing the composition or pH value [26]. On molecular level, the interactions between cellulose and most derivatization solvents is clear. Simultaneously, both of the derivatization and dissolution processes occur for cellulose. However, it is still unclear whether the hydrogen bonds of cellulose participate in the derivatization and dissolution processes. Table 3.3 lists some of the derivatization solvents and the corresponding derivative substituents.

3.4.4 Preparation and characterization of cellulose derivates

There are three active hydroxyl groups on the D-glucopyranose unit (AGU) of cellulose, including one primary hydroxyl group (C-6 position) and two secondary hydroxyl groups (C-2 and C-3 positions respectively). Thus, a series of derivatization reactions may happen theoretically with hydroxyl, such as oxidation, crosslinking, etherification, esterification, graft copolymerization, etc., to produce various kinds of

Table 3.3: The derivatization solvents and corresponding derivative substituents for cellulose.

Solvent	Substituents
H_3PO_4 aqueous solution (>85%)	$-PO_3H_2$
$CF_3COOH/CF_3(CO)_2O$	$-COCF_3$
Me_3SiCl/pyridine	$-SiMe_3$
CCl_3CHO/DMSO/TEA	$-CH(OH)-CCl_3$
NaOH/urea	$-CONH_2$
$HCOOH/ZnCl_2$	$-CH=O$
N_2O_4/DMF	$-O-N=O$
$(CH_2O)_x$/DMSO	$-CH_2OH$
$CS_2/NaOH/H_2O$	$-C-(S)SNa$

cellulose derivates [131], as listed in Table 3.4. The cellulose derivates with the functional groups (substitutional group), degree of substitution (DS), and solubility are illustrated. It is known that the derivatization of cellulose improves the solubility, introduces new functions and expands application in food, medicine, chemical, construction, environmental protection, petrochemicals, etc. It is a significant way to resolve the key problems of melting and dissolution of cellulose.

3.4.4.1 Esterification of cellulose

Cellulose ester refers to the esterification reaction of cellulose hydroxyl group with acid, acid anhydride, acid halide, etc., which may be the earliest research and production product in cellulose chemistry [136]. The hydroxyl groups of cellulose are polar groups, which can be substituted by nucleophilic groups or nucleophilic compounds in strong acid solution to generate corresponding cellulose esters via nucleophilic substitution reaction. As depicted in Figure 3.12, there are three main mechanisms for the esterification of cellulose. Firstly, as shown in equation (1), hydronium ions are generated and then substituted, which happens for the reaction of cellulose and inorganic acid. Secondly, cellulose and organic acid interact through the nucleophilic addition reaction, which proceeds according to equation (2). According to equation (3), acid catalysis can promote the esterification reaction of cellulose. A proton is added to the electronegative oxygen of carboxyl group in advance, leading to positive carbon atoms. This is beneficial to the activity of nucleophilic alcohol molecules to generate H_2O. The all reactions are reversible, indicating that the esterification reaction is typical equilibrium for cellulose. Theoretically, cellulose may react with both the inorganic and organic acids to produce mono-, di-, and tri-substituted cellulose esters.

3.4.4.2 Cellulose inorganic acid ester

Cellulose inorganic acid ester produced by the reactions between cellulose and nitric acid, sulfuric acid, phosphoric acid, xanthate, etc. The main inorganic acid esters are cellulose nitrate and xanthate (an important intermediate for producing regenerated cellulose). Cellulose nitrate (CN) is an important industry material from the nitration reaction using natural cellulose [137]. When it is used in producing goods, CN presents good solubility, transparency, viscosity, and stability, especially a narrow range of ammonia content. The mixed acids of HNO_3/H_2SO_4 are most used to synthesize CN. In this process, H_2SO_4 plays the role of dehydrating agent to remove the water produced during the reaction. H_2SO_4 cannot penetrate into crystalline microfibers of cellulose, while it may be used as a swelling agent to facilitate the penetration of HNO_3 for fibrils, as well as accelerate the esterification reaction. The maximum DS of CN ester produced by the mixed acid system can reach 2.9 (nitrogen content 13.8%). During the esterification process, a small amount of sulfuric acid ester is also generated. It may affect the stability of CN. Furthermore, sulfuric acid and cellulose produce unstable sulfuric acid

Table 3.4: The main cellulose derivates and their characters [24, 34, 132–135].

Category	Material	Functional groups (substitutional group)	DS	Solubility
Cellulose ester	Cellulose acetate (CA)	$-C(O)CH_3$	0.6–0.9	H_2O
			1.2–1.8	Ethyl alcohol
			2.2–2.7	Acetone
			2.8–3.0	Chloroform
	Cellulose acetate propionate (CAP)	$-C(O)CH_3/-C(O)CH_2CH_3$	2.4/0.2	Acetone, ethyl acetate
	Cellulose acetate butyrate (CAB)	$-C(O)CH_3/-C(O)(CH_2)_2CH_3$	0.2/2.7	Acetone, valerone
			1.1/1.6	Acetone
	Cellulose nitrate (CN)	$-NO_2$	1.8–2.0	Ethyl alcohol
			2.0–2.3	Methyl alcohol, acetone, methyl ethyl ketone
			2.2–2.8	Acetone
	Cellulose sulfate (CS)	$-OSO_3Na$	> 1.0	H_2O
	Cellulose xanthate ester (CXE)	$-C(S)SNa$	0.5–0.6	NaOH aqueous
Cellulose ether	Methylcellulose (MC)	$-CH_2$	0.1–0.4	4–8% NaOH aqueous (low temperature)
			0.4–0.6	4–8% NaOH aqueous
			1.3–2.6	H_2O (sizing agent)
			2.1–2.6	Ethyl alcohol
			2.4–2.7	Organic solvent
			2.6–2.8	Hydrocarbon solvent
	Ethylcellulose (EC)	$-CH_2-CH_3$	0.7–1.7	H_2O
			>1.5	Organic solvent
	Hydroxyethyl cellulose (HEC)	$-CH_2-CH_2-OH$	0.11–0.35	6–8% NaOH aqueous
			0.66–1.66	H_2O
	Hydroxy propyl cellulose (HPC)	$-CH_2-CHOH-CH_3$	3.5	H_2O
	Carboxymethyl cellulose (CMC)	$-CH_2-COONa$	0.1–3.0	H_2O

Table 3.4: (continued)

Category	Material	Functional groups (substitutional group)	DS	Solubility
	Quaternary ammonium salt of cellulose (QC)	$-CH(OH)-CH_2-$ $N(CH_3)_3{}^+$	0.01–0.1	H_2O
	Cyanoethyl cellulose (CEC)	$-CH_2-CH_2-CN$	~2.0	Acetone, acrylonitrile, nitromethane, zinc chloride solution
	Ethyl hydroxyethyl cellulose (EHEC)	$-CH_2-CH_3$ $-CH_2-CH_2-OH$	<3	H_2O
	Hydroxyethylmethyl cellulose (HEMC)	$-CH_2-CH_2-OH$ $-CH_3$	1.4–1.9	H_2O
	Hydroxyethyl carboxymethyl cellulose (HECMC)	$-CH_2-CH_2-OH$ $-CH_2-COONoa$	0.3–1.1	H_2O
	Quaternary hydroxyethyl cellulose (QHEC)	$-CH_2-CH_2-OH$ $-CH(OH)-CH_2-$ $N(CH_3)_3{}^+$	0.49	H_2O

ester by-products and nitric acid mixtures during the reaction, which may cause degradation, spontaneous combustion, and even explosion of CN. Sulfuric acid may degrade cellulose, decreasing the viscosity of the esterification product, and even destroying the fiber morphology. Moreover, sulfuric acid is corrosive to steel and do harm to the environment. Therefore, it is essential to explore new industrialization methods to prepare CN. Cellulose sulfate (CS) with water-solubility is prepared by introduction of sulfate ester groups to the cellulose AGU. The introduction of the sulfate group breaks the original intermolecular and intramolecular hydrogen bonds of cellulose, increasing the solubility in water greatly. When DS is higher than 1.0, CS presents the unique antienzymatic property. It is negatively charged. The mutual repulsion between the charges induces the stretching of the cellulose molecular. It has been widely applied as drilling fluid additives, thickener, and slow-releasing capsule membranes in petrochemical, coating, and pharmaceutical industry, respectively. Cellulose xanthate ester (CXE) is an important intermediate for the production of regenerated cellulose. It was first discovered by Cross and Beven in 1892 [138]. The principle for preparing CXE is the reaction between alkali cellulose and CS_2 via viscose

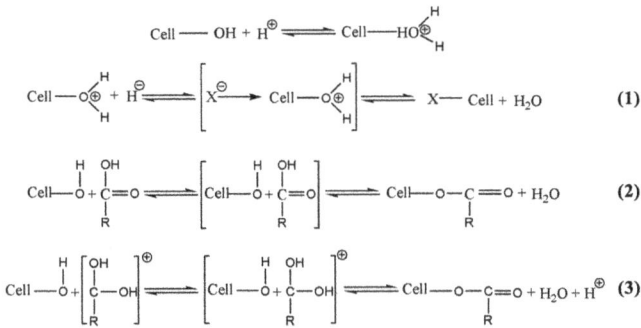

Figure 3.12: The mechanism for esterification of cellulose.

process. Firstly, CS with low reactivity reacts with NaOH to produce dihiocarbonate with ionized water-solubility in high reactivity. And then, the dihiocarbonate spontaneously reacts with cellulose to produce CXE. However, there are usually various side reactions in this process, forming trithiocarbonate, thio-acid compound, carbon disulfide, carbonate, etc. [139]

3.4.4.3 Cellulose organic esters

Cellulose organic esters can be prepared by reaction between cellulose and organic acids, acid anhydrides, and acid chlorides respectively. The more important products are cellulose acetate and its mixed esters (such as cellulose acetate, butyl acetate, etc.).

Cellulose acetate (CA) is mainly acetylated from cotton fiber or wood fiber, which is also called acetyl cellulose [140]. The most widely used is cellulose diacetate with a DS of 2–3, as listed in Table 3.4. In 1864, Schutzenberger used acetic anhydride to successfully acetylate cellulose, opening up a way to produce CA. It is the earliest commercialized cellulose derivative with a history more than 150 years [26]. Until now, the preparation methods can be divided into multiphase system and solution process with acetylation for commercial CA. Inert diluents, such as benzene, toluene, pyridine, etc., are employed to replace (partially replace) acetic acid to accomplish the acetylation. The cellulose maintains a fibrous structure throughout the process. The perchloric acid is commonly used as catalyst due to the strong catalytic ability and irresponsive to cellulose to form acid esters. Notably, this fibrous heterogeneous ethyl acetate is also suitable for acetylation of acetic acid vapor with zinc chloride as a catalyst [141]. Except for the fibrous triacetin, almost all cellulose acetates are acetylated by solution process, in which acetic acid, sulfuric acid, and liver vinegar are used as solvent, catalyst, and esterification agent respectively. The acetylation of cellulose with the catalyst of sulfuric acid occurs from the amorphous domains of cellulose, and then to the crystalline domains. The bulk structure of cellulose changes from heterogeneous to single homogeneous phase via the cycle reactions of layer-by-layer reaction of the cellulose fibers, dissolving, and exposure. Therefore, this process is also known as quasi-homogeneous reaction with the character of a gradual transition from heterogeneous to homogeneous. Owing to the insolubility of cellulose triacetate in acetone, it is often partly hydrolyzed to reduce the degree of esterification, transforming it into cellulose diacetate [142].

Cellulose acetate propionate (CAP) is synthesized by activating cellulose using acid, following by catalyzing with sulfuric acid [143]. As a catalyst, there are different mechanisms for catalyzing. On the one hand, sulfate esters are produced with the reaction between sulfuric acid and cellulose. At the beginning of esterification, sulfonation, acetylation, and propionylation compete with each other for the hydroxyl groups on cellulose chains. Although the reaction of sulfonation happens faster than acylation, most of the sulfonic acid groups are replaced by acetyl and propionyl groups to yield CAP at the end of the esterification due to the poor stability and easy decomposition of the sulfate ester. On the other hand, the acetyl sulfuric acid and propionyl sulfuric acid generated by the reaction between sulfuric acid and acid anhydride react with cellulose to form sulfuric acid ester. Finally, the sulfuric acid ester reacts with the corresponding acid to produce CAP.

Cellulose acetate butyrate (CAB) possesses the excellent UV resistance, heat resistance of cellulose acetate and the excellent moisture resistance, stability of cellulose butyrate [142]. Meanwhile, CAB presents a high degree of flexibility, broad solubility, and miscibility with other resins due to the presence of butyryl. CAB with butyryl content of 25–38% may be soluble with plasticizers and resins easily, indicating great application in coatings. CAB with butyryl content of 38–55% is easily soluble in alcohol solvents, which may greatly improves the poor fluidity, poor leveling property and undurability in wood and plastic coatings [144].

Cellulose fatty acid ester refers to the cellulose fatty esters with more than four C for the acid ester [145]. The advantages of cellulose fatty acid esters are low processing temperature, high impact strength, excellent solubility in nonpolar solvents and great compatibility with hydrophobic polymers. It can be molded without plasticizers. Therefore, it has great potential application as biodegradable plastics, coating, and films industries [146].

3.4.4.4 Etherification of cellulose

Cellulose ethers, which are produced by the reaction between hydroxyl groups of cellulose and etherification reagents, are an important branch of cellulosic derivatives. In 1905, Suida prepared methyl cellulose by using dimethyl sulfate and cellulose [147]. After that, the industrialization of carboxymethyl cellulose was realized in the 1920s for the first time. Cellulose ether has been one of the most important water-soluble polymers with a wide variety and excellent performance, as listed in Table 3.4. Compared with cellulose, one of the most important advantages of cellulose ethers is the excellent solubility, which can be controlled by the type of substituent and the degree of substitution. The hydrophilic substituents (such as hydroxyethyl, quaternary ammonium groups, etc.) and polar substituents may endow cellulose ethers with water solubility even in a low degree of substitution. For hydrophobic substituents (such as methyl, ethyl, etc.), cellulose ethers can only be swelled or dissolved in dilute alkali solutions, dissolved in water and organic solvents with an increasing degree of substitution [148].

The mechanisms for etherification of cellulose are as follows [149].

(1) Williamson etherification

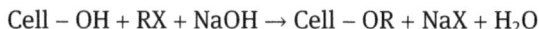

$$Cell - OH + RX + NaOH \rightarrow Cell - OR + NaX + H_2O$$

where R is an alkyl group, and X is generally a halogen element. Methyl cellulose, ethyl cellulose and carboxymethyl cellulose are prepared according to this reaction.

(2) Ring-opening addition reaction of alkoxy via base-catalyzed

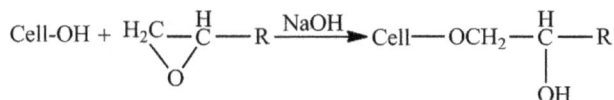

$$Cell\text{-}OH + H_2C\overset{H}{\underset{O}{\diagdown\!C}}\!-R \xrightarrow{\ NaOH\ } Cell-OCH_2-\overset{H}{\underset{OH}{C}}-R$$

where R is an alkyl group. Hydroxyethyl cellulose, propyl cellulose, and T-based cellulose are prepared according to this reaction. In order to suspend and disperse cellulose effectively, a large amount of organic diluent should be added.

(3) Alkali-catalyzed Michael addition reaction

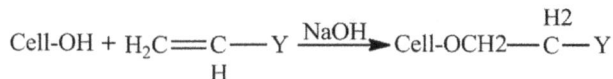

$$Cell\text{-}OH + H_2C\!=\!\overset{}{\underset{H}{C}}\!-Y \xrightarrow{\ NaOH\ } Cell-OCH2-\overset{H2}{C}-Y$$

where Y is an electron withdrawing group. In this process, an addition reaction happens between an activated vinyl and cellulose hydroxyl groups. The most typical reaction is for acrylic resin with alkali cellulose to produce cyanoethyl cellulose. This reaction is temperature-sensitive and is generally carried out under milder conditions and low alkali concentrations of 1–6% with the temperature ca. 30 °C.

Owing to the insolubility of cellulose in water and general solvents, cellulose etherification reactions are mainly carried out in a heterogeneous system. The large numbers of hydroxyl groups on cellulose molecules yield a large amount of intramolecular and intermolecular hydrogen bonds, forming crystalline fibril structure in the solid state. These hydroxyl groups are blocked in the crystal regions, hindering the approach of etherifying agent and the etherification reactions. Thus, it is difficult to obtain an ideal etherified product. Pretreatment is often required to improve the hydroxyl group activity of cellulose before etherification reaction, such as alkali swelling, high-energy electron radiation, microwave, and steam explosion technology. Among them, NaOH is the most effective and widely used pretreatment agent, as well as the catalyst for cellulose derivatization reactions.

3.4.4.5 Alkyl cellulose ether

Methylcellulose (MC) is one of the most widely used alkyl cellulose ether. It is obtained by partially or fully methylating the three hydroxyl groups on the cellulose AGU. The physical and chemical properties of MC are greatly affected by the DS, as listed in Table 3.4. MC is nonionic cellulose ether with good heat resistance and salt tolerance. The aqueous solution presents surface activity, which can be used as a thickener for water-soluble adhesives. The prominent feature of MC is the thermally reversible gelation behavior [150]. MC with a certain degree of substitution can be well dissolved in water to form a transparent and uniform solution. As the temperature increases, the solution will be partially dehydrated with an enhancement of viscosity, presenting a turbid solution. Finally, the system will form a condensation gel structure. When the MC solution is heated to a certain temperature, it shows phase separate and becomes solid. On the contrary, the system may gradually turns to the initial solution state from the solid state with the temperature decreases, as shown in Figure 3.13 [151]. This phenomenon indicates that there is a solution–gel (Sol–Gel) transition accompanied by a microphase transition from unaggregated to aggregated state. At low temperature, the solution shows high transparency and light transmittance of almost 100%, suggesting that MC dissolves well in water. When the temperature increases, the solution becomes turbid suddenly with a sharply dropped light transmittance [152]. The solution forms an opaque gel structure in a very narrow temperature range. These phenomena suggest that the gel of MC is a thermally reversible physical gel.

Ethyl cellulose (EC) is the ether product of cellulose which is partially ethylated. The ethyl group in EC makes it to be more hydrophobic than MC. The solubility of EC is also affected by the degree of substitution. When the DS of ethyl is in the range of 0.7–1.7, EC is soluble in water. When the DS is higher than 1.5, it can be dissolved in organic solvents, as listed in Table 3.4. EC and its modified ethers possess versatility that they

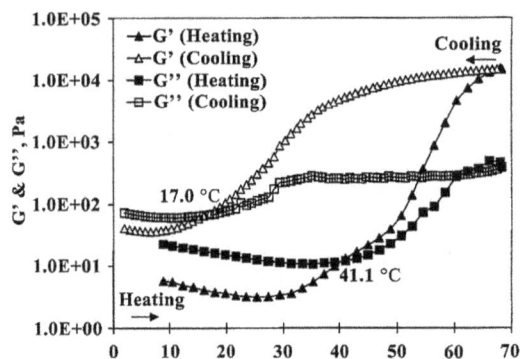

Figure 3.13: The change of rheological behavior as the temperature variation [151]. @2001 American Chemical Society.

maintain significant mechanical strength and flexibility in a very wide temperature range, especially at relatively low temperatures. It also presents low flammability, strong heat resistance, and good cold resistance. Therefore, it may improve the toughness and strength in composites as filler. It also shows good compatibility with most resins and plasticizers, holding great potential application as plastics, inks, films, adhesives, etc. Notably, EC can effectively eliminate surface adhesion, which is conducive to the molding of hard and soft plastics [153].

3.4.4.6 Hydroxyalkyl cellulose ether

The hydroxyalkyl cellulose is obtained by replacing the hydrogen atom of cellulose hydroxyl group with an alkyl group. The most representative is hydroxyethyl cellulose (HEC) and hydrixypropyl cellulose (HPC). In the hydroxyalkylation of cellulose, there are two main differences with alkylation and carboxymethylation of cellulose. Firstly, the used alkali acts as a catalyst without participating in chemical reactions. Secondly, hydroxyalkylation may happen to both of the hydroxyl group of the cellulose chains and substituted alkyl side chains [154]. Therefore, besides the DS, there is also the concept of molar substitution (MS), which is the average number of hydroxyalkyl groups contained on the hydroxyl group per mole of hydroxyalkyl cellulose. The ratio of MS/DS is defined as the length of the side chain oligomer ether. The hydroxyalkyl cellulose contains a large number of active hydroxyl groups, showing the characteristics of typical polyols. Therefore, it is often used to prepare cellulose derivatives with special functions [149].

HEC is white or light yellow powder with odorless and tasteless. It shows alkali solubility with MS of 0.05–0.5. It can be dissolved in water when MS is above 1.3. As listed in Table 3.4, the DS of most water-soluble HEC is 0.8–1.2. Actually, most of HEC products are soluble in both water and a mixed solvent composed of water and water-soluble organic solvents [155]. The aqueous solution of HEC is stable without ionization for HEC is a nonionic material. Thus, there will be no change in viscosity, as well as the precipitation or precipitation of residues, causing by the high concentration of salt in solution [149]. HEC also presents thickening effect on many monovalent or divalent high-concentration electrolyte solutions. HEC is insoluble in hydrocarbon solvents, but can be slightly soluble in solvents such as ethylene glycol, propylene glycol, glycerin, and N-acetylethanolamine when heated. It can be dissolved in some strong polar solvents such as dimethyl sulfoxide, phenol, dimethyl formamide, and formic acid. HEC possess the characteristics of thickening, suspending, bonding, emulsifying, dispersing, and retaining moisture. For example, HEC can be used as a protective agent for the emulsion polymerization of vinyl acetate with a good stability in a wide pH range. It can also be used as a dispersant for suspension polymerization of styrene, acrylate, and propylene. HEC can prevent the gelation of pigments in emulsified coatings, contributing to the dispersion of pigments, the stability of latex, enhancement of the viscosity [156].

HEC can be modified to produce cationic surfactant by cationization, which is valuable in the field of washing and cosmetics. The introduction of hydrophobic groups into hydroxyethyl cellulose molecule presents hydrophobic property. It also presents significant thickening property, temperature, and salt resistance, and shear stability [157]. It holds a great potential application as controlling agents for water fluid flowing, paint additives, and petroleum exploitation auxiliaries.

3.4.4.7 Anionic cellulose ether

Owing to the anionic side groups, anionic cellulose is an anionic polyelectrolyte of the key ionic cellulose ether. Anionic cellulose ethers mainly include carboxymethyl cellulose (CMC), ethyl cellulose sulfonate, and various carboxymethyl cellulose derivatives, among which CMC is the most widely used. As listed in Table 3.4, CMC is an important water-soluble cellulose derivative. CMC usually refers to the sodium salt of carboxymethyl cellulose due to the poor water solubility of the acid form [149]. In commercially, CMC is white or light yellow fibrous powder with the properties of nontoxic, tasteless, insoluble in acid and organic solvents (such as methanol, ethanol, ether, acetone, chloroform, and benzene), and easily soluble in water with good film-forming, significant adhesion, dispersibility, emulsification, diffusion, and viscosity [61]. CMC also presents good biocompatibility and biodegradability. The main indicators to measure the quality of CMC are DS and viscosity [158]. Generally, the transparency and stability of the aqueous solution improves with the increase of DS. When the DS is between 0.7 and 1.2, the transparency of CMC is relatively good with a largest viscosity under a pH in the range of 6–9. CMC presents a significant acid resistance and salt resistance with the DS above 0.8 [135]. The pH value of dilute CMC solution is ca. 7, in which more than 99% of the carboxyl groups appear in the form of sodium salt with a very small part of free carboxyl groups.

Because of its excellent properties including thickening, bonding, film-forming, water retention, emulsification, and suspension, CMC is commonly used as binder, thickener, suspending agent, emulsifier, dispersant, stabilizer, sizing agent, etc., in industry. For example, CMC may replace HEC in coating industry due to the uniform distribution of substituents, strong antibacterial properties, relatively stable viscosity, and low cost. Meanwhile, it shows low splashing, good film formation, good flow and leveling, low sag, high brush viscosity, excellent pigment compatibility and excellent biological stability. Furthermore, owing to the significant biocompatibility, good affinity and loading capacity for active substances such as drugs, it has been widely used in the field of drug delivery and sustained release [159].

3.4.4.8 Cationic cellulose ether

Cationic cellulose ethers is a product obtained by reacting cellulose or cellulose ether with cationic etherification reagents. It is known that the AGU of cellulose contains three active hydroxyl groups, which can react with cationic etherification reagents to prepare cationic cellulose ethers [160]. Meanwhile, there are also reactive hydroxyl groups on the molecular chain of other cellulose ethers including HEC and HPC. Since cellulose is insoluble in water and tends to aggregates into crystalline fibril, the degree of substitution is usually not high by directly cationized, as well as little change of macrostructure. Commercial cationic cellulose ethers are mainly prepared with water-soluble hydroxyalkyl cellulose to accomplish high DS and good water solubility. The commonly used cationic etherification reagents are mainly glycidyl trimethyl ammonium chloride (GTAC), dimethyl diallyl ammonium chloride (DDAC) and various acrylic acid cation derivatives. Among them, GTAC is most widely studied and used. Because of the unstable epoxy group of GTAC, it is usually made into the form of 3-chloro-2-hydroxypropyl trimethyl ammonium chloride [161]. It is cyclized with alkali before used. The converted glycidyl quaternary ammonium salt reacts with the cellulose hydroxyl group to yield cationic cellulose ether via ring-opening nucleophilic process under alkaline conditions [162].

The procedures for cationization of HEC are as follows. Firstly, HEC is dispersed in isopropanol or tert-butyl alcohol aqueous solution, following by addition of lye for alkalization and cationic etherifying agent. The temperature is increased for the cationization of HEC. Finally, hydrochloric acid, sulfuric acid, etc. is added for neutralization. The final product is post-processed in acetone aqueous solution [163]. This product of cation HEC is also known as "Polymerized Quaternary Ammonium Salt-10" (PQ-10) by the American Cosmetics, Toiletry and Fragrance Industry Association (CTFA).

3.4.4.9 Cyanoethyl cellulose

Cyanoethyl cellulose (CEC) is prepared by the reaction of alkali cellulose and acrylonitrile under mild conditions with temperature of ca. 30 °C under the alkali concentration of 1–2%. CEC with low DS (0.3–0.5) can effectively inhibit the growth of mold and bacteria, which has been used in textiles [164]. CEC that is partially substituted possesses good heat and acid resistance, which may avoid degradation. It may also be used in insulators due to the high insulation. CEC with a DS about 2.0 is soluble in some solvents, such as lactone, acrylonitrile, nitromethane, and zinc chloride solution of a constant concentration. CEC with high DS is presents high dielectric constant with good solubility in organic solvents. When the residual hydroxyl groups of CEC are removed (such as acetylation), the dielectric loss of it is minimized with significantly increased dielectric properties. Thus, it can be used as a pigment component in electroluminescent devices [165].

In addition to the above-mentioned cellulose ethers, there are other cellulose ethers, such as aralkyl cellulose, aryl cellulose, and cellulose silicone ethers. Benzyl cellulose or benzyl cellulose, dibenzyl cellulose or benzyl cellulose, and tribenzyl cellulose, belonging to the aralkyl cellulose ethers, can be prepared by one-step reaction. The aryl cellulose and its derivatives are usually indirectly synthesized by a two-step reaction. Owing to the introduction of bulky groups, the reaction usually occurs on the C-6 hydroxyl group with the least steric hindrance for the cellulose ethers. Recently, it is reported that the trityl chloride derivatives like methoxy-substituted triphenylmethyl chlorides exhibit higher efficiency for blocking hydroxyl groups at C-6 location of cellulose, which has shorten the reaction time largely [166].

3.4.5 Grafting copolymerization of cellulose

Grafting copolymerization is an important method of cellulose modification. In this strategy, functional copolymers are employed to modify cellulose by keeping the main advantages of cellulose. The characteristic of grafting copolymerization is that monomers are grafted onto the cellulose backbone through covalent bonds through polymerization. A large number of cellulose derivatives, which can be used in absorbent materials, ion exchange fibers, permanent dyed fabrics, and molded sheets, have been designed and synthesized through graft copolymerization between cellulose and acrylic acid, acrylonitrile, methyl methacrylate, and other polymer monomers. In addition, these derivatives can also be used to improve the thermal stability and antifouling properties of cotton fabrics, the wear resistance, and chemical stability of fibers, and the adhesion properties of viscose fibers and rubber.

3.4.5.1 Mechanisms and methods

The grafting of monomers onto the cellulose backbone is mainly following the three methods [167].
(1) The "grafting to…" method. Active functional groups are firstly introduced into cellulose molecules. And then, it reacts with the active groups terminated polymer to obtain the cellulose grafting copolymer.
(2) The "grafting from…" method. Active center point is firstly co-polymerized onto cellulose molecules. And then, the monomer polymerization is initiated to obtain the cellulose grafting copolymer.
(3) The "grafting by…" method. In this method, the modified cellulose is commonly used as macromolecule to react with selected small molecules to obtain the cellulose grafting copolymer.

Among them, the most commonly used is the "grafting from…" method. Because small monomers are the main participants in the reaction with low steric hindrance. In this way, it is easy to prepare cellulose grafting copolymer with a high grafting rate. On the contrary, a polymer with relative high molecule weight may participate in the reaction via the "grafting to…" method. The steric hindrance may be large, hindering the polymerization to yield cellulose derivatives with a high grafting rate. For the "grafting by…" method, it is difficult to synthesize cellulose macro-monomers. Thus, this method is less employed.

Currently, free radical polymerization is one of the main ways for grafting modification of cellulose. Owing to the wide selection of monomers, mild conditions, low cost, and easy availability of initiators and reaction media (such as water, etc.), it has attracted huge attention. Tang et al. grafted poly[2-(dimethylamino)ethyl methacrylate] (PDMAEMA) onto the surface of cellulose nanocrystals through free radical polymerization. The resultant suspension of PDMAEMA-grafted-cellulose nanocrystals (PDMAEMA-g-CNC) possessed pH-responsive properties [168].

It should be noted that the traditional free radical polymerization is still slow in initiation, faster in chain growth, and prone to chain transfer and chain termination. Thus, it is hard to control the polymerization, resulting in uncontrollable molecular weight and molecular structure, broad molecular weight distribution and even branching of polymers. It may seriously affect the performance of the product. The living polymerization has been considered to makes up for the shortcomings of free radical polymerization, and may realize to synthesize fine polymers with a certain composition, molecular design, and functionality [169]. The main methods are atom transfer radical polymerization (ATRP), reversible addition-fragmentation chain transfer polymerization (RAFT) and nitroxide stable radical polymerization (NMP). In these processes, a large number of reactive free radicals are devitalized to a dormant state. A fast dynamic balance between a small amount of growing free radicals and a large number of dormant free radicals can be established to reduce the concentration of reactive free radicals greatly [170]. Therefore, the occurrence of double radical termination and chain transfer can be reduced. NMP is the earliest living free radical polymerization technology used for grafting cellulose [171]. However, it is mainly applied in controlling styrene and its derivatives. It is essential to design a nitrogen oxide compound with special structure in the polymerization of methacrylate and methacrylamide monomers. In contrast, ATRP shows the advantages of a wide range of applicable monomers, mild reaction conditions, controllable molecular weight, and facile molecular design [170]. Therefore, it has been widely used in the field of cellulose graft modification [172, 173]. RAFT has also attracted attention due to the unique merits (wide range of applicable monomers, mild reaction conditions, diverse implementation methods, etc.) in cellulose graft modification [174].

3.4.5.2 Heterogeneous phase grafted copolymer of cellulose

The graft copolymerization of cellulose, which mainly occurs on the surface of the cellulose and the amorphous area, is generally carried out in a heterogeneous medium. The growth of the side chain increases the solubility of the copolymer, which may gradually form a local heterogeneous-homogeneous equilibrium to promote the polymerization reaction. The grafting rate is generally not high (<10%) for the copolymer yielded by the general polymerization. In addition, the polymerization controllability of the side chain is not high, resulting in wide molecular weight distribution for copolymers. In 2002, Carlmark [175] successfully grafted polymethyl methacrylate on the cellulose matrix via RAFT method. The molecular weight distribution of the product is narrower with significant controllability of graft density and molecular weight. Therefore, the controlling of molecular weight and graft density of grafted cellulose copolymers has been focused onto the living polymerization methods. However, it is still difficult to increase the grafting rate of copolymers due to the heterogeneous reaction. Furthermore, the graft copolymer yielded by the heterogeneous reaction is insoluble, whose structure can only be analyzed by indirect means such as infrared and elemental analysis. This may limit the development of this heterogeneous synthesis strategy. Thus, it may be an inevitable direction to develop homogeneous graft copolymerization of cellulose [172].

3.4.5.3 Grafted copolymer from cellulose derivatives

The grafted copolymer from cellulose derivatives presents better solubility, higher decomposition temperature, and weaker hydrogen bond effect than cellulose. It can be dissolved in some organic solvents (such as chloroform, tetrahydrofuran, etc.) [176]. Cellulose derivatives still keep the structural characteristics and excellent mechanical properties of cellulose. Thus, it is easier to carry out homogeneous graft modification for cellulose derivatives than cellulose. The ring-opening polymerization (ROP) in homogeneous mild conditions for grafting cellulose derivatives have been proposed and developed. The amphiphilic copolymers with mechanical and/or barrier properties, and even interfacial properties can be achieved, anticipated to be utilized for applications such as encapsulation and release [177]. Furthermore, the graft copolymers of cellulose derivatives show diverse structures and multifunctions. For example, cellulose diacetate (CDA), as an organic acid ester, is widely used in textiles, membrane separation, plastics, coatings, etc. [178]. However, the glass transition temperature of CDA is too high with poor melting handling performance. When it is modified by the grafting method, it can not only overcome the shortcomings, but also introduce the grafting groups with an improvement of performance [179]. When the poly(ethylene terephthalate) was grafted onto CDA, it would improve heat-resistant quality of CDA and help to improve the ability of thermo processing with a significant application as tissues scaffold [180].

3.5 Concluding remarks and future trends

In the twenty-first century, biomass has become an important renewable resource of human society due to the serious pollution, difficult biodegradation, and irreparable depletion of petroleum resources. The use of biomass resources to synthesize high-value materials is a great trend nowadays, which may facilitate the transition from fossil-based industrial to the biomass-based industry [181]. Because biomass is abundant in storage of organic carbon, it stores organic carbon in the form of materials to avoid the generation and emission of CO. Therefore, the utilization of biomass is not only in line with the sustainable development, but also promotes the establishment of the low-carbon model in industries [16, 182–184]. As one of the most important biomass materials, cellulose may become the main chemical raw material with widely application in the future. The research about cellulose is conducive to promoting the development and application of biomass materials, and it is in line with the sustainable development strategy and the "green" process. The water system to dissolve cellulose for constructing series of functional materials is of great attractive. Various cellulose-based materials can be prepared directly from the cellulose solution through "green" process in aqueous solution with great application. These materials include fiber, membranes, catalyst, aerogels, bioplastics, hydrogels, microparticles and hybrid materials with excellent biocompatibility, biodegradability and mechanical properties [38, 61, 88, 185–191]. Moreover, they are rich in raw materials. As a result, they will form a very promising "new industry of cellulose-based materials." It is believed that the full and effective use of cellulose will play an incalculable role in global economy and social progress.

Author contributions: All the authors have accepted responsibility for the entire content of this submitted manuscript and approved submission.

Research funding: The authors acknowledge the financial support provided by the National Natural Science Foundation of China (Grant No. 51802259), the Natural Science Foundation of Shaanxi (Grant No. 2019JQ-510), Xi'an and Xi'an Beilin District Programs for Science and Technology Plan (Grant Nos. 201805037YD15CG21(18) and GX1913), the Promotion Program for Youth of Shaanxi University Science and Technology Association (Grant No. 20190415), Fund of Key Laboratory of Processing and Quality Evaluation Technology of Green Plastics of China National Light Industry Council (Grant No. PQETGP2019003), the PhD Start-up fund project (Grant No. 108-451118001) of Xi'an University of Technology.

Conflict of interest statement: The authors declare no conflicts of interest regarding this article.

References

1. Sanderson K. From plant to power. Nature 2009;461:710–1.
2. Raupach MR, Canadell JG. Carbon and the anthropocene. Curr Opin Environ Sustain2010;2:210–8.
3. Hubbe MA, Buehlmann U. A continuing reverence for wood. BioResources 2010;5:1–2.
4. Tsien TH, Needham J. Science and civilization in China. Cambridge: Cambridge University Press; 1985.

5. Zhu HL, Luo W, Ciesielski PN, Fang ZQ, Zhu JY, Henriksson G, et al. Wood-derived materials for green electronics, biological devices, and energy applications. Chem Rev 2016;116:9305–74.
6. Gallezot P. Conversion of biomass to selected chemical products. Chem Soc Rev 2012;41:1538–58.
7. Araki J, Miyayama M. Wet spinning of cellulose nanowhiskers; fiber yarns obtained only from colloidal cellulose crystals. Polymer 2020;188:122116.
8. Payen A. Memoir on the composition of the tissue of plants and of woody. Compt Rend 1838;7: 1052–125.
9. Heuser E. The chemistry of cellulose. New York: John Wiley & Sons; 1944.
10. Kassig H, Kennedy J, Phillips G, Williams P. In cellulose and its derivatives. New York: Ellis Norwood; 1985: 3–25pp.
11. Moon RJ, Martini A, Nairn J, Simonsen J, Youngblood J. Cellulose nanomaterials review: structure, properties and nanocomposites. Chem Soc Rev 2011;40:3941–94.
12. Tu H, Xie K, Ying DF, Luo LB, Liu XY, Chen F, et al. Green and economical strategy for spinning robust cellulose filaments. ACS Sustain Chem Eng 2020;8:14927–37.
13. Sun B, Zhang M, Chen J, He Z, Fatehi P, Ni Y. Applications of cellulose-based materials sustained drug delivery systems. Curr Med Chem 2019;26:2485–501.
14. Moran JI, Alvarez VA, Cyras VP, Vazquez A. Extraction of cellulose and preparation of nanocellulose from sisal fibers. Cellulose 2008;15:149–59.
15. Jung YH, Chang TH, Zhang HL, Yao CH, Zheng QF, Yang VW, et al. High-performance green flexible electronics based on biodegradable cellulose nanofibril paper. Nat Commun 2015;6:7170.
16. Zhou X, Deng J, Yang R, Zhou D, Fang C, He X, et al. Facile preparation and characterization of fibrous carbon nanomaterial from waste polyethylene terephthalate. Waste Manag 2020;107: 172–81.
17. Lei W, Zhou X, Fang C, Li Y, Song Y, Wang C, et al. New approach to recycle office waste paper: reinforcement for polyurethane with nano cellulose crystals extracted from waste paper. Waste Manag 2019;95:59–69.
18. Mogosanu GD, Grumezescu AM. Natural and synthetic polymers for wounds and burns dressing. Int J Pharm 2014;463:127–36.
19. Ramamoorthy SK, Skrifvars M, Persson A. A review of natural fibers used in biocomposites: plant, animal and regenerated cellulose fibers. Polym Rev 2015;55:107–62.
20. Emam HE, Mowafi S, Mashaly HM, Rehan M. Production of antibacterial colored viscose fibers using in situ prepared spherical Ag nanoparticles. Carbohydr Polym 2014;110:148–55.
21. Michud A, Tanttu M, Asaadi S, Ma YB, Netti E, Kaariainen P, et al. Ionic liquid-based cellulosic textile fibers as an alternative to viscose and lyocell. Textil Res J 2016;86:543–52.
22. Schild G, Sixta H. Sulfur-free dissolving pulps and their application for viscose and lyocell. Cellulose 2011;18:1113–28.
23. Cai J, Zhang L, Liu SL, Liu YT, Xu XJ, Chen XM, et al. Dynamic self-assembly induced rapid dissolution of cellulose at low temperatures. Macromolecules 2008;41:9345–51.
24. Song YB, Sun YX, Zhang XZ, Zhou JP, Zhang LN. Homogeneous quaternization of cellulose in NaOH/urea aqueous solutions as gene carriers. Biomacromolecules 2008;9:2259–64.
25. Chang CY, Zhang LZ, Zhou JP, Zhang LN, Kennedy JF. Structure and properties of hydrogels prepared from cellulose in NaOH/urea aqueous solutions. Carbohydr Polym 2010;82:122–7.
26. Wang S, Lu A, Zhang LN. Recent advances in regenerated cellulose materials. Prog Polym Sci 2016; 53:169–206.
27. Cai J, Zhang L. Rapid dissolution of cellulose in LiOH/urea and NaOH/urea aqueous solutions. Macromol Biosci 2005;5:539–48.
28. Cai J, Zhang L. Unique gelation behavior of cellulose in NaOH/urea aqueous solution. Biomacromolecules 2006;7:183–9.

29. Cai J, Zhang L, Zhou J, Qi H, Chen H, Kondo T, et al. Cellulose in NaOH/urea aqueous solution: structure and properties. Adv Mater 2007;19:821–5.
30. Guo Y, Zhou JP, Song YB, Zhang LN. An efficient and environmentally friendly method for the synthesis of cellulose carbamate by microwave heating. Macromol Rapid Commun 2009;30: 1504–8.
31. Kosan B, Michels C, Meister F. Dissolution and forming of cellulose with ionic liquids. Cellulose 2008;15:59–66.
32. Cuissinat C, Navard P, Heinze T. Swelling and dissolution of cellulose. Part IV: free floating cotton and wood fibres in ionic liquids. Carbohydr Polym 2008;72:590–6.
33. Pinkert A, Marsh KN, Pang SS, Staiger MP. Ionic liquids and their interaction with cellulose. Chem Rev 2009;109:6712–28.
34. Wang H, Gurau G, Rogers RD. Ionic liquid processing of cellulose. Chem Soc Rev 2012;41:1519–37.
35. Cai Y, Zhang H, Guo Q, Shao H, Hu X. Structure and properties of cellulose fibers from ionic liquids. J Appl Polym Sci 2010;115:1047–53.
36. Lei W, Zhou X, Fang C, Song Y, Li Y. Eco-friendly waterborne polyurethane reinforced with cellulose nanocrystal from office waste paper by two different methods. Carbohydr Polym 2019;209: 299–309.
37. Yu SC, Liu MH, Ma M, Qi M, Lu ZH, Gao CJ. Impacts of membrane properties on reactive dye removal from dye/salt mixtures by asymmetric cellulose acetate and composite polyamide nanofiltration membranes. J Membr Sci 2010;350:83–91.
38. Zhao DW, Chen CJ, Zhang Q, Chen WS, Liu SX, Wang QW, et al. High performance, flexible, solid-state supercapacitors based on a renewable and biodegradable mesoporous cellulose membrane. Adv Energy Mater 2017;7:1700739.
39. Kesting RE. Phase inversion membranes. ACS Symp Ser 1985;269:131–64.
40. Laity PR, Glover PM, Hay JN. Composition and phase changes observed by magnetic resonance imaging during non-solvent induced coagulation of cellulose. Polymer 2002;43:5827–37.
41. Thakur VK, Voicu SI. Recent advances in cellulose and chitosan based membranes for water purification: a concise review. Carbohydr Polym 2016;146:148–65.
42. Saljoughi E, Amirilargani M, Mohammadi T. Effect of poly (vinyl pyrrolidone) concentration and coagulation bath temperature on the morphology, permeability, and thermal stability of asymmetric cellulose acetate membranes. J Appl Polym Sci 2009;111:2537–44.
43. Nguyen TPN, Yun ET, Kim IC, Kwon YN. Preparation of cellulose triacetate/cellulose acetate (CTA/CA)-based membranes for forward osmosis. J Membr Sci 2013;433:49–59.
44. Wang B, Kang HL, Yang HG, Xie JJ, Liu RG. Preparation and dielectric properties of porous cyanoethyl cellulose membranes. Cellulose 2019;26:1261–75.
45. Zhang L, Mao Y, Zhou J, Cai J. Effects of coagulation conditions on the properties of regenerated cellulose films prepared in NaOH/urea aqueous solution. Ind Eng Chem Res 2005;44:522–9.
46. Mao Y, Zhou J, Cai J, Zhang L. Effects of coagulants on porous structure of membranes prepared from cellulose in NaOH/urea aqueous solution. J Membr Sci 2006;279:246–55.
47. Ruan D, Zhang L, Mao Y, Zeng M, Li X. Microporous membranes prepared from cellulose in NaOH/thiourea aqueous solution. J Membr Sci 2004;241:265–74.
48. Liu S, Zhang L. Effects of polymer concentration and coagulation temperature on the properties of regenerated cellulose films prepared from LiOH/urea solution. Cellulose 2009;16:189–98.
49. Saljoughi E, Sadrzadeh M, Mohammadi T. Effect of preparation variables on morphology and pure water permeation flux through asymmetric cellulose acetate membranes. J Membr Sci 2009;326: 627–34.
50. Lei W, Fang C, Zhou X, Yin Q, Pan S, Yang R, et al. Cellulose nanocrystals obtained from office waste paper and their potential application in PET packing materials. Carbohydr Polym 2018;181: 376–85.

51. Lagerwall JPF, Schutz C, Salajkova M, Noh J, Park JH, Scalia G, et al. Cellulose nanocrystal-based materials: from liquid crystal self-assembly and glass formation to multifunctional thin films. NPG Asia Mater 2014;6:e80.
52. Kistler SS. Coherent expanded aerogels and jellies. Nature 1931;127:741–2.
53. Kistler SS. Coherent expanded-aerogels. J Phys Chem 1932;36:52–64.
54. Weatherwax R, Caulfield D. Cellulose aerogels: an improved method preparing a highly expanded form of dry cellulose. TAPPI (Tech Assoc Pulp Pap Ind) 1971;54:985–6.
55. Tan C, Fung B, Newman J, Vu C. Organic aerogels with very high impact strength. Adv Mater 2001; 13:644–6.
56. Duchemin BJ, Staiger MP, Tucher N, Newman RH. Aerocellulose based on all-cellulose composites. J Appl Polym Sci 2010;115:216–21.
57. Wang Z, Liu S, Matsumoto Y, Kuga S. Cellulose gel and aerogel from LiCl/DMSO solution. Cellulose 2012;19:393–9.
58. Tingaut P, Zimmermann T, Sèbe T. Cellulose nanocrystals and microfibrillated cellulose as building blocks for the design of hierarchical functional materials. J Mater Chem 2012;22: 20105–11.
59. Kettunen M, Silvennoinen RJ, Houbenov N, Nykanen A, Ruokolainen J, Sainio J, et al. Photoswitchable superabsorbency based on nanocellulose aerogels. Adv Funct Mater 2011;21: 510–7.
60. Jin H, Kettunen M, Laiho A, Pynnonen H, Paltakari J, Marmur A, et al. Superhydrophobic and superoleophobic nanocellulose aerogel membranes as bioinspired cargo carriers on water and oil. Langmuir 2011;27:1930–4.
61. Chang CY, Zhang LN. Cellulose-based hydrogels: present status and application prospects. Carbohydr Polym 2011;84:40–53.
62. Qiu XY, Hu SW. Smart materials based on cellulose: a review of the preparations, properties, and applications. Materials 2013;6:738–81.
63. Galiano F, Briceño K, Marino T, Molino A, Christensen KV, Figoli A. Advances in biopolymer-based membrane preparation and applications. J Membr Sci 2018;564:562–86.
64. Li B, Konecke S, Wegiel LA, Taylo LS, Edgar KJ. Both solubility and chemical stability of curcumin are enhanced by solid dispersion in cellulose derivative matrices. Carbohydr Polym 2013;98: 1108–16.
65. Wang Q, Cai J, Zhang L, Xu M, Cheng H, Han C, et al. A bioplastic with high strength constructed from a cellulose hydrogel by changing the aggregated structure. J Mater Chem 2013;1:6678–86.
66. Guan Q, Yang H, Han Z, Ling Z, Yu S. An all-natural bioinspired structural material for plastic replacement. Nat Commun 2020;11:5401.
67. Zhou X, Zhang X, Wang D, Fang C, Lei W, Huang Z, et al. Preparation of cellulose nanocrystal film from waste papers and synthesis of waterborne polyurethane nanocomposite films. J Renew Mater 2020;8:631–45.
68. Trache D, Hussin MH, Haafiz MKM, Thakur VK. Recent progress in cellulose nanocrystals: sources and production. Nanoscale 2017;9:1763–86.
69. Favier V, Chanzy H, Cavaille J. Polymer nanocomposites reinforced by cellulose whiskers. Macromolecules 1995;28:6365–7.
70. Angles MN, Dufresne A. Plasticized starch/tunicin whiskers nanocomposites. 1. Structural analysis. Macromolecules 2000;33:8344–53.
71. Céline C, Nicholas M, Ryo K, Rowan SJ. Development, processing and applications of bio-sourced cellulose nanocrystal composites. Prog Polym Sci 2020;103:101221.
72. Youssef H, Lucian AL, Orlando JR. Cellulose nanocrystals: chemistry, self-assembly, and applications. Chem Rev 2010;110:3479–500.

73. Roman M, Winter WT. Effect of sulfate groups from sulfuric acid hydrolysis on the thermal degradation behavior of bacterial cellulose. Biomacromolecules 2004;5:1671–7.

74. D'Acierno F, Hamad WY, Michal CA, MacLachlan MJ. Thermal degradation of cellulose filaments and nanocrystals. Biomacromolecules 2020;21:3374–86.

75. Thomas B, Raj MC, Athira KB, Rubiyah MH, Joy J, Moores A, et al. Nanocellulose, a versatile green platform: from biosources to materials and their applications. Chem Rev 2018;118:11575–625.

76. Reid MS, Villalobos M, Cranston ED. Benchmarking cellulose nanocrystals: from the laboratory to industrial production. Langmuir 2017;33:1583–98.

77. Grishkewich N, Mohammed N, Tang JT, Tam KC. Recent advances in the application of cellulose nanocrystals. Curr Opin Colloid Interface 2017;29:32–45.

78. Elazzouzi-Hafraoui S, Nishiyama Y, Putaux JL, Heux L, Dubreuil F, Rochas C. The shape and size distribution of crystalline nanoparticles prepared by acid hydrolysis of native cellulose. Biomacromolecules 2008;9:57–65.

79. Rosa MF, Medeiros ES, Malmonge JA, Gregorski KS, Wood DF, Mattoso LHC, et al. Cellulose nanowhiskers from coconut husk fibers: effect of preparation conditions on their thermal and morphological behavior. Carbohydr Polym 2010;81:83–92.

80. Angles MN, Dufresne A. Plasticized starch/tunicin whiskers nanocomposite materials. 2. Mechanical behavior. Macromolecules 2001;34:2921–31.

81. Diddens I, Murphy B, Krisch M, Muller M. Anisotropic elastic properties of cellulose measured using inelastic X-ray scattering. Macromolecules 2008;41:9755–9.

82. Iwamoto S, Kai W, Isogai A, Iwata T. Elastic modulus of single cellulose microfibrils from tunicate measured by atomic force microscopy. Biomacromolecules 2009;10:2571–6.

83. Wagner R, Moon R, Pratt J, Shaw G. Uncertainty quantification in nanomechanical measurements using the atomic force microscope. Nanotechnology 2011;22:455703.

84. Nishino T, Matsuda I, Hirao K. All-cellulose composite. Macromolecules 2004;37:7683–7.

85. Fleming K, Gray DG, Matthews S. Cellulose crystallites. Chem Eur J 2001;7:1831–5.

86. Dong XM, Kimura T, Revol JF, Gray DG. Effects of ionic strength on the isotropic–chiral nematic phase transition of suspensions of cellulose crystallites. Langmuir 1996;12:2076–82.

87. Orts WJ, Godbout L, Marchessault RH, Revol JF. Enhanced ordering of liquid crystalline suspensions of cellulose microfibrils: a small angle neutron scattering study. Macromolecules 1998;31:5717–25.

88. Rol F, Belgacem MN, Gandini A, Bras J. Recent advances in surface-modified cellulose nanofibrils. Prog Polym Sci 2019;88:241–64.

89. Nogi M, Yano H. Transparent nanocomposites based on cellulose produced by bacteria offer potential innovation in the electronics device industry. Adv Mater 2008;20:1849–52.

90. Shah N, Ul-Islam M, Khattak WA, Park JK. Overview of bacterial cellulose composites: a multipurpose advanced. Mater Carbohydr Polym 2013;98:1585–98.

91. Huang Y, Zhu CL, Yang JZ, Nie Y, Chen CT, Sun DP. Recent advances in bacterial cellulose. Cellulose 2014;21:1–30.

92. Czaja WK, Young DJ, Kawecki M, Brown RM. The future prospects of microbial cellulose in biomedical applications. Biomacromolecules 2007;8:1–12.

93. Nakagaito AN, Yano H. Novel high-strength biocomposites based on microfibrillated cellulose having nano-order-unit web-like network structure. Appl Phys A 2005;80:155–9.

94. Lin SP, Calvar IL, Catchmark JM, Liu JR, Demirci A, Cheng KC. Biosynthesis, production and applications of bacterial cellulose. Cellulose 2013;20:2191–219.

95. Romling U, Galperin MY. Bacterial cellulose biosynthesis: diversity of operons, subunits, products, and functions. Trends Microbiol 2015;23:545–57.

96. Klemm D, Schumann D, Udhardt U, Maesch S. Bacterial synthesized cellulose-artificial blood vessels for microsurgery. Prog Polym Sci 2011;26:1561–603.

97. Astley OM, Chanliaud E, Donald AM, Gidley MJ. Tensile deformation of bacterial cellulose composites. Int J Biol Macromol 2003;32:28–35.
98. Chawla PR, Bajaj IB, Survase SA, Singhal RS. Microbial cellulose: fermentative production and applications. Food Technol Biotechnol 2009;47:107–24.
99. Menriksson M, Berglund LA, Isaksson P, Lindstrom T, Nishino T. Cellulose nanopaper structures of high toughness. Biomacromolecules 2008;9:1579–85.
100. Yoon SH, Jin HJ, Kook MC, Pyun YR. Electrically conductive bacterial cellulose by incorporation of carbon nanotubes. Biomacromolecules 2006;7:1280–4.
101. de Moura MR, Mattoso LHC, Zucolotto V. Development of cellulose-based bactericidal nanocomposites containing silver nanoparticles and their use as active food packaging. J Food Eng 2012;109:520–4.
102. Nehls I, Wagenknecht W, Philipp B, Stscherbina D. Characterization of cellulose and cellulose derivatives in solution by high resolution ^{13}C-NMR spectroscopy. Prog Polym Sci 1994;19:29–78.
103. Cocinero EJ, Gamblin DP, Davis BG, Simons JP. The building blocks of cellulose: the intrinsic conformational structures of cellobiose, its epimer, lactose, and their singly hydrated complexes. J Am Chem Soc 2009;131:11117–23.
104. Chen X, Burger C, Wan F, Zhang J, Rong L, Hsiao B, et al. Structure study of cellulose fibers wet-spun from environmentally friendly NaOH/urea aqueous solutions. Biomacromolecules 2007;8:1918–26.
105. Morgan JLW, Strumillo J, Zimmer J. Crystallographic snapshot of cellulose synthesis and membrane translocation. Nature 2013;493:181–6.
106. Payne CM, Knott BC, Mayes HB, Hansson H, Himmel ME, Sandgren M, et al. Fungal cellulases. Chem Rev 2015;115:1308–448.
107. Park S, Baker JO, Himmel ME, Parilla PA, Johnson DK. Cellulose crystallinity index: measurement techniques and their impact on interpreting cellulase performance. Biotechnol Biofuels 2010;3:1–10.
108. Sugiyama J, Okano T, Yamamoto H, Horii F. Transformation of valonia cellulose crystals by an alkaline hydrothermal treatment. Macromolecules 1990;23:3196–8.
109. Wada M, Kondo T, Okano T. Thermally induced crystaltransformation from cellulose I$_\alpha$ to I$_\beta$. Polym J 2003;35:155–9.
110. Wada M, Heux L, Sugiyama J. Polymorphism of cellulose I Family: reinvestigation of cellulose IV$_I$. Biomacromolecules 2004;5:1385–91.
111. Siro I, Plackett D. Microfibrillated cellulose and new nanocomposite materials: a review. Cellulose 2010;17:459–94.
112. Cherhal F, Cousin F, Capron I. Influence of charge density and ionic strength on the aggregation process of cellulose nanocrystals in aqueous suspension, as revealed by small-angle neutron scattering. Langmuir 2015;31:5596–602.
113. French AD. Idealized powder diffraction patterns for cellulose polymorphs. Cellulose 2014;21:885–96.
114. Heinze T, iebert T. Unconventional methods in cellulose functionalization. Prog Polym Sci 2001;26:1689–762.
115. Isogai A, Ishizu A, Nakano J. Dissolution mechanism of cellulose in SO_2-amine-dimethylsulfoxide. J Appl Polym Sci 1987;33:1283–90.
116. Lindman B, Karlstrom G, Stigsson L. On the mechanism of dissolution of cellulose. J Mol Liq 2010;156:76–81.
117. Swatloski RP, Spear SK, Holbrey JD, Rogers RD. Dissolution of cellose with ionic liquids. J Am Chem Soc 2002;124:4974–5.
118. Zhang H, Wu J, Zhang J, He J. 1-Allyl-3-methylimidazolium chloride room temperature ionic liquid: a new and powerful nonderivatizing solvent for cellulose. Macromolecules 2005;38:8272–7.

119. Wu J, Zhang J, Zhang H, Ren Q, Guo M. Homogeneous acetylation of cellulose in a new ionic liquid. Biomacromolecules 2004;5:266–8.
120. Zavrel M, Bross D, Funke M, Buchs J, Spiess AC. High-throughput screening for ionic liquids dissolving (ligno-) cellulose. Bioresour Technol 2009;100:2580–7.
121. Gadd KA. New solvent for cellulose. Polymer 1982;23:1867–9.
122. Jayme G, Lang F. Über das celluloselösende system eisen(III)-Weinsäure-Kaliumhydroxyd. Colloid & Polymer 1955;144:75–81.
123. Abe M, Fukaya Y, Ohno H. Fast and facile dissolution of cellulose with tetrabutylphosphonium hydroxide containing 40 wt% water. Chem Commun 2012;48:1808–10.
124. Zheng XY, Gandour RD, Edgar KJ. Remarkably regioselective deacylation of cellulose esters using tetraalkylammonium salts of the strongly basic hydroxide ion. Carbohydr Polym 2014;111:25–32.
125. Cai J, Kimura S, Wada M, Kuga S. Nanoporous cellulose as metal nanoparticles support. Biomacromolecules 2009;10:87–94.
126. Patwardhan PR, Satrio JA, Brown RC, Shanks BH. Influence of inorganic salts on the primary pyrolysis products of cellulose. Bioresour Technol 2010;101:4646–55.
127. Zhou J, Zhang L. Solubility of cellulose in NaOH/urea aqueous solution. Polym J 2000;32:866–70.
128. Zhang L, Ruan D, Gao S, Polym J. Dissolution and regeneration of cellulose in NaOH/thiourea aqueous solution. J Polym Sci B Polym Phys 2002;40:1521–9.
129. Cai J, Zhang L, Chang C, Cheng G, Chen X, Chu B. Hydrogen-bond-induced inclusion complex in aqueous cellulose/LiOH/urea solution at low temperature. ChemPhysChem 2007;8:1572–9.
130. Ca J, Liu Y, Zhang L. Dilute solution properties of cellulose in LiOH/urea aqueous system. J Polym Sci B Polym Phys 2006;44:3093–101.
131. Wang SR, Guo XJ, Liang T, Zhou Y, Luo ZY. Mechanism research on cellulose pyrolysis by py-GC/MS and subsequent density functional theory studies. Bioresour Technol 2012;104:722–8.
132. Eyley S, Thielemans W. Surface modification of cellulose nanocrystals. Nanoscale 2014;6:7764.
133. Kim GY, Lee HD, Kim YH. Preparation and thermoresponsive properties of 2-hydroxy-3-butoxypropyl hydroxyethyl cellulose and its hydrogel crosslinked with epichlorohydrin. Polymer-Korea 2020;44:495–504.
134. Arai K, Shikata T. Hydration/dehydration behavior of cellulose ethers in aqueous solution. Macromolecules 2017;50:5920–8.
135. Kamel S, Ali N, Jahangir K, Shah SM, El-Gendy AA. Pharmaceutical significance of cellulose: a review. Express Polym Lett 2008;2:758–78.
136. Nevell TP, Zeronian SH. Cellulose chemistry and its application. New York: E. Horwood. John Wiley: Halsted Press; 1985.
137. Roy D, Semsarilar M, Guthrie JT, Perrier S. Cellulose modification by polymer grafting: a review. Chem Soc Rev 2009;38:2046–64.
138. Winding CC, Hiatt GD. Polymeric materials. New York: McGraw-Hill; 1961.
139. Fleet R, McLeary JB, Grumel V, Weber WG, Matahwa H, Sanderson RD. RAFT mediated polysaccharide copolymers. Eur Polym J 2008;44:2899–911.
140. Zhang S, Wang KY, Chung TS, Chen HM, Jean YC, Amy G. Well-constructed cellulose acetate membranes for forward osmosis: minimized internal concentration polarization with an ultra-thin selective layer. J Membr Sci 2010;360:522–35.
141. Liu ZT, Fan X, Wu J, Zhang L, Song L, Gao Z, et al. A green route to prepare cellulose acetate particle from ramie fiber. React Funct Polym 2007;67:104–12.
142. Hornig S, Heinze T. Efficient approach to design, stable water-dispersible nanoparticles of hydrophobic cellulose esters. Biomacromolecules 2008;9:1487–92.
143. Schilling M, Bouchard M, Khanjian H, Learner T, Phenix A, Rivenc R. Application of chemical and thermal analysis methods for studying cellulose ester plastics. Acc Chem Res 2010;43:888–96.

144. Gindl W, Keckes J. Tensile properties of cellulose acetate butyrate composites reinforced with bacterial cellulose. Compos Sci Technol 2004;64:2407–13.
145. Lcrépy L, MiriL V, Joly N, Martin P, Lefebvre J. Effect of side chain length on structure and thermomechanical properties of fully substituted cellulose fatty esters. Carbohydr Polym 2011; 83:1812–20.
146. Thiebaud S, Borredon M. Solvent-free wood esterification with fatty acid chlorides. Bioresour Technol 1995;52:169–73.
147. Suida W. Über Den Einflu\Der Aktiven Atomgruppen in Den Textilfasern Auf das Zustandekommen Von Färbungen. Monatshefte Fur Chemie 1905;26:413–27.
148. Donges R. Non-ionic cellulose ethers. Br Polym J 1990;23:315.
149. Klemm D, Philipp B, Heinze T, Heinze U, Wagenknecht W. Comprehensive cellulose chemistry. Weinheim: Wiley-VCH Vecrlag GmbH; 1998.
150. Siepmann J, Peppas NA. Modeling of drug release from delivery systems based on hydroxypropyl methylcellulose (HPMC). Adv Drug Deliv Rev 2012;64:163–74.
151. Li L, Thangamathesvaran PM, Yue CY, Tam KC, Hu X, Lam YC. Gel network structure of methylcellulose in water. Langmuir 2001;17:8062–8.
152. Xu Y, Li L. Thermoreversible and salt-sensitive turbidity of methylcellulose in aqueous solution. Polymer 2005;46:7410–7.
153. Crabbe-Mann M, Tsaoulidis D, Parhizkar M, Edirisinghe M. Ethyl cellulose, cellulose acetate and carboxymethyl cellulose microstructures prepared using electrohydrodynamics and green solvents. Cellulose 2018;25:1687–703.
154. Kohler S, Liebert T, Heinze T, Vollmer A, Mischnick P, Mollmann E, et al. Interactions of ionic liquids with polysaccharides 9. Hydroxyalkylation of cellulose without additional inorganic bases. Cellulose 2010;17:437–48.
155. Angadi SC, Manjeshwar LS, Aminabhavi TM. Interpenetrating polymer network blend microspheres of chitosan and hydroxyethyl cellulose for controlled release of isoniazid. Int J Biol Macromol 2010;47:171–9.
156. Chu ML, Feng NR, An H, You GL, Mo CS, Zhong HY, et al. Design and validation of antibacterial and pH response of cationic guar gum film by combining hydroxyethyl cellulose and red cabbage pigment. Int J Biol Macromol 2020;162:1311–22.
157. Svensson AV, Huang LG, Johnson ES, Nylander T, Piculell L. Surface deposition and phase behavior of oppositely charged polyion/surfactant ion complexes. 1. Cationic guar versus cationic hydroxyethylcellulose in mixtures with anionic surfactants. ACS Appl Mater Interfaces 2009;1:2431–42.
158. Benchabane A, Bekkour K. Rheological properties of carboxymethyl cellulose (CMC) solutions. Colloid Polym Sci 2008;286:1173–80.
159. Hebeish A, Hashem M, Abd El-Hady MM, Sharaf S. Development of CMC hydrogels loaded with silver nano-particles for medical applications. Carbohydr Polym 2013;92:407–13.
160. Lu BL, Xu AR, Wang JJ. Cation does matter: how cationic structure affects the dissolution of cellulose in ionic liquids. Green Chem 2014;16:1326–35.
161. Pahimanolis N, Salminen A, Penttila PA, Korhonen JT, Johansson LS, Ruokolainen J, et al. Nanofibrillated cellulose/carboxymethyl cellulose composite with improved wet strength. Cellulose 2013;20:1459–68.
162. Tucker I, Petkov J, Penfold J, Thomas RK. Adsorption of nonionic and mixed nonionic/cationic surfactants onto hydrophilic and hydrophobic cellulose thin films. Langmuir 2010;26:8036–48.
163. Yang RM, Shi RH, Peng SH, Zhou D, Liu H, Wang YM. Cationized hydroxyethylcellulose as a novel, adsorbed coating for basic protein separation by capillary electrophoresis. Electrophoresis 2008;29:1460–6.
164. Frey MW. Electrospinning cellulose and cellulose derivatives. Polym Rev 2008;48:378–91.

165. Zhou JP, Li QA, Song YB, Zhang LN, Lin XY. A facile method for the homogeneous synthesis of cyanoethyl cellulose in NaOH/urea aqueous solutions. Polym Chem 2010;1:1662–8.
166. Liu D, Xia K, Yang R. Synthetic pathways of regioselectively substituting cellulose derivatives: a review. Curr Org Chem 2012;16:1838–49.
167. Kang HL, Liu RG, Huang Y. Graft modification of cellulose: methods, properties and applications. Polymer 2015;70:A1–16.
168. Tang JT, Lee MFX, Zhang W, Zhao BX, Berry RM, Tam KC. Dual responsive pickering emulsion stabilized by poly[2-(dimethylamino) ethyl methacrylate] grafted cellulose nanocrystals. Biomacromolecules 2014;15:3052–60.
169. Zoppe JO, Habibi Y, Rojas OJ, Venditti RA, Johansson LS, Efimenko K, et al. Poly(N-isopropylacrylamide) brushes grafted from cellulose nanocrystals via surface-initiated single-electron transfer living radical polymerization. Biomacromolecules 2010;11:2683–91.
170. Joubert F, Musa OM, Hodgson DRW, Cameron NR. The preparation of graft copolymers of cellulose and cellulose derivatives using ATRP under homogeneous reaction conditions. Chem Soc Rev 2014;43:7217–35.
171. Malmstrom E, Carlmark A. Controlled grafting of cellulose fibres – an outlook beyond paper and cardboard. Polym Chem 2012;3:1702–13.
172. Meng T, Gao X, Zhang J, Yuan JY, Zhang YZ, He JS. Graft copolymers prepared by atom transfer radical polymerization (ATRP) from cellulose. Polymer 2009;50:447–54.
173. Hansson S, Ostmark E, Carlmark A, Malmstrom E. Arget ATRP for versatile grafting of cellulose using various monomers. ACS Appl Mater Interfaces 2009;1:2651–9.
174. Roy D, Knapp JS, Guthrie JT, Perrier S. Antibacterial cellulose fiber via RAFT surface graft polymerization. Biomacromolecules 2008;9:91–9.
175. Carlmark A, Malmström E. Atom transfer radical polymerization from cellulose fibers at ambient temperature. J Am Chem Soc 2002;124:900–1.
176. Thakur VK, Thakur MK, Gupta RK. Graft copolymers from cellulose: synthesis, characterization and evaluation. Carbohydr Polym 2013;97:18–25.
177. Carlmark A, Larsson E, Malmstrom E. Grafting of cellulose by ring-opening polymerisation – a review. Eur Polym J 2012;48:1646–59.
178. Li YX, Liu RG, Huang Y. Synthesis and phase transition of cellulose-graft-poly (ethylene glycol) copolymers. J Appl Polym Sci 2008;110:1797–803.
179. Vlcek P, Janata M, Latalova P, Dybal J, Spirkova M, Toman L. Bottlebrush-shaped copolymers with cellulose diacetate backbone by a combination of ring opening polymerization and ATRP. J Polym Sci Polym Chem 2008;46:564–73.
180. Liang WC, Hou J, Fang XC, Bai FD, Zhu TH, Gao FF, et al. Synthesis of cellulose diacetate based copolymer electrospun nanofibers for tissues scaffold. Appl Surf Sci 2018;443:374–81.
181. Kalia S, Kaith BS, Kaur I. Cellulose fibers: bio- and nano-polymer composites. Spring-Verlag Berlin Heidelberg; 2011.
182. Uihlein A, Schebek L. Environmental impacts of a lignocellulose feedstock biorefinery system: an sssessment. Biomass Bioenergy 2009;33:793–802.
183. Zhou X, Su J, Wang C, Fang C, He X, Lei W, et al. Design, preparation and measurement of protein/CNTs hybrids: a concise review. J Mater Sci Technol 2020;46:74–87.
184. Li QQ, McGinnis S, Sydnor C, Wong A, Renneckar S. Nanocellulose life cycle assessment. ACS Sustain Chem Eng 2013;1:919–28.
185. Kontturi E, Laaksonen P, Linder MB, Nonappa, Groschel AH, Rojas OJ, et al. Advanced materials through assembly of nanocelluloses. Adv Mater 2018;30:1703779.
186. Azeredo HMC, Rosa MF, Mattoso LHC. Nanocellulose in bio-based food packaging applications. Ind Crop Prod 2017;97:664–71.

187. Chen MY, Kang HL, Gong YM, Guo J, Zhang H, Liu RG. Bacterial cellulose supported gold nanoparticles with excellent catalytic properties. ACS Appl Mater Interfaces 2015;7:21717–26.
188. Olsson RT, Samir MASA, Salazar-Alvarez G, Belova L, Strom V, Berglund LA, et al. Making flexible magnetic aerogels and stiff magnetic nanopaper using cellulose nanofibrils as templates. Nat Nanotechnol 2010;5:584–8.
189. Brodin M, Vallejos M, Opedal MT, Area MC, Chinga-Carrasco G. Lignocellulosics as sustainable resources for production of bioplastics-A review. J Clean Prod 2017;162:646–64.
190. Kolakovic R, Laaksonen T, Peltonen L, Laukkanen A, Hirvonen J. Spray-dried nanofibrillar cellulose microparticles for sustained drug release. Int J Pharm 2012;430:47–55.
191. Zhong LX, Fu SY, Zhou XS, Zhan HY. Effect of surface microfibrillation of sisal fibre on the mechanical properties of sisal/aramid fibre hybrid composites. Compos Appl Sci Manuf 2011;42: 244–52.

Guoqiang Zhu, Chengguo Liu* and Chaoqun Zhang

4 Plant oil-based polymers

Abstract: Polymer materials derived from natural resources have gained increasing attention in recent years because of the uncertainties concerning petroleum supply and prices in the future as well as their environmental pollution problems. As one of the most abundant renewable resources, plant oils are suitable starting materials for polymers because of their low cost, the rich chemistry that their triglyceride structure provides, and their potential biodegradability. This chapter covers the structure, modification of triglycerides and their derivatives as well as synthesis of polymers therefrom. The remarkable advances during the last two decades in organic synthesis using plant oils and the basic oleochemicals derived from them are selectively reported and updated. Various methods, such as condensation, radical/cationic polymerization, metathesis procedure, and living polymerization, have also been applied in constructing oil-based polymers. Based on the advance of these changes, traditional polymers such as polyamides, polyesters, and epoxy resins have been renewed. Partial oil-based polymers have already been applied in some industrial areas and recent developments in this field offer promising new opportunities.

Keywords: fatty acids; living polymerization; olefin metathesis; plant oil; thermoplastic polymers; thermosetting polymers.

4.1 Introduction

Modern chemical industry of polymeric materials has been well developed based on fossil feedstocks. Due to the uncertainties concerning petroleum supply and prices in the future as well as their environmental pollution problems, it is important and urgent to utilize renewable raw materials in industrial applications for a sustainable development [1–5]. Natural oils, which can be derived from both plant and animal origin, are abundant in most parts of the world. Approximately 80% of the global oil and fat production is plant oil, whereas 20% is of animal origin (share decreasing) [6]. In 2019–2020 the production of plant oils amounted to over 200 million tones. For more detailed information of the commodity plant oils production, the readers are referred to a website from United States Department of Agriculture [7]. Natural oils have been used for centuries in the production of coatings, inks, plasticizers, lubricants, and

*Corresponding author: Chengguo Liu, Institute of Chemical Industry of Forest Products, Chinese Academy of Forestry, Nanjing, P. R. China, E-mail: liuchengguo@icifp.cn
Guoqiang Zhu, Institute of Chemical Industry of Forest Products, Chinese Academy of Forestry, Nanjing, P. R. China
Chaoqun Zhang, South China Agricultural University, Guangzhou, P. R. China

This article has previously been published in the journal Physical Sciences Reviews. Please cite as: G. Zhu, C. Liu and C. Zhang "Plant oil-based polymers" *Physical Sciences Reviews* [Online] 2021, 7. DOI: 10.1515/psr-2020-0070 | https://doi.org/10.1515/9781501521942-004

agrochemicals, as reported in recent books and reviews [1–5, 8–10]. As one of the most widely applied renewable resource, natural oils are suitable starting materials for polymers because of their low cost, the rich chemistry that their triglyceride structure provides, and their potential biodegradability. Hence, aside from being large-scale utilized as biodiesel [11, 12], the development of polymers from this green alternative has gain intensive attention.

The development of oil-based polymers relies greatly on the fundamental progress in oleochemistry and polymerization methods. Classical and well-established oleochemical transformations occur preferentially at the ester functionality of the native triglycerides, such as hydrolysis to free fatty acids and transesterification to fatty acid methyl esters [4]. Fatty acids are transformed by reactions at the carboxy group to soaps, esters, amides, or amines. Hydrogenation of both fatty acids and their methyl esters gives fatty alcohols, which are used for the production of surfactants. During the last two decades, a variety of new modification methods, such as "click" reactions, metathesis reactions, and transition-metal-catalyzed reactions, have been applied from petroleum chemistry to oleochemistry and renewed the traditional polymers from these basic oleochemicals. Meanwhile, various polymerization methods including condensation, radical/cationic polymerization, metathesis procedures, and living polymerization, from uncontrolled polymerization to controlled polymerization, have been employed to construct oil-based polymers [2]. As a result, numerous thermo-plastic and thermosetting polymers as well as elastomers, additives, and surfactants from natural oils have been successfully developed. Some of these polymers have good physical and chemical properties which can compete with petroleum-based materials. Hence, aside from those traditional areas mentioned above, oil-based polymers have also been partially employed in structural application such as vehicles, vessels, and building materials.

The present chapter deals mainly with the use of plant oils and their derivatives as sources of polymeric materials. The diminishing demand for fats in these materials is linked to the presence in these fats of high levels of saturated acids, *trans* acids, and cholesterol, and to low levels of polyunsaturated fatty acids.

4.2 Structure and modification of plant oils

4.2.1 Structure of plant oils

Plant oils are predominantly made up of triglyceride molecules, which are composed of three fatty acids joined at a glycerol junctures, as shown in Figure 4.1a. The most common oils contain fatty acids that vary from 14 to 24 carbons in length with 0–3 double bonds per fatty acid chain. The fatty acids of bulk oils currently used in oleo-chemistry are rather uniform.

(a)

R=fatty acid chain

(b)

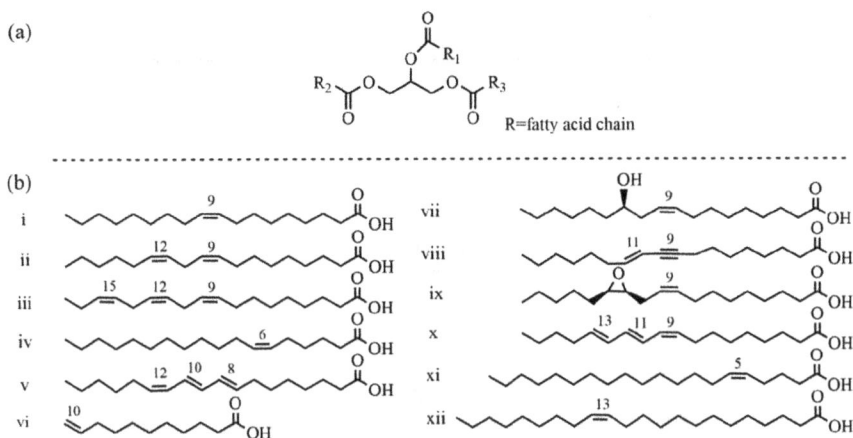

Figure 4.1: (a) General triglyceride molecule, the major component of natural oils. (b) Fatty acids commonly used in polymer chemistry: (i) oleic acid, (ii) linoleic acid, (iii) linolenic acid, (iv) petroselinic, (v) calendic acid, (vi) 10-undecenoic acid, (vii) ricinoleic acid, (viii) santalbic acid, (ix) vernolic acid, (x) α-eleostearic acid, (xi) 5-eicosenoic acid, and (xii) erucic acid.

Table 4.1 shows the fatty acid distributions of several common oils [8, 9, 13]. The industrially-utilized fatty acids includes saturated ones with an even number of carbon atoms (C8–C18) and unsaturated C18 fatty acids such as oleic acid, linoleic acid, and linolenic acid, as shown in Figure 4.1b. The cultivation of the respective plants for the production of these oils would increase the agricultural biodiversity, an important aspect of a sustainable utilization of renewable feedstocks. Moreover, classic breeding and genetic engineering are necessary to improve the oil yield and modulate the fatty acid composition for chemical utilization [14, 15].

The most important parameters affecting the physical and chemical properties of plant oils are the stereochemistry of the double bonds of the fatty acid chains, their degree of unsaturation as well as the length of the carbon chain of the fatty acids. The degree of unsaturation, which can be expressed by the iodine value (amount of iodinein g that can react with double bonds present in 100 g of sample under specified conditions) can be used as a simple parameter to divide oils into three classes: drying (iodine value > 170; e.g. linseed oil), semi-drying (100 < iodine value < 170; e.g. sunflower or soy oils) and non-drying (iodine value < 100; e.g. palm kernel oil) oils [2]. Iodine values of common fatty acids and their triglycerides were given in a book [16].

4.2.2 Modification of oils and their derivatives

Triglyceride molecules contain active sites amenable to chemical reaction: the unsaturated carbon-carbon bonds, the ester groups, epoxy groups, hydroxyl groups, the allylic carbon, and the α-carbons to the ester group (see Figure 4.1b). Meanwhile,

Table 4.1: Fatty acid distribution mainly in various plant oils [8].

Fatty acid	n_C:$n_{C=C}$*	Canola	Corn	Linseed	Olive	Palm	Rapeseed	Soybean	High oleic
Myristic	14:0	0.1	0.1	0.0	0.0	1.0	0.1	0.1	0.0
Palmitic	16:0	4.1	10.9	5.5	13.7	44.4	3.0	11.0	6.4
Palmitoleic	16:1	0.3	0.2	0.0	1.2	0.2	0.2	0.1	0.1
Stearic	18:0	1.8	2.0	3.5	2.5	4.1	1.0	4.0	3.1
Oleic	18:1	60.9	25.4	19.1	71.1	39.3	13.2	23.4	82.6
Linoleic	18:2	21.0	59.6	15.3	10.0	10.0	13.2	53.2	2.3
Linolenic	18:3	8.8	1.2	56.6	0.6	0.4	9.0	7.8	3.7
Arachidic	20:0	0.7	0.4	0.0	0.9	0.3	0.5	0.3	0.2
Gadoleic	20:1	1.0	0.0	0.0	0.0	0.0	9.0	0.0	0.4
Behenic	22:0	0.3	0.1	0.0	0.0	0.1	0.5	0.1	0.3
Erucic	22:1	0.7	0.0	0.0	0.0	0.0	49.2	0.0	0.1
Lignoceric	24:0	0.2	0.0	0.0	0.0	0.0	1.2	0.0	0.0

*The number of carbon atoms (n_C) and C=C double bonds ($n_{C=C}$) per fatty acid.

fatty acids and their esters can be easily obtained either by simple hydrolysis or alcoholysis of triglycerides. Hence, a variety of reactions including oxidation, C–C-bonding additions to the C–C double bond, metathesis reactions, and C–H activation have been employed to functionalized oils and their derivatives [4, 5]. Here a few typical methods to prepare monomers and oligomers using natural oils and their derivatives are to be briefly introduced, and more examples of their applications will be reviewed in the next sections.

4.2.2.1 Conventional functionalization

This modification method is defined as incorporating polymerizable groups (e.g. epoxy and vinyl groups) onto the active sites of triglycerides or their derivatives *via* one-step reaction or multi-step reactions to form polymerizable macromonomers or oligomers. The chemical reactions used in this method mainly include epoxidation, alcoholysis, transesterification, esterification, amidation, etc. Figure 4.2 illustrates several pathways developed by Wool and co-workers to prepare functionalized triglycerides [8]. For example, the epoxidized triglycerides (Figure 4.2a) can be achieved in a straightforward fashion by reaction of unsaturated fatty acid chains with, e.g., molecular oxygen, hydrogen peroxide as well as by chemo-enzymatic reactions. If these functionalized oils react with acrylic acid by ring opening and esterification of the epoxy groups, acrylated epoxidized oils can be obtained (Figure 4.2e). Monoglycerides can be converted from triglycerides through a glycerolysis or an amidation reaction (Figure 4.2d). One key step for this modification is to reach a higher level of molar mass and functionality, as well as to incorporate chemical functionalities known to impart special properties in a polymer network (e.g. aromatic or cyclic structures). These oil-based macromonomers can be used in the synthesis of epoxy resins, alkyd resins, polyurethanes, polyesters, polyamides, and other polymers.

4.2.2.2 Olefin metathesis

As mentioned above, either by simple hydrolysis or alcoholysis of triglycerides, fatty acids, fatty acid esters, and other oil derivatives can be easily obtained. Recent contributions show a growing interest in the use of oil derivatives as precursors of monomers, not only because of their renewability, but also due to the properties they can provide to the final polymers. Metathesis of oil derivatives is one of the most important contributions that should be highlighted.

Olefin metathesis, first described by Boelhouwer and co-workers [17, 18], is a facile approach to derivatise the double bonds present in unsaturated fatty acid derivatives in order to obtain new polymerizable monomers. Olefin metathesis have made great progress in the last decade [5]. Figure 4.3 lists the two metathesis reactions of oil derivatives: self-metathesis (SM) and cross-metathesis (CM). The research on these reactions has been accelated during the last few years due to the development of functional-group-tolerant metathesis catalysts by Grubbs and others [19–25]. For instance, Warwel et al. [26] showed that the ethenolysis of fatty acid derivatives is possible with very low amounts of Grubbs first generation catalyst (0.01 mol% or less) and that the resulting 9-decenoic acid methyl ester can be dimerized with the same

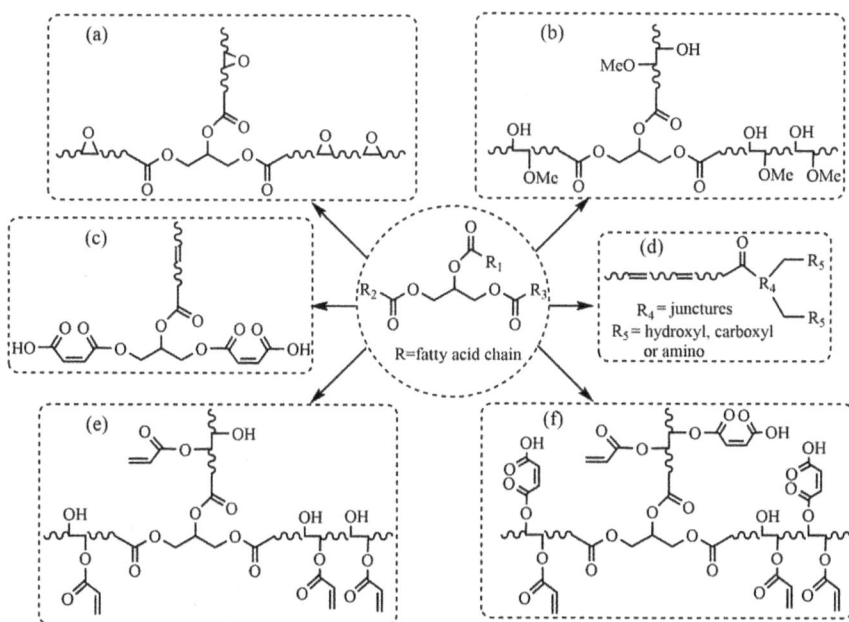

Figure 4.2: Chemical pathways leading to macromonomers from triglycerides.

catalyst to yield a long-chain α, ω-diester for polyester synthesis. Dixneuf et al. [27, 28] reported the CM of fatty acid derived unsaturated esters, acids, and aldehydes with acrylonitrile and fumaronitrile in the presence of Hoveyda-Grubbs second generation catalyst. Grubbs and Hoveyda-Grubbs second generation catalysts displayed very good activities with high conversions and CM selectivities. For more detailed information about the possibilities of SM and CM with heterogeneous or homogeneous catalyst systems, product yields, and structure of fatty acid based monomers in oleochemistry, the reader is referred to some review articles [4, 5].

4.2.2.3 Transition-metal-catalyzed addition

The transition-metal-catalyzed addition provides an efficient and highly regio- as well as chemoselective conversion of an internal double bond into a terminal functional group via an isomerization of the double bond along the fatty acid chain and an exclusive trapping of the ω-double bond [29, 30]. Figure 4.4a lists an example leading to ω-functionalized fatty acid esters for polyester synthesis [31, 32].

4.2.2.4 Thiol-ene addition

The century-old thiol-ene addition has attracted renewed interest because of its being highly efficient and orthogonal to a wide range of functional groups, as well as for

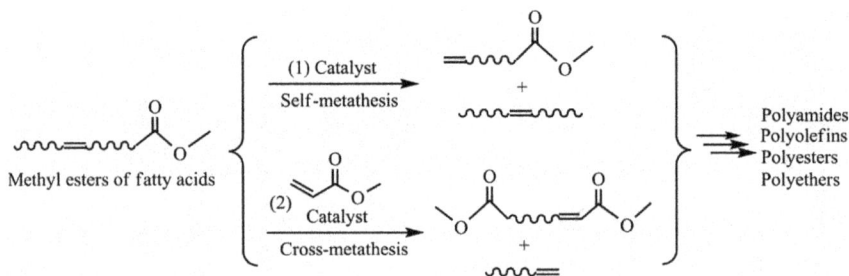

Figure 4.3: General scheme of catalytic metathesis reactions.

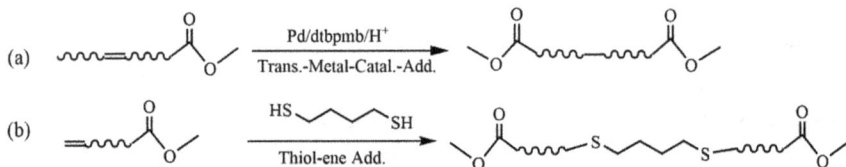

Figure 4.4: Examples of the synthesis of oil-based monomers via (a) transition-metal-catalyzed additions and (b) thiol-ene additions.

being compatible with water and oxygen [33]. The photochemically/thermally-induced version of this reaction is known to proceed by a radical mechanism to give an anti-Markovnikov-type thioether [33, 34]. Several examples have been reported to deal with the synthesis of fatty acid derived monomers and/or to their polymerization for polyurethane, epoxy resin, polyester, etc. [35–40]. Figure 4.4b lists one of the examples from a metathesis-yielding monomer to a new fatty acid monomer for polyester synthesis [35].

4.3 Oil-based thermosetting polymers

Plant oils are one of the oldest feedstocks for polymerization in human history, especially in the areas such as coatings, rubber substitutes, polyamides, and alkyd resins. In the past 80 years, although they have been substituted by petroleum in most chemical industry, some types of polymers for the production of coatings and nylon still remain the traditional feedstock of natural oils. As the emergence of so many modification methods mentioned above, more and more polymers and materials are returned to use this promising resource. In the following sections, the preparation, properties as well as the applications of oil-based polymers are discussed in more detail.

4.3.1 Direct-polymerized polymers

The most straightforward modification of natural oils to get polymer materials is based on the direct-polymerized reactions of functionalities on the triglycerides. These reactions commonly include direct cross-linking of unsaturated double bonds as well as epoxy groups which is initiated by light, heat, and catalyst. For example, drying oils (see above for a definition) are characterization by their ability to form resins due to autoxidation, peroxide formation and subsequent radical polymerization and are therefore applied as binders and film formers in paint and coating formulation [2].

Up to now two kinds of direct-polymerization methods are employed to synthesize polymers and resins: free radical polymerization (FRP) and cationic polymerization (CP). The FRP of triglycerides is always subsequent to the autooxidation or peroxide formation. The studies in early years [41, 42] revealed that after the initial reaction step of an abstraction of a bisallylic hydrogen atom and trapping of the radical by oxygen followed by hydrogen abstraction leads to the formation of hydroperoxides, the curing proceeds by isomerization of the double bonds, C–C bond scission giving many functionalities (e.g. alcohol). Subsequently, the curing continues as evidenced by a rapid decrease in the double bond concentrations as well as epoxide formation, meanwhile cross-linking reactions between the formed radicals (e.g. alkyl, alkoxyl, and peroxyl radicals) occurs, accompanying the increase in viscosity and molar mass.

A simplified scheme of the overall cross-linking reaction of drying oils is provided in Figure 4.5. The CP includes direct cationic polymerization of unsaturated carbon-carbon double bonds, cationic copolymerization with vinyl monomers, and ring opening polymerization of epoxy groups. Good initiators, catalysts, and/or mediums such as supercritical CO_2 fluid will benefit the cationic polymerization due to the reduction of incompatibility and other effects.

Coatings are the first direct-polymerized materials from natural oils. About 2500 years ago, Chinese wrote down records of preparing coatings from Chinese tung oil (TO) *via* heat treatment [43]. The principal constituent of this oil is a glyceride composed of alpha-elaeostearic acid (*cis*-9, *trans*-11, *trans*-13-octadecatrienoic acid). This highly unsaturated, conjugated triene system makes it an excellent drying oil at room temperature, used principally in the preparation of paints, varnishes, and related materials. Not until the 13th century, English began to prepare paints from linseed oil in a similar way like Chinese, after that the plant oil-based coatings were gradually known to the world. Linseed oil (LO) mainly consists of glycerides based on linolenic (all-*cis*-9, 12, 15-octadecatrienoic acid) and linoleic acid (all-*cis*-9, 12- octadecadienoic acid). Compared to TO, it has a lower unsaturated and conjugated system, thus leading to a slower curing rate and less rigidity. Castor oil (CO) is a non-drying oil which mainly contains ricinoleic acid ((9Z, 12R)-12-hydroxy-9-octadecenoic acid). This oil can be converted into conjugated or non-conjugated diene structures *via* vacuum dehydration, namely dehydrated castor oil (DCO), which can be used for preparing high-performance vanishes. At the beginning of 19th century, oil-based varnish was prepared by adding natural resins into dry oils, and hardness as well as brightness of this varnish were improved. By reacting with liquid S_2Cl_2 as well as sulfur powder, LO, CO, rapeseed oil (RO) or other monoglycerides were widely used in preparation of rubber substitutes in the beginning of the 20th century [16]. Materials from these drying or semi-drying oils were basically cured by FRP. However, because of the flexibility of fatty acid chains, the chain transfer process of the yielding radicals can readily occurs, leading to polymers with low molar masses and poor properties. Vinyl monomers were copolymerized to improve this. For instance, it was reported that styrene-oil copolymerization can lead to improved film properties. Generally, free radical type initiator,

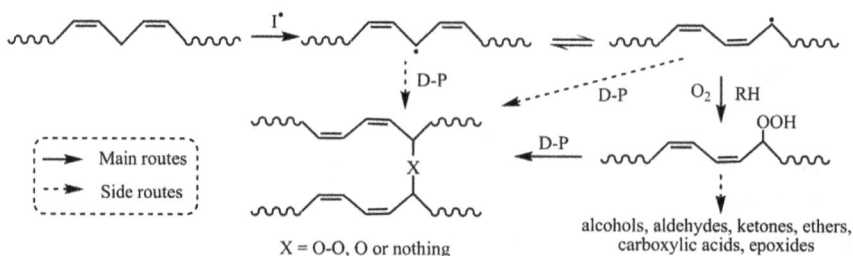

Figure 4.5: Simplified scheme of the overall direct-polymerization reaction.

such as benzoyl peroxide and ditertiarybutyl peroxide, has been used to accelerate this copolymerization reaction. Hewitt and Armitage [44] proposed two types of reaction mechanism for conjugated and non-conjugated oils.

Additionally, by direct cationic copolymerization of many plant oils with vinyl monomers (e.g. styrene), Larock and co-workers [45, 46] have developed a series of polymers with different properties characteristic, ranging from hard and brittle to relatively soft and ductile plastics. Epoxidized oils can also be polymerized by cationic catalysts, such as iodonium salt, sulfonium salt, diazonium salt, ammonium salt, boron trifluoride, etc. Chen et al. [47] reported that a novel norbornyl epoxidized linseed oil was polymerized with divinyl ethers *via* photo-induced curing: the epoxy conversion reached 93.3% at 600 s and the contents of natural oils in the formed polymers exceed 80%. Moreover, cationic ring-opening polymerization of epoxidized soybean oil (ESO) in scCO$_2$ was performed to obtain polymers with molar masses (M_n) ranging from 1 to 25 kDa and application possibilities as surfactants, lubricants or hydraulic fluids [48, 49].

4.3.2 Alkyd resins

Alkyd resins is another oldest polymers prepared from triglyceride oils which produced by the esterification of polyhdroxy alcohols with polybasic acids and fatty acids [1]. Alkyd resins have acquired a good reputation because of their economy and ease of application. Additionally, they are to a greater extent biologically degradable polymers because of the oil and glycerol parts. As a result, they become the most widely-used surface coatings in many areas such as vehicles, boats, electronic devices, and so on.

Two approaches were developed to form alkyd resins [1, 9]. The former approach is based on the alcoholysis of the oil by part of polyols, followed by the esterification of the remaining hydroxy functions using a polyacid. The latter approach is instead based on a one-step process consisting of the direct reaction among the fatty acid, the polyol and a polyacid. Today, the latter approach is more often used to prepare alkyd resins due to the lower cost of manufacturing and the better properties such as high viscosity, good drying and hardness. Several types of alkyd resins, including liquid crystalline type, high-solid content type and water-soluble type were reviewed by Güner et al. [1]. This study will not address these materials in further detail, but refers the reader to a recent article and a book focusing on alkyd resins from renewable raw materials [50, 51].

4.3.3 Unsaturated polyester resins

Unsaturated polyesters (UPEs) are usually linear polymers obtained by condensation of saturated and unsaturated diacids/anhydrides with diols. The unsaturation present in this type of polyesters provides a site for subsequent cross-linking with vinyl monomers, such as styrene [52]. Usually the viscous blends of UPEs with dilute vinyl monomers are called as unsaturated polyester resins (UPRs). Petroleum-derived UPR is currently one of the most widely utilized thermosetting polymers because of its low cost, ease of handling, and good balance of mechanical, electrical, and chemical properties. The reinforced composite materials of UPRs have been broadly employed in aerospace, automotive, marine, infra-structure, military, sports, and other industrial fields.

Several methods have been reported for chemically converting plant oils into UPRs. The first method is converting plant oils into feedstocks like diols, which can generate linear UPEs *via* direct polycondensation. For example, Qin et al. [53] firstly synthesized soybean oil monoglyceride (SOMG) by the alcoholysis of soybean oil with glycerol, then synthesized soybean oil-based UPE *via* polycondensation of SOMA, propanediol, and SOMG. The synthesis route is shown in Figure 4.6. The resulting UPR with 20 wt% of SOMG as diols showed tensile strength and modulus of about 17.5 MPa and 0.4 GPa, respectively, and glass transition temperature (T_g) of 70 °C.

In fact, the first method is not commonly used, possibly because plant oils are not easily converted into high-purity feedstocks, and the content of the resulting materials is low if only one kind of oil-based feedstock is employed. As a result, the second method, converting plant oils into unsaturated ester (UE) macromonomers or oligomers containing active C=C groups, are more preferred by researchers. A series of unsaturated ester (UE) resins were developed by wool and co-workers (Figure 4.2) [8, 54, 55]. These UEs include maleate half esters (Figure 4.2c), vinyl esters (VEs) or acrylates (Figure 4.2e). All of these UE resins share a similar utilization way like UPRs, for instance, copolymerizing with styrene or reinforced by fibers. However, most of the obtained UEs possessed a low C=C functionality, thus leading to low cross-link density and inferior performance in stiffness and heat resistance for the resulting materials. For instance, acrylated epoxidized soybean oil (AESO), synthesized through ring-opening reaction of epoxidized soybean oil with acrylic acid, usually possessed a C=C

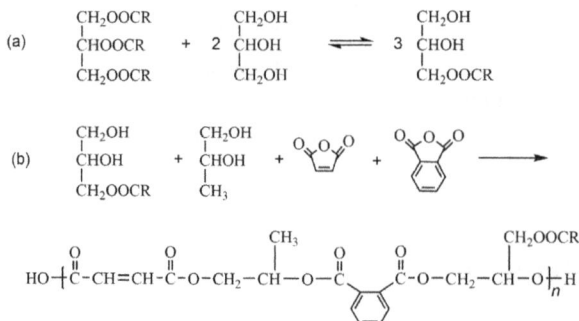

Figure 4.6: Synthetic route for soybean oil-based (a) monoglyceride and (b) unsaturated polyester.

functionality of 2–4. Consequently, the resultant AESO resin with 40 wt% of styrene only demonstrated the tensile strength of 21 MPa, Young's modulus of 1.6 GPa, and glass transition temperature (T_g) of 65 °C [56]. In order to improve the C=C functionality of AESO, maleic anhydride was used to further react with the generated hydroxyl groups on AESO (Figure 4.2f) [55]. The obtained MA-modified AESO (MAESO) could reach a C=C functionality up to 5.9, and the resulting materials with 38 wt% of styrene showed the tensile strength of 41–44 MPa, Young's modulus of 2.2–2.5 GPa, and T_g of 115–130 °C. As the new UE monomer comprised two types of active C=C moieties, it can be named as unsaturated co-ester (Co-UE). In our group, we also developed two types of new Co-UEs from tung oil and soybean oil, which also demonstrated good mechanical and thermal properties after curing with 40% of styrene [57, 58].

Physically blending of plant oil derivatives with petroleum-based UPRs is another efficient way to prepare oil-based UPRs. The plant oil derivatives are usually added into UPR during the polycondensation process of UPE or the curing process of UPR. Mehta et al. [59] used methyl ester of soybean oil and epoxidized methyl linseedate (EML) to modify UPR, and prepared biocomposites containing the modified UPR and hemp fiber mat. The notched Izod impact strength of these resulting biocomposites was enhanced by 90% compared to that of the pure UPR composites. Miyagawa et al. [60, 61] partially replaced UPRs with epoxidized methyl soyate (EMS) or EML to prepare novel biobased resins. These novel bioplastics showed relatively high elastic modulus values and notched Izod impact strength. Ghorui et al. [62] utilized maleated castor oil (COMA) as biomodifier in UPR/fly ash composites. By the incorporation of 5 wt% COMA, impact strength of the obtained matrix increased by 52% as compared to that of the neat UPR/fly ash matrix. In our group, we also reported that UPR terminated with dicyclopentadiene (DCPD-UPR) was modified by tung oil triglyceride *via* intermolecular Diels–Alder reaction occurring at the later stage of melt polycondensation [63]. Compared with the neat DCPD-UPR matrix, the obtained bioplastic with a tung oil content of about 14.8 wt% has a maximum increase of 373% in impact strength.

Aside from the structural factors of the macromonomer itself, the crosslink density effect of polymer matrixes was important to ultimate properties of the designed thermosetting polymers [59, 60, 64]. Nevertheless, in our group, we found the phase-separation effect was also an important effect affecting the mechanical properties of the TO-based UE thermosets [65]. The phase separation effect also plays an important role in the UPRs prepared *via* physically blending. The resulting UPRs usually possess good toughness due to the occurrence of phase separation [59, 60, 63, 64, 66]. The phase separation in solutions or matrixes can be attributed to the incompatibility among oil-based modifiers, general UPEs, and the diluting monomers.

Plant oils can also be converted into vinyl monomers. For instance, Scala et al. [67] and Can et al. [68] have synthesized fatty acid-based monomers (Figure 4.7) as styrene replacements from oleic acid to decrease the volatile organic compound emissions caused by styrene. Fatty acids from oils and fats can be used directly as blocking agents to improve the corrosion resistance of UPR materials [69].

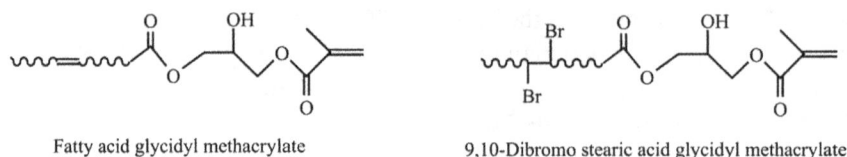

Fatty acid glycidyl methacrylate

9,10-Dibromo stearic acid glycidyl methacrylate

Figure 4.7: Fatty acid-based monomers as styrene replacements.

4.3.4 Epoxy resins

Epoxy resins are commonly known as thermosetting materials for which the resin precursors contain at least one epoxy function. These epoxy functions are highly reactive towards the active hydrogen such as those present in alcohol, amine and carboxylic acid, leading to extremely versatile materials that range from laminated circuit board, structural carbon fiber composites, electronic component encapsulations and adhesives, etc. Almost of 90% of commercial epoxy resins were produced from raw materials of bisphenol A (BPA) and epichlorohydrin. However, the potential health risk cannot be neglected when using BPA resins to produce some materials [70]. The public concern over BPA has led to renewed efforts to developing sustainable epoxy resins. In the formation of sustainable epoxy resins, epoxidized plant oils and fatty acids have been largely utilized as reported in the literature [1, 2, 10]. Most plant oils do not contain natural epoxy groups exception the oils containing vernolic acid obtained from *Vernonia galamensis* and related Vernonia species [71]. Usually, the epoxidation of unsaturated fatty acids or triglycerides can be achieved straightly using reactions with molecular oxygen, hydrogen peroxide as well as chemo-enzymatic reactions, etc. [2, 4].

In industry the unsaturated fatty compounds are converted into epoxidized plant oils by the performic acid procedure (Figure 4.8). The epoxy rate constants of formic acid react with several plant oils were reported by Wool et al. [72]. They found that the level of unsaturation and concentration of formic acid had a significant effect on the value of the rate constant. The double bonds of oleic acid had similar reactivity with double bounds of linoleic acid, but approximately three times less reactive than double bonds of linolenic acid due to steric and electronic effects. Many epoxidized plant oils could be prepared used this method such as epoxidized canola oil, epoxidized grapeseed oil, and epoxidized castor oil. However, the low selectivity and side reactions (Figure 4.9) were also found in reactions, ultimately low epoxy value plant oils were obtained [70, 73–75]. Besides, strong acid catalysts always corrode equipments and they should be removed from final products by neutralization [76].

Lipase-catalyzed chemoenzymatic oxidations overcome the defects of strong acid catalysts and always have the following important advantages: (1) mild reaction conditions, i.e., neutral pH of the reaction mixture; (2) formation of stable hydroperoxides directly from FA, i.e., no need for acetic or formic acid addition; (3)

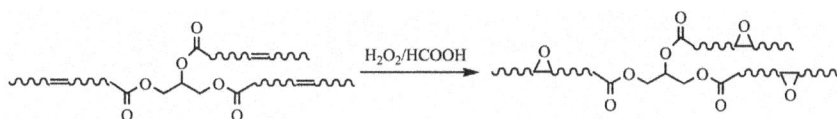

Figure 4.8: Generic scheme of epoxidation of plant oils by the performic acid process.

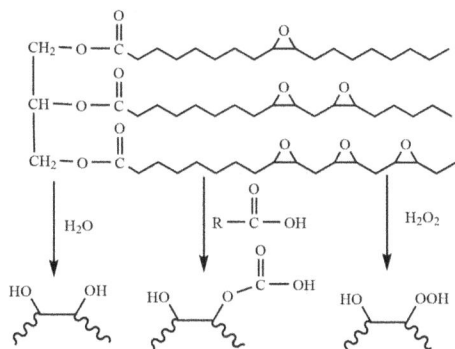

Figure 4.9: Side reactions during epoxidation.

high regio- and stereo-selectivity, (4) significant suppression of side reactions, and (5) high conversion [76–78]. It was reported that the conversion of double bonds to epoxy groups could exceed 90% by Novozyme 435 at optimal conditions for *in situ* epoxidation of soybean oil [76]. In the experiments 110 wt% toluene and 25 mol% free fatty acid were used to suppress hydrolysis of TG.

Epoxidized plant oils can be polymerized with various hardeners, such as diamines, diacids, and anhydrides [70]. A flame-retardant epoxy resin was synthesized *via* bis(*m*-aminophenyl)methylphosphine oxide reacting with epoxidized 10-undecenoyl triglyceride. This epoxidized resin showed enhanced flammability because of the presence of phosphorous increased the limiting oxygen index (LOI) values [79]. Epoxidized fatty acids derived from plant oils can be modified by amidification and thiol-ene reaction as hardeners, which can be further cross-linked with epoxidized plant oils to produce rigid epoxidized resins. Lapinte et al. [80] synthesized novel aliphatic diamines as hardeners for epoxidized linseed oil by amidation of grapeseed oil-based fatty acids, followed by thiol-ene coupling with cysteamine hydrochloride (Figure 4.10). They also found that the bio-diamines more reactive with epoxidized plant oils than petroleum diamines. Some polycarboxylic acid like citric acid (CA) and rosin derivative-fumaropimaric acid (FPA) were chosen as hardeners to cross-link ESO [81, 82]. Those networks possessed excellent self-healing, shape memory, and reprocessing properties due to the presence of dynamic covalent bond exchange and further confirmed as transesterification reactions through stress relaxation [82]. Epoxidized plant oils can also be polymerized with anhydrides, aromatic anhydrides like phthalic anhydride (PA) generally used to synthesis rigid resins, aliphatic anhydrides usually used to produce flexible and rubbery resins [83, 84].

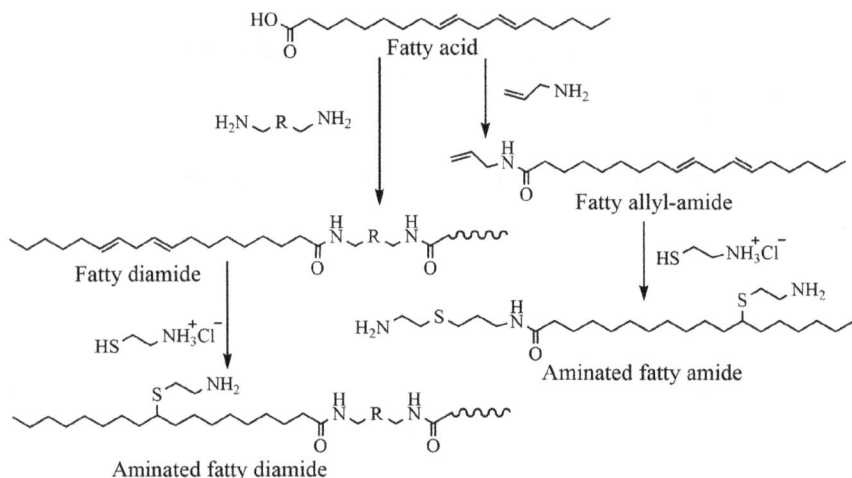

Figure 4.10: Synthesis of fatty diamines as hardeners.

Traditionally, epoxidized plant oils are used as plasticizers and stabilizers for poly(vinyl chloride), but also in the painting and coating formulations [85]. Epoxidized castor oil (ECO), ESO, and other epoxidized plant oils are currently used in epoxy resin applications [2, 10, 78, 86–89]. A comparison of ECO and ESO polymerized with these catalysts revealed that ECO provided a higher T_g value and a lower coefficient of thermal expansion, which was attributed to a higher number of intermolecular interactions, such as hydrogen bonding, in the ECO system [78]. Moreover, ESO cured with cyclic acid anhydrides in the presence of tertiary amines also resulted in thermosetting properties [88]. It was observed that the higher degrees of epoxidation led to higher T_g and hardness of the resins. Therefore, it should be possible to obtain epoxy resins with improved properties from oils with even higher iodine values.

Some other researchers preferred to mechanically reinforce epoxy resins derived from plant oil with different inorganic or organic fillers such as glass, carbon and mineral fibers. For instance, starting from epoxidized linseed oil (ELO), novel nanocomposites with better stiffness were prepared from surface treated alumina nanowhiskers, and layered silicates, and 3-glycidylpropylheptaisobutyl-T8-polyhedral oligomeric silsesquioxane (g-POSS) as nanofillers [90, 91]. Improvements on mechanical and thermal properties were provided by such nanofillers with a low content of 10 wt%, while its flammability, heat evolution, and viscosity during processing were reduced. This is due to their high surface specificity in contrast to "microfillers". In the case of POSS, the resulting nanocomposites were prepared by curing g-POSS and ELO as comonomers with 1 wt% of *N*-benzylpyrazium hexafluoroantimoniate as a thermally latent cationic catalyst.

Fatty acid derivatives can also be employed as epoxy resin applications. Compared to triglyceride oils, the resulting epoxidized fatty acid-derived oxazoline exhibits lower viscosity and higher reactivity in the design of interpenetrating networks [92]. Novel phosphorous-containing derivatives obtained from fatty acid have been prepared as valuable monomers in the design of epoxy thermosetting materials with good flame-retardant properties [79, 93].

Vernonia seed oil contains naturally epoxy acids in its structure (Figure 4.1b(ix)). A UV-curable resin was synthesized *via* transesterification of vernonia fatty acids with a hyperbranched hydroxy functional polyether [94]. Many other epoxidized plant oils were also widely used for the synthesis of cationic UV-curable coatings [1].

4.3.5 Polyurethane resins

4.3.5.1 Isocyanate-based PUs

The reactions of oil-based polyols and diisocyanates to form polyurethanes (PUs) are regarded as one of the most important routes in the development of PUs. The most widely used diisocyanates are toluenediisocyanate (TDI), isophoronediisocyanate (IPDI), hexamethylenediisocyanate (HDI), methylene-4, 4'-diphenyldiisocyanate (MDI), etc., as shown in Figure 4.11 [1, 95]. However, up to now, no literatures are working on the synthesis of new diisocynates from natural oils and derivatives. Hence, the synthesis of oil-based polyols attracts the major concern of researchers. The oil-based polyols can be achieved by introducing hydroxyls onto the unsaturated fatty acid chains. These products have low price but high functionalities, thus providing PUs excellent properties comparable to the petroleum-based PUs. These merits also make the oil-based PUs grow quickly and widely be used as coatings, foams, molding resins, and composites in the areas of packaging, vehicles, agricultural devices, building materials, etc. [96].

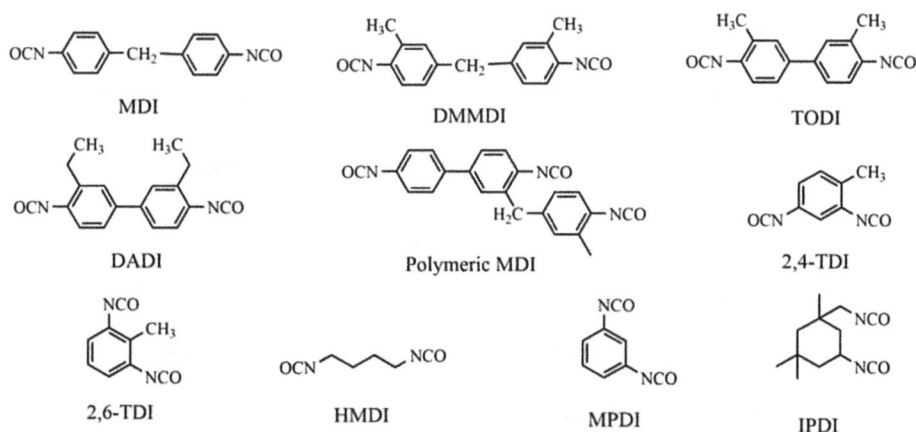

Figure 4.11: Chemical structures of common isocyanates used in polyurethane production.

4.3.5.1.1 Polyols prepared by epoxidation ring-opening for PUs

One method to prepare oil-based polyols is based on the ring opening reactions of epoxy groups, such as alcoholysis, acidolysis, and aminolysis (Figure 4.12) [97]. Recently, various epoxidized plant oils (canola, midoleic sunflower, soybean, linseed, sunflower and corn) were employed to prepare polyols [2, 98]. These oil-based polyols gave PUs similar cross-link density (and thus similar T_g) values as well as mechanical properties despite the somewhat different distributions of fatty acids. An interesting approach to obtain novel polyols from plant oil renewable materials was recently described by Lligadas et al. [99] who combined acid-catalyzed ring-opening polymerization of epoxidized methyl oleate with a subsequent partial reduction of ester groups to yield oligomeric polyether polyols (Figure 4.13).

The kinetics of ring-opening reactions were investigated by several researchers. The activation energy of ring-opening between the epoxy and different nucleophilic reagents was studied by zhang et al. [100]. The activation energy of the ring-opening reaction between the epoxy and the carboxylic groups was 72.2 kJ/mol, while the activation energy of the ring-opening reaction between epoxy and hydroxyl groups with protons as catalysts was 75.1 kJ/mol (Figure 4.14). Steric hindrance has a negative effect on the ring-opening reactions. Therefore, the higher activation energy can be caused by long carbon chains of fatty acids.

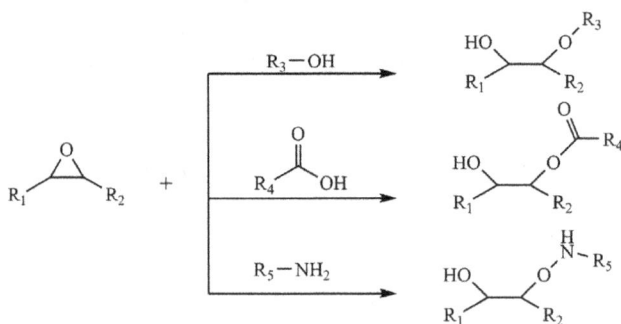

Figure 4.12: Ring-opening reactions of epoxidized plant oils initiated by alcohol, acid, or amine.

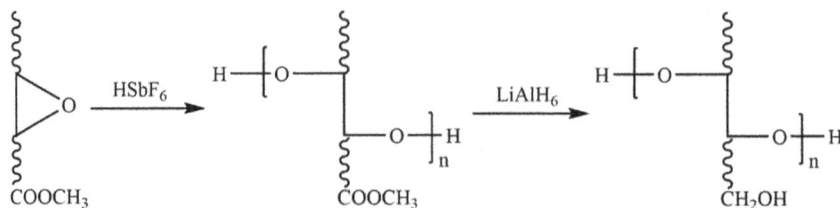

Figure 4.13: Polyols prepared by ring opening polymerization of epoxidized methyl oleate and subsequent partial reduction of the esters.

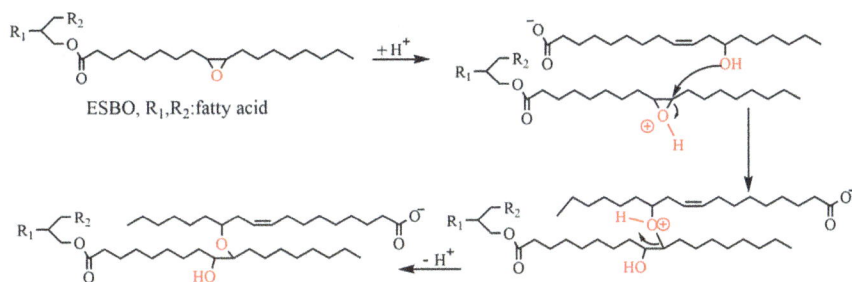

Figure 4.14: Mechanism of ring-opening reactions between epoxy and hydroxyl groups with acid as catalyst.

The hydrogenated triglycerides with one secondary hydroxyl group were obtained by epoxides group which could be directly hydrogenated by dihydrogen [97]. Some nucleophilic reagents such as water, monohydroxy alcohol, polyhydroxy alcohol were used to open the epoxy groups of plant oils catalyzed by inorganic acids [101]. The polyols may be complex blends of products and side reactions during epoxy opening cannot be ignored. The side-reactions of transesterification and opening polymerization of epoxides group during triglyceride functionalization were studied by many researchers (Figure 4.15) [97, 99, 102]. High viscosity polyols resulting from the presence of the oligomers caused by above side-reactions. Thus, ring opening reaction of epoxidized plant oils leads easily to prepare biobased polyols with high viscosity and odorless. However, in almost of all above mentioned polyols, the small compounds employed in ring-opening of epoxidized plant oils are based on petroleum. Several research groups have dedicated considerable effort to the development of the solvent-free/catalyst-free method and 100% biobased polyols to prepare PUs [103, 104]. The plant oils were oxidized into epoxidized oils with formic acid and hydrogen peroxide, followed by ring-opening reaction with castor oil fatty acid (Figure 4.16) [103]. However, high temperature is used in this reaction and some oligomers can be obtained due to the high temperature and long reaction time. The PUs from those green polyols showed better thermal resistance and thermal mechanical properties than others PUs from castor oil with methoxylated soybean oil polyol [104].

Amines are often used to open the oxirane ring of epoxidized plant oils. The polyol mixture was prepared by isopropanolamine as a nucleophilic reagent with ESO. The results suggested that both ester groups and epoxy groups in ESO reacted with amino group of isopropanolamine through simultaneous ring-opening and amidation reactions [105]. The PUs which backbone with quaternary ammonium salts were synthesized by ESO and diethylamine and the resultant PUs indicated excellent antibacterial activity and bio-compatibility properties (Figure 4.17) [106].

Some researchers tried to find out the relationship between the OH number of polyols and the thermal mechanical properties of PUs in order to improve the

1. Transesterification

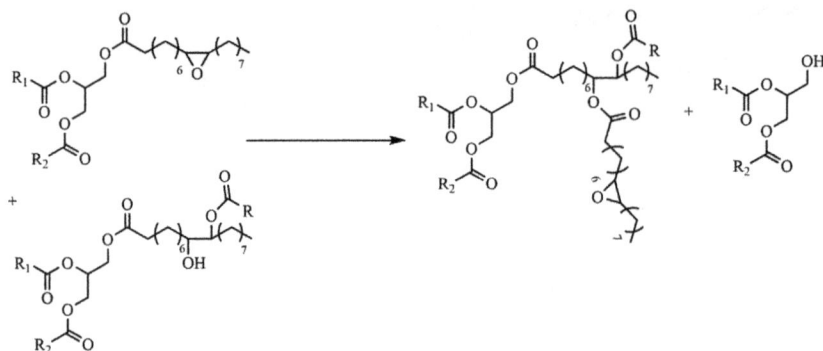

2. Ring opening of epoxide groups by alcohol

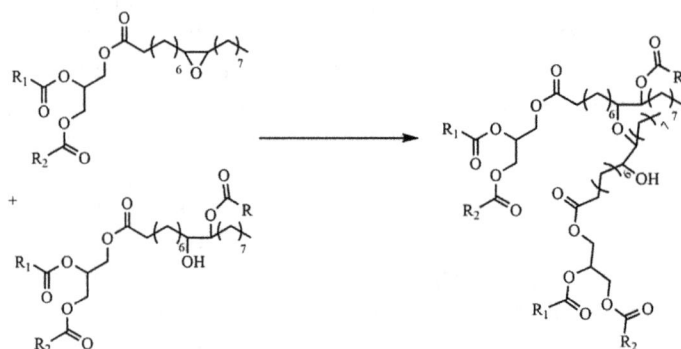

3. Epoxide oligomerization

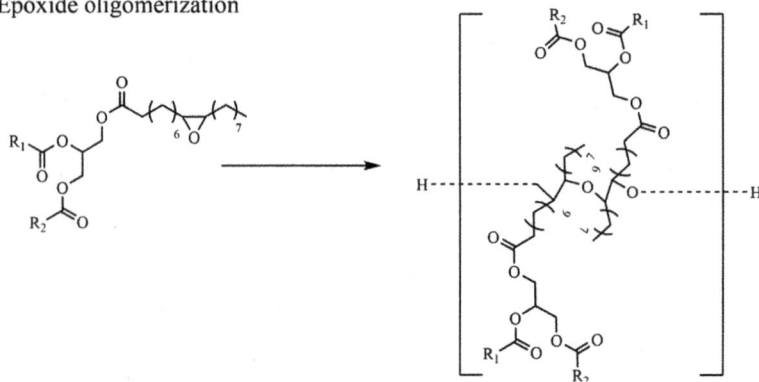

Figure 4.15: Possible side-reactions yielding oligomers during triglyceride epoxide ring-opening reaction with acids.

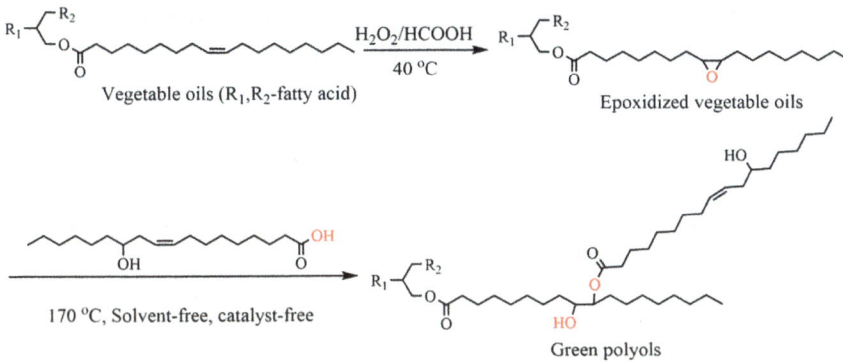

Figure 4.16: Green polyols prepared by solvent-free/catalyst-free method.

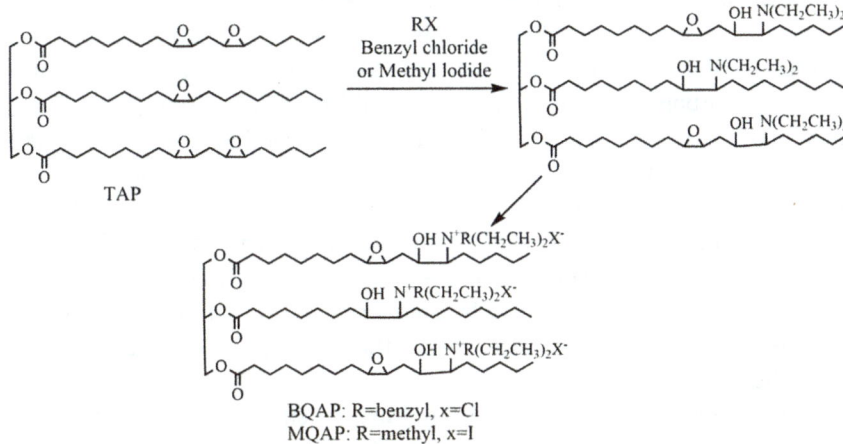

BQAP: R=benzyl, x=Cl
MQAP: R=methyl, x=I

Figure 4.17: Synthesizes route for preparation of functional polyols from ESO.

mechanical and thermal mechanical performance [107]. A series of polyols with a range of OH numbers based on SBO and ESO were prepared by Wang et al. [101]. The PUs prepared from polyols with OH numbers larger than 170 mg of KOH/g were glassy, whereas those prepared from polyols with OH numbers less than 170 were rubbery. Glassy PUs displayed decent mechanical strength, whereas rubbery samples showed relatively poor elastic properties characterized by lower strengths. The tensile strength of the PUs increased when the OH number was increasing. However, the elongation and thermal stability of the PUs decreased with the OH number increasing. This could be caused by the reduced crosslinking density of the resins due to the lower OH value of the polyols.

4.3.5.1.2 Polyols prepared by hydroformylation/hydrogenation and their resulting PUs

The polyols with the primary hydroxy groups could be prepared by the method of hydroformylation/hydrogenation (Figure 4.18) [70, 107]. The polyols with primary hydroxy groups usually react with isocyanates more easily than the second hydroxy groups of polyols obtained by process of epoxidation/ring opening [108]. The syngas and catalyst such as rhodium and cobalt carbonyl complex are commonly used to synthesize vegetable-oil based polyols by the method of hydroformylation/hydrogenation. The rhodium and cobalt demonstrated different effects on the mechanical properties of PUs. The PUs prepared by rhodium catalysis polyols tend to form rigid structures due to the high conversion of triglycerides into polyols. However, the cobalt-catalyzed hydroformylation resulted in a lower polyol conversion and ultimately the rubbery PUs were obtained with lower mechanical properties [109].

4.3.5.1.3 Polyols prepared by ozonolysis/reduction and their resulting PUs

Ozonolysis was used to prepare polyols with terminal primary hydroxyl groups and different functionalities from trilinolein (or triolein), SO, and low-saturation canola oil [110–113]. The carbon-carbon double bonds could be broken during ozonolysis, leading to a length reduction of fatty acid chains at the same time (Figure 4.19). As a result, the low-molecular-weight polyols, although approximately 60% of the polyols obtained by the method of epoxidation/oxirane ring opening or hydroformylation/reduction, could be prepared usually with low viscosity at room temperature [108, 111]. Influence of aprotic and protic solvents on the formation of polyols during the ozonolysis process of canola oil were studied by Omonov et al. [108]. It was observed that ethyl acetate as aprotic solvent tends to accelerate the formation of oligomeric polyols. In addition, the formation of carbolic acids also observed in the early stage of ozonolysis when ethyl acetate as solvent. In contrary, when methanol or ethanol and their mixtures with ethyl acetate as protic solvent, the high-molecular oligomeric polyols and carbolic acids could be forbidden or reduced during the ozonolysis

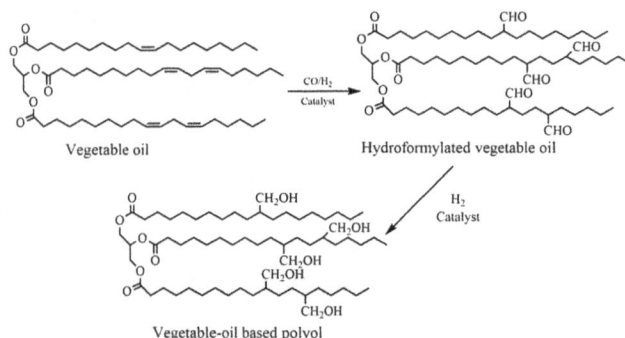

Figure 4.18: Typical route to synthesize polyols *via* hydroformylation/hydrogenation reaction.

of canola oil. The author also further found that a clear correlation between the ozonolysis time, product yields, and the reaction exothermicity.

The resulting PUs displayed excellent mechanical properties and higher glass transition temperatures compared to the PUs from epoxidized and hydroformylated polyols with the same functionality, presumably due to the absence or lower content of dangling chains in the former. We can also use the different feed gas to alert the mechanical and thermal properties. The polyols with varying OH number could be obtained by air and oxygen as feed gas [107]. The PU plastic sheets were synthesized from canola oil by ozonolysis/reduction to investigated the OH/NCO molar ratio effects of PUs properties, reported by Kong et al. [114]. The T_g values for PUs with OH/NCO molar ratios of 1.0/1.0, 1.0/1.1, 1.0/1.2 were found to be 23, 41, and 43 °C, respectively. It was also found that PU with 1.0/1.1 of OH/NCO molar ratio had the excellent mechanical properties with a yield point of 18 MPa and elongation at break of 39%.

The ozonolysis method also accompany with some drawbacks such as the need to use toxic solvents, relatively high reaction temperature, and sophisticated equipment for synthesis of polyols [115].

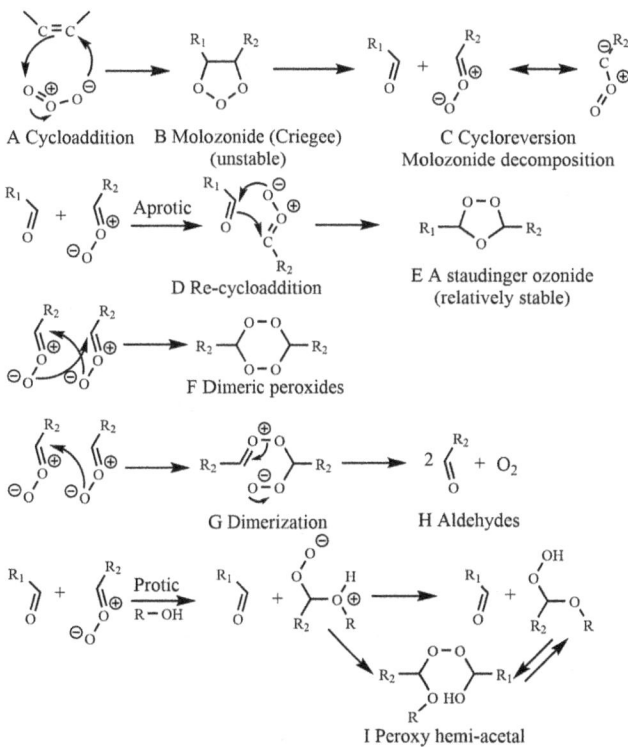

Figure 4.19: Proposed mechanisms of ozonolysis and compounds formed by the ozonolysis of unsaturated fatty acids.

4.3.5.1.4 Polyols prepared by transesterification/amidation and their resulting PUs

It is difficult to prepare the plant oil-polyols with OH number over 250 mg KOH/g by epoxidized methods [116]. However, the OH of polyols formed by epoxidation and transesterification of plant oils can easily exceed 250 mg KOH/g and those polyols with low viscosity about 2.4–13.5 Pa·s (Table 4.2). Kong et al. firstly established the production of polyols and their corresponding PUs from plant oil starting materials whose glycerol backbone was removed explicitly during the polyol synthesis reaction (Figure 4.20) [116]. The OH number of two polyols (Liprol™ 270, Liprol™ 320) synthesized by transesterification with 1,3-propanediol and 1,2-propanediol were about 275 and 323 mg KOH/g, respectively. Besides, they possessed low viscosities about 2.976 and 3.068 Pa·s at 25 °C. Hence, the resulting PUs indicated rigid structures, excellent hydrolytic stability and the alkali resistance due to removal of the glycerol from triacylglyceride backbone. The monoglycerides were also synthesized by transesterification of sunflower oil, and the according PUs were found to be cytocompatible, immunocompatible, hemocompatible, and histocompatible, which supported the adherence and proliferation of cells [117].

The polyols with primary hydroxyl groups can also be prepared by amidation with different amines similarly as the transesterification method. A variety of polyols with different structures were synthesized from rapeseed oil (RO) with diethanolamine (DEA), triethanolamine (TEA) and glycerol (GL) at different molar ratios (Figure 4.21). The PU networks showed high tensile strength and hardness but low elongation at break due to the high OH number of RO/DEA polyols ranged from 298–412 mg KOH/g. The polyols prepared by transamidation were further modified by condensation polymerization with dimer fatty acid and those PUs can be used as wood finishes [118].

Although transesterification and amidation can convert plant oils into polyols with high OH functionalities, one obvious defect of this method could be the absence of any hydroxyl groups in long acid chains of the polyols. Therefore, the PUs prepared by transesterification and amidation are usually with inferior mechanical and thermal properties. However, the PUs from castor oil by method of transesterification or amidation without the mentioned problems as a result of the natural hydroxyl groups in ricinoleic fatty acid chains [70].

4.3.5.1.5 Polyols prepared by thiol-ene reaction and their resulting PUs

Thiol-ene reaction classified as "click chemistry" provides a platform to prepare plant oils-based polyols with high yields, high reaction rates, and fewer side products [70, 107, 119]. The mechanism of thiol-ene reaction is presented in Figure 4.22, and the comparison of polyols from different method is shown in Table 4.3. Thiol-ene reactions proceed through four steps: (1) Formation of free radicals by thermal initiation or photo-irradiation; (2) Formation of thiyl radicals by transferring free radicals at the sulfur atom; (3) Anti-Markovnikov addition of thiyl radicals to double bonds; (4) New transfer reactions with thiol groups to final products and new thiyl radicals [70, 120].

Table 4.2: Mechanical and thermal properties PUs synthesized by using polyols produced through transesterification/amidation reaction [70, 110].

Polyols				PU			
Seed oil (s)	Reaction mechanism	OH number	Isocyanate	Glass transition temperature T_g (°C)	Tensile strength (MPa)	Elongation at break (%)	Young's modulus (MPa)
Rapeseed	Transesterification	330–997	MDI	93.4	11.4–36	420–1189	NF[a]
Oleic acid	Transesterification with trimethylol propane	27–130	MDI	−33 to −58	0.6–6.9	47–110	NF
Neem	Transesterification with glycerol/anhydride	14–32	TDI	NF	NF	NF	NF
Castor	Transesterification with glycerol	NF	HDI and IPDI	49–100	6.1–18.4	3.4–92.3	NF
Castor	Transesterification with pentaerythritol	190–234	Tolonate	−21 to 34 and 62–72	11.1–17.4	140–209	NF
Sunflower	Transesterification	NF	TDI	NF	21.9–27.3	500–863	NF
Sunflower	Transesterification with glycerol	NF	TDI	NF	19–23	700–750	NF
Canola	Transesterification with propane diol	275–323	MDI	70–85	61–67	4.6–5.9	1430–1700
Canola, sunflower, and camelina	Transesterification with propane diol	259–286	MDI	69–94	55–64	5.4–6.4	1500–1900
Mesua ferrea, castor, and sunflower	Transesterification	NF	TDI	NF	5.0–7.5	595–620	NF
Rapeseed	Transamidation with diethanolamine	310–460	PMDI	89–98	48–53	140–209	19–25
Rapeseed	Transamidation with triethanolamine	302–374	PMDI	39–48	14–24	12.6–26.2	496–931
Oleic acid	Transamidation with diethanolamine	69–132	TDI	NF	NF	NF	NF

[a] The required information has not been reported or could not be found from the corresponding paper.

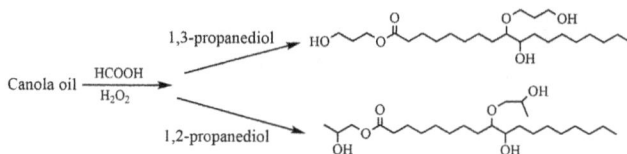

Figure 4.20: Synthesis of poly(ether ester) polyols from canola oil.

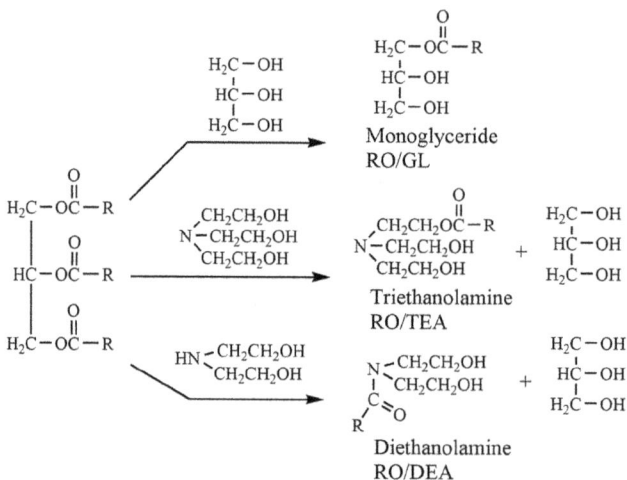

Figure 4.21: Idealized transamidation reactions of RO with DEA, TEA, and GL.

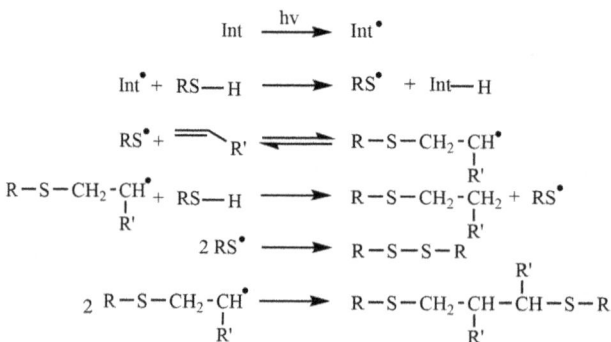

Figure 4.22: Mechanism of thiol-ene reaction.

The polyols were prepared by one step chemical conversion of plant oils at 80 °C and only 65% of carbon-carbon double bonds were changed into hydroxyl groups. Therefore, the obtained polyols usually have a hydroxy value of <200 mg KOH/g and the correspond PUs always indicated poor mechanical and thermal mechanical

Table 4.3: Comparison of polyols prepared by different methods [70].

	Epoxidation/ring-opening	Ozonolysis/reduction	Hydroformylation/hydrogen	Transesterification/admidation	Thiol-ene
Number of steps	2	2	2	1	1
Functionality	Secondary, tunable	Terminal, 3	Primary, tunable	Terminal, 2-3	Primary, tunable
Functionality/OH	70–320	228–260	150–200	190–400	Around 223
Molecular weight	Variable (>1000)	500–700	900–1100	350–400	1070–1440
Viscosity	High	Low	Medium	Low	Medium
Reaction temperature	Variable (50–170 °C)	Room temperature	High (120 °C)	High (140–220 °C)	Low or room temperature
Reaction time	Long	Medium	–	–	–

properties [40]. However, Alagi et al. demonstrated this reaction at low temperature (−10 or −20 °C) and more than 99% carbon–carbon double bond conversion and more than 95% hydroxyl functionalization were achieved [121]. The intermediating radicals are more stable at low temperature as a result of high conversation and functionalization. To best of our knowledge, the double bonds of plant oils are internal (1,2-disubstituted) double bonds and those double bonds are always with low reactivity in thiol-ene reactions at mild conditions. Thus the high excess of mercaptans, longer reaction time, and high concentration of initiators were used to improve the reactivities of thiol-ene reactions. Ionescu et al. changed the methodologies of synthesis and used mercaptanized soybean oil (SBO) instead of plant oils to overcome the defects of common thiol-ene reactions with plant oils (Figure 4.23). Plant oil polyols, polyamines, epoxides, isocyanates, isothiocyanates, and silicon compounds useful for making bio-based polymers were successfully prepared using this method in a single step and without by-products [120].

4.3.5.1.6 Polyols prepared by other reactions and their resulting PUs
Transition-metal-catalyzed reactions are also adopted as one of the useful methods to obtain polyols. For instance, the epoxy groups in triglycerides or fatty acid derivatives can be hydrogenated *via* transition-metal-catalyzed addition [122]. Moreover, the unsaturated C=C bonds in fatty acid chain were converted to carbonyl groups by transition-metal-catalyzed oxidation, then the formed groups were reduced to polyols in the presence of Raney-Ni catalyst [123]. Dimethyl nonadecanedioate (Figure 4.4a) can also be reduced to diols for PU synthesis.

More recently, Rio et al. [124] prepared novel amorphous and semi-crystalline PUs from polyols obtained by ADMET polymerization of a CO-based diene and the semicrystalline polymers showed good shape memory properties. Moreover, novel

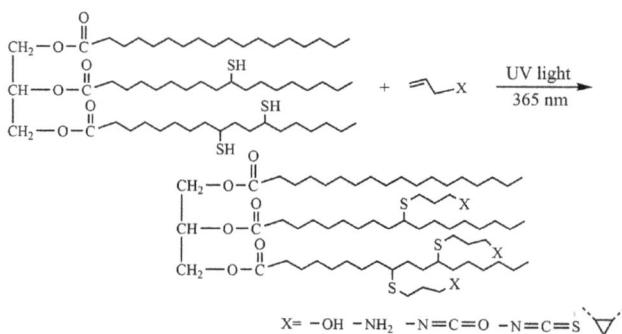

Figure 4.23: General thiol -ene reaction between mercaptanized soybean oil and ally compounds with reactive groups.

silicon-containing PUs based on plant oils were prepared and characterized [99]. In particular, a silicon-fatty acid OH-terminated triglyceride (Si-oil-polyol) was prepared and reacted with MDI.

4.3.5.2 Non-isocyanate-derived PUs

Due to the threat of environment pollution (e.g. usually phosgene is used for synthesis of isocyanate), the reaction of cyclic carbonates with amines has been emerging as a promising non-isocyanate route to PUs [125–128]. The plant oil-based cyclic carbonate can be prepared with oxiranes and CO_2 in the presence of a catalyst, such as alkali metal halides, quaternary ammonium halides, and polystyrene bound quaternary ammonium salts [70]. The non-isocyanate-derived PUs (NIPUs) were synthesized followed by reacting with amine. For example, ESO was converted to carbonated soybean oil (CSBO) with CO_2 in 99% conversion using tintetrachloride/tetrabutylammonium bromide as catalyst [129]. Subsequently, CSBO was reacted with ethylene diamine to yield well-performing NIPUs. The liner or network would be obtained due to intermolecular hydrogen bonds among the new formed hydrogen groups at β-carbons of the urethane moiety. NIPUs usually show comparable mechanical properties, and better water absorption, thermal stability, and chemical resistance than conventional PUs [70, 130]. Various amines and molar ratios of carbonate to amine were used to prepare NIPUs and their mechanical and thermal mechanical properties were reported by many researchers [131–133]. The carbonated soybean oil reacting with different amines to produce the NIPUs were reported by Javni. The reactants 1,2-ethylenediamine, 1,4-butylenediamine, and 1,6-hexamethylenediamine were used with the carbonate to amine ratio of 1:0.5, 1:1, and 1:2, respectively. During the reaction, amides resulting from the reaction of amines with ester groups were found (Figure 4.24). The amides as side products have a negative effect on mechanical and thermal properties, thus the concentration of amides in reaction system must be reduced. The stoichiometric carbonate-to-amine ratio endowed the resultant polymer networks with the highest density, T_g, hardness, and tensile strengths due to the low concentration of amides [131]. On the other hand, the mechanical and thermal mechanical properties were dominantly controlled by the carbonate-to-amine ratio and the structure of amines. The aromatic and cycloaliphatic diamines were used to react with carbonated soybean oil (CSBO) for the purpose of increase the rigidity and tensile strength of NIPUs [132].

Recently, Cramail and co-workers investigated the synthesis of linear PUs from cyclic dicarbonates based on methyl oleate [134]. However, low molar mass PUs were achieved due to the presence of secondary hydroxyl functions and the formation of macrocycles.

Transurethanization is another route worth mentioning to synthesize NIPUs. For instance, Jayakannan and Deepa [135, 136] have described the PUs synthesis *via* a

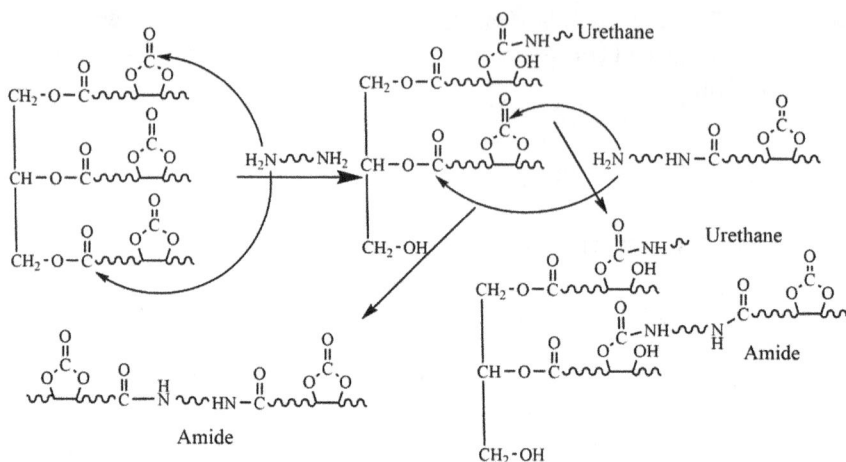

Figure 4.24: Chemical routes for the urethane and amide formation in the reaction of CSBO with diamines.

transurethane reaction of biscarbamate and diol under solvent-free and melt conditions. However, NIPUs based on fatty acid-based monomers using this approach have not been reported.

More recently, Cramail and co-workers [137, 138] have presented a novel AB-type self-polyaddition method to obtain fully biobased NIPUs from methyl oleate, ricinoleic acid, naturally occurring oleic acid as well as undecylenic acid derivatives. Firstly, a set of novel AB-type monomers was prepared *with* the employment of environmentally-benign thiol-ene chemistry and well characterized, subsequently a series of new semi-crystalline PUs were synthesized *via* the AB-type polycondensation approach. The obtained PUs reached a high M_n to 10 kDa and all the PUs exhibited fair thermal stability with no significant weight loss below 200 °C.

4.3.6 Other thermosetting polymers

In order to combine the merits of different polymers or resins, numerous copolymers and interpenetrating polymer networks (IPNs) have been developed. Many of them (e.g. polyester-amide, urethane alkyd, polyurethane acrylate, or IPN, etc.) were derived from natural oils and fatty acid derivatives [1, 2, 9, 10]. For example, three novel polyurethane acrylates (PUAs) were prepared by the reaction of soy polyols with IPDI and hydroxyethylacrylate [139]. When compared with PUs, the tensile strengths of PUAs were improved by 118, 39, and 9%, respectively, with the increase of OH number of three soy polyols. The development of all these composites are accompanied with the advance of every component, thus we will not discuss these in detail.

4.4 Oil-based thermoplastic polymers

4.4.1 Polyamides

The study of polyamides (PAs) prepared from natural oils is an old field of research which seems to be processed slowly in the last several decades. AB-monomers (i.e., ω-amino carboxylic acids), AA- and BB-monomers (dicarboxylic acids with diamines), or the ring opening polymerization of lactams were usually used to synthesis PAs [140]. One of the most commercially used polymers, Nylon 11, was developed from CO [141]. The product had a wide range of flexibility, excellent dimensional stability and electrical properties, good chemical resistance, and low cold brittleness temperatures. Recently, BASF relaunched PA 6, 10, a polymer that is derived from sebacic acid, which can be produced from CO [142]. Another CO-based product that BASF introduced to the market is Lupranol Balance, a polyol made of up to 31 % of CO by weight [142]. France-based Arkema has also been working with castor derivatives since the early 1940s. It produces the 100% CO-based Rilsan (PA 11), which is mainly used to make synthetic fibers as an alternative to other oil-based PA [142]. Today, Rilsan is used in high-value-added applications such as fuel lines in cars, offshore pipelines, gas distribution piping systems and in traditional markets such as electronics, sports equipment, furniture and automobile components. Polyamides from TO and soybean oil (SO) were used mainly in the paint industry because of their thixotropic character [141]. In addition, SO-based dimer-acids were used as co-monomers in the preparation of co-polyamides with different α-amino acids [143]. The introduction of the amino acids into the polymers did not enhance the biodegradation ability but did improve the swelling properties of the co-polyamides in water at a relative high pH and temperature. The co-polyamide with L-tyrosine as co-monomer was discussed as easily de-inkeable copy toner revealing that images of these soy-based toners were similar to those of commercially available toners.

PA 11 were first synthesized from amino acid monomer and commercial application since 1955 [144]. However, toxic and corrosive gasses such as NH_3 and HBr were used in this method, which is incompatible with the guidelines of green chemistry (Figure 4.25). For linear PAs synthesis, AB-monomers can be prepared *via* cross-metathesis of primary fatty acid with suitably functionalized olefins theoretically [140]. The monomer 12-aminododecanoic acid would be synthesized through 10-undecenoic acid cross-metathesis with allyl chloride or acrylonitrile and further subsequent nucleophilic substitution and hydrogenation or hydrogenation. However, PA 12 synthesized from this monomer has not been described (Figure 4.26). Besides, the high toxicity of functionalized olefins should be high concerned in this research [145]. The precursor of PA 12 were also synthesized with a different method, in which the hydroformylation of 10-undecenenitrile, a substrate readily prepared from renewable castor oil, to a linear aldehyde in high yields and regio-selectivities with a (dicarbonyl) rhodium acetoacetonate-biphephos [Rh (acac)(CO)$_2$-biphephos] catalyst (Figure 4.27a) [146]. The safety of this method should also be considered as above due to the use of syngas (i.e., CO). Besides, the thiol-ene addition reaction was successfully used to prepare monomer of PA 14, thioether was introduced into the structure of monomer at the same time (Figure 4.27b) [147]. Equimolar methyl-10-undecenoate and cysteamine hydrochloride reacted via thiol-ene addition at 75 °C, and a high yield of 85% was obtained ultimately. The melting point of PA 14 obtained from this method (138 °C) was observed lower than that of common PA 14 as a result of the thiol atom replace one −CH$_2$− moiety in PA 14 monomer structure [140]. It can be concluded that the thiol-ene addition is a straightforward and sustainable method to obtain AB-monomers, and more complex PAs can be synthesized from this method in future.

The AA- and BB- monomers, azelaic obtained from oleic oil derivatives usually was used to synthesis PA 6 and 9 for the applications in food packaging, wire coatings, or in

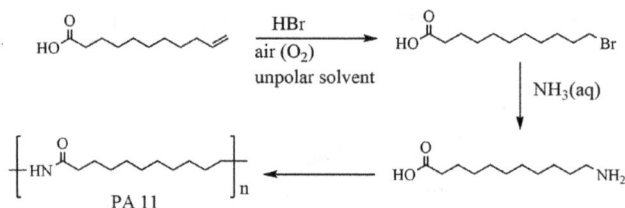

Figure 4.25: Synthesis route to commercially available PA 11.

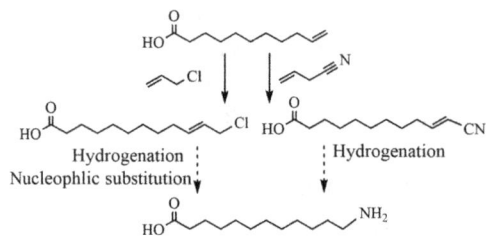

Figure 4.26: Potential synthesis route to PA 12 monomer via olefin cross metathesis.

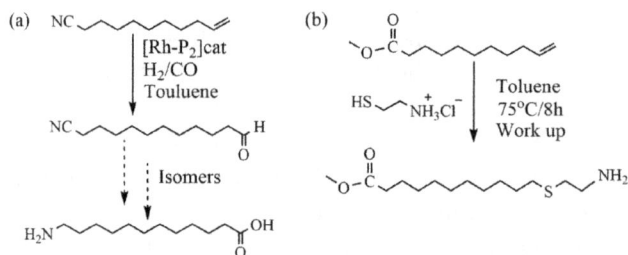

Figure 4.27: (a) The precursor of PAs prepared by hydroformylation of 10-undecenenitrile; (b) The PA 14 monomer synthesized from thiol-ene addition reaction.

the automotive and electronic industries [140]. Except dicarboxylic acids, α,ω-diesters synthesized *via* cross-metathesis of fatty acid methyl esters with methyl acrylate or themselves in the presence of metathesis catalysts, usually were utilized to prepare PAs. The middle to long chain unsaturated α, ω-diesters (1.8-, 1.11-, 1.12-, 1.15-, and 1.20-) could be obtained from cross-metathesis reaction, which are valuable renewable monomers for PAs [148]. Meier and co-workers [149] reported the synthesis of unsaturated PA X and 20 from oil derivatives *via* two different approaches, each involving a metathesis step (Figure 4.28). The results showed that the first approach from methyl 10-undecenoate cannot produce polymers with high molar masses, while the second approach from dimethyl(E)-icos-10-enedioate (a self-metathesis monomer) produced renewable polyamides X and 20 with molar masses up to 14,700 Da in the presence of 5 mol% of 1, 5, 7-triazabicyclo[4.4.0]dec-5-ene. This method presents several advantages over traditional techniques like the avoidance of acid chlorides.

4.4.2 Polyesters

Polyhydroxyalkanoates (PHAs), naturally synthesized by a large variety of bacteria, are alternative polyesters to alleviate problems associated with degreasing organic solvent amount in paint formulation. Plant oils are among the different feeding sources for the bacteria. The readers are referred to a book chapter [9] discussing these materials.

Olefin metathesis were often employed to develop non-crosslinked polyesters. Warwel et al. [26] developed two novel methods consisting of metathesis reactions to prepare UPEs with molar masses (M_w) up to 70 kDa, as shown in Figure 4.29. These UPEs with internally unsaturated bonds can hardly be crosslinked with vinyl monomers as thermosets, thus can only be used as special surfactants.

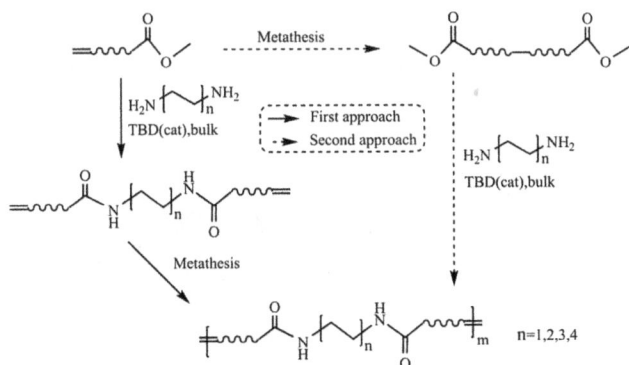

Figure 4.28: Schematic representation of the two approaches to obtained PA X, 20 from oil-based derivatives.

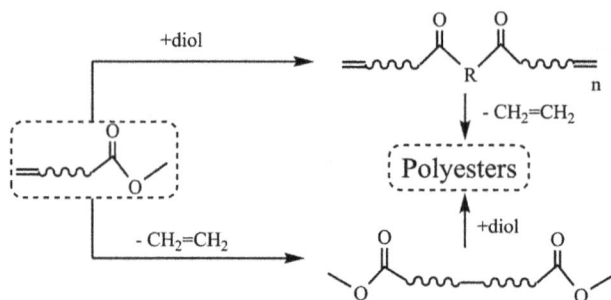

Figure 4.29: Polyesters based on ω-unsaturated fatty acid methyl ester.

Figure 4.30 shows some relatively-new metathesis schemes for polymerization, which were termed as acyclic diene metathesis (ADMET) polymerization, acyclic triene metathesis (ATMET) polymerization. The ADMET polymerization is a step-growth polymerization, usually performed on α, ω-dienes, which is driven by the release of ethylene as condensate (Figure 4.30a) [150]. This polymerization is really a promising approach to obtain high molar mass because it demonstrates the versatility of trigycerides for direct polymer synthesis. For instance, the ADMET polymerization of 10-undecenyl-10-undecenoate resulted in high-molar-mass unsaturated polyesters with polyethylene-like structure, and it was possible to prepare telechelics and ABA triblock copolymers with this monomer in a one-step procedure [151]. Analogously, if α, ω-dienes was changed to model triglycerides or high-oleic sunflower oil, the ATMET polymerization occurred (Figure 4.30b). The cross-linking of the resulting polymers could be completely avoided in the presence of a chainstopper, and hyperbranched polymers were obtained in a one-step procedure [152].

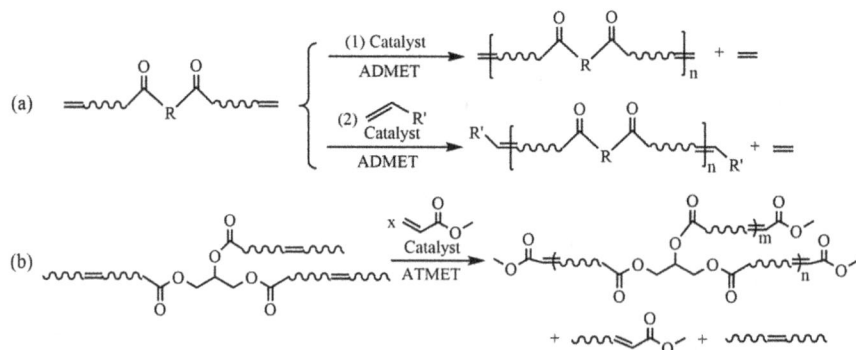

Figure 4.30: General catalytic metathesis reactions for polymers:
(a) ADMET polymerization without (1) and with (2) a generic chain-stopper, (b) ATMET polymerization with a chain-stopper.

Estolides, which result from ester bond formation between a hydroxyl or olefinic group of one fatty acid and the terminal carboxyl group of a second fatty acid, are biodegradable and applicable as lubricants or in cosmetic products [153–155]. In addition, Meier and co-workers [35] prepared a set of 10-undecenoic acid derived monomers *via* thiol-ene addition for the synthesis of linear or hyperbranched polyesters. The obtained hyperbranched polyesters with molar masses (M_n) of 4–10 kDa were stable up to 300 °C and showed T_m values ranging from 50 to 70 °C. Other approaches, e.g. lipase-catalyzed synthesis approach, can also be employed to produce polyesters [156].

4.4.3 Other thermoplastic polymers

Mono-functional (meth)acrylate monomers can be synthesized from plant oils derivatives such as fatty acids, fatty alcohol, and fatty acid methyl esters, as shown in Figure 4.31 [157–159]. Thermoplastic polymers with pendant fatty chains can then be prepared by uncontrollable polymerization like free radical polymerization. For instance, Wang and co-workers [160, 161] synthesized high-purity methyl methacrylate monomers from soybean oil *via* amidation and esterification, then prepared thermoplastic polymers *via* free radical polymerization, as depicted in Figure 4.32.

Controllable/living polymerization methods, such as atom transfer radical polymerization (ATRP) [95, 162] cationic ring opening polymerization (CROP) [163, 164], group transfer polymerization (GTP) [165, 166] as well as living anionic polymerization [167], can also be applied to control the polymerization of the above fatty acid-based monomers. Besides, Ring-opening metathesis polymerization (ROMP) of norbornenyl-functionalized fatty acid based monomers was studied by Xia et al. and Mutlu et al. [168, 169], as shown in Figure 4.33. Polymers with controlled molecular weight and architecture can be obtained by the controllable/living polymerization.

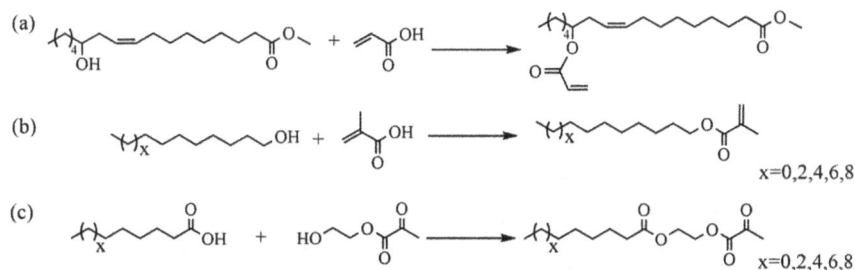

Figure 4.31: (a) Fatty acid methyl ester-derived (meth)acrylate monomers; (b) Fatty alcohol-derived (meth)acrylate monomers; (c) Fatty acid -derived (meth)acrylate monomers.

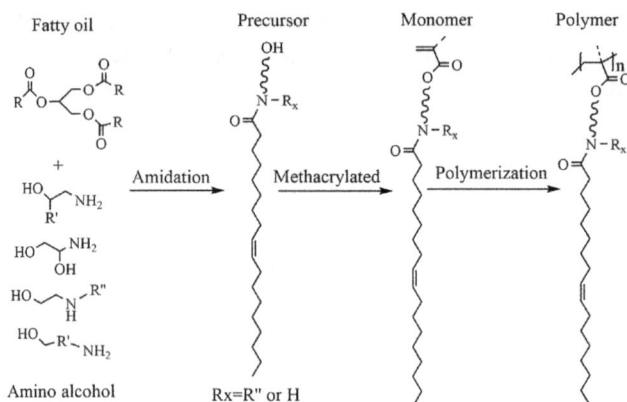

Figure 4.32: Synthesis of (methyl) methacrylate monomers from soybean oil [160, 161].

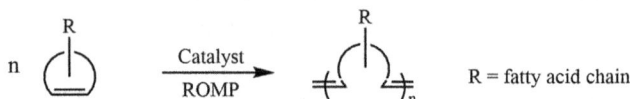

Figure 4.33: A general example of ROMP.

4.5 Conclusions

This chapter summarizes the recent advances in the realm of the modification of plant oils and fatty acid derivatives as well as their use in the preparation of polymers. In the modification methods, conventional functionalization, olefin metathesis, transition-metal-catalyzed addition, and thiol-ene addition were introduced or renewed. These modification methods as well as some novel polymerizations, e.g. ADMET, ATRP or CROP, directly or indirectly contributed to form novel polymers. Based on the advances of these modification and polymerization methods, the areas of traditional polymers (e.g. PUs) are greatly enlarged. Some resultant polymers demonstrate excellent ultimate properties that can compete with petroleum-based products. All of these advances will make plant oils as essential raw materials to be used in various applications in the future.

Author contributions: All the authors have accepted responsibility for the entire content of this submitted manuscript and approved submission.

Research funding: The authors are grateful to the financial support from National Natural Science Foundation of China (31822009 and 31770615), the Major State Research & Development Program of China (2016YFD0600802), and the Fundamental Research Funds of CAF (CAFYBB2020QA005).

Conflict of interest statement: The authors declare no conflicts of interest regarding this article.

References

1. Güner FS, Yağcı Y, Erciyes AT. Polymers from triglyceride oils. Prog Polym Sci 2006;31:633–70.
2. Meier MAR, Metzger JO, Schubert US. Plant oil renewable resources as green alternatives in polymer science. Chem Soc Rev 2007;36:1788–802.
3. Lligadas G, Ronda JC, Galia M, Cadiz V. Plant oils as platform chemicals for polyurethane synthesis: current state-of-the-art. Biomacromolecules 2010;11:2825–35.
4. Biermann U, Bornscheuer U, Meier MAR, Metzger JO, Schäfer HJ. Oils and fats as renewable raw materials in chemistry. Angew Chem Int Ed 2011;50:3854–71.
5. de Espinosa LM, Meier MAR. Plant oils: the perfect renewable resource for polymer science?! Eur Polym J 2011;47:837–52.
6. Metzger JO, Bornscheuer U. Lipids as renewable resources: current state of chemical and biotechnological conversion and diversification. Appl Microbiol Biotechnol 2006;71:13–22.
7. https://www.fas.usda.gov/data/oilseeds-world-markets-and-trade, 2020.
8. Wool RP, Sun XS. Bio-based polymers and composites. Amesterdam: Elsevier; 2005.
9. Belgacem MN, Gandini A. Materials from vegetable oils: major sources, properties and applications. In: Belgacem MN, Gandini A, editors. Monomers, polymers and composites from renewable resources. Amsterdam: Elsevier; 2008:39–66 pp.
10. Raquez JM, Deléglise M, Lacrampe MF, Krawczak P. Thermosetting (bio)materials derived from renewable resources: a critical review. Prog Polym Sci 2010;35:487–509.
11. Melero JA, Iglesias J, Morales G. Heterogeneous acid catalysts for biodiesel production: current status and future challenges. Green Chem 2009;11:1285–308.
12. Lestari S, Maki-Arvela P, Beltramini J, Lu GQM, Murzin DY. Transforming triglycerides and fatty acids into biofuels. ChemSusChem 2009;2:1109–19.
13. Liu K. Soybeans: chemistry, technology, and utilization. New York: Chapman and Hall; 1997.
14. Carlsson AS. Plant oils as feedstock alternatives to petroleum – a short survey of potential oil crop platforms. Biochimie 2009;91:665–70.
15. Dyer JM, Stymne S, Green AG, Carlsson AS. High-value oils from plants. Plant J 2008;54:640–55.
16. Shahidi F. Bailey's industrial oil and fat products, 6th ed. New York: Wiley & Sons; 2005.
17. Dam PBV, Mittelmeijer MC, Boelhouwer C. Metathesis of unsaturated fatty acid esters by a homogeneous tungsten hexachloride–tetramethyltin catalyst. J Chem Soc, Chem Commun 1972; 22:1221–2.
18. Boelhouwer C, Mol JC. Metathesis of fatty acid esters. J Am Oil Chem Soc 1984;61:425–30.
19. Schwab P, France MB, Ziller JW, Grubbs RH. A series of well-defined metathesis catalysts-synthesis of [RuCl2(CHR')(PR3)2] and its reactions. Angew Chem Int Ed Engl 1995;34:2039–41.
20. Schwab P, Grubbs RH, Ziller JW. Synthesis and applications of RuCl2(CHR')(PR3)2: the influence of the alkylidene moiety on metathesis activity. J Am Chem Soc 1996;118:100–10.
21. Scholl M, Ding S, Lee CW, Grubbs RH. Synthesis and activity of a new generation of ruthenium-based olefin metathesis catalysts coordinated with 1,3-dimesityl-4,5-dihydroimidazol-2-ylidene ligands§. Org Lett 1999;1:953–6.
22. Scholl M, Trnka TM, Morgan JP, Grubbs RH. Increased ring closing metathesis activity of ruthenium-based olefin metathesis catalysts coordinated with imidazolin-2-ylidene ligands. Tetrahedron Lett 1999;40:2247–50.

23. Garber SB, Kingsbury JS, Gray BL, Hoveyda AH. Efficient and recyclable monomeric and dendritic Ru-based metathesis catalysts. J Am Chem Soc 2000;122:8168–79.

24. Grela K, Harutyunyan S, Michrowska A. A highly efficient ruthenium catalyst for metathesis reactions. Angew Chem Int Ed 2002;41:4038–40.

25. Forman GS, Bellabarba RM, Tooze RP, Slawin AMZ, Karch R, Winde R. Metathesis of renewable unsaturated fatty acid esters catalysed by a phoban-indenylidene ruthenium catalyst. J Organomet Chem 2006;691:5513–6.

26. Warwel S, Bruse F, Demes C, Kunz M, Klaas MRG. Polymers and surfactants on the basis of renewable resources. Chemosphere 2001;43:39–48.

27. Malacea R, Fischmeister C, Bruneau C, Dubois JL, Couturier JL, Dixneuf PH. Renewable materials as precursors of linear nitrile-acid derivatives viacross-metathesis of fatty esters and acids with acrylonitrile and fumaronitrile. Green Chem 2009;11:152–5.

28. Bruneau C, Fischmeister C, Miao XW, Malacea R, Dixneuf PH. Cross-metathesis with acrylonitrile and applications to fatty acid derivatives. Eur J Lipid Sci Technol 2010;112:3–9.

29. Beller M. A personal view on homogeneous catalysis and its perspectives for the use of renewables. Eur J Lipid Sci Technol 2008;110:789–96.

30. Behr A, Gomes JP. The refinement of renewable resources: new important derivatives of fatty acids and glycerol. Eur J Lipid Sci Technol 2010;112:31–50.

31. Quinzler D, Mecking S. Linear semicrystalline polyesters from fatty acids by complete feedstock molecule utilization. Angew Chem Int Ed 2010;49:4306–8.

32. Cole-Hamilton DJ. Nature's polyethylene. Angew Chem Int Ed 2010;49:8564–6.

33. Hoyle CE, Bowman CN. Thiol-ene click chemistry. Angew Chem Int Ed 2010;49:1540–73.

34. Firdaus M, Montero de Espinosa L, Meier MAR. Terpene-based renewable monomers and polymers via thiol-ene additions. Macromolecules 2011;44:7253–62.

35. Turunc O, Meier MAR. Fatty acid derived monomers and related polymers via thiol-ene (click) additions. Macromol Rapid Commun 2010;31:1822–6.

36. Dondoni A. The emergence of thiol-ene coupling as a click process for materials and bioorganic chemistry. Angew Chem Int Ed 2008;47:8995–7.

37. Stemmelen M, Pessel F, Lapinte V, Caillol S, Habas JP, Robin JJ. A fully biobased epoxy resin from vegetable oils: from the synthesis of the precursors by thiol-ene reaction to the study of the final material. J Polym Sci Polym Chem 2011;49:2434–44.

38. Campos LM, Meinel I, Guino RG, Schierhorn M, Gupta N, Stucky GD, et al. Highly versatile and robust materials for soft imprint lithography based on thiol-ene click chemistry. Adv Mater 2008; 20:3728–33.

39. Killops KL, Campos LM, Hawker CJ. Robust, efficient, and orthogonal synthesis of dendrimers via thiol-ene "click" chemistry. J Am Chem Soc 2008;130:5062–4.

40. Desroches M, Caillol S, Lapinte V, Auvergne RM, Boutevin B. Synthesis of biobased polyols by thiol–ene coupling from vegetable oils. Macromolecules 2011;44:2489–500.

41. Frankel EN. Review. Recent advances in lipid oxidation. J Sci Food Agric 1991;54:495–511.

42. Mallegol J, Gardette JL, Lemaire J. Long-term behavior of oil-based varnishes and paints I. Spectroscopic analysis of curing drying oils. J Am Oil Chem Soc 1999;76:967–76.

43. Situ Y, Huang H, Hu J, Fu H, Chen H. Research progress of polymer synthesis based on plant oil. Fine Chem 2006;23:1041–7.

44. Hewitt DH, Armitage F. Styrene copolymers in surface coatings. J Oil Colour Chem Assoc 1946;29: 109–28.

45. Li F, Larock RC. Synthesis, structure and properties of new tung oil–styrene–divinylbenzene copolymers prepared by thermal polymerization. Biomacromolecules 2003;4:1018–25.

46. Andjelkovic DD, Valverde M, Henna P, Li F, Larock RC. Novel thermosets prepared by cationic copolymerization of various vegetable oils-synthesis and their structure-property relationships. Polymer 2005;46:9674–85.
47. Chen JX, Soucek MD, Simonsick WJ, Celikay RW. Synthesis and photopolymerization of norbornyl epoxidized linseed oil. Polymer 2002;43:5379–89.
48. Liu Z, Sharma BK, Erhan SZ. From oligomers to molecular giants of soybean oil in supercritical carbon dioxide medium: 1. Preparation of polymers with lower molecular weight from soybean oil. Biomacromolecules 2007;8:233–9.
49. Liu Z, Doll KM, Holser RA. Boron trifluoride catalyzed ring-opening polymerization of epoxidized soybean oil in liquid carbon dioxide. Green Chem 2009;11:1774–80.
50. Haveren J, Oostveen EA, Miccichè F, Noordover BAJ, Koning CE, Benthem RATM, et al. Resins and additives for powder coatings and alkyd paints, based on renewable resources. J Coating Technol Res 2007;4:177–86.
51. Deligny P, Tuck N. Resins for surface coatings: alkyds and polyesters (vol. II). New York: Wiley; 2000.
52. Mighani H. Synthesis of thermally stable polyesters. In: Saleh HM, editor. Polyester, synthesis of thermal stable polyesters. London: InTech Sec 1–1; 2012:3–17 pp.
53. Qin Y, Jia J, Zhao L, Huang Z, Zhao S, Zhang G, et al. Synthesis and characterization of soybean oil based unsaturated polyester resin. In: Chen R, Sung WP, editors. Advanced materials research. Switzerland: Trans Tech Publications Ltd; 2012, vol 393–395:349–53 pp.
54. Can E, Wool RP, Kusefoglu S. Soybean and castor oil based monomers: synthesis and copolymerization with styrene. J Appl Polym Sci 2006;102:2433–47.
55. Lu J, Khot S, Wool RP. New sheet molding compound resins from soybean oil. I. Synthesis and characterization. Polymer 2005;46:71–80.
56. Khot SN, Lascala JJ, Can E, Morye SS, Williams GI, Palmese GR, et al. Development and application of triglyceride-based polymers and composites. J Appl Polym Sci 2001;82:703–23.
57. Liu CG, Shang QQ, Jia PY, Dai Y, Zhou YH, Liu ZS. Tung oil-based unsaturated co-ester macromonomer for thermosetting polymers: synergetic synthesis and copolymerization with styrene. ACS Sustainable Chem Eng 2016;4:3437–49.
58. Liu CG, Dai Y, Hu Y, Shang QQ, Feng GD, Zhou J, et al. Highly functional unsaturated ester macromonomer derived from soybean oil: synthesis and copolymerization with styrene. ACS Sustainable Chem Eng 2016;4:4208–16.
59. Mehta G, Mohanty AK, Misra M, Drzal LT. Biobased resin as a toughening agent for biocomposites. Green Chem 2004;6:254–8.
60. Miyagawa H, Mohanty AK, Burgueno R, Drzal LT, Misra M. Novel biobased resins from blends of functionalized soybean oil and unsaturated polyester resin. J Polym Sci B Polym Phys 2007;45: 698–704.
61. Miyagawa H, Mohanty AK, Burgueno R, Drzal LT, Misra M. Development of biobased unsaturated polyester containing functionalized linseed oil. Ind Eng Chem Res 2006;45:1014–8.
62. Ghorui S, Bandyopadhyay NR, Ray D, Sengupta S, Kar T. Use of maleated castor oil as biomodifier in unsaturated polyester resin/fly ash composites. Ind Crop Prod 2011;34:893–9.
63. Liu CG, Lei W, Liu ZS, Hu LH, Zhou YH. Use of tung oil as a reactive toughening agent in dicyclopentadiene-terminated unsaturated polyester resins. Ind Crop Prod 2013;49:421–8.
64. Haq M, Burgueño R, Mohanty AK, Misra M. Bio-based polymer nanocomposites from UPE/EML blends and nanoclay: development, experimental characterization and limits to synergistic performance. Compos Appl Sci Manuf 2011;42:41–9.
65. Liu C, Dai Y, Wang C, Xie H, Zhou Y, Lin X, et al. Phase-separation dominating mechanical properties of a novel tung-oil-based thermosetting polymer. Ind Crop Prod 2013;43:677–83.

66. Cui JF, Tan WH, Liu CG, Zhou YH. Synthesis and property characterization of tung oil based dicyclopentadiene unsaterated polyester resin for coatings. Paint Coating Ind 2012;42:51–5.
67. La Scala JJ, Sands JM, Orlicki JA, Robinette EJ, Palmese GR. Fatty acid-based monomers as styrene replacements for liquid molding resins. Polymer 2004;45:7729–37.
68. Can E, La Scala JJ, Sands JM, Palmese GR. The synthesis of 9–10 dibromo stearic acid glycidyl methacrylate and its use in vinyl ester resins. J Appl Polym Sci 2007;106:3833–42.
69. Pan BF, Tan KF. Research and development of a new kind of humidity &heat resistant resin modified with α-linolenic acid. Fiber Compos (in Chinese) 2005;3:3–5.
70. Zhang YH, Yuan L, Guan QB, Liang GZ, Gu AJ. Developing self-healable and antibacterial polyacrylate coatings with high mechanical strength through crosslinking by multi-amine hyperbranched polysiloxane via dynamic vinylogous urethane. J Mater Chem 2017;5:16889–97.
71. Al-Mulla EAJ, Suhail AH, Aowda SA. New biopolymer nanocomposites based on epoxidized soybean oil plasticized poly(lactic acid)/fatty nitrogen compounds modified clay: preparation and characterization. Ind Crop Prod 2011;33:23–9.
72. La Scala J, Wool RP. Effect of FA composition on epoxidation kinetics of TAG. J Am Oil Chem Soc 2002;79:373–8.
73. Çolak S, Küsefoğlu SH. Synthesis and interfacial properties of aminosilane derivative of acrylated epoxidized soybean oil. J Appl Polym Sci 2007;104:2244–53.
74. Köckritz A, Martin A. Oxidation of unsaturated fatty acid derivatives and vegetable oils. Eur J Lipid Sci Technol 2008;110:812–24.
75. Kim JR, Sharma S. The development and comparison of bio-thermoset plastics from epoxidized plant oils. Ind Crop Prod 2012;36:485–99.
76. Vlek T, Petrovi ZS. Optimization of the chemoenzymatic epoxidation of soybean oil. J Am Oil Chem Soc 2006;83:247–52.
77. Uyama H, Kuwabara M, Tsujimoto T, Kobayashi S. Enzymatic synthesis and curing of biodegradable epoxide-containing polyesters from renewable resources. Biomacromolecules 2003;4:211–5.
78. Park S-J, Jin F-L, Lee J-R. Synthesis and thermal properties of epoxidized vegetable oil. Macromol Rapid Commun 2004;25:724–7.
79. Lligadas G, Ronda JC, Galià M, Cádiz V. Development of novel phosphorus-containing epoxy resins from renewable resources. J Polym Sci Polym Chem 2006;44:6717–27.
80. Stemmelen M, Lapinte V, Habas J-P, Robin J-J. Plant oil-based epoxy resins from fatty diamines and epoxidized vegetable oil. Eur Polym J 2015;68:536–45.
81. Altuna FI, Pettarin V, Williams RJJ. Self-healable polymer networks based on the cross-linking of epoxidised soybean oil by an aqueous citric acid solution. Green Chem 2013;15:3360–6.
82. Chen J, Yu Y, Zhu B, Han J, Liu C, Liu C, et al. Synthesis of biocompatible and highly fluorescent N-doped silicon quantum dots from wheat straw and ionic liquids for heavy metal detection and cell imaging. Sci Total Environ 2020;765:142754.
83. Omonov TS, Curtis JM. Biobased epoxy resin from canola oil. J Appl Polym Sci 2014;131:40142.
84. Carbonell-Verdu A, Bernardi L, Garcia-Garcia D, Sanchez-Nacher L, Balart R. Development of environmentally friendly composite matrices from epoxidized cottonseed oil. Eur Polym J 2015;63: 1–10.
85. Crivello JV, Narayan R, Sternstein SS. Fabrication and mechanical characterization of glass fiber reinforced UV-cured composites from epoxidized vegetable oils. J Appl Polym Sci 1997;64: 2073–87.
86. Liang G, Chandrashekhara K. Cure kinetics and rheology characterization of soy-based epoxy resin system. J Appl Polym Sci 2006;102:3168–80.
87. Park SJ, Jin FL, Lee JR, Shin JS. Cationic polymerization and physicochemical properties of a biobased epoxy resin initiated by thermally latent catalysts. Eur Polym J 2005;41:231–7.

88. Gerbase AE, Petzhold CL, Costa APO. Dynamic mechanical and thermal behavior of epoxy resins based on soybean oil. J Am Oil Chem Soc 2002;79:797–802.

89. Park SJ, Jin FL, Lee JR. Effect of biodegradable epoxidized castor oil on physicochemical and mechanical properties of epoxy resins. Macromol Chem Phys 2004;205:2048–54.

90. Miyagawa H, Misra M, Drzal LT, Mohanty A. Biobased epoxy/layered silicate nanocomposites: thermophysical properties and fracture behavior evaluation. J Polym Environ 2005;13:87–96.

91. Miyagawa H, Mohanty A, Drzal LT, Misra M. Effect of clay and alumina-nanowhisker reinforcements on the mechanical properties of nanocomposites from biobased epoxy: a comparative study. Ind Eng Chem Res 2004;43:7001–9.

92. Trumbo DL, Otto JT. Epoxidized fatty acid-derived oxazoline in thermoset coatings. J Coating Technol Res 2008;5:107–11.

93. Lligadas G, Ronda JC, Galia M, Cadiz V. Synthesis and properties of thermosetting polymers from a phosphorous-containing fatty acid derivative. J Polym Sci Polym Chem 2006;44:5630–44.

94. Samuelsson J, Sundell PE, Johansson M. Synthesis and polymerization of a radiation curable hyperbranched resin based on epoxy functional fatty acids. Prog Org Coating 2004;50:193–8.

95. Cayli G, Meier MAR. Polymers from renewable resources: bulk ATRP of fatty alcohol-derived methacrylates. Eur J Lipid Sci Technol 2008;110:853–9.

96. Babb DA. Polyurethanes from renewable resources. In: Rieger B, Kunkel A, Coates GW, Reichardt R, Dinjus E, Zevaco TA, editors. Synthetic biodegradable polymers. Berlin: Springer; 2012, vol 245:315–60 pp.

97. Caillol S, Desroches M, Boutevin G, Loubat C, Auvergne R, Boutevin B. Synthesis of new polyester polyols from epoxidized vegetable oils and biobased acids. Eur J Lipid Sci Technol 2012;114: 1447–59.

98. Zlatanic A, Lava C, Zhang W, Petrovic ZS. Effect of structure on properties of polyols and polyurethanes based on different vegetable oils. J Polym Sci B Polym Phys 2004;42:809–19.

99. Lligadas G, Ronda JC, Galia M, Biermann U, Metzger JO. Synthesis and characterization of polyurethanes from epoxidized methyl oleate based polyether polyols as renewable resources. J Polym Sci Polym Chem 2006;44:634–45.

100. Zhang C, Li Y, Chen R, Kessler MR. Polyurethanes from solvent-free vegetable oil-based polyols. ACS Sustainable Chem Eng 2014;2:2465–76.

101. Wang C-S, Yang L-T, Ni B-L, Shi G. Polyurethane networks from different soy-based polyols by the ring opening of epoxidized soybean oil with methanol, glycol, and 1,2-propanediol. J Appl Polym Sci 2009;114:125–31.

102. Wu Q, Hu Y, Tang J, Zhang J, Wang C, Shang Q, et al. High-performance soybean-oil-based epoxy acrylate resins: "green" synthesis and application in UV-curable coatings. ACS Sustainable Chem Eng 2018;6:8340–9.

103. Zhang C, Madbouly SA, Kessler MR. Biobased polyurethanes prepared from different vegetable oils. ACS Appl Mater Interfaces 2015;7:1226–33.

104. Zhang C, Xia Y, Chen R, Huh S, Johnston PA, Kessler MR. Soy-castor oil based polyols prepared using a solvent-free and catalyst-free method and polyurethanes therefrom. Green Chem 2013;15: 1477–84.

105. Miao S, Zhang S, Su Z, Wang P. Synthesis of bio-based polyurethanes from epoxidized soybean oil and isopropanolamine. J Appl Polym Sci 2013;127:1929–36.

106. Bakhshi H, Yeganeh H, Mehdipour-Ataei S, Shokrgozar MA, Yari A, Saeedi-Eslami SN. Synthesis and characterization of antibacterial polyurethane coatings from quaternary ammonium salts functionalized soybean oil based polyols. Mater Sci Eng C Mater Biol Appl 2013;33:153–64.

107. Ghasemlou M, Daver F, Ivanova EP, Adhikari B. Polyurethanes from seed oil-based polyols: a review of synthesis, mechanical and thermal properties. Ind Crop Prod 2019;142:111841.

108. Pfister DP, Xia Y, Larock RC. Recent advances in vegetable oil-based polyurethanes. ChemSusChem 2011;4:703–17.
109. Guo A, Demydov D, Zhang W, Petrovic ZS. Polyols and polyurethanes from hydroformylation of soybean oil. J Polym Environ 2002;10:49–52.
110. Petrovic ZS, Guo A, Zhang W. Structure and properties of polyurethanes based on halogenated and nonhalogenated soy–polyols. J Polym Sci Polym Chem 2000;38:4062–9.
111. Omonov TS, Kharraz E, Curtis JM. Ozonolysis of canola oil: a study of product yields and ozonolysis kinetics in different solvent systems. J Am Oil Chem Soc 2011;88:689–705.
112. Petrovic ZS. Polyurethanes from vegetable oils. Polym Rev 2008;48:109–55.
113. de Souza VHR, Silva SA, Ramos LP, Zawadzki SF. Synthesis and characterization of polyols derived from corn oil by epoxidation and ozonolysis. J Am Oil Chem Soc 2012;89:1723–31.
114. Narine SS, Kong X. Physical properties of polyurethane plastic sheets produced from polyols from canola oil. Biomacromolecules 2007;8:2203–9.
115. Tran P, Graiver D, Narayan R. Ozone-mediated polyol synthesis from soybean oil. J Am Oil Chem Soc 2005;82:653–9.
116. Kong X, Liu G, Curtis JM. Novel polyurethane produced from canola oil based poly(ether ester) polyols: synthesis, characterization and properties. Eur Polym J 2012;48:2097–106.
117. Das B, Chattopadhyay P, Mandal M, Voit B, Karak N. Bio-based biodegradable and biocompatible hyperbranched polyurethane: a scaffold for tissue engineering. Macromol Biosci 2013;13:126–39.
118. Rajput SD, Mahulikar PP, Gite VV. Biobased dimer fatty acid containing two pack polyurethane for wood finished coatings. Prog Org Coating 2014;77:38–46.
119. Xia Y, Larock RC. Vegetable oil-based polymeric materials: synthesis, properties, and applications. Green Chem 2010;12:1893–909.
120. Ionescu M, Radojčić D, Wan X, Petrović ZS, Upshaw TA. Functionalized vegetable oils as precursors for polymers by thiol-ene reaction. Eur Polym J 2015;67:439–48.
121. Alagi P, Choi YJ, Seog J, Hong SC. Efficient and quantitative chemical transformation of vegetable oils to polyols through a thiol-ene reaction for thermoplastic polyurethanes. Ind Crop Prod 2016;87:78–88.
122. Guo A, Cho YJ, Petrovic ZS. Structure and properties of halogenated and nonhalogenated soy-based polyols. J Polym Sci Polym Chem 2000;38:3900–10.
123. Guo A, Javni I, Petrovic Z. Rigid polyurethane foams based on soybean oil. J Appl Polym Sci 2000;77:467–73.
124. Rio ED, Lligadas G, Ronda JC, Galià M, Meier MAR, Cádiz V. Polyurethanes from polyols obtained by ADMET polymerization of a castor oil-based diene: characterization and shape memory properties. J Polym Sci Polym Chem 2011;49:518–25.
125. Bahr M, Bitto A, Mulhaupt R. Cyclic limonene dicarbonate as a new monomer for non-isocyanate oligo- and polyurethanes (NIPU) based upon terpenes. Green Chem 2012;14:1447–54.
126. Pyo SH, Hatti-Kaul R. Selective, green synthesis of six-membered cyclic carbonates by lipase-catalyzed chemospecific transesterification of diols with dimethyl carbonate. Adv Synth Catal 2012;354:797–802.
127. Ma ZW, Hong Y, Nelson DM, Pichamuthu JE, Leeson CE, Wagner WR. Biodegradable polyurethane ureas with variable polyester or polycarbonate soft segments: effects of crystallinity, molecular weight, and composition on mechanical properties. Biomacromolecules 2011;12:3265–74.
128. Bahr M, Mulhaupt R. Linseed and soybean oil-based polyurethanes prepared via the non-isocyanate route and catalytic carbon dioxide conversion. Green Chem 2012;14:483–9.
129. Li Z, Zhao Y, Yan S, Wang X, Kang M, Wang J, et al. Catalytic synthesis of carbonated soybean oil. Catal Lett 2008;123:246–51.

130. Guan J, Song Y, Lin Y, Yin X, Zuo M, Zhao Y, et al. Progress in study of non-isocyanate polyurethane. Ind Eng Chem Res 2011;50:6517–27.

131. Javni I, Hong DP, Petrović ZS. Soy-based polyurethanes by nonisocyanate route. J Appl Polym Sci 2008;108:3867–75.

132. Javni I, Hong DP, Petrović ZS. Polyurethanes from soybean oil, aromatic, and cycloaliphatic diamines by nonisocyanate route. J Appl Polym Sci 2013;128:566–71.

133. Lee A, Deng Y. Green polyurethane from lignin and soybean oil through non-isocyanate reactions. Eur Polym J 2015;63:67–73.

134. Foltran S, Maisonneuve L, Cloutet E, Gadenne B, Alfos C, Tassaing T, et al. Solubility in CO_2 and swelling studies by in situ IR spectroscopy of vegetable-based epoxidized oils as polyurethane precursors. Polym Chem 2012;3:525–32.

135. Deepa P, Jayakannan M. Polyurethane-oligo(phenylenevinylene) random copolymers: π-conjugated pores, vesicles, and nanospheres via solvent-induced self-organization. J Polym Sci Polym Chem 2008;46:5897–915.

136. Deepa P, Jayakannan M. Solvent-free and nonisocyanate melt transurethane reaction for aliphatic polyurethanes and mechanistic aspects. J Polym Sci Polym Chem 2008;46:2445–58.

137. Palaskar DV, Boyer A, Cloutet E, Alfos C, Cramail H. Synthesis of biobased polyurethane from oleic and ricinoleic acids as the renewable resources via the AB-type self-condensation approach. Biomacromolecules 2010;11:1202–11.

138. More AS, Gadenne B, Alfos C, Cramail H. AB type polyaddition route to thermoplastic polyurethanes from fatty acid derivatives. Polym Chem 2012;3:1594–605.

139. Wang C, Chen X, Chen J, Liu C, Xie H, Cheng R. Synthesis and characterization of novel polyurethane acrylates based on soy polyols. J Appl Polym Sci 2011;122:2449–55.

140. Meier MAR. Plant-oil-based polyamides and polyurethanes: toward sustainable nitrogen-containing thermoplastic materials. Macromol Rapid Commun 2019;40:e1800524.

141. Oldring PKT, Turk N. Polyamides. In: Resins for surface coatings. New York: Wiley; 2000, vol III: 131–97 pp.

142. Mutlu H, Meier MAR. Castor oil as a renewable resource for the chemical industry. Eur J Lipid Sci Technol 2010;112:10–30.

143. Deng YL, Fan XD, Waterhouse J. Synthesis and characterization of soy-based copolyamides with different?-amino acids. J Appl Polym Sci 1999;73:1081–8.

144. Genas M. Rilsan (polyamid 11), synthese und eigenschaften. Angew Chem 1962;74:535–40.

145. Ameh Abel G, Oliver Nguyen K, Viamajala S, Varanasi S, Yamamoto K. Cross-metathesis approach to produce precursors of nylon 12 and nylon 13 from microalgae. RSC Adv 2014;4:55622–8.

146. Ternel J, Couturier J-L, Dubois J-L, Carpentier J-F. Catalyzed tandem isomerization/hydroformylation of the bio-sourced 10-undecenenitrile: selective and productive catalysts for production of polyamide-12 precursor. Adv Synth Catal 2013;355:3191–204.

147. Türünç O, Firdaus M, Klein G, Meier MAR. Fatty acid derived renewable polyamides via thiol–ene additions. Green Chem 2012;14:2577–83.

148. Rybak A, Meier MAR. Cross-metathesis of fatty acid derivatives with methyl acrylate: renewable raw materials for the chemical industry. Green Chem 2007;9:1356–61.

149. Mutlu H, Meier MAR. Unsaturated PA X,20 from renewable resources via metathesis and catalytic amidation. Macromol Chem Phys 2009;210:1019–25.

150. Mutlu H, de Espinosa LM, Meier MAR. Acyclic dienemetathesis: a versatile tool for the construction of defined polymer architectures. Chem Soc Rev 2011;40:1404–45.

151. Rybak A, Meier MAR. Acyclic diene metathesis with a monomer from renewable resources: control of molecular weight and one-step preparation of block copolymers. ChemSusChem 2008;1:542–7.

152. Biermann U, Metzger JO, Meier MAR. Acyclic triene metathesis oligo- and polymerization of high oleic sun flower oil. Macromol Chem Phys 2010;211:854–62.

153. Cermak SC, Isbell TA, Evangelista RL, Johnson BL. Synthesis and physical properties of petroselinic based estolide esters. Ind Crop Prod 2011;33:132–9.
154. Lin JT, Chen GQ. Identification of minor acylglycerols less polar than triricinolein in castor oil by mass spectrometry. J Am Oil Chem Soc 2012;89:1773–84.
155. Zhang HX, Olson DJH, Van D, Purves RW, Smith MA. Rapid identification of triacylglycerol-estolides in plant and fungal oils. Ind Crop Prod 2012;37:186–94.
156. Tsujimoto T, Uyama H, Kobayashi S. Enzymatic synthesis and curing of biodegradable crosslinkable polyesters. Macromol Biosci 2002;2:329–35.
157. Brister EH, Jarrett W, Thames SF. Castor-acrylated monomer ^{1}H- and ^{13}C-nuclear magnetic resonance spectral assignments. J Appl Polym Sci 2001;82:1850–4.
158. Maiti B, De P. RAFT polymerization of fatty acid containing monomers: controlled synthesis of polymers from renewable resources. RSC Adv 2013;3:24983–90.
159. Maiti B, Kumar S, De P. Controlled RAFT synthesis of side-chain oleic acid containing polymers and their post-polymerization functionalization. RSC Adv 2014;4:56415–23.
160. Wang ZK, Yuan L, Trenor NM. Sustainable thermoplastic elastomers derived from plant oil and their "click-coupling" via TAD chemistry. Green Chem 2015;17:3806–18.
161. Yuan L, Wang Z, Trenor NM, Tang C. Amidation of triglycerides by amino alcohols and their impact on plant oil-derived polymers. Polym Chem 2016;7:2790–8.
162. Raghunadh V, Baskaran D, Sivaram S. Efficiency of ligands in atom transfer radical polymerization of lauryl methacrylate and block copolymerization with methyl methacrylate. Polymer 2004;45:3149–55.
163. Hoogenboom R, Schubert US. Microwave-assisted cationic ring-opening polymerization of a soy-based 2-oxazoline monomer. Green Chem 2006;8:895–9.
164. Huang HY, Hoogenboom R, Leenen MAM, Guillet P, Jonas AM, Schubert US, et al. Solvent-induced morphological transition in core-cross-linked block copolymer micelles. J Am Chem Soc 2006; 128:3784–8.
165. Sogah DY, Hertler WR, Webster OW, Cohen GM. Group transfer polymerization – polymerization of acrylic monomers. Macromolecules 1987;20:1473–88.
166. Sannigrahi B, Wadgaonkar PP, Sehra JC, Sivaram S. Copolymerization of methyl methacrylate with lauryl methacrylate using group transfer polymerization. J Polym Sci Polym Chem 1997;35: 1999–2007.
167. Raghunadh V, Baskaran D, Sivaram S. Living anionic polymerization of lauryl methacrylate and synthesis of block copolymers with methyl methacrylate. J Polym Sci Polym Chem 2004;42:875–82.
168. Xia Y, Lu YS, Larock RC. Ring-opening metathesis polymerization (ROMP) of norbornenyl-functionalized fatty alcohols. Polymer 2010;51:53–61.
169. Mutlu H, Meier MAR. Ring-opening metathesis polymerization of fatty acid derived monomers. J Polym Sci Polym Chem 2010;48:5899–906.

Xing Zhou*, Xin Zhang, Pu Mengyuan, Xinyu He and
Chaoqun Zhang*

5 Bio-based polyurethane aqueous dispersions

Abstract: With the advances of green chemistry and nanoscience, the synthesis of green, homogenous bio-based waterborne polyurethane (WPU) dispersions with high performance have gained great attention. The presented chapter deals with the recent synthesis of waterborne polyurethane with the biomass, especially the vegetable oils including castor oil, soybean oil, sunflower oil, linseed oil, jatropha oil, and palm oil, etc. Meanwhile, the other biomasses, such as cellulose, starch, lignin, chitosan, etc., have also been illustrated with the significant application in preparing polyurethane dispersions. The idea was to highlight the main vegetable oil-based polyols, and the isocyanate, diols as chain extenders, which have supplied a class of raw materials in WPU. The conversion of biomasses into active chemical agents, which can be used in synthesis of WPU, has been discussed in detail. The main mechanisms and methods are also presented. It is suggested that the epoxide ring opening method is still the main route to transform vegetable oils to polyols. Furthermore, the nonisocyanate WPU may be one of the main trends for development of WPU using biomasses, especially the abundant vegetable oils.

Keywords: biomass; structure; vegetable oil; waterborne polyurethane dispersion.

5.1 Introduction

Aqueous polymer dispersions are important materials consisting of very small polymer particles dispersed in water with a variety of application in industrial processes. Their particles are barely visible macroscopically when finally processed and providing the function for which they were selected, appearing as milky fluids with slightly blue or red light occasionally in various kinds of micromorphologies [1]. The essential effects achieved by polymer dispersions are mainly protecting, binding, and finishing. They are commonly used to protect metal, wood, and leather against water and microorganisms, and are used as binders for pigments, fillers, and fibers and to finish the surfaces of metal, wood or paper, etc. In general the term "dispersion" characterizes a

*Corresponding authors: Xing Zhou, PhD, Faculty of Printing, Packaging Engineering and Digital Media Technology, Xi'an University of Technology, Xi'an 710048, P. R. China; and School of Materials Science and Engineering, Xi'an University of Technology, Xi'an 710048, P. R. China; and Chaoqun Zhang, College of Materials and Energy, South China Agricultural University, Guangzhou 510642, P. R. China, E-mail: zdxnlxaut@163.com (X. Zhou), zhangcq@scau.edu.cn (C. Zhang)
Xin Zhang, Pu Mengyuan and Xinyu He, Faculty of Printing, Packaging Engineering and Digital Media Technology, Xi'an University of Technology, Xi'an 710048, P. R. China

This article has previously been published in the journal Physical Sciences Reviews. Please cite as: X. Zhou, X. Zhang, P. Mengyuan, X. He, and C. Zhang "Bio-based polyurethane aqueous dispersions" *Physical Sciences Reviews* [Online] 2021, 6. DOI: 10.1515/psr-2020-0075 | https://doi.org/10.1515/9781501521942-005

two phase system consisting of finely dispersed solid particles in a continuous liquid phase [2]. Polyurethane (PU) dispersed in aqueous under the assistant of hydrophilic chain extender (such as dimethylol propionic acid) can be assigned to the colloid dispersion. It is a kind of multipurpose resin with various product forms, is of great interest with regard to the increased use of multifunctional and adaptable materials [3–10]. Polyurethanes differ from other synthetic resins in raw materials, formula combination, product forms, and application fields. Indeed the waterborne polyurethane (WPU) dispersion is perhaps the most significant example of useful polymers with the combination of environment-friendly and high performance in aqueous, even with the natural raw materials of biomass [1, 11–14]. It has been widely used as coatings for various fibers, adhesives for alternative substrates, binders in printing inks, primers for metals, caulking materials, emulsion polymerization media for different monomers, paint additives, defoamers, associate thickeners, pigment pastes, and textile dyes and so on in the fields of transportation, construction, machinery, electronic equipment, furniture, clothing, textile, printing and packaging, food processing, biomedical, etc. [15–17]. Polyurethanes are synthesized by polyols (diol), polyisocyanate (diisocyanate), and other additives via step-growth or emulsion polymerization process. Organic polyisocyanate and oligomer polyols are the main raw material of synthetic polyurethane, accounting for more than 80% of the total weight, as shown in Figure 5.1 [1]. Polyurethane can be assigned as a kind of block copolymer from the aspect of heterogeneously molecular structure. Polyols consists of the soft segment of PU, whose characteristic is bearing plurality of hydroxyl groups, deciding the main properties. It is well-known that highly branched polyols may result in rigid PU with good heat and chemical resistance, less branched polyols give PU with good flexibility (at low temperature) and low chemical resistance. Similarly, low molecular weight polyols produce rigid PU while high molecular weight long chain polyols yield flexible PU [18].

However, all the raw materials of PU come from petroleum, which may be limited by the price, availability, and even the global political and institutional tendencies. Notably, the renewable resources, which can be the raw materials of polyols (diols) and even isocyanate, have been strongly linked to the synthesis of PU.

An excellent example of naturally occurring polyols is castor oil. Other vegetable oils (VO) can be used to prepare polyols by chemical transformations as well. They are not only relatively inexpensive, accessible, produced in large quantities, but also can avoid the limitation caused by petrochemical resources [19]. At present, Several

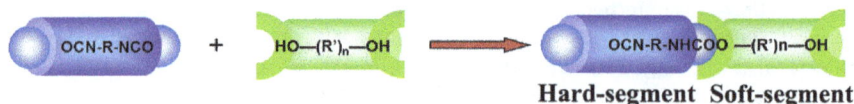

OCN-R-NCO + HO—(R')n—OH ⟶ OCN-R-NHCOO—(R')n—OH

Hard-segment Soft-segment

Figure 5.1: The structure of polyurethane of blocks copolymer [1] @Copyright 2020 Elsevier.

biorenewable materials are the focus as feedstocks for the preparation of bio-based polyols or diols as the raw materials of PU, including vegetable oil, nut shell, plant, lignin, sucrose and starch, poly(lactic acid), and poly(ε-caprolactone) [20, 21]. The bio-based polyols are used to react with isocyanate to give the chemical nature of PU with excellent thermal–physical and mechanical properties similar to those of traditional petroleum-based polyols. Meanwhile, some biomass materials can be used to modify PU to improve its all aspects of functionality. PU, especially WPU has been greatly employed in the development of bio-based applications. The intent of this chapter is to give a short overview of the WPU dispersions synthesized from different bio-based polyols from biomass. It should be noted that bio-based sourcing does not entail biodegradability. Bio-based sourcing could be measured by renewable carbon content owing to D6866ASTM standards [22].

5.2 WPU synthesized from biomass resources

5.2.1 Properties of polymer dispersion

To understand and prepare WPU dispersions, the properties of polymer dispersion should be introduced firstly. In this case, the selection of raw materials and polymerization methods, the controlling of the molecule chains, and the additives can be valid. It is well known that the aggregation state of polymer dispersion is commonly thermodynamically unstable, which means it is a dispersion rather than a solution. The phase separation and coagulation should be hindered significantly due to the large internal surface area of dispersed polymer particles. Thus, stabilization agent is usually essential. Driving force for the agglomeration of particles is the gain of energy by reducing the internal surface [2]. Lastly, in the aqueous phase, a polymer block and a substantially polymer-free water phase are yielded to form polymer particles with free-standing or non selfsupporting. Notably, the coagulation can be accelerated by salts (NaCl, KCl, etc.), acids (HCl, H_2SO_4), solvents, freezing, shear, etc. For example, the electrolytic stability of WPU dispersion can be detected by using aqueous NaCl solution [17]. Indisputably, polymer dispersion with high stability is usually attractive, which can be accomplished by providing ionic groups, such as the incorporation of ionic groups into the polyurethane chains by dimethylol propionic acid (DMPA) or the use of anionic or cationic surfactants in emulsion. To the morphology of the polymer dispersion, it normally consists of spherical particles in micro or nano dimensions, showing a translucent appearance. The polymer content, particle size, particle size distribution, electrolyte content, and the viscosity are affected by dissolved constituents in the aqueous phase. When the WPU dispersion is prepared via prepolymer method, the chain extending, dispersing, and emulsifying processes of the prepolymer by water is of great important [23–25]. Thus, in preparation of WPU dispersion, these properties are all key points that are necessary to be considered.

5.2.2 Vegetable oil-based polyols

Vegetable oils, which comes from different plants (soybean, palm, rapeseed, etc.,), are considered as the most important renewable resources and can be used as a reliable starting material for obtaining new products with a wide range of chemical conversion possibilities [22]. The chemical compositions of vegetable oils are triglyceride and three long chain fatty acids of different compositions depending on oil resource, and reactivity depends on carbon–carbon double bonds (C=C) and ester active sites [26]. In addition, some vegetable oils can be directly used as polyols with natural hydroxyl groups. But some oils do not have hydroxyl groups in them, which are introduced hydroxyl group into molecular structure by chemical modification of active groups such as double bond and ester group. The main method is through the hydrolysis of natural triglycerides in vegetable oils (esters of glycerol with fatty acids with C6 to C22 carbon atoms) [27]. The main vegetable oils and the structures are illustrated in Figure 5.2. The fatty acid composition and unsaturation of the common vegetable oil, which are the main indexes to distinguish vegetable oil, are listed in Table 5.1 [11, 28]. The resources are abundant and vegetable oils has actually become the most promising and attractive raw materials in WPU industry because of its rich, inherent biodegradability and low price.

The vegetable oil composition helps to improve the elasticity and mechanical properties of polyurethane. Whilst, vegetable oil-based WPU has performance defects, such as low chemical resistance, low heat resistance, and physicomechanical stability by reason of the presence of hydrophilic carboxylate group. Apart from the castor oil, the other main vegetable oils are hard to use directly for preparation of WPU because of

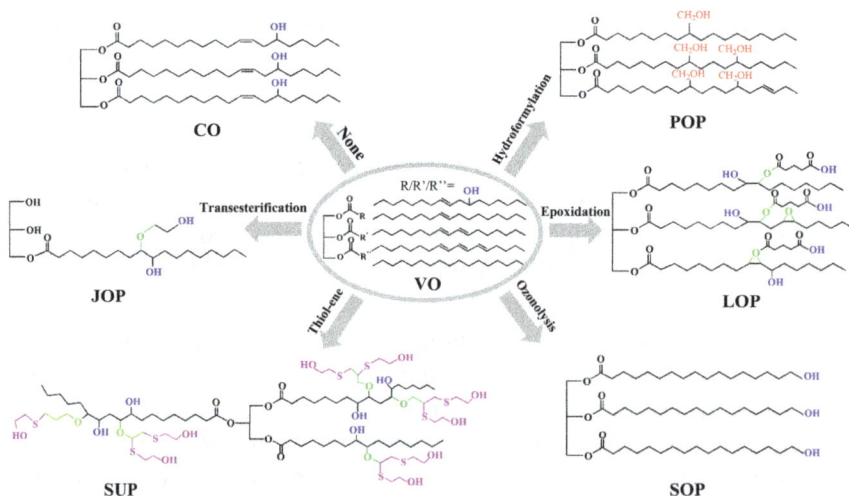

Figure 5.2: The molecule structure of the main used VO in polyurethane dispersions.

Table 5.1: Degree of unsaturation and composition of common vegetable oils [11, 28].

Oil	Double bonds[a]	Fatty acid composition (%)					Iodine value (mg/100 g)
		C16:0	C18:0	C18:1	C18:2	C18:3	
Castor	3.0	1.5	0.5	5.0	4.0	0.5	81–91
Coconut	–	9.8	3.0	6.9	2.2	–	–
Corn	4.5	10.9	2.0	25.4	59.6	1.2	118–128
Cottonseed	3.9	21.6	2.6	18.6	54.4	0.7	98–118
Linseed	6.6	6.6	5.5	3.5	19.1	15.3	>177
Olive	2.8	13.7	2.5	71.1	10.1	0.6	76–88
Palm	1.7	42.8	4.2	40.5	10.1	0.6	50–55
Palm kernel	–	8.8	2.4	13.6	1.1	–	–
Peanut	3.4	11.4	2.4	48.3	31.9	–	84–100
Rapeseed	3.8	4	2	56	26	10	100–115
Canola	3.9	4.1	1.8	60.9	21.0	8.8	100–115
Sesame	3.9	9	6	41	43	1	–
Soybean	4.6	11.0	4.0	23.4	53.3	7.8	123–139
Sunflower	4.7	5.2	2.7	37.2	53.8	1.0	125–140
Cottonseed	3.9	21.6	2.6	18.6	54.4	0.7	98–118

[a]Number of C–C double bonds per triglyceride

the absence of active hydroxyl groups. The C=C bonds and the ester functionality presented in triglycerides of the main vegetable oils allow for the introduction of such groups to react with isocyanate [28]. Notably, the considerable number of C=C bonds of greater than 2.5 double bonds per triglyceride are inclined to meet the reactions, as listed in Table 5.1. The main used vegetable oils in polyurethane dispersions are illustrated in Figure 5.2 including castor oil, soybean oil, sunflower oil, linseed oil, jatropha oil, and palm oil. This section will introduce the main reaction routes of the main vegetable oils to prepare polyols and their main structures.

Castor oil, in possession of carbon–carbon double and hydroxyl bonds naturally in their fatty acid chains, is a fatty acid glyceride in which the typical hydroxyl value is 163 mg KOH/g and iodine value of 81–91 mg/100 g. Castor oil can directly react with isocyanates to form urethane bonds as natural vegetable oil polyols without any prior modification [29]. It is the uniform distribution of hydroxyl groups on the castor oil chain that makes the synthesis of WPU with large degree of crosslinking, higher mechanical properties and thermal stability. Castor oil can be directly used in the manufacture of polyurethane adhesives, coatings, foam plastics, and can also be modified for use. Victoria et al. conducted studies of WPU synthesized by unmodified castor oil, isophorone diisocyanate (IPDI) and tartaric acid as internal emulsifiers, by replacing the tartaric acid, derived from the wine-grape process, and is more flexible and thermally stable in the high temperature range. Moreover, the special hydrophobic triglyceride structure of castor oil can also improve the mechanical properties, water resistance and chemical resistance of the prepared WPU films [30, 31]. The anionic

WPU prepared by Luong et al. exhibits the addition of castor oil as a soft segment in polymer chains significantly changed the mechanical properties and flexibility of thin films [32]. Additionally, by adding low molecular polyols such as ethylene glycol, glycerol, pentaerythritol, sorbitol to alcoholysis, and transesterification, castor oil derivatives polyols with different hydroxyl values and molecular weights can be generated. These derivatives have been widely used in PU coatings due to the increase of primary hydroxyl group content, which improves the reactivity of the products [26].

However, the absence of natural hydroxyl groups exists in most vegetable oil, such as soybean oil, sunflower oil, linseed oil, tung oil, jatropha oil, and palm oil. So chemical modification is used to require the introduction of hydroxyl groups on the C=C bonds reactive sites. A series of vegetable oil polyols have been developed by making full use of low-price vegetable oil and mainly intended as raw material for PU. They are generally yielded by epoxidation of the double bond followed by ring opening of the epoxide with an alcohol, which may generate secondary hydroxyl functions [33]. Currently, there are mainly five synthetic routes that take advantage of the properties of C=C bonds or esters in the molecular structure of vegetable oils to convert these cheap and renewable feedstocks into corresponding vegetable oil polyols. They are epoxidation/oxirane ring-opening reactions, thiol–ene reaction, hydroformylation/hydrogenation, ozonolysis, and transesterification/transamidation, respectively. The main reactions are as illustrated in Table 5.2. The reactions have been described by Desroches et al. in detail [22]. In this section, these routes will be introduced briefly to realize the functions of these reactions. The first route for functionalizing triglycerides processes in two procedures: epoxidation of C=C bonds of unsaturated fatty acids (reaction A, Table 5.2) followed by a nucleophilic ring opening of epoxide group (reaction B, Table 5.2). In this case, one or more hydroxyl groups can be added onto the fatty acid aliphatic chain because of the nucleophiles of the groups [34–38]. Epoxide groups can also be directly hydrogenated (reaction C, Table 5.2) from dihydrogene addition with Raney nickel catalyst, in which the obtained triglycerides exhibit only secondary hydroxyl group [39, 40]. Reaction D (Table 5.2) depicts the Oxidative cleavage of C=C bonds by ozone, leading to the aldehyde group [41]. In this case, an unsaturated vegetable oil with this reaction may yield a maximum of three aldehyde groups per triglyceride molecule (one terminal aldehyde on each fatty acid chain). Reaction E (Table 5.2) gives the hydroformylation of vegetable oils under the function of CO and H_2, leading to aldehyde functions that can be hydrogenated to yield primary alcohols [42]. This route can also be combined with reaction N (Table 5.2) to complete hydroaminomethylation of fatty esters with various amines. It proceeds in three procedures through hydroformylation (reaction E, Table 5.2) followed by condensation of the aldehyde function with the primary or the secondary amine, and then hydrogenation (reaction N, Table 5.2) yielding the secondary or the tertiary amine [43]. Reaction F (Table 5.2) shows an important functionalization method called thiol–ene coupling [44, 45]. It expands the path for the functional routs by using the rich electrons of double bonds of vegetable oil, which allows radical addition of various molecules proceeded through a

Table 5.2: Mains reactions used on vegetable oils [22].

Reaction	Name	Scheme
A	Epoxidation	$R_1R_2C=CR_3R_4 \longrightarrow$ epoxide $R_1R_2C(-O-)CR_3R_4$
B	Epoxide ring opening	epoxide $\xrightarrow{\text{NuH}}$ $R_1R_2C(OH)-CR_3R_4(Nu)$ + $R_1R_2C(Nu)-CR_3R_4(OH)$ NuH=RNH$_2$, RCO$_2$H, RSH, ROH
C	Epoxides hydrogenation	epoxide $\xrightarrow{\text{H}_2,\ \text{cat}}$ $R_1R_2C(OH)-CR_3R_4(H)$
D	Ozonolysis–hydrogenation	$R_1R_2C=CR_3R_4$ $\xrightarrow[\text{H}_2,\ \text{cat}]{\text{O}_3}$ $R_1R_2CH(OH)$ + $R_3R_4CH(OH)$
E	Hydroformylation	epoxide $\xrightarrow{\text{CO, H}_2,\ \text{cat}}$ product (HO, R$_1$R$_2$R$_3$R$_4$)
F	Thiol–ene coupling	$R_1R_2C=CR_3R_4$ $\xrightarrow{\text{R}_5\text{SH}}$ $R_1R_2CH-CR_3R_4(SR_5)$
G	Oxidation	$R_1R_2C=CR_3(CH_2R_4)$ $\longrightarrow$ $R_1R_2C(OOH)-CR_3=CR_4$
H	Reduction	$R_1C(=O)O-R_2$ $\longrightarrow$ R_1CH_2OH
I	(Trans)esterification	$R_1C(=O)O-R_2$ $\xrightarrow{\text{R}_3\text{OH}}$ $R_1C(=O)O-R_3$ + R_2OH
J	Amidification	$R_1C(=O)O-R_2$ $\xrightarrow{\text{R}_3\text{R}_4\text{NH}}$ $R_1C(=O)N(R_3)R_4$ + R_2OH
K	Diels aldercyclization	diene + dienophile $\longrightarrow$ cyclohexene (R$_1$, R$_2$, R$_3$, R$_4$)

Table 5.2: (continued)

Reaction	Name	Scheme
L	Ethoxylation/propoxylation	
M	Metathesis	
N	Hydrogenation	
O	Carbonatation	

photoreaction. This reaction can also be used to enhance the antiwear properties of vegetable oils with poorly sensitive to oxygen. Double bonds can also be oxidized (Reaction G, Table 5.2) into allyl hydroperoxides, which were reduced in a second step into secondary hydroxyl groups through reaction H (Table 5.2). Reaction H can also be used to reduce fatty acid dimers into fatty diol dimers in order to confer high hydrophobic, stretching, and chemical stability properties to polyurethane. Reaction I (Table 5.2) illustrates the transesterifification of the ester group under the function methoxide such as sodium. Mostly, it happens for the side reaction for partial transesterification of triglycerides. For example, vegetable oil can be epoxidation using H_2O_2 in acidic conditions, following by an epoxide ring opening reaction with polyols or alcoholamines. Commonly, more than 80% of epoxide groups can be transformed into primary or secondary hydroxyl groups, in which the product is the mixture of polyester and polyether polyols. This mixture is yielded by the partial transesterification of triglycerides (reaction I, Table 5.2) and epoxide ring opening (reaction B, Table 5.2), respectively. Reaction J shows the amide formation (Table 5.2), whose reaction with triglycerides esters may compete with that of epoxide ring opening (reaction B, Table 5.1). In this case, the ester groups of triglycerides are usually essential. Reaction K (Table 5.2) involves the reaction concerning linoleic acid. Firstly, a Diels–Alder reaction happens to the conjugated linoleic acid molecule. And then, the products react with another linoleic acid molecule to yield a cyclohexene adduct. In this process, conjugated linoleic acid is reposefully generated from isomerization of linoleic acid or by extraction and isolation from some natural oils [46]. The ethoxylation or propoxylation reactions (reaction L, Table 5.2) usually happen in terms of long chain fatty alcohols, leading to interesting surfactant properties. In this route, the reaction between alcohols and chlorosulfonic acid or phosphoric anhydride happens, significantly yielding sulfates or phosphates surfactants, respectively [47]. Reaction M (Table 5.2) depicts the epoxidation of triolein after metathesis reaction. It can be employed as the first step for preparation of terminal hydroxyl groups according to ring opening of epoxide groups by methanol in the presence of fluoroboric acid. This can directly produce the polyols which may react with diisocyanates to yield polyurethanes with short dangling chains [48]. Reaction N (Table 5.2) shows the hydrogenation step realized under hydrogen flow with Raney nickel [42, 49]. The polyols obtained from this process can be directly used in polyurethane synthesis or be functionalized by further reaction such as esterification [50]. Reaction O (Table 5.2) depicts the direct carbonatation of epoxidized soybean oil [51, 52].

Soybean oil, a widely grown edible oil agricultural product, is a polyunsaturated or linoleic and highly susceptible to lipid oxidation type of oil. But soybean oil does not contain natural hydroxyl group, it needs to be obtained prior to its use for WPU. Meanwhile, because of the polyfunctionality of soybean oil based polyols, it is easy to appear high cross-linking and gelation when the polyols are used in PU synthesis. It is preferred in PU foam and elastomer, but not in WPU dispersions. The decrease of hydroxyl functionalities may be the effective way to adjust the polyols for preparation of WPU dispersion. Lu et al. reported that a series of methoxylated soybean oil polyols (MSOLs) with different hydroxyl functionalities ranging from 2.4 to 4.0 had been used to prepare WPU dispersion with a uniform particle size [53]. At present, soybean oil polyols are prepared by epoxidation and hydroxylation with a functional degree of 1–6 and a hydroxyl value of 50–700 mg KOH/g. Epoxide ring opening reaction with acetic acid is the main route to transform soybean oil to polyols for preparation of PU. Zhang et al. significantly investigated reaction kinetics of ring-opening between epoxidized soybean oil and castor oil fatty acid (Figure 5.3) with 1,8-diazabicycloundec-7-ene as catalyst. It is suggested that the activation energy of the ring-opening reaction between the epoxy and the carboxylic groups was 72.2 kJ/mol, while that by using protons as catalysts was 75.1 kJ/mol. The higher energy may be attributed to the steric effect of long carbon chains of castor oil [54]. Feng et al. developed a solvent-free and scalable method for the preparation of soybean-oil-based polyols by thiol–ene reaction with 2-mercaptoethanol and the resulting polyols posed high reactivity towards diisocyanates because of the primary hydroxyl groups [55]. Additionally, the use of this epoxidized soybean oil and glutaric acid leads to a fully bio-based emulsifier, which can potentially replace typical petroleum-based emulsifiers for the production of WPU [56].Furthermore, the acrylic monomers, butyl acrylate, and methyl methacrylate, have undergone emulsion polymerization in soybean oil-based anionic WPU dispersions to give urethane–acrylic hybrid latexes. This contributes to the improvement in thermal and mechanical properties [3, 4].

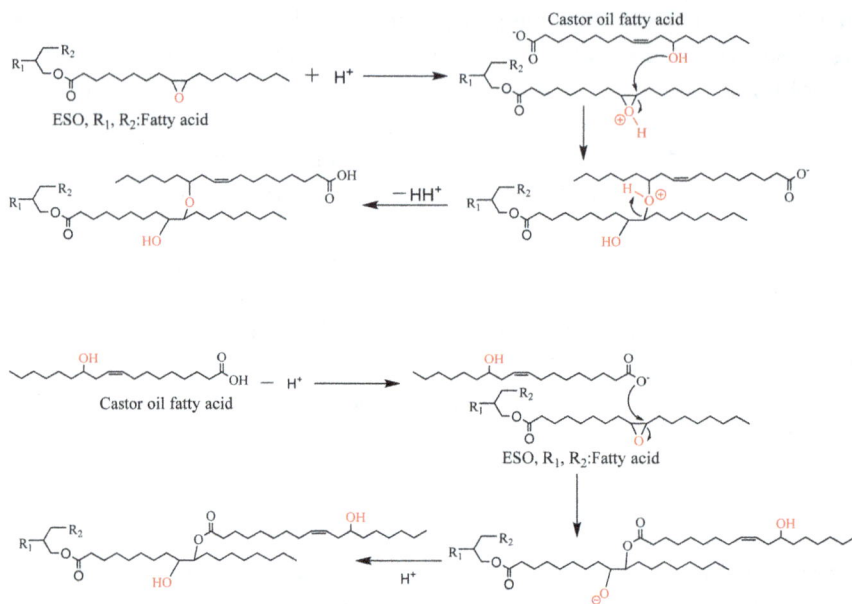

Figure 5.3: The reaction mechanism between epoxidized soybean oil and castor oil fatty acid. Top: ring-opening initiated by OH. Bottom: ring-opening initiated by acid [54] @Copyright 2014 American Chemical Society.

Sunflower oil is important cooking oil. It has been reported to be used in polyols, isocyanate, and chain extender in preparing PU. The epoxidation of the C=C bonds and the ring-opening action of the epoxy group are to introduce the hydroxyl group into the structure of vegetable oil to form polyols [57]. Ismail et al. synthesized high primary hydroxyl functionality polyol from sunflower oil by a thiol–yne reaction [58]. Owing to the high hydroxyl polyols in ultraviolet curing, polyurethane synthesis, polyester synthesis, and other fields have a wide range of applications [59–61]. Interestingly, sunflower oil may also play a role in preparing isocyanate, which is relatively difficult to prepare due to high reactivity and toxicity. Hojabri et al. synthesized 1,7-heptane diisocyanate utilizing oleic acid as a renewable source from sunflower oil, which was reacted with petroleum- and canola-oil-based polyols, respectively, to synthesize PU. The bio-based PU showed comparable performance with that from petroleum-derived 1,6- hexane diisocyanate [62]. For synthesis of WPU, dihydroxy acids are also an important raw material. Among the known dihydroxy acids, dimethylolpropionic acid is the most widely used in the study for the stabilization of polyurethane colloids in aqueous phases. However, it is complex pathway, and is also an expensive material in industry. Hasan et al. proposed a novel carboxylic acid group containing polyol was prepared from sunflower oil via epoxidation, ring opening, and saponification reactions and applied as a chain extender in preparation of WPU [63].

Linseed oil, extracted from flax seed, is one of the cheap renewable resources and is mainly produced in Canada, China, Argentina, and the United States. Recent studies have shown that linseed oil-based polyols was synthesized by transesteration of glycerin/linseed oil and then used as chain extender to synthesize WPU with diisocyanates [64, 65]. Cheng et al. introduced the WPU dispersions synthesized from linseed oil by using a transesterification process and the linseed oil-based films had excellent adhesion, durability, and lightfastness for wood coatings [64]. Moreover, linseed oil was successfully used to replace dimethylol propionic acid as a chain extender in anionic polyurethane dispersion system. Chen et al. reported in preparation of anionic waterborne polyurethane dispersions that epoxidized linseed oil subjected to ring-open by glycol and HCl, followed by saponification to polyhydroxy fatty acid [66].

Jatropha oil, obtained from the seed of jatropha curcas, contains a high concentration of toxic ingredients of phorbol esters resulting inedible [67]. Hence, jatropha oil is an excellent competitor among other vegetable oils in the manufacture of polyurethane dispersants because its price has not been affected by the growth of the food industry [68]. The oil content of the jatropha oil is about 40–60%, which makes good economic sense for extraction [69]. Sariah et al. prepared jatropha oil-based polyol from jatropha oil by epoxidation and oxirane ring opening process and investigated the effect of hard segment, hydroxyl number, and ionic emulsifier content in jatropha oil-based WPU on colloidal stability and rheological properties [70, 71].

Palm oil, produced from oil palm, consists of triglycerides and diglycerides [72]. In order to the preparation of PU, it is crucial to introduction of hydroxyl into polyols in palm oil backbone. Recent studies have shown that epoxidation and transesterification with small molecular polyols are alternative methods for introducing hydroxyl groups to become palm oil-based polyol. Yeoh et al. prepared palm oil-based polyester polyol derived from epoxidized palm olein and glutaric acid and reacted with isophorone diisocyanate to produce aliphatic polyurethane, without the incorporation of any commercial petrochemical-based polyol by a one-shot foaming method [73]. Meanwhile, Sittinun et al. researched palm olein-based polyol was synthesized by the transesterification, epoxidation, and reaction with ethylene glycol and ultimately was used as a starting material for producing polyurethane foam [74].

In general, a series of vegetable oil-based polyols were synthesized from various green natural vegetable resources as raw materials for different applications of polyurethane. The vegetable oils consisting of various functional sites such as ester groups and unsaturated carbon–carbon double bonds makes it easy to convert to different functional monomers such as diols, polyols, and -OH containing polymer, which can be used to synthesize polyurethane. Hence, it can be used as an alternative to petroleum-based polymers as green precursors for bio-based PU production.

5.2.3 Vegetable oil based isocyanates

Besides the vegetable oil based polyols, researchers have devoted to the green-ization of isocyanates instead of petroleum based isocyanates by using vegetable oils simultaneously. However, it is still a challenge to synthesize suitable diiso-cyanates derived from vegetable oils, which are suitable for PU because of the high reactivity and toxicity of isocyanates. Notably, owing to the inherently aliphatic of isocyanates with the absence of unsaturation from vegetable oils, it is exactly suitable for the WPU synthesis for potential application in coatings [28]. It is re-ported that there is commercially available fatty-acid based diisocyanate known as dimer diisocyanate supplied by Henkel Corporation Co. or General Mills Co [75]. Soybean oil is expected to synthesize diisocyanate by functionalized with isocya-nate moieties. Two strategies are developed in which one is to brominate triglyc-eride at the allylic positions using N-bromosuccinimide, following by the reaction with AgNCO to yield isocyanate-containing triglyceride [76]. Secondly, as depicted in Figure 5.4, iodine isocyanate adduct of soybean oil containing 3.1 isocyanate groups per triglyceride in one-step could be prepared by the reaction of iodine isocyanate and soybean oil at room temperature. These prepared isocyanates are suitable for preparation of WPU significantly [77]. In addition, oleic acid is also potential for preparation of 1,7-heptamethylene diisocyanate in three steps by combination oxidation of the dicarboxylic acid with Curtius rearrangement [62]. Unsaturated α,ω-dicarboxylic acid synthesized from oleic acid by metathesis was also used to prepare a novel long chain unsaturated diisocyanate using a similar procedure [78]. It should be noted that the limited methods have been used to prepare isocyanate from vegetable oil. It is still a challenge for this conception to obtain isocyanate from vegetable oil because of the high reactivity of isocyanate groups. Non isocyanate methods may be a great potential path for preparation of PU, while this strategy is still immature in WPU dispersions because of the demanding of hydrophilic polyurethane chains.

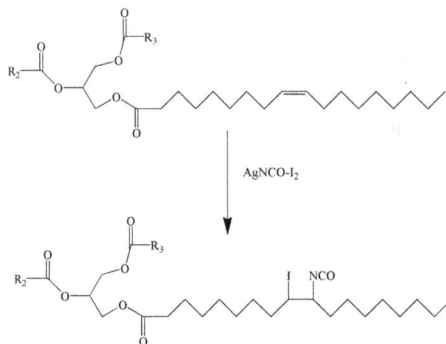

Figure 5.4: Synthesis of iodine-containing plant oil-based polyisocyanates [77] @Copyright 2010 WILEY-VCH.

5.2.4 Lignin based polyols

Lignin, originate from wood, is the renewable and abundant substance of aromatic polymers in nature [79]. Lignin is a crosslinked network structure compound made from three monomers (coniferyl alcohol, *p*-coumaryl alcohol, and sinapyl alcohol) [80] The three-dimensional network of lignin can provide sufficient strength and hardness, and can be directly mixed into the polymer matrix to obtain antibacterial, oxidation resistance, ultraviolet resistance, flame retardant, and other functions [81–84]. Therefore, it has great potential in industrial application. The main functional groups of lignin are hydroxyl, methoxy, ether, carbonyl, and carboxyl groups [85]. These active groups provide possibilities for modification of lignin to introduce new chemically active sites on lignin structure, for example hydroxyl functionalization developed multifunctional lignin copolymers as the raw material of PU production [86, 87]. Since lignin contains a large amount of hydroxyl groups (phenolic and aliphatic), lignin and modified lignin could be used as polyol derivatives to prepare lignin-based polyurethane [88].

Owing to the presence of hydroxyl groups, lignin can directly react with isocyanates as polyol precursor for the PU synthesis rather mild condition [89]. Lignin possesses a complex heterogeneous structure with wide molecular weight distribution. Therefore, solvent fractionation was widely applied as an efficient pretreatment pathway to extract homogeneous lignin fractions for clear structural features and purity [90, 91] Wang et al. explored that the lignin-based polyurethane was fabricated with unmodified softwood Kraft lignin fractions as primary hydroxyl groups [92]. The results show that the increase of the molecular weight of lignin can improve the rigidity or deformation resistance of lignin-based polyurethane. However, a variety of polyurethane materials have been developed using lignin as a single polyol, which is usually brittle and hard. Therefore, polyether, poly(ethylene glycol), and polyester, poly(trimethylene glutarate), are incorporated as secondary polyols to be added to polyurethane synthesis [93]. Lignin acts as not only a cross-linking agent but also a hard segment because of high content of aromatic rings and hydroxyl functionality to form the PU polymers. Renewable flexible lignin-based PU was also synthesized by lignin with long flexible aliphatic chains and HDI as hard segment, and biodegradable poly (ε-caprolactone) as soft segment [93].

In addition, lignin can be chemically modified to improve its reactivity. The chemical modification of lignin methods mainly includes isocyanate modification, halogenated alkyl modification, esterification, oxypropylation, and hydroxymethylation [94]. Oxidation of lignin was widely studied as a method for improving the amount of alcoholic hydroxyl groups and the functionality of lignin [95, 96]. At present, Zhang and his group studied ozone oxidized lignin replaced part of polyethylene glycol to react with isocyanate to prepare lignin-based polyurethane [89]. The results indicated that ozone oxidation led to the higher hydroxyl content and is an efficient modification method of lignin with higher reactivity. Furthermore, owing to increased hydroxyl group content, the oxidized lignin-based PU presented the higher degree of crosslinking with excellent overall performance than lignin-based PU. Ren te al. researched that epichlorohydrin modified lignin were used into the synthesis of WPU and analyzed the effect of lignin on particle size distribution, thermal stability and mechanical properties, and water resistance [88].

5.2.5 Cashew nut shell liquid based polyols

Cashew nut shell liquid, extracted from cashew nut, include many reactive phenol derivatives with highly hydrophobic fatty chains, such as cardanol, cardol, 2-methyl cardanol, and bitter citric acid [97]. Prior art literature advocates that cashew nut shell liquid is routinely used for the development of sustainable PU. For instance, Tuhin and Niranjan reported the synthesis of self-healable anticorrosive PU coating using cashew nut shell liquid as chain terminating agent [98]. Moreover, a series of polyols derived from cashew nut shell liquid as a component for PU preparation [98]. Cardanol, containing a characteristic long unsaturated aliphatic chain, is also used as polyols for the synthesis of sustainable PU with desired properties [99, 100]. Use of cardanol in PU production has been well developed. Wang et al. introduced epoxidized cardanol polyols by thiol-based click reaction with relative high hydroxyl number building block for PU (Figure 5.5) [101].

5.2.6 Plant straw based polyols

The production of biopolyols from lignocellulosic biomass has been of interest to the polyurethane industry. Among them, there are abundant lignocellulose in agricultural crop residue, such as soybean straw [101], rape straw [102], wheat straw [103], and corn straw [104]. The conversion of these plant straws into liquid biopolyols can be achieved

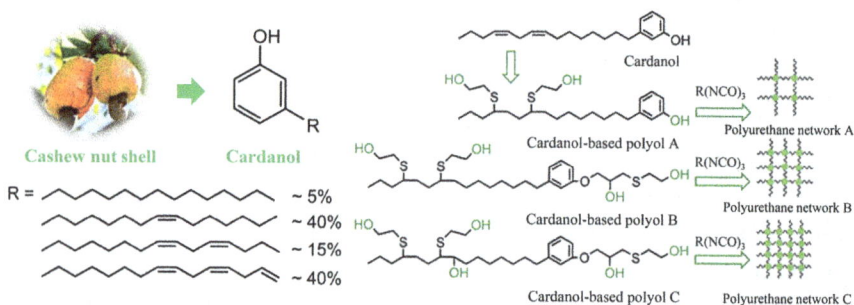

Figure 5.5: Chemical structure of cardanol with the unsaturated alkyl phenolic structure, Cashew nut shell-derived polyols are presented here for applications as sustainable building blocks of polyurethanes [101] @Copyright 2012 Elsevier.

through a liquefaction process. The process of biomass liquefaction is composed of decomposition, esterification, condensation polymerization and a series of other reactions, and finally, resulting in the final formation of biopolyols [105]. It was found that the polyurethane synthesized from biopolyols extracted from plant straw showed good performance, which was comparable to that of petroleum-based products.

Now researchers have made a lot of work in polyurethane synthesized from biopolyols such aspects as raw materials. Various plant straw materials have been liquefied in different solvents to generate polyols and to subsequently produce the PU. The liquefaction products of polyhydric-alcohol as solvent are polyesters or polyether polyols rich in hydroxyl groups, which can be reacted with isocyanates to obtain polyurethane materials with excellent properties [106]. For example, Hu et al. proposed the feasibility of high quality biopolyols extracted from soybean straw using crude glycerol for synthesis of polyurethane. The polyols containing hydroxyl numbers from 440 to 540 mg KOH/g and produced well-behaved PU compared with traditional petroleum based [107]. Serrano researched the preparation of wheat straw polyols in glycerol solvent by a liquefaction reaction, which has been employed to synthesize biodegradable PU [108]. Therefore, biopolyols extracted from plant straw studies of the prepared PU have been developed to meet the requirements of the environment.

5.2.7 Terpene based polyols

Terpenes, a class of natural hydrocarbons obtained from many plants, are widely distributed in nature, including pinene, limonene, pirocarvone, myrcene, alloocimene, ocimene, and farnesene [109]. The structural diversity (i.e., linear, cyclic, or polycyclic) and the presence of double bonds make terpenes with multifunction and high reactivity [110]. In recent years, additional advances have been made in the synthesis of different terpene polymers, especially polyurethane [111].

Limonene, a monoterpene consisting of two distinct double bonds, is an inexpensive natural compound as an economically feasible renewable feedstock isolated from citrus fruits [114]. Because of this special structure, limonene can be easily modified into other valuable materials, such that epoxidation or thiol–ene click reactions allows the introduction of hydroxyl groups, which can be used to obtain polyols (Figure 5.6) [113]. Thus, the limonene-based polyols have been widely used for preparation of polyurethane. In this aspect, Gupta et al. reported the synthesis of new Mannich polyol with aromatic structure based on limonene which was further used for the preparation of PU. Additionally, limonene dioxide, the double epoxidation product of limonene, is a key precursor for the preparation of bio-based polycarbonates or non isocyanate polyurethanes, as shown in Figure 5.6B [114, 115]. Moreover, pinene, analicyclic epoxy resin with endocyclic structure, was extracted from turpentine. Wu and coworkers reported that pinene serve as precursors for the synthesis of polyols which could react with polyisocyanate to produce polyurethane composite polymers by crosslinking [116].

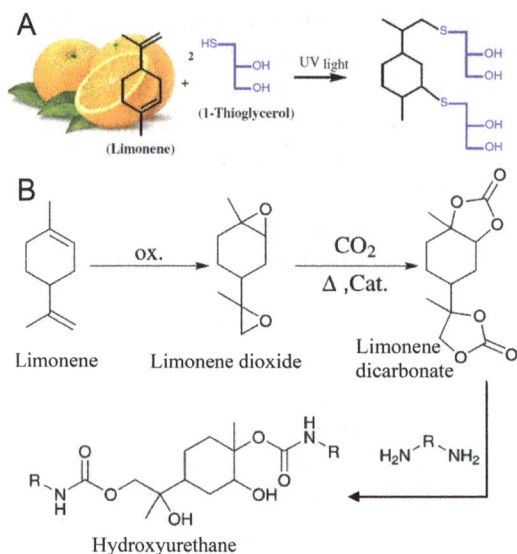

Figure 5.6: A: Synthesis of polyols based on limonene by thiol–ene "click" chemistry [112] @Copyright 2014 Springer. B: Isocyanate free strategy from limonene to hydroxyurethane [113] @Copyright 2012 RSC.

5.2.8 Rosin based polyols

Rosin, mainly composed of 90% rosin acid, is a kind of brittle and transparent solid natural resin secreted from the trunk of pine tree with excellent properties of insulation, adhesion, and emulsification, which has been widely used in materials, chemistry, medicine and cosmetics, etc. [117, 118] Rosin acid is a naturally rigid small molecule with a rigid hydrophenanthrene ring structure, and can make the polymer has a high mechanical rigidity [119]. Owing to rosin molecular structure of conjugated bonds, carboxyl groups, and other reactive groups, different derivatives can be derived through addition, isomerization, polymerization, esterification, and other ways. Hence, rosin and its derivatives have been used as alternatives for chemical raw materials in the field of polymer synthesis, such as PU [120, 121]. Liu et al. researched that rosin-based nonisocyanate PU coatings were successfully prepared by rosin-based cyclic carbonate with amines [122].

In the process of synthesis of polyurethane, a rosin-based diol chain extender can be obtained by modifying the conjugated double bonds and carboxyl groups on the rosin, which can be introduced into the hard segment structure of polyurethane to obtain the bio-based polyurethane. Xu et al. reported that fumaropimaricacid polyester polyol from rosin acid was successfully applied in the synthesis of series of novel rosin-based WPU, which improved mechanical properties, thermal stability, and water resistance [121]. In addition, hydrophenanthrene ring structure is a nonplanar ring structure, which can undergo conformational transformation under the action of external pressure, and can undergo huge deformation and store elastic energy during stretching [123, 118]. So, it can synthesize the raw materials of shape memory polyurethane with high recovery rate and excellent elastic function. Zhang et al. prepared thermoplastic shape memory PU with a rosin-based diol as the rosin-based chain extender, which improved stability of the hard phase and recovery of the corresponding polyurethanes with a more than 1000% strain [124]. In general, with the content of rosin-based chain extender increasing, the phase separation degree and the stability of PU gradually enhanced, all of which were due to the hydrophenanthrene ring structure.

5.3 WPU modified by biomass resources

Owing to the low solid content, weak intermolecular force , poor thermal stability, and mechanical properties in WPU, its wide application has been restricted [125]. To this end, the modification of WPU is necessary, such as the addition of multifunctional materials is an effective strategy. Safety and environmentally friendly consideration, various biomass resources derived from animals and plants such as cellulose, starch, chitosan, and lignin have been employed with WPU to develop versatility for practical application, i.e., food packaging [126], degradable plastics [127], and biomedical materials [128] applications. They have a variety of structural and functional groups, easy to chemical and physical modification. Therefore, being a major contributor to PU industry with the advantages of low cost, renewable, nontoxicity, easy availability, and biodegradable benign proved to be a convincing alternative.

5.3.1 Cellulose

Cellulose, among the most widely distributed and abundant natural polymers, is comprised of a linear, long-chain glucose polymer that is rich in oxygen, particularly hydroxyl groups [129]. Cellulose is ubiquitous in plants, rich in about a third of higher plants, 40% of cellulose content in dry wood, and more than 90% in cotton and flax [130]. Since the high aspect ratio renewability, large specific surface area, low density, excellent biodegradability and mechanical properties, cellulose has attracted great attention in WPU modification [131]. However, crystalline state and amorphous state exist in natural cellulose structure. Amorphous cellulose affects the properties of the material to some extent, so was removed by mechanical, enzymatic or chemical processes to obtain nanocelluloses, including cellulose nanofiber (CNF), cellulose nanocrystal (CNC), and bacterial cellulose (BC) [132]. Owing to its small size, large specific surface area and excellent mechanical properties and other special properties, nanocelluloses can be widely used in reinforcement agent for composite materials, biodegradable film, packaging material, functional film, medical dressing, and other fields [133–135].

Recently, some research reported that cellulose nanofibers, with higher aspect ratio, can be used as reinforcing nanofillers improving the thermal and mechanical properties of the WPU for widely application [136–138]. Kong et al. introduced CNF into the WPU by chemical grafting (derived from nanofibers by the 2,2,6,6-tetramethylpiperidine-1-oxyl (TEMPO) oxidation) and physical blending [139]. The results showed, comparing the two methods at 0.1 wt% nanocellulose addition resulted in the maximum improvement of the comprehensive properties especially the tensile strength. Additionally, CNC with highly crystalline rod-like nanostructures is inherently strong material and reinforce hard segments of WPU through hydrogen bonds in the interfacial area [140]. Lei et al. researched that CNC extracted from office waste paper was used to reinforce WPU [141]. CNC have been incorporated into castor oil-WPU matrix to lead to higher hardness, elastic recovery and high compatibility in suspensions to be applied as coatings were also investigated [142].

More than that, it was found that the effective introduction of modified CNC resulted in a significant increase for the wide application of CNC-WPU, such as water resistance [143], electrical and antimicrobial properties [144, 145]. Zhang et al. introduced that highly compressible WPU was endowed by the brittle cellulose nanocrystal with the assistance of carbon nanotube to prepare high-performance piezoresistive sensor for artificial electronic skin application [145]. Finally, bacterial cellulose with 3D nanofiber network structure is produced by some bacterial strains, which favors the strong interchain hydrogen bonds with high mechanical properties in humid environment [146]. Urbina et al. developed a BC/WPU based water-activated shape memory nanocomposites in the form of a gelatinous and translucent membrane can be developed for biomedical applications and highest BC content reached a shape recovery ratio of $92.8 \pm 6.3\%$ as shown in [147].

5.3.2 Starch

Starch, one of the most abundant polysaccharides, is produced as a storage polymer by many granules of plants (potatoes, corn and so on) formed as water in-soluble granules, and its output is second only to cellulose as a pure natural renewable resource [148]. Owing to its good biodegradability, stable supply, and low price, starch is widely used in polymer materials and products with good environment, including paper making, food, textile, medicine, and petrochemical industry, etc. [149–151] Starch have been introduced in synthetic WPU by chemically bonded [152], and most notably in reinforcement of mechanical properties by obtaining a filler effect, increasement of elongation to break and modulus by increasing starch content [153, 154]. Moreover, starch can be alcoholized or hydrolyzed to produce alkyl glucosides (α-methyl glucoside and hydroxyalkyl glucosides) and glucose, which are used as polyolic starters for synthesizing rigid polyether polyols [155], the main routes for using starch in illustrated in Figure 5.7.

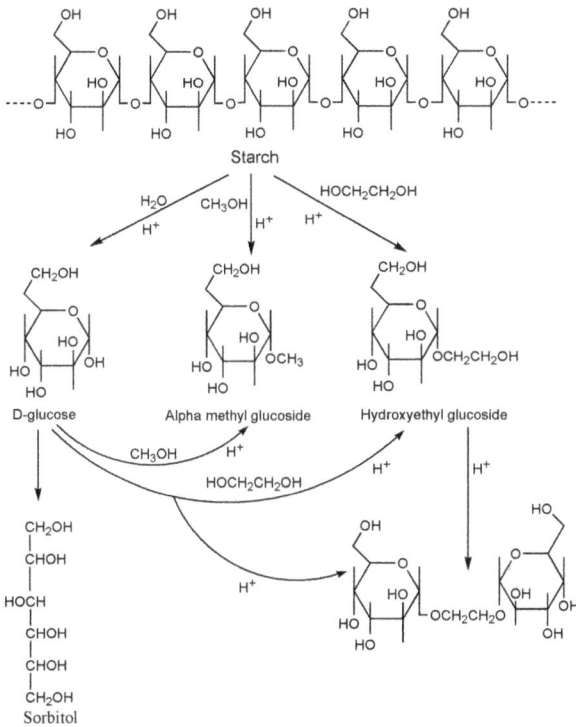

Figure 5.7: Chemical transformations of starch and ᴅ-glucose into polyols that can be used as polyfunctional starters for polyether polyols synthesis [155] @Copyright 2005 Rapra Technology.

Recently, N.L. Tai reported physical mixing of PU ionomer with starch, in which ionic act as internal emulsifier and lead to high compatible starch-PU blends without obvious phase separation [154]. As shown in Figure 5.8, interaction among the charged particles in starch-PU dispersion and enhanced the intermolecular forces and the emergence of physical crosslinking. Consequently, the surface topography of the anionic starch-PU was smoother and improved the ductility and tensile strength of the starch-PU. Additionally, starch nanocrystals, with an average equivalent diameter of 25–40 nm, is another good option to modified WPU [156, 157]. It is worth noting that starch nanocrystals through strong hydrogen bonding interaction with WPU to improve thermal stability and mechanical properties [158]. Moreover, incorporation of modified starch into the WPU to improve mechanical properties also appears in some literatures, such as oxidized starch [159] and vinyltrimethoxysilane starch [126].

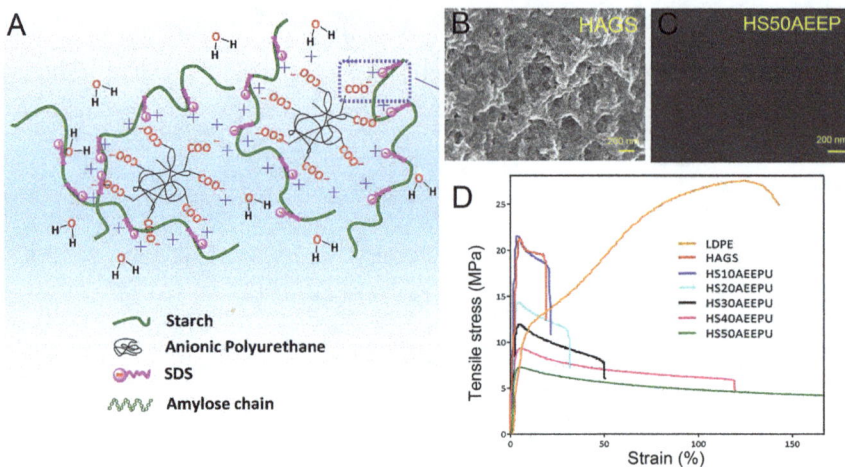

Figure 5.8: WPU modified by starch. A: Schematic diagram showing the proposed/postulated interaction between HAGS and AEEPU; B: Scanning electron microscope micrographs of surface, HAGS = high amylose glycerol plasticized starch film, AEEPU = 50 indicate the ratio of Anionic polyurethane in the HAGS film; C: Stress-strain curves for HAGS, LDPE and HAGS-AEEPU films [154] @Copyright 2012 Elsevier.

5.3.3 Chitosan

Chitosan, by the removal of acetylated molecular groups from natural chitin, is a kind of polysaccharide and an excellent candidate for wide application in biomedical field [160, 161]. Chitosan has active groups such as hydroxyl group and amino group, which can easily change the mechanical and physical chemical properties of chitosan by chemical modification [162]. Owing to its nontoxicity, antibacterial, thermal stability, cytocompatibility, and antibacterial hemocompatibility, chitosan has been used as modifier in polyurethane copolymers for drug delivery, medical hydrogel scaffolds, and antibacterial fabrics [163–165]. Lin reported functionalized biodegradable PU with chitosan crosslinkers via the dynamic Schiff reaction, which successfully produce novel self healing hydrogels for biomedical applications [166]. Thus, the soft segments of the chitosan crosslinker can be altered to develop biodegradable WPU based hydrogels and cryogels. In addition, chitooligosaccharide, as products by chemical and enzymatic hydrolysis of chitosan, has applied in the WPU modification for enhancing the mechanical properties due to the reactive amino and hydroxyl groups [167].

5.3.4 Lignin

Lignin is an important biomass raw material for polyurethane synthesis as mentioned above. However, it can also modify the properties of polyurethane for wider application

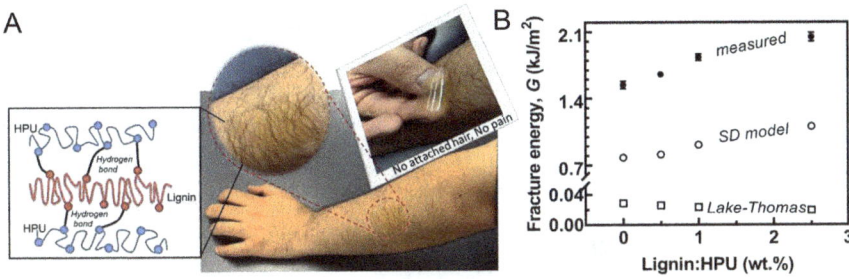

Figure 5.9: WPU modified by lignin. A: Schematic diagram showing the proposed/postulated interaction between lignin and HPU, lignin-HPU patch on arm; B: Comparison between calculated and measured fracture energies of lignin-HPU at different concentrations of lignin. @Copyright 2012 American Chemical Society.

because of good features, such as excellent stability and aging resistance property [168–170]. Zhang et al. researched that WPU modified by lignin as antioxidant agent, aiming to enhance waterproof-breathable properties and weather accelerated aging performance for application of aging-resistant fabric [170]. The result demonstrated that lignin mainly affects the morphology, surface chemistry structure of WPU and allows for the development of satisfactory aging resistance. Moreover, since abundant polar sites on its backbone, lignin can be used for physical crosslinking of polymers resulted in developments in mechanical properties. Oveissi et al. showed that lignin-PU hydrogel construct with considerable superior mechanical properties to produce tough, thin, printable hydrogels for 3D printing, and fiber spinning (Figure 5.9) [171].

5.3.5 Sodium alginate

Sodium alginate, a natural biodegradable polysaccharide carbohydrate extracted from brown algae, is widely applied in fields of food additives [171], biomedicine [172], tissue engineering [173], printing industry, and so on. Wang et al. introduced sodium alginate into WPU through *in-situ* interaction and the formation of interpenetrating networks structure [174]. Since hydrogen bonding and physical crosslinking between the two polymers, it has positive effects on the thermal and mechanical properties of composite film [175].

5.3.6 Natural phenolic acid

Natural phenolic acid, extracted from fruits, nuts and flowers, is containing phenolic hydroxyl group and carboxyl group, which is an ideal substitute for petroleum resources. Recently, Zhang et al. conducted systematic studies various natural phenolic acids as neutralizers in preparation of castor oil-based cationic WPU. Due to the

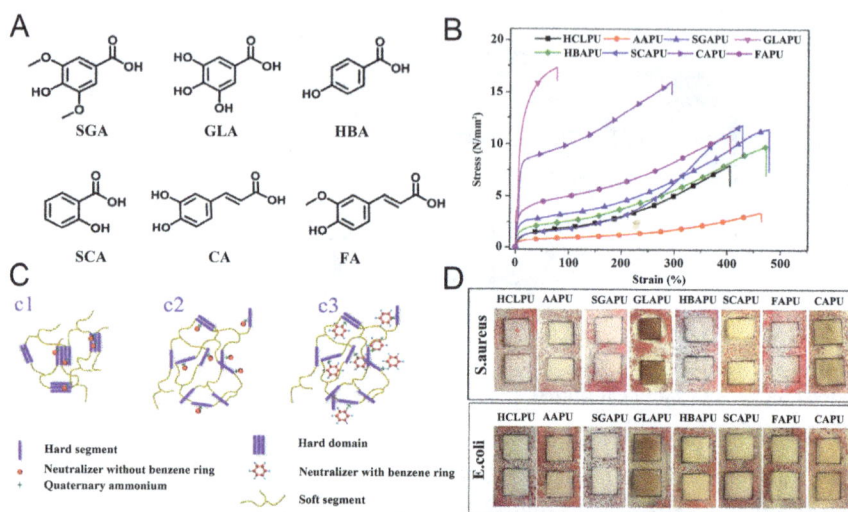

Figure 5.10: WPU modified by natural phenolic acid. A: Structures of natural phenolic acid neutralizers: caffeic acid (CA), ferulic acid (FA), syringic acid (SGA), gallic acid (GLA), 4-hydroxybenzoic acid (HBA), and salicylic acid (SCA); B: Stress–strain curves of films based on PU with different neutralizers; C: Phase separation scheme (c1–c3) of PUs with different neutralizers. (c1) High phase separation (acetic acid/PU); (c2) low phase separation (hydrochloric acid/PU); (c3) WPU neutralized with phenolic acids [176] @Copyright 2020 Elsevier; D: Contact-active antibacterial activity of WPU films against *S. aureus* and *E. coli* [175] @Copyright 2020 American Chemical Society.

introduction of the hydrophobic, rigid, conjugated structure of phenolic acids in WPU hard segment leading to an increase degree of phase mixing, the WPU endowed mechanical properties, antibacterial activities, as shown in Figure 5.10 [175]. Moreover, Ren et al. introduced environmentally friendly application of gallic acid with three hydroxyl groups modified WPU, which led to the formation of crosslinking that improved the water resistance of WPU [176].

On the whole, biomass resources can play an important role in the synthesis and modification of WPU. Therefore, the feasibility of using all kinds of natural raw materials to prepare WPU instead of petroleum-based raw materials is proved. From the perspective of green and environmental protection, biomass as raw material has a very broad development and investment prospects. Presently facing the cost of the problem, the development of technology and market development is not perfect, but with the science and technological progress, the biomass energy in WPU industry will get rapid development.

5.4 Application

During the past few years, WPU has demonstrated many important applications in coating, adhesive, binders, textile dyes, and even drug deliver particles, etc. What should be noted is the printing ink. Inspired by the environmental pollution and threatening human health of solvent-based printing ink, water-based ink has a broad development prospect, especially suitable for tobacco, alcohol, food, beverage, medicine, children's toys, and other packaging and printing products with strict hygienic conditions. Notably, WPU exhibit excellent properties such as high tensile and tear strength, low viscosity, thermal stability, good compatibility, which can act as the binder applied in ink, especially flexographic ink and UV-curable ink [14, 177]. However, compared to traditional solvent-based ink, the low curing speed and water resistance of WPU binder in inks restrict the printing effects. To improve these performances, composite inorganic materials through physical action in WPU dispersion have undergone rapid development [178]. Zhang et al. illustrated the synthesis of waterborne UV-curable WPU/SiO$_2$ flexographic ink, resulting in the ink accompanied by thermal stability, water resistance, good rheological behavior and abrasion resistance performances in applications of food packages [179]. Zhang et al. introduced sodium lignosulfonate as functional filler into castor oil-based WPU, which could find excellent UV absorption of the composites because of the conjugated structures of lignin macromolecular [180].

Biodegradable polyurethane based on block copolymer have undergone rapid development in recent years because of its potential for use in food packaging application, especially polylactic acid (PLA) and polycaprolactone (PCL) [181]. PLA, derived from renewable resources, is an aliphatic polyester with excellent biodegradability and high strength [182]. PCL is widely used in the processing of drug carriers, plasticizers, degradable plastics, nanofiber spinning, and plastic materials because of its good biocompatibility and biodegradability, and good shape memory and temperature control properties [183]. Significantly, PLA and PCL blend with PU by means of the chain-extension method can provide PU specific properties, such as biocompatibility, good mechanical strength, and low permeability. For example, Fathilah et al. developed the compositions of the hard PLA and the soft PCL diols in PU polymers to optimize mechanical properties, flexibility and gas barrier property, demonstrating its potential utility as biodegradable packaging materials [184]. The result revealed that the increase in PCL content leads to the generation of PU polymers with higher molecular weight, which improved tensile strength and elongation at break, and was accompanied by an increase in degradation rate. Meanwhile, Arrieta et al. synthesized PU based on triblock copolymer of PLA and PCL loaded with catechin, were developed with thermally-activated shape memory effective antioxidant activity for biodegradable food packaging purpose [185].

PU has been used as adhesives as it was born. With the combination of biomass, WPU adhesives have attracted great attention in research and innovation due to the safety, widely source, and even biodegradability. The strength of such adhesion can be traced back to the formation of covalent bonding to the substrate, Van der Waals forces and hydrogen bonding and mechanical interlocking [186, 187]. For the aqueous phase, the unavoidable problem is the low water resistance of adhesives. For example, the

WPU dispersion from soy protein isolates shows limited mechanical behavior in dry and wet states. Lopez et al. [188] employed acrylic to modify WPU dispersion to yield WPU-acrylic hybrid nanoparticles for application as pressure-sensitive adhesives (PSAs) via one-step miniemulsion polymerization. It is reported that there is a large increase in the cohesive strength and hence a much higher shear holding time (greater than seven weeks at room temperature), which is a very desirable characteristic for PSAs. Liu et al. [189] prepared a bio WPU by using soy oil to improve the wet strength of a soy protein adhesive. They modified the hydrophilicity of the proteins that leads to low wet strength by mixing them with soy oil-based WPU (optimum concentration of about 50%) and developing covalent and non covalent bonds that lead to the increase of the adhesive's water resistance. It is suggested that the good penetration and developed adhesion of the small molecules of the protein to wood and the beneficial effect of the WPU hinder the penetration of water in the interphase region between the adhesive and wood [186]. It is a fact that the key problem is still the aqueous phase in WPU dispersion for application as adhesives.

WPU dispersions have also been widely used in textile area. The PU aqueous emulsion shape with recovery effect was used to finish cotton fabrics through a padding → predrying → curing process (the conventional method) [190]. During the curing process at a high temperature, the isocyanates released reacted to form cross-linking structures bonding the polymer to cotton fabric. Moreover, the wool fabrics can also be treated by this dispersion. Zhang et al. have also prepared the bio-based WPU dispersion with significant recovery effect in textile and hair styling agent [191].

5.5 Conclusion and outlook

Growing concerns over long-term waste recovery and environmental pollution issues direct the use of renewable and biodegradable resources for sustainable polyurethane in the industrial production. Herein, the recent achievements in synthesis of WPU dispersions from biomass resources for sustainable materials have been systematically introduced. Traditionally, the industrial production of polyurethane deeply depends on petroleum stocks. Recently, bio-based polyurethanes have been further explored for material functionalization application in many fields, including printing, packaging, and shape-memory polymer. The biomass, especially the vegetable oils and cellulose-based polymers have been paid a large amount of attention in last decade, providing solutions to the increasing environmental and energy concerns.

It should be noted that there are still many problems in bio-based WPU dispersions. For example, the transition of the vegetable oils to polyols (or diols, isocyanates) is limited. The most used route may be the epoxide ring opening method, which may decrease the purity of the products. The nonisocyanate WPU dispersion is still confused even though the vegetable oil-based isocyanate has been designed. Moreover, WPU chains with the controllable architectures are still explored to be built based on vegetable oil-based monomers, especially in employing living polymerization methods to expand the applications of these bio-based polymers. In a word, the bio-based monomers for WPU aqueous dispersions should be designed and developed in a larger scale to meet the needs of functional and smart polymer technology.

Author contributions: All the authors have accepted responsibility for the entire content of this submitted manuscript and approved submission.

Research funding: The authors acknowledge the financial support provided by the National Natural Science Foundation of China [Grant No. 51802259], the Natural Science Foundation of Shaanxi [Grant No. 2019JQ-510], Xi'an and Xi'an Beilin District Programs for Science and Technology Plan [Grant No. 201805037YD15CG21(18) and GX1913], the Promotion Program for Youth of Shaanxi University Science and Technology Association [Grant No. 20190415], Fund of Key laboratory of Processing and Quality Evaluation Technology of Green Plastics of China National Light Industry Council [Grant No. PQETGP2019003], the Ph.D. Start-Up Fund Project [Grant No. 108-451118001] of Xi'an University of Technology.

Conflict of interest statement: The authors declare no conflicts of interest regarding this article.

References

1. Zhou X, Hao YY, He XY, Zhou D, Xie L, Liu SL, et al. Protean morphology of waterborne polyurethane dispersion: an overview of nanoparticles from sphere to irregular elongated shape. Prog Org Coating 2020;146:105742.
2. Urban D, Takamura K. Polymer Dispersions and Their Industrial Applications. Weinheim: Wiley-VCH; 2002.
3. Xia Y, Larock RC. Vegetable oil-based polymeric materials: synthesis, properties, and applications. Green Chem 2010;12:1893–909.
4. Lu YS, Larock RC. New hybrid latexes from a soybean oil-based waterborne polyurethane and acrylics via emulsion polymerization. Biomacromolecules 2007;8:3108–14.
5. Zhou X, Fang C, Yu Q. Synthesis of polyurethane dispersions in nanoparticles and their properties that depend on aging time. J Dispersion Sci Technol 2015;36:1178–89.
6. Zhou X, Fang C, Chen J, Li S, Li Y, Lei W. Correlation of raw materials and waterborne polyurethane properties by sequence similarity analysis. J Mater Sci Technol 2016;32:687–94.
7. Zhou X, Fang C, Lei W, Su J, Li L, Li Y. Thermal and crystalline properties of waterborne polyurethane by in situ water reaction process and their potential application as biomaterial. Prog Org Coating 2017;104:1–10.
8. Zhou X, Fang C, He X, Wang Y, Yang J, Yang L, et al. The morphology and structure of natural clays from Yangtze River and their interactions with polyurethane elastomer. Compos Appl Sci Manuf 2017;96:46–56.
9. Zhou X, Su J, Wang C, Fang C, He X, Lei W, et al. Design, preparation and measurement of protein/CNTs hybrids: a concise review. J Mater Sci Technol 2020;46:74–87.

10. Zhou X, Deng J, Fang C, Yu R, Lei W, He X, et al. Preparation and characterization of lysozyme@carbon nanotubes/waterborne polyurethane composite and the potential application in printing inks. Prog Org Coating 2020;142:105600.

11. Zhang C, Garrison TF, Madbouly SA, Kessler MR. Recent advances in vegetable oil-based polymers and their composites. Prog Polym Sci 2017;71:91–143.

12. Zhang C, Liang HY, Liang DS, Lin ZR, Chen Q, Feng PJ, et al. Renewable castor-oil-based waterborne polyurethane networks: simultaneously showing high strength, self-healing, processability and tunable multishape memory. Angew Chem Int Ed 2020. https://doi.org/10.1002/anie.202014299.

13. Wang X, Liang HY, Jiang JZ, Wang QW, Luo Y, Feng PJ, et al. A cysteine derivative-enabled ultrafast thiol-ene reaction for scalable synthesis of a fully bio-based internal emulsifier for high-toughness waterborne polyurethanes. Green Chem 2020;22:5722–9.

14. Zhou X, Li Y, Fang CQ, Li SJ, Cheng YL, Lei WQ, et al. Recent advances in synthesis of waterborne polyurethane and their application in water-based ink: a review. J Mater Sci Technol 2015;31: 708–22.

15. Engels HW, Pirkl HG, Albers R, Albach RW, Krause J, Hoffmann A, et al. Polyurethanes: versatile materials and sustainable problem solvers for today's challenges. Angew Chem Int Ed 2013;52: 9422–41.

16. Zhou X, Deng J, Wang D, Fang C, Song R, Zhang W, et al. Growth of polypyrrole conductive and integrated hybrids with lysozyme nanolayer and the thermal properties. Compos Appl Sci Manuf 2020;137:105975.

17. Fang CQ, Zhou X, Yu Q, Liu SL, Guo DG, Yu RE, et al. Synthesis and characterization of low crystalline waterborne polyurethane for potential application in water-based ink binder. Prog Org Coating 2014;77:61–71.

18. Zafar F, Sharmin E. Polyurethane: an introduction. Croatia: InTech; 2012.

19. Höfer R, Daute P, Grützmacher R, Westfechtel A. Oleochemical polyols - a new raw material source for polyurethane coatings and floorings. J Coat Technol 1997;69:65–72.

20. Heinrich LA. Future opportunities for bio-based adhesives-advantages beyond renewability. Green Chem 2019;21:1866–88.

21. Noreen A, Zia KM, Zuber M, Tabasum S, Zahoor AF. Bio-based polyurethane: an efficient and environment friendly coating systems: a review. Prog Org Coating 2016;91:25–32.

22. Desroches M, Escouvois M, Auvergne R, Caillol S, Boutevin B. From vegetable oils to polyurethanes: synthetic routes to polyols and main industrial products. Polym Rev 2012;52: 38–79.

23. Duer M, Veis A. Water brings order. Nat Mater 2013;12:1081–2.

24. Lattuada M, Sandkühler P, Wu H, Sefcik J, Morbidelli M. Kinetic modeling of aggregation and gel formation in quiescent dispersion of polymer colloids. Macromol Symp 2004;206:307–20.

25. Zhou X, Fang CQ, Lei WQ, Du J, Huang T, Li Y, et al. Various nanoparticles morphology of polyurethane dispersions controlled by water. Sci Rep 2016;6:34574.

26. Fu CQ, Zheng ZT, Yang Z, Chen YW, Shen L. A fully bio-based waterborne polyurethane dispersion from vegetable oils: from synthesis of precursors by thiol-ene reaction to study of final material. Prog Org Coating 2014;77:53–60.

27. Ionescu M, Petrović ZS, Wan X. Ethoxylated soybean polyols for polyurethanes. J Polym Environ 2007;15:237–43.

28. Daniel P, Pfister DP, Xia Y, Larock RC. Recent advances in vegetable oil-based polyurethanes. ChemSusChem 2011;4:703–17.

29. Hormaiztegui MEV, Aranguren MI, Mucci VL. Synthesis and characterization of a waterborne polyurethane made from castor oil and tartaric acid. Eur Polym J 2018;102:151–60.

30. Gurunathan T, Chung JS. Physicochemical properties of amino–silane-terminated vegetable oil-based waterborne polyurethane nanocomposites. ACS Sustain Chem Eng 2016;4:4645–53.

31. Liang HY, Wang SW, He H, Wang MQ, Liu LX, Lu JY, et al. Aqueous anionic polyurethane dispersions from castor oil. Ind Crop Prod 2018;122:182–9.
32. Luong ND, Sinh LH, Minna M, Jurgen W, Torsten W, Matthias S, et al. Synthesis and characterization of castor oil-segmented thermoplastic polyurethane with controlled mechanical properties. Eur Polym J 2016;81:129–37.
33. Palaskar DV, Boyer A, Cloutet E, Le Meins JF, Gadenne B, Alfos C, et al. Original diols from sunflower and ricin oils: synthesis, characterization, and use as polyurethane building blocks. J Polym Sci, Part A: Polym Chem 2012;50:1766–82.
34. Zhang Y, Liu BY, Huang KX, Wang SY, Quirino RL, Zhang ZX, et al. Eco-friendly castor oil-based delivery system with sustained pesticide release and enhanced retention. ACS Appl Mater Interfaces 2020;12:37607–18.
35. Zhang Y, Zhang WB, Wang X, Dong QW, Zeng XY, Quirino RL, et al. Waterborne polyurethanes from castor oil-based polyols for next generation of environmentally-friendly hair-styling agents. Prog Org Coating 2020;142:105588.
36. Shen YB, He JL, Xie ZX, Zhou X, Fang CQ, Zhang C. Synthesis and characterization of vegetable oil based polyurethanes with tunable thermomechanical performance. Ind Crop Prod 2019;140: 111711.
37. Liang B, Li RP, Zhang Q, Yang ZH, Yuan T. Synthesis and characterization of a novel tri-functional bio-based methacrylate prepolymer from castor oil and its application in UV-curable coatings. Ind Crop Prod 2019;135:170–8.
38. Zhao MH, Wang YQ, Liu LX, Liu LX, Chen M, Zhang Q, et al. Green coatings from renewable modified bentonite and vegetable oil based polyurethane for slow release fertilizers. Polym Compos 2018; 39:4355–63.
39. Petrovic Z, Guo A, Zhang W. Structure and properties of polyurethanes based on halogenated and nonhalogenated soy-polyols. J Appl Polym Sci 2000;38:4062–9.
40. Guo A, Cho Y, Petrovic ZS. Structure and properties of halogenated and nonhalogenated soy-based polyols. J Appl Polym Sci 2000;38:3900–10.
41. Ramani N, Graiver D, Farminer KW, Tran PT, Tran T. Novel modified fatty acid esters and method of preparation thereof. US patent 0084603, 2010.
42. Sharma V, Kundu PP. Condensation polymers from natural oils. Prog Polym Sci 2008;33:1199–215.
43. Behr A, Fiene M, Buß C, Eilbracht P. Hydroaminomethylation of fatty acids with primary and secondary amines: a new route to interesting surfactant substrates. Eur J Lipid Sci Technol 2000; 102:467–71.
44. Bantchev GB, Kenar JA, Biresaw G, Han MG. Free radical addition of butanethiol to vegetable oil double bonds. J Agric Food Chem 2009;57:1282–90.
45. Koenig NH, Swern D. Organic sulfur derivatives. I. Addition of mercaptoacetic acid to long-chain monounsaturated compounds. J Am Chem Soc 1957;79:362–5.
46. Larock R, Dong X, Chung S, Reddy C, Ehlers L. Preparation of conjugated soybean oil and other natural oils and fatty acids by homogeneous transition metal catalysis. J Am Oil Chem Soc 2001;78: 447–53.
47. Gunstone FD. Chemical reactions of fatty acids with special reference to the carboxyl group. Eur J Lipid Sci Technol 2001;103:307–14.
48. Zlatanic A, Petrovic ZS, Dusek K. Structure and properties of triolein-based polyurethane networks. Biomarcomolecules 2002;3:1048–56.
49. Lyon C, Garrett V, Frankel E. Rigid urethane foams from hydroxymethylated castor oil, safflower oil, oleic safflower oil, and polyol esters of castor acids. J Am Oil Chem Soc 1974;51:331–4.
50. Petrovic Z, Guo A, Javni I, Cvetkovic I, Hong DP. Polyurethane networks from polyols obtained by hydroformylation of soybean oil. Polym Int 2008;57:275–81.

51. Li ZR, Zhao YH, Yan SR, Wang XK, Kang MQ, Wang JW, et al. Catalytic synthesis of carbonated soybean oil. Catal Lett 2008;123:246–51.
52. Parzuchowski PG, Jurczyk-Kowalska M, Ryszkowska J, Rokicki G. Epoxy resin modified with soybean oil containing cyclic carbonate groups. J Appl Polym Sci 2006;102:2904–14.
53. Lu Y, Larock RC. Soybean-oil-based waterborne polyurethane dispersions: effects of polyol functionality and hard segment content on properties. Biomacromolecules 2008;9:3332–40.
54. Zhang C, Li Y, Chen R, Kessler MR. Polyurethanes from solvent-free vegetable oil-based polyols. ACS Sustain Chem Eng 2014;2:2465–76.
55. Feng YC, Liang HY, Yang ZM, Yuan T, Luo Y, Li PW, et al. A solvent-free and scalable method to prepare soybean-oil-based polyols by thiol-ene photo-click reaction and biobased polyurethanes therefrom. ACS Sustain Chem Eng 2017;5:7365–73.
56. Liu LX, Lu JY, Zhang Y, Liang HY, Liang DS, Jiang JZ, et al. Thermosetting polyurethanes prepared with the aid of a fully bio-based emulsifier with high bio-content, high solid content, and superior mechanical properties. Green Chem 2019;21:526–37.
57. Babanejad N, Farhadian A, Omrani I, Nabid MR. Design, characterization and in vitro evaluation of novel amphiphilic block sunflower oil-based polyol nanocarrier as a potential delivery system: raloxifene-hydrochloride as a model. Mater Sci Eng C 2017;78:59–68.
58. Omrani I, Farhadian A, Babanejad N, Shendi HK, Ahmadi A, Nabid MR. Synthesis of novel high primary hydroxyl functionality polyol from sunflower oil using thiol-yne reaction and their application in polyurethane coating. Eur Polym J 2016;82:220–31.
59. Hajirahimkhan S, Xu CC, Ragogna PJ. Ultraviolet uv curable coatings of modified lignin. ACS Sustain Chem Eng 2018;6:14685–94.
60. Guo L, Huang S, Qu JQ. Synthesis and properties of high-functionality hydroxyl-terminated polyurethane dispersions. Prog Org Coating 2018;119:214–20.
61. Valverde C, Lligadas G, Ronda JC, Galià M, Cádiz V. Hydroxyl functionalized renewable polyesters derived from 10-undecenoic acid: polymer structure and postpolymerization modification. Eur Polym J 2018;105:68–78.
62. Hojabri L, Kong X, Narine SS. Fatty acid-derived diisocyanate and biobased polyurethane produced from vegetable oil: synthesis, polymerization, and characterization. Biomacromolecules 2009;10:884–91.
63. Shendi HK, Omrani I, Ahmadi A, Farhadian A, Babnejad N, Nabid MR. Synthesis and characterization of a novel internal emulsifier derived from sunflower oil for the preparation of waterborne polyurethane and their application in coatings. Prog Org Coating 2017;105:303–9.
64. Cheng Z, Li QT, Yan Z, Liao GF, Zhang BX, Yu YM, et al. Design and synthesis of novel aminosiloxane crosslinked linseed oil-based waterborne polyurethane composites and its physicochemical properties. Prog Org Coating 2019;127:194–201.
65. Lu KT, Chang JP. Synthesis and antimicrobial activity of metal-containing linseed oil-based waterborne urethane oil wood coatings. Polymers 2020:12.
66. Chen R, Zhang C, Kessler MR. Anionic waterborne polyurethane dispersion from a bio-based ionic segment. RSC Adv 2014;4:35476–83.
67. Openshaw K. A review of jatropha curcas: an oil plant of unfulfilled promise. Biomass Bioenergy 2000;19:1–15.
68. Kumar A, Sharma S. An evaluation of multipurpose oil seed crop for industrial uses (Jatropha curcas L.): a review. Ind Crop Prod 2008;28:1–10.
69. Ling CK, Aung MM, Rayung M, Abdullah LC, Lim HN, Noor ISM. Performance of ionic transport properties in vegetable oil-based polyurethane acrylate gel polymer electrolyte. ACS Omega 2019;4:2554–64.
70. Sariah S, Chuah AL, Aung MM, Salleh MZ, Biak DRA, Basri M, et al. Physicochemical properties of jatropha oil-based polyol produced by a two steps method. Molecules 2017;22:551.

71. Saalah S, Abdullah LC, Aung MM, Salleh MZ, Biak DRA, Basri M, et al. Colloidal stability and rheology of jatropha oil-based waterborne polyurethane (JPU) dispersion. Prog Org Coating 2018; 125:348–57.
72. Tanaka R, Hirose S, Hatakeyama H. Preparation and characterization of polyurethane foams using a palm oil-based polyol. Bioresour Technol 2008;99:3810–6.
73. Su YP, Lin H, Zhang ST, Yang ZH, Yuan T. One-step synthesis of novel renewable vegetable oil-based acrylate prepolymers and their application in uv-curable coatings. Polymers 2020;12:1165.
74. Sittinun A, Pisitsak P, Manuspiya H, Thiangtham S, Chang YH. Utilization of palm olein-based polyol for polyurethane foam sponge synthesis: potential as a sorbent material. J Polym Environ 2020;28:3181–91.
75. Lligadas G, Ronda JC, Galia M, Cadiz V. Plant oils as platform chemicals for polyurethane synthesis: current state-of-the-art. Biomacromolecules 2010;11:2825–35.
76. Cayli G, Kusefoglu S. Biobased polyisocyanates from plant oil triglycerides: synthesis, polymerization, and characterization. J Appl Polym Sci 2010;109:2948–55.
77. Cayli G, Kusefoglu S. A simple one-step synthesis and polymerization of plant oil triglyceride iodo isocyanates. J Appl Polym Sci 2010;116:2433–40.
78. Hojabri L, Kong X, Narine SS. Novel long chain unsaturated diisocyanate from fatty acid: synthesis, characterization, and application in bio-based polyurethane. J Polym Sci, Part A: Polym Chem 2010;48:3302–10.
79. Zakzeski J, Bruijnincx PCA, Jongerius AL, Weckhuysen BM. The catalytic valorization of lignin for the production of renewable chemicals. Chem Rev 2013;110:3552–99.
80. Ma XZ, Chen J, Zhu J, Yan N. Lignin-based polyurethane: recent advances and future perspectives. Macromol Rapid Commun 2020:2000492.
81. Dong X, Dong MD, Lu YJ, Turley A, Lin T, Wu CQ. Antimicrobial and antioxidant activities of lignin from residue of corn stover to ethanol production. Ind Crop Prod 2011;34:1629–34.
82. Domenek S, Louaifi A, Guinault A, Baumberger S. Potential of lignins as antioxidant additive in active biodegradable packaging materials. J Polym Environ 2013;21:692–701.
83. El Salamouny S, Shapiro M, Ling KS, Shepard BM. Black tea and lignin as ultraviolet protectants for the beet armyworm nucleopolyhedrovirus. J Entomol Sci 2009;44:50–8.
84. Liu LN, Qian MB, Song PA, Huang GB, Yu YM, Fu SY. Fabrication of green lignin-based flame retardants for enhancing the thermal and fire retardancy properties of polypropylene/wood composites. ACS Sustain Chem Eng 2016;4:2422–31.
85. Bernardini J, Cinelli P, Anguillesi I, Coltelli MB, Lazzeri A. Flexible polyurethane foams green production employing lignin or oxypropylated lignin. Eur Polym J 2015;64:147–56.
86. Cinelli P, Anguillesi I, Lazzeri A. Green synthesis of flexible polyurethane foams from liquefied lignin. Eur Polym J 2013;49:1174–84.
87. Liu J, Liu HF, Deng L, Liao B, Guo QX. Improving aging resistance and mechanical properties of waterborne polyurethanes modified by lignin amines. J Appl Polym Sci 2013;130:1736–42.
88. Ren LF, Zhao YX, Qiang TT, He QQ. Synthesis of a biobased waterborne polyurethane with epichlorohydrin-modified lignin. J Dispersion Sci Technol 2019;40:1499–506.
89. Cheradame H, Detoisien M, Gandini A, Pla F, Roux G. Polyurethane from kraft lignin. Br Polym J 1989;21:269–75.
90. Li H, Liang Y, Li PC. Conversion of biomass lignin to high-value polyurethane: a review. A review. J Bioresour Bioprod 2020;5:163–79.
91. Griffini G, Passoni V, Suriano R, Levi M, Turni S. Polyurethane coatings based on chemically unmodified fractionated lignin. ACS Sustain Chem Eng 2015;3:1145–54.
92. Wang Y, Wyman C, Cai C, Ragauskas AJ. Lignin-based polyurethanes from unmodified kraft lignin fractionated by sequential precipitation. ACS Appl Polym Mater 2019;1:1672–9.

93. Zhang Y, Liao JJ, Fang XC, Bai FD, Qiao K, Wang LM. Renewable high performance polyurethane bioplastics derived from lignin-poly(ε-caprolactone). ACS Sustain Chem Eng 2017;5:4276–84.
94. Laurichesse S, Avérous L. Chemical modification of lignins: towards biobased polymers. Prog Polym Sci 2014;39:1266–90.
95. Lora JH, Glasser WG. Recent industrial applications of lignin: a sustainable alternative to nonrenewable materials. J Polym Environ 2002;10:39–48.
96. Sadeghifar H, Cui C, Argyropoulos DS. Toward thermoplastic lignin polymers. Part I. selective masking of phenolic hydroxyl groups in kraft lignins via methylation and oxypropylation chemistries. Ind Eng Chem Res 2012;51:16713–20.
97. Ghosh T, Karak N. Cashew nut shell liquid terminated self-healable polyurethane as an effective anticorrosive coating with biodegradable attribute. Prog Org Coating 2020;139.
98. Kathalewar M, Sabnis A. Preparation of novel CNSL-based urethane polyol via nonisocyanate route: curing with melamine-formaldehyde resin and structure-property relationship. J Appl Polym Sci 2015;132:41391.
99. Suresh KI. Rigid polyurethane foams from cardanol: synthesis, structural characterization, and evaluation of polyol and foam properties. ACS Sustain Chem Eng 2012;1:232–42.
100. Wang HR, Zhou QX. Synthesis of cardanol-based polyols via thiol-ene/thiol-epoxy dual click-reactions and thermosetting polyurethanes therefrom. ACS Sustain Chem Eng 2018;6:12088–95.
101. Hu SJ, Wan CX, Li YB. Production and characterization of biopolyols and polyurethane foams from crude glycerol based liquefaction of soybean straw. Bioresour Technol 2012;103:227–33.
102. Zhang J, Hori N, Takemura A. Thermal and time regularities during oilseed rape straw liquefaction process to produce bio-polyol. J Clean Prod 2020;277:124015.
103. Chen FG, Lu ZM. Liquefaction of wheat straw and preparation of rigid polyurethane foam from the liquefaction products. J Appl Polym Sci 2010;111:508–16.
104. Wang TP, Zhang LH, Li D, Yin J, Wu S, Mao ZH. Mechanical properties of polyurethane foams prepared from liquefied corn stover with PAPI. Bioresour Technol 2008;99:2265–8.
105. Jason D, Yan N. Producing bark-based polyols through liquefaction: effect of liquefaction temperature. ACS Sustain Chem Eng 2013;1:534–40.
106. Ye LY, Zhang JM, Zhao J, Tu S. Liquefaction of bamboo shoot shell for the production of polyols. Bioresour Technol 2014;153:147–53.
107. Wang D, Zhou X, Song R, Fang C, Wang Z, Wang C, et al. Freestanding silver/polypyrrole composite film for multifunctional sensor with biomimetic micropattern for physiological signals monitoring. Chem Eng J 2020;404:126940.
108. Serrano L, Rincón E, García A. Bio-degradable polyurethane foams produced by liquefied polyol from wheat straw biomass. Polymers 2020;12:2646.
109. Monica FD, Kleij AW. From terpenes to sustainable and functional polymers. Polym Chem 2020;11:5109–27.
110. Touaibia M, Boutekedjiret C, Perino S, Chemat F. Natural terpenes as building blocks for green chemistry. Singapore: Springer; 2019.
111. Liu GF, Wu GM, Jin C, Kong ZW. Preparation and antimicrobial activity of terpene-based polyurethane coatings with carbamate group-containing quaternary ammonium salts. Prog Org Coating 2015;80:150–5.
112. Gupta RK, Ionescu M, Radojcic D, Wan X, Petrovic ZS. Novel renewable polyols based on limonene for rigid polyurethane foams. J Polym Environ 2014;22:304–9.
113. Bhr M, Bitto A, Mülhaupt R. Cyclic limonene dicarbonate as a new monomer for non-isocyanate oligo- and polyurethanes (NIPU) based upon terpenes. Green Chem 2012;14:1447–54.
114. Firdaus M, Meier MAR. Renewable polyamides and polyurethanes derived from limonene. Green Chem 2013;15:370–80.

115. Luc C, Xavier F, Serge K. Ultrasonic and catalyst free epoxidation of limonene and other terpenes using dimethyl dioxirane in semi-batch conditions. ACS Sustain Chem Eng 2018;6:12224–31.

116. Wu GM, Kong ZW, Chen J, Huo SP, Liu GF. Preparation and properties of waterborne polyurethane/epoxy resin composite coating from anionic terpene-based polyol dispersion. Prog Org Coating 2014;77:315–21.

117. Maiti S, Ray SS, Kundu AK. Rosin: a renewable resource for polymers and polymer chemicals. Prog Polym Sci 1989;14:297–338.

118. Wilbon PA, Chu FX, Tang CB. Progress in renewable polymers from natural terpenes, terpenoids, and rosin. Macromol Rapid Commun 2013;34:8–37.

119. Li TT, Liu XQ, Jiang YH, Ma SQ, Zhu J. Bio-based shape memory epoxy resin synthesized from rosin acid. Iran Polym J 2016;25:1–9.

120. Xu X, Song ZQ, Shang SB, Cui SQ, Rao XP. Synthesis and properties of novel rosin-based waterborne polyurethane. Polym Int 2011;60:1521–6.

121. Hsieh CC, Chen YC. Synthesis of bio-based polyurethane foam modified with rosin using an environmentally-friendly process. J Clean Prod 2020:276.

122. Liu GF, Wu GM, Chen J, Kong ZW. Synthesis, modification and properties of rosin-based non-isocyanate polyurethanes coatings. Prog Org Coating 2016;101:461–7.

123. Vevere L, Fridrihsone A, Kirpluks M. A Review of wood biomass-based fatty acids and rosin acids use in polymeric materials. Polymer 2020;12:2706.

124. Zhang L, Jiang Y, Xiong Z, Liu X, Na H, Zhang R, et al. Highly recoverable resin-based shape memory polyurethenes. J Mater Chem 2013;1:3263.

125. Wang YX, Zhang LN. High-strength waterborne polyurethane reinforced with waxy maize starch nanocrystals. J Nanosci Nanotechnol 2008;8:5831–8.

126. Li YY, Jing WW, Wang JH, Li JF. Elucidating the relationship between structure and property of waterborne polyurethane-cellulose nanocrystals nanocomposite films. Sci Adv Mater 2020;12:1213–24.

127. Omrani I, Babanejad N, Shendi HK, Nabid MR. Fully glutathione degradable waterborne polyurethane nanocarriers: preparation, redox-sensitivity, and triggered intracellular drug release. Mater Sci Eng C 2017;70:607–16.

128. Shin EJ, Choi SM. Advances in waterborne polyurethane-based biomaterials for biomedical applications. Adv Exp Med Biol 2018;107:251–83.

129. Åkerholm M, Hinterstoisser B, Salmén L. Characterization of the crystalline structure of cellulose using static and dynamic FT-IR spectroscopy. Carbohydr Res 2004;339:569–78.

130. Siró I, Plackett D. Microfibrillated cellulose and new nanocomposite materials: a review. Cellulose 2010;17:459–94.

131. Liu H, Song J, Shang SB, Song ZQ, Wang D. Cellulose nanocrystal/silver nanoparticle composites as bifunctional nanofillers within waterborne polyurethane. ACS Appl Mater Interfac 2012;4:2413–19.

132. Klemm D, Kramer F, Moritz S, Lindstrom T, Ankerfors M, Gray D, et al. Nanocelluloses: a new family of nature-based materials. Angew Chem Int Ed 2011;50:5438–66.

133. Lee KY, Aitomaki Y, Berglund LA, Oksman K, Bismarck A. On the use of nanocellulose as reinforcement in polymer matrix composites. Compos Sci Technol 2014;105:15–27.

134. Azeredo HMC, Rosa MF, Mattoso LHC. Nanocellulose in bio-based food packaging applications. Ind Crop Prod 2016;97:664–71.

135. Jorfi M, Foster EJ. Recent advances in nanocellulose for biomedical applications. J Appl Polym Sci 2015;132.

136. Chen RD, Huang CF, Hsu SH. Composites of waterborne polyurethane and cellulose nanofibers for 3D printing and bioapplications. Carbohydr Polym 2019;212:75–88.

137. Dutta GK, Karak N. Waste brewed tea leaf derived cellulose nanofiber reinforced fully bio-based waterborne polyester nanocomposite as an environmentally benign material. RSC Adv 2019;9: 20829–40.
138. Choi SM, Lee MW, Shin EJ. One-pot processing of regenerated cellulose nanoparticles/ waterborne polyurethane nanocomposite for eco-friendly polyurethane matrix. Polymers 2019; 11.
139. Kong L, Xu D, He Z, Wang F, Gui S, Fan J, et al. Nanocellulose-reinforced polyurethane for waterborne wood coating. Molecules 2019;24.
140. Zhou X, Zhang X, Wang D, Fang C, Lei W, Huang Z, et al. Preparation and characterization of waterborne polyurethane/cellulose nanocrystal composite membrane from recycling waste paper. J Renew Mater 2020;8:631–45.
141. Lei W, Zhou X, Fang C, Song Y, Li Y. Eco-friendly waterborne polyurethane reinforced with cellulose nanocrystal from office waste paper by two different methods. Carbohydr Polym 2019;209: 299–309.
142. Hormaiztegui MEV, Daga B, Aranguren MI, Mucci V. Bio-based waterborne polyurethanes reinforced with cellulose nanocrystals as coating films. Prog Org Coating 2020;144:105649.
143. Zhang P, Lu Y, Fan M, Jiang P, Dong Y. Modified cellulose nanocrystals enhancement to mechanical properties and water resistance of vegetable oil-based waterborne polyurethane. J Appl Polym Sci 2019;136.
144. Zhang SD, Sun K, Liu H, Chen XY, Zheng YJ, Shi XZ, et al. Enhanced piezoresistive performance of conductive WPU/CNT composite foam through incorporating brittle cellulose nanocrystal. Chem Eng J 2020;387:124045.
145. Cheng LS, Ren SB, Lu XN. Application of eco-friendly waterborne polyurethane composite coating incorporated with nano cellulose crystalline and silver nano particles on wood antibacterial board. Polymers 2020;12:407.
146. Hu WL, Chen SY, Yang JX, Li Z, Wang HP. Functionalized bacterial cellulose derivatives and nanocomposites. Carbohydr Polym 2014;101:1043–60.
147. Urbina L, Alonso-Varona A, Saralegi A, Palomares T, Eceiza A, Corcuera MA, et al. Hybrid and biocompatible cellulose/polyurethane nanocomposites with water-activated shape memory properties. Carbohydr Polym 2019;216:86–96.
148. Zia F, Zia KM, Zuber M, Kamal S, Aslam N. Starch based polyurethanes: a critical review updating recent literature. Carbohydr Polym 2015;134:784–98.
149. Le Corre D, Bras J, Dufresne A. Starch nanoparticles: a review. Biomacromolecules 2010;11:1139.
150. Chemelli A, Gomernik F, Thaler F, Huber A, Hirn U, Bauer W, et al. Cationic starches in paper-based applications-a review on analytical methods. Carbohydr Polym 2020;235:115964.
151. de Azeredo HMC. Nanocomposites for food packaging applications. Food Res Int 2009;42: 1240–53.
152. Chum HL. Polymers from biobased materials. USA: Noyes Data Corporation; 1991.
153. Tai NL, Adhikari R, Shanks R, Halley P, Adhikari B. Flexible starch-polyurethane films: effect of mixed macrodiol polyurethane ionomers on physicochemical characteristics and hydrophobicity. Carbohydr Polym 2018;197:312–25.
154. Lee SJ, Kim BK. Covalent incorporation of starch derivative into waterborne polyurethane for biodegradability. Carbohydr Polym 2012;87:1803–9.
155. Ionescu M. Chemistry and technology of polyols for polyurethanes. United Kingdom: Rapra Technology; 2005.
156. Wang Y, Zhang L. High-strength waterborne polyurethane reinforced with waxy maize starch nanocrystals. J Nanosci Nanotechnol 2008;8:5831–8.

157. Chang PR, Ai FJ, Chen Y, Dufresne A, Huang J. Effects of starch nanocrystal-graft-polycaprolactone on mechanical properties of waterborne polyurethane-based nanocomposites. J Appl Polym Sci 2008;111:619–27.

158. Chen GJ, Wei M, Chen JH, Huang J, Dufresne A, Chang PR. Simultaneous reinforcing and toughening: new nanocomposites of waterborne polyurethane filled with low loading level of starch nanocrystals. Polymer 2008;49:1860–70.

159. Yang DY, Zhang HQ, Rong XS, Qiu FX. Investigations on oxidised starch based waterborne polyurethane nanocomposites. Plast Rub Compos 2012;41:425–9.

160. Rinaudo M. Chitin and chitosan-properties and applications. ChemInform 2007;31:603–32.

161. Younes I, Rinaudo M. Chitin and chitosan preparation from marine sources, structure, properties and applications. Mar Drugs 2015;13:1133–74.

162. Kean T, Thanou M. Biodegradation, biodistribution and toxicity of chitosan. Adv Drug Deliv Rev 2010;62:3–11.

163. Bankoti K, Rameshbabu AP, Datta S, Maity PP, Goswami P, Datta P, et al. Accelerated healing of full thickness dermal wounds by macroporous waterborne polyurethane-chitosan hydrogel scaffolds. Mater Sci Eng C 2017;81:133–43.

164. Naz F, Zuber M, Zia KM, Salman M, Chakraborty J, Nath I, et al. Synthesis and characterization of chitosan-based waterborne polyurethane for textile finishes. Carbohydr Polym 2018;200:54–62.

165. Liu YX, Zou YL, Wang J, Wang S, Liu XF. A novel cationic waterborne polyurethane coating modified by chitosan biguanide hydrochloride with application potential in medical catheters. J Appl Polym Sci 2020:e50290. https://doi.org/10.1002/app.50290.

166. Lin TW, Hsu SH. Self-healing hydrogels and cryogels from biodegradable polyurethane nanoparticle crosslinked chitosan. Adv Sci 2020;7:1901388.

167. Xu WW, Xiao MH, Yuan LT, Zhang J, Hou ZS. Preparation, physicochemical properties and hemocompatibility of biodegradable chitooligosaccharide-based polyurethane. Polymers 2018; 10:580.

168. Hu LS, Guang CY, Liu Y, Su ZQ, Gong SD, Yao YJ, et al. Adsorption behavior of dyes from an aqueous solution onto composite magnetic lignin adsorbent. Chemosphere 2020;246:125757.

169. Bhagavathi LR, Deshpande AP, Ram GDJ, Panigrahi SK. Hygrothermal aging, fatigue and dynamic mechanical behavior of cellulosic particles reinforced one-component moisture curable polyurethane adhesive joints. Int J Adhes Adhes 2021;105:102771.

170. Oveissi F, Naficy S, Le TYL, Fletcher DF, Dehghani F. Tough and processable hydrogels based on lignin and hydrophilic polyurethane. ACS Appl Bio Mater 2018;1:2073–81.

171. Yang Q, Hu G, Ma Z. Review of characteristics of sodium alginate and its application in meat products. China Food Addit 2010:164–8.

172. Daemi H, Barikani M, Barmar M. Compatible compositions based on aqueous polyurethane dispersions and sodium alginate. Carbohydr Polym 2013;92:490–6.

173. Drury JL, Mooney DJ. Hydrogels for tissue engineering: scaffold design variables and applications. Biomaterials 2003;24:4337–51.

174. Wang X, Zhang Y, Liang HY, Zhou X, Fang CQ, Zhang CQ, et al. Synthesis and properties of castor oil-based waterborne polyurethane/sodium alginate composites with tunable properties. Carbohydr Polym 2019;208:391–7.

175. Zhang Y, Zhang WB, Deng HH, Zhang WH, Kang J, Zhang C. Enhanced mechanical properties and functional performances of cationic waterborne polyurethanes enabled by different natural phenolic acids. ACS Sustain Chem Eng 2020;8:17447–57.

176. Ren LF, Ma XD, Zhang J, Qiang TT. Preparation of gallic acid modified waterborne polyurethane made from bio-based polyol. Polymer 2020;194:122370.

177. Xu HP, Qiu FX, Wang YY, Wu WL, Yang DY, Guo Q. UV-curable waterborne polyurethane-acrylate: preparation, characterization and properties. Prog Org Coating 2012;73:47–53.

178. Chen Y, Zhou S, Gu G, Wu L. Microstructure and properties of polyester-based polyurethane/titania hybrid films prepared by sol-gel process. Polymer 2006;47:1640.
179. Zhang JY, Xu HP, Hu L, Yang Y, Li HB, Huang C, et al. Novel waterborne uv-curable hyperbranched polyurethane acrylate/silica with good printability and rheological properties applicable to flexographic ink. ACS Omega 2017;2:7546–58.
180. Zhang WB, Zhang Y, Liang HY, Liang DS, Cao HY, Liu CG, et al. High bio-content castor oil based waterborne polyurethane/sodium lignosulfonate composites for environmental friendly UV absorption application. Ind Crop Prod 2019;142:111836.
181. Johannson C. Bio-nanocomposites for food packaging applications. Oxford: Oxford University Press; 2011.
182. Avnesh K, Sudesh KY, Subhash C. Biodegradable polymeric nanoparticles based drug delivery systems. Colloids Surf B Biointerfaces 2009;75:1–18.
183. Woodruff MA, Hutmacher DW. The return of a forgotten polymer-polycaprolactone in the 21st century. Prog Polym Sci 2010;35:1217–56.
184. Ali FB, Kang DJ, Kim MP, Cho CH, Kim BJ. Synthesis of biodegradable and flexible, polylactic acid based, thermoplastic polyurethane with high gas barrier properties. Polym Int 2014;63:1620–6.
185. Arrieta MP, Sessini V, Peponi L. Biodegradable poly(ester-urethane) incorporated with catechin with shape memory and antioxidant activity for food packaging. Eur Polym J 2017;94:111–24.
186. Mucci VL, Hormaiztegui MEV, Aranguren MI. Plant oil-based waterborne polyurethanes: a brief review. J Renew Mater 2020;8:579–601.
187. Zhou X, Fang C, Yu Q, Yang R, Xie L, Cheng Y, et al. Synthesis and characterization of waterborne polyurethane dispersion from glycolyzed products of waste polyethylene terephthalate used as soft and hard segment. Int J Adhesion Adhes 2017;74:49–56.
188. Lopez A, Contraires ED, Canetta E, Creton C, Keddie JL, Asua JM. Waterborne polyurethane-acrylic hybrid nanoparticles by miniemulsion polymerization: applications in pressure-sensitive adhesives. Langmuir 2011;27:3878–88.
189. Liu H, Li C, Sun XS. Soy-oil-based waterborne polyurethane improved wet strength of soy protein adhesives on wood. Int J Adhesion Adhes 2017;73:66–74.
190. Hu J, Meng H, Li G, Ibekwe SI. A review of stimuli-responsive polymers for smart textile applications. Smart Mater Struct 2012;21:053001.
191. Liang HY, Li YC, Huang SY, Huang KX, Zeng XY, Dong QW, et al. Tailoring the performance of vegetable oil-based waterborne polyurethanes through incorporation of rigid cyclic rings into soft polymer networks. ACS Sustain Chem Eng 2020;8:914–25.

Samy Madbouly*, Sean Edlis and Nicolas Ionadi

6 Soybean-based polymers and composites

Abstract: Development and evaluation of new bio-based sustainable plastics to replace the petroleum-based materials in different industrial applications has both environmental and economic benefits. Bio-based polymers can be widely used in biomedical and agriculture applications due to their excellent biodegradability and biocompatibility. Soy protein is a natural material that can be isolated from soybean, which is a major agricultural crop in the U.S. The viability of soybean-based polymers and composites is questioned due to their high-water absorption and poor mechanical properties. There have been many environmentally friendly attempts to improve the properties of soybean polymers as soybeans and their extracts are widely available worldwide. Soy protein, hulls, and oils all find use in the development of different biodegradable polymers. While the development looks promising, there is still more work to do to make the soybean polymers useful and economically viable. Blending soy protein with other biodegradable polymers, such as polylactide (PLA) and polyurethane dispersion is a valid approach to improve the mechanical properties of soy protein and reduce its water sensitivity.

Keywords: bio-based; biodegradable; composite; horticulture crop containers; polymer; soybean.

6.1 Introduction

Since the mid-1900s there has been a dramatic increase in the amount of plastic that gets used. The substance that was once only found in nature, now relatively dictates our way of life. Everything that surrounds us either contains plastics or took advantage of the use of plastics for their fabrication. In 2018 alone the world produced a volume of 359 million metric tons of plastic, growing by a near 240% since the 1950s [1].

While plastic has transformed the way we live today, there is one crucial flaw with the material that can be seen worldwide. When the disposal of plastic gets considered by the average person, there is usually no consideration and an assumption is made. The assumption being is that this material will breakdown in naturally, just how paper and metal products decompose back into the Earth. This is not the case; plastic pollution is flooding our planet. While there are commercial ways to recycle some

*Corresponding author: Samy Madbouly,** Behrend College, School of Engineering, Pennsylvania State University, Erie, PA 16563, USA, E-mail: sum1541@psu.edu
Sean Edlis and Nicolas Ionadi, Behrend College, School of Engineering, Pennsylvania State University, Erie, PA 16563, USA

This article has previously been published in the journal Physical Sciences Reviews. Please cite as: S. Madbouly, S. Edlis and N. Ionadi "Soybean-based polymers and composites" *Physical Sciences Reviews* [Online] 2021, 7. DOI: 10.1515/psr-2020-0069 | https://doi.org/10.1515/9781501521942-006

plastics, not all plastics can be recycled. Nor would recycling be a viable idea for most companies to invest into, until recently.

In the past couple years, more and more people are catching onto the problem that plastic pollution is causing more companies to take up the problem seriously. To solve this problem, oddly enough, today's top minds are looking into the past to revolutionize the future. This may seem odd; however, Henry Ford was one of the earliest pioneers of the biodegradable plastic. The area that interested him and his scientists the most was the used of soy by-products to help reduce costs during the great depression.

In the 1930s the Ford Motor Company started to use soybean oil as the base for their paints, instead of the previous lacquer that was used [2]. By 1933, some of Ford's vehicles even came with an option for a trim package that included parts produced with a soy protein plastic [2]. This soy protein plastic came from blending a phenol-formaldehyde resin with soybean meal to create a biodegradable plastic [3]. By 1941 the Ford Motor Company even produced a concept car where its outer body was almost completely made of soybean plastic.

While it's been almost 90 years since the start of using soy by-products as plastic substitutes, there has not been any many major breakthroughs with this technology. A majority of all the soy plastics that are produced today have relatively the same properties as they did in the 1940s. However, in the past few years that has been changing. Plastic pollution is now a seriously threat to our planet and much research is going into the study of how to dispose of them properly or completely negate the need for recycling (biodegradable plastics). Once a thing of the past, soy plastics just might lead the way to the future.

Soy plastics can have two different meanings, either conventual polymers, like polyethylene or polypropylene, that are mixed with soybean by-products as an additive or filler. Or the use of soybean proteins, that are extracted out of soy by-products, to produce natural soy protein plastics. Soy protein plastics are commonly derived from three different soybean by-products: soy flour (SF), soy protein concentrate (SPC), and soy protein isolate (SPI). SF is the by-product of dehulled soybeans that have been solvent-defatted, then meal ground into flour after being dried. SF is fat-free, low fiber flour with an average protein percentage of 50% by weight [3]. SPC is the by-product of SF that has had all of its water and alcohol-soluble sugars leeched out, through heat, increasing the protein percentage to 65 wt.% [3]. While SPI is the by-product of SPC that has gone through alkali and reprecipitated acidification near its isoelectric point, pH 4.5 [3]. This process isolates the protein molecules within the concentrate increasing the protein percentage up to or even more than 90 wt.% [3]. SPI and SPC are available in two common forms, namely isoelectric (water insoluble) and neutralized (sodium salt) form [3].

SPIs are the most common soy protein extract. They consist of four main proteins called globulins that exist in all the soybean extracts. The four globulins are 2S, 7S, 11S, and 15S. The 7S and 11S occur the most frequently, making up approximately 60% of the total storage protein of soybeans [4]. The 7S themselves make up about 70% of the total protein in SPI. This results in the 7S having a large effect on the properties of the SPI. The main factors that take effect are the molecular weight (M_W) and the reaction to water. The 7S have a M_W of 200,000 and are hydrophilic. On the other hand, 11S have a M_W of 350,000 mostly due to the tighter orientation of their functional groups and are hydrophobic due to the bonds between their subunits [5]. However, the higher percentage of the 7S drives the properties of the SPI. The most notable is its hydrophilic nature. This causes the soybean-based polymer derived from the SPI to absorb moisture which is one of the main issues regarding soybean-based polymers.

To make soybean-based polymers commercially viable, the water sensitivity issue needs to be addressed. Research has shown that water "acts as a spacer between chains" in other, more common polymers such as polycarbonate and nylon 6 [6]. In tests and applications at and above room temperature, absorbed moisture reduces the modulus of a polymer as well as its glass transition temperature (T_g) [6]. Both of these factors combine to result in a reduced stiffness of the polymer. The moisture likely has the same effect on soybean-based polymers. This issue often leads soybean-based polymers to be blended with other bio-based polymers that are hydrophobic. Moisture in the proteins may also have an impact on the processing of the polymers as well. If not properly dried, the moisture could cause bubbles and delamination in the molded parts.

One blend that addresses this issue of water absorption is a SPI, poly(ε-caprolactone) (PCL), and toluene 2,4 diisocyanate (TDI) blend [5]. PCL is a semicrystalline polymer that belongs to the polyester family. It is known for its ability to provide good mechanical properties while being biodegradable and miscible with a wide range of polymers. This makes it a perfect complement to SPI and other soybean extracts as it reduces the sensitivity to water while remaining biodegradable. Depending on its environment and M_W, PCL can biodegrade in anywhere from several months to several years [7]. In addition to the water resistance, this blend provides a workable tensile strength that lies between 14.8 and 16.3 MPa [5].

Soybean polymers have a lot of potential in various applications. A main use for these polymers is for films and packaging. A common gripe with consumers is the over usage of plastic in packaging. Soybean biodegradable packaging addresses these concerns, while still performing an adequate job and is not harmful to the food it encloses. Soybean polymers also find use as a bonding agent for plywood and other wood composites, insulating foams, and coatings for cars and agricultural equipment [8].

Soybean-based polymers are typically compounded using twin screw extruders. This is due to twin screw extruders offering a more homogeneous and plasticized mix than a single screw extruder can achieve [5]. After compounding, the polymer is commonly compression or injection molded. Compression molding requires less expensive machinery and does not introduce the polymer to potential degradation due to the high shear heating and pressures experienced in injection molding. However, injection molding can offer more complex geometries. Additionally, injection molding offers more consistent parts as compression molding can lead to weld lines in parts due to poor interaction between the compounded pellets [5].

6.2 Production of DSF, SPC, and SPI

To be able to turn soybeans and soybean waste into a usable material for polymer production, the raw beans must go through several processing steps. These processing steps facilitate the removal of soluble nonprotein matter from the soybean, increasing the overall protein percentage in the soybean. This is important for producing plastic materials out of soybean, since the protein chains make up the primary chain structure of soybean-based plastics.

As mentioned above, soy protein comes in three general forms for plastic production, defatted soy flower (DSF), SPC, and SPI. Each type of soy protein increases the overall protein percentages; DSF ($\approx$54 wt.% protein), SPC ($\approx$70 wt.% protein), and SPI ($\approx$90 wt.% protein) [9]. The increase of protein percentage for each type of soy protein is done by continually refining the soybean material. DSF is the first by-product of refining raw soybeans through solvent extraction.

Before the soybeans can be turned into DSF through solvent extraction, the raw beans must first be prepared for this process. First, the raw beans are dried until they reach a moisture content of 10% or lower [9], this allows for an easier dehulling process of the soybean meat. Next, the beans are allowed to temper for 2–5 days or until they reach a moisture equilibrium point [9]. After the soy material is allowed to equilibrate, it is cleaned, all remaining seeds are cracked, and the soybean material is put into a second tempering process. The tempering process is facilitated by steam, either from direct or indirect contact, through heating the material to 65–70 °C and to 10.5–11% moisture content [9]. Tempering the soybean material allows for an easier flaking process or flattening the soybean material between two rolling cylindrical dies to 0.2–0.35 mm thickness.

After this process is completed the flaked soybean material is ready to be turned into DSF through solvent extraction. Two different methods of solvent extraction can be used to turn flaked soybeans into DSF, through the percolation or flood method [9]. Percolating the solvent over the soybean material is favorable for thin flakes with increased surfaces areas that have little resistance to diffusion [9]. While in the flood method, the solid material is totally immersed into the solvent. Immersing the soybean material completely in the solvent is beneficial for extracting the protein out of thicker, denser, soybean flakes [9]. The recommended solvent medium for protein extraction in soybeans is polar solvents, like ethanol and isopropyl alcohols [9].

The two described methods of solvent extraction, percolation and flood, can be performed in three different manners. First, the solvent extraction process can be done through batch solvent extraction [9], where a certain amount of flakes and volume of solvent are in contact with each other. In this process, the solvent is continually extracted, filtered, and recirculated into the flakes until the desired percentage of protein in the flakes are met. Second, the process can be facilitated through semi-continuous solvent extraction [9]. In semicontinuous extraction systems, a series of batch extraction processes are linked together. Within this process the solvent flows from one extractor to the next after the flakes have reached their specified protein percentage. Fresh solvent and flakes are added separately after each has been depleted from use. Third, the solvent extraction process can be completed through continuous solvent extraction [9]. In this process, fresh solvent and soybean flakes are continually fed into and through the extractor to extract the oil and increase the protein content of the flakes.

Continuous solvent extraction of soybean flakes to defatted soybean flakes, is the most suitable process for industrial applications. It allows for large, continuous, amounts of soybean flakes to be defatted. While continuous solvent extraction is the best method for industrial applications, batch processes are best suited for laboratory and experimental applications. Since only a finite amount of defatted soybean flakes can be produced at once. As well as, the constant monitoring of the process allows for increased control over the final product.

After the solvent extraction process the resulting soybean flakes and residual solvent must be dealt with, to do so they must go through the post extraction process. The soybean flakes and solvent are split into two different streams. One of the streams is used for the refinement of the soybean oil rich solvent, which is known as full micella [9]. Within this stream the full micella goes through a distillation process to extract the soybean oil from the solvent. Common methods of distillation for full micella are flash evaporation, vacuum distillation, and steam stripping [9]. While the other stream caring the solvent saturated soybean flakes, leads to a meal desolventizing process. In this process the saturated soybean flakes are exposed to heat, allowing for the solvent to be evaporated away and recollected for further use.

Once the post extraction process is completed, the dried solvent extracted flakes are finally ready to be turned in to a DSF. For this to be completed the dried flakes are ground in a mill, just like how flower is produced, to the desired consistency. The now ground DSF can be further refined through the use of sieves. By passing the flour through continually finer meshed sieves, the protein percentage of the flower is increased. This is done by the sieves trapping the larger carbohydrate latent material, while allowing for the finer protein rich flower to pass through.

While DSF can be used to produce plastic products, like films [9], they are rarely used on their own and are more often used as a filler for other plastics. To increase the usability of DSF it must be refined into SPC. There are three different methods of turning DSF into SPC, each of which results in an SPC that has varying drawbacks. To turn DSF into SPC all soluble nonprotein materials, like carbohydrates (mono, di, and oligosaccharides) must be removed [9]. During this process, the protein structures suffer some denaturization from the harsh process of removing the carbohydrates. Each of the three processes vary in the degree and the nature of how the protein structures are affected.

The aqueous alcohol wash process is the first of the three different processes to be discussed. In this process aqueous solutions of lower aliphatic alcohols (methanol, ethanol, and isopropyl alcohol) are used to extract soluble sugars from the DSF [9]. The optimal percentage of alcohol for this process, with minimalized denaturing to the proteins, is 60%wt [9]. While the alcohol wash process retains most of the functional properties, the overall solubility of the protein concentrates is decreased when compared to the other two processes. This process also fallows the same principal processes as the previous mentioned solvent extraction process for DSF.

The second process of converting DSF into SPC is the acid wash process. This process relies on the pH sensitivity of proteins, as the pH level nears the isoelectric point (pH 4.5) of the soy protein the solubility of the protein increases [9]. The DSF is mixed with the acidic water solution by drops until the pH value reaches the isoelectric point. Once this is accomplished the mixture is centrifuged until separation and the soy concentrate is then dried [9]. This results in isoelectric SPC (60–70 wt.% protein), which can be rewashed to increase protein content further. While the altered pH level increases the solubility of the proteins, the additional wash stage needed to neutralize the now acidic SPC allows for considerable water absorption. To properly remove the excess moisture from the concentrate the material must be rotary vacuum filtered or decanting centrifuges must be utilized, which results in SPC (70–75 wt.% protein) [9].

6.3 Recent advancements

In an experiment by Fernandez–Espada et al. [10], the processing conditions for the injection molding of a soy protein biopolymer were optimized to find the best conditions that would produce parts that will absorb water while still being able to have workable mechanical properties. They used a 50% soy protein-50% glycerol mixture. The reference values for the processing conditions were a 40 °C cylinder temperature, a 500-bar injection pressure, and a 70 °C mold temperature as these settings are deemed to be the most influential [11–14]. The cylinder and mold temperatures had a greater effect on the tensile and water absorption properties than the pressure [10]. Processing protein-based polymers at high temperature should be done with caution as there is not much data on the injection molding of these materials, and high temperatures in the cylinder may cause premature crosslinking in the polymer [11, 15]

The authors found that, the increased crosslinking at the higher temperatures will make the polymer more stiff and brittle which will lower its load bearing capability. If the polymer crosslinks prematurely, it will not be able to orient correctly. In addition, the highest and lowest pressure values yielded similar elongation values, but the highest setting yielded a higher stress capability. The highest stress values are achieved at the highest injection pressure. The modulus is also the highest at the highest pressure setting. The difference in the stress bearing capabilities is that the higher pressure results in a higher degree of orientation in the polymer [10]. This leads to less potential flaws.

The tensile data for the final stage were found to be mold temperature dependent. The maximum stress and elongation increased at each temperature interval. However, the maximum stress fell at 130 °C, but the elongation increased greatly. These findings were observed by other research groups [12, 16]. The increase of the properties seen in this stage of the experiment can be explained by the higher molding temperatures allowing more chain mobility in the polymer. This allows the chains to flow more freely and gain improves alignment and orientation [16, 17].

The highest cylinder temperature value results in the lowest water uptake. This is due to the phase separation of the glycerol being more apparent at the lower temperatures. This better orientation leads to a more plasticized polymer that does not absorb as much moisture. The values for the injection pressure are the highest at 500 bar and then lower slightly at 900 bar. More pressure results in a higher density which yields samples with higher void fraction after the immersion time, but the 900 bar pressure allows for better packing of the protein which limits the water uptake. The mold temperature has the greatest effect on the water uptake of the polymer. The higher mold temperatures result in more crosslinking in the protein. This is the ideal time for the protein to crosslink as opposed to in the cylinder. The degree of crosslinking is so great that the polymers only lost the glycerol portion at the highest temperature shown by the sub 50% soluble matter loss in the table [10].

A study by Balla et al. [18] found that abundant and inexpensive soybean hulls can be modified with dilute acid hydrolysis to be used in natural fiber reinforced composites. The hulls were used to enhance thermoplastic copolyester (TPC) composites through the 3-D printing process of fused filament fabrication (FFF). Physical and chemical treatments were done to the fibers to determine their effect on the composites. Some of the factors that had influence on the fiber treatment were the interlayer bonding of the filaments, surface quality, and the number of printing defects. The dilute acid hydrolysis treatment of the fibers resulted in nearly a 100% increase in composite density as it reduced pore sizes by half [18].

3-D printed parts lack mechanical strength compared to conventionally molded parts, and therefore require fibers to make up for that disparity. Typically, they are reinforced with synthetic fibers. However, these synthetic fibers require high amounts of energy to be produced [19]. This energy requirement leads some to use natural fibers [20]. Natural fibers such as wood, hemp, and jute are lower cost, biodegradable, and lead to easier recycling [21–25]. The downside to these fibers is their hydrophilicity, clumping, and lack of thermal stability. These fibers also lead to increased porosity as the parts increase in size. The dilute acid hydrolysis processed soybean hulls were tested as a better replacement for the natural fibers due to their downfalls [26].

Table 6.1 shows the different types of materials tested in this experiment. A neat polymer was made without fiber treatment to act as the control. Composites were created with as received hulls both dry and wet shear mixed. Lastly, two sets of composites were created with two different stages of the dilute acid hydrolyzed hulls. Table 6.1 also shows the moisture percentage of each of the composites as well as the printing parameters used to make them.

Figure 6.1 shows the mechanical testing data for each of the five composite make-ups. The samples were tensile tested at a cross head speed of 100 mm/s. The composites were tested at two fiber weight percentages: 5 and 10%. In all four of the graphs show in Figure 6.1, the hydrolyzed composites performed better than the neat TPC and unhydrolyzed hull treated composites. This data shows the viability of the dilute acid hydrolyzed soybean hulls as a natural fiber reinforcement for TPC composites.

Figure 6.2 demonstrates the surface morphology along the build direction for typical TPC-soybean hull composite in comparison with unfilled TPC prepared by FFF fabrication. Obviously, a good interlayer bonding can be observed with few isolated defects. The high viscosity of TPC compared to the TPC-soybean hull leads to a straight interface for TPC and uneven interface with lateral flow in the composite. Unbonded regions or debonding between the layers was also observed for TPC as a result for the high viscosity in comparison to the low viscosity of TPC-soybean hull composite as clearly seen in Figure 6.2. Increasing the printing speed can eliminate this issue. In a conclusion, the interface between layers, the layer-to-layer and bead-to-dead bonding were found to be strong, tight, and diffuse in the composite compared to the unfilled TPC due to the difference in viscosity as seen in Figure 6.2.

In addition to the proteins and hulls being used, the soybean oil also finds use in the polymer industry. An experiment done by Mauck et al. [27] found that the use of acrylated epoxidized soybean oil (AESO) increased the mechanical properties of polylactide (PLA) polymers while being able to maintain the glass transition temperature of neat PLA. Derived from plant sugars, PLA is a biodegradable thermoplastic used in disposable packaging, consumer products, textiles, and other similar applications [28–30]. However, it has low tensile elongation at break and fracture toughness. This causes it to be brittle which this limits its abilities in use. The AESO improves these properties while not

Table 6.1: Different composites and FFF parameters used in the present investigation. Reproduced with permission from [18].

Material	Treatment	Moisture (%)	FFF printing parameters	
AR-Dry	Dry blended as-received hulls	10.05 ± 0.37	Layer thickness	200 µm
AR-Shear	Wet shear mixed as-received hulls	3.16 ± 0.26	Printing speed	30 mm/s
ST1	Single-stage hydrolyzed hulls	2.75 ± 0.52	Extrusion width	550 µm
ST2	Two-stage hydrolyzed hulls	2.96 ± 0.31	Printing temperature	220 °C
TPC	Neat TPC		Bed temperature	65 °C
			Nozzle diameter	500 µm
			Fill angle	±45°

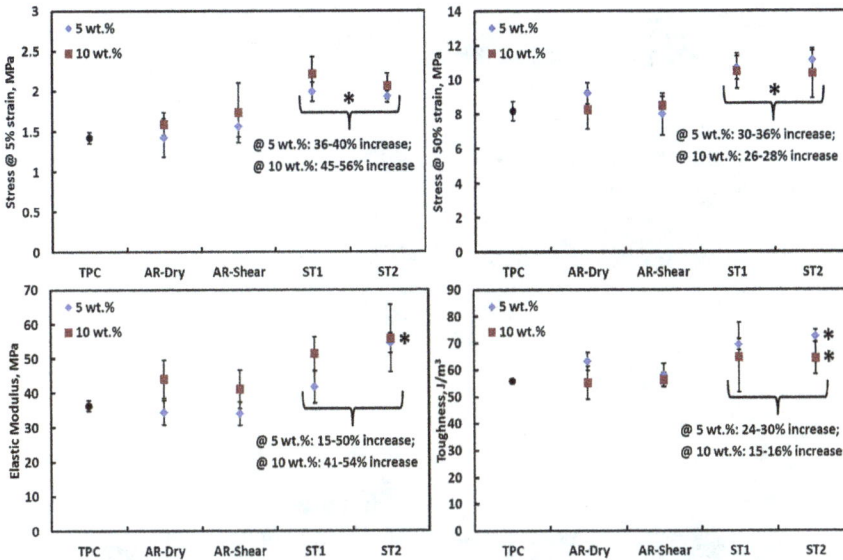

Figure 6.1: Comparison of tensile mechanical properties of pure TPC with TPC-soybean hull fiber composites showing the influence of fiber treatment and concentration. *$p < 0.05$ compared to pure TPC. Adapted with permission from [18].

adding too much of a plasticizing effect. AESO is also biodegradable and renewable which considers the environmental impact.

AESO is modified soybean oil that is commercially available. It is produced by epoxidizing the soybean oil and then opening the epoxide group rings using acrylic acid [31]. This causes the carbon–carbon double bonds bond to hydroxyl and acrylate groups which cause the AESO to be very reactive. This helps it crosslink with the PLA when blended.

Table 6.2 shows the different compositions of each blend tested in the experiment. The first was the PLA blended with unaltered soybean oil (SYBO). The second was a blend with the AESO. The third was a mixture of both the SYBO and the AESO with the PLA. The final combination also included both SYBO and AESO, but it was blended with a PLA star polymer that had an AESO core. This polymer was synthesized by having the hydroxyl groups on the AESO polymerize L-lactide through ring opening. Neat PLA was used as the control.

Table 6.3 shows the tensile properties of each blend in the experiment. The PLA/ AESO and the fourth blend with each of the four components PLA/AESO/SYBO/STAR provided the best improvements in both elongation at break and toughness while maintaining a similar modulus. The fourth blend provides the most viable mechanical properties as the PLA/AESO blend has a large deviation on its data. This is thought to be due to the effects of processing on this particular blend. The PLA/SYBO blend also had

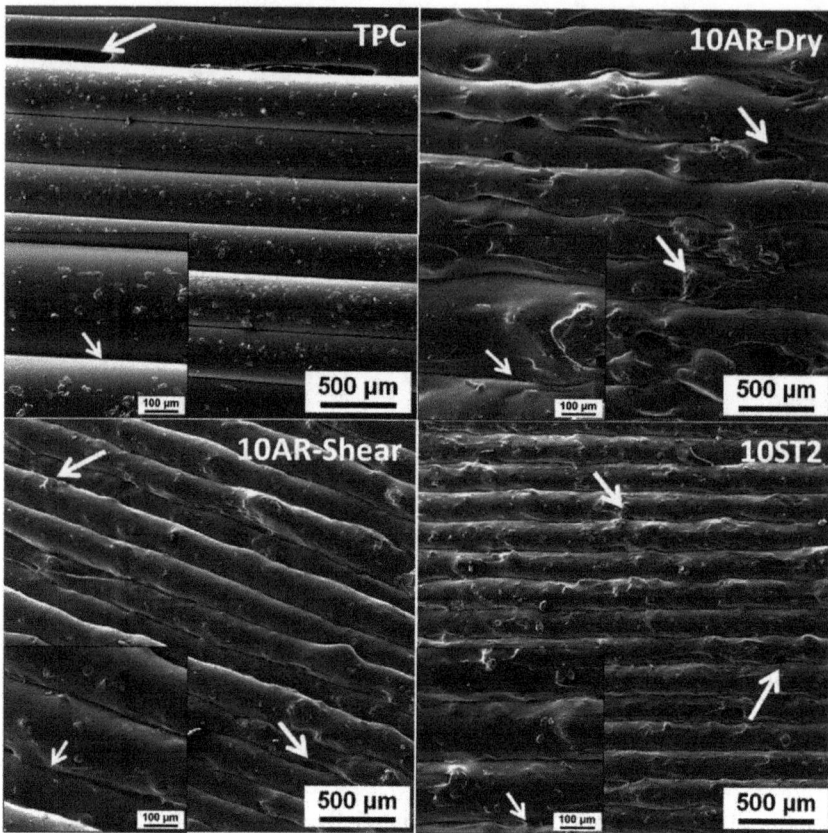

Figure 6.2: Surface morphology for TPC and TPC/soybean hull composite. The interlayer interface and defects showed by the arrows. Reproduced with permission from [18].

Table 6.2: Blend compositions (wt.%). Reproduced with permission from [27].

Blend	PLA	AESO	SYBO	STAR
PLA/SYBO	95	0	5	0
PLA/AESO	95	5	0	0
PLA/AESO/SYBO	95	2.5	2.5	0
PLA/AESO/SYBO/STAR	90	2.5	2.5	5

a noticeable variability. This is likely due to the immiscibility of SYBO and PLA. The two are at different viscosities during melt mixing which results in an uneven dispersion of the SYBO droplets in the PLA [32]. The PLA/AESO blend relies on the crosslinking of the oil during processing to deter the wide disparity in droplets throughout the polymer matrix. However, this crosslinking is also believed to be the reason for the variability in the PLA/AESO blend as the crosslinking hindered the ability to prepare defect-free samples. These shortcomings were righted by the addition of the PLA star polymer. The star polymer reduced the oil droplet size distribution and the average drop diameter which led to more reproducible mechanical results [27].

Liu et al. [33] also employed AESO in their study. In their experiment, they made environmentally green composites from AESO resin and microcrystalline cellulose (MCC). The MCC forms hydrogen bonds due to its hydroxyl groups bonding to the hydroxyl, C=O, and epoxy groups of the AESO. The MCC loading was tested in between 20 and 40% weight percentage. The addition of the MCC had increasing effects on the density, hardness, flexural modulus, and flexural strength of the AESO resin. However, the MCC is hydrophilic and diminished the water resistance of the composites at room temperature.

MCC is a hydrolyzed cellulose with amorphous regions in between the cellulose microcrystals. It is much stiffer and stronger than both amorphous and plain cellulose [34]. It is used in various applications due to its biodegradability, stability, low density, and nontoxicity [35–38]. MCC is commonly used with petroleum-based polymers. However, due to its hydrophilic nature, it requires coupling agents [34, 39, 40], copolymerization [41, 42], and chemical treatment [43–45] to bond to the polymers. Bio-based cellulose reinforced composites such as that with AESO takes advantage of chemical treatment to bond the MCC to the polymer. AESO composites often use styrene to reduce viscosity of the resin. Along with the hazards accompanying styrene [46], it makes the resin highly hydrophobic which would make it incompatible with the cellulose. This study relied on the hydrogen bonding previously stated to replace the reactivity introduced by the styrene in other AESO composites.

Figure 6.3 shows change in flexural properties of the composites at different MCC loadings. The percent strain gradually decreased at each preceding level. The flexural modulus and strength values increase at 20 and 30% loading, but the strength value tapers off at the highest loading of 40%. This is likely due to the build-ups of the MCC in the AESO resin not completely wetting out and forming voids and cracks similar to the

Table 6.3: Tensile properties of PLA/OIL blends. Reproduced with permission from [27].

Blend	Tensile strength (MPa)	Elongation at break (%)	Toughness (MPa)	Stress at break (MPa)	Modulus (GPa)
PLA	64 ± 2	4.1 ± 0.9	1.9 ± 0.5	58 ± 2	3.1 ± 0.1
PLA/SYBO	44 ± 3	24 ± 14	8 ± 4	25 ± 6	2.9 ± 0.1
PLA/AESO	62 ± 1	31 ± 39	10 ± 12	33 ± 11	2.9 ± 0.1
PLA/AESO/ SYBO	44 ± 4	20 ± 9	6 ± 3	29 ± 3	2.9 ± 0.1
PLA/AESO/ SYBO/STAR	44 ± 1	39 ± 6	11 ± 2	30 ± 1	2.90 ± 0.02

issues in the previous study [27, 47]. The consistent increase in modulus makes sense as the increase in MCC crystal structure provides more and more stiffness to the composite [48]. This is also the reason for the decrease in flexural strain. The pure AESO resin is much more flexible than those with the added MCC which allows it to prevent stress concentrations and handle loadings better [49].

Figure 6.4 shows the water absorption of the composites at different MCC loadings. At both room temperature and 100 °C, the water absorption increased as the MCC loading increased. The neat AESO has some water absorption as the hydroxyl groups of the resin attract the water molecules. As stated, the MCC is very hydrophilic and contains much more hydroxyls groups. Therefore, the increased loadings result in more water absorption. The increase is not necessarily linear as the MCC conglomerates that build up at the 40% loading allow for more passageways for water as they are not completely wet out by the resin. The composites absorb more water at the higher temperature. This is due to the higher diffusivity of the boiling water [50]. The higher diffusivity makes the water more active, forming a larger amount of microcracks in the composite for more water to bond to [51]. The viability of this composite is reduced due to the high-water intake displayed by the data.

Environmentally friendly polyurethane aqueous dispersion (PUDs) was mixed with soybean protein to develop shape-memory degradable polyurethane/SP blends

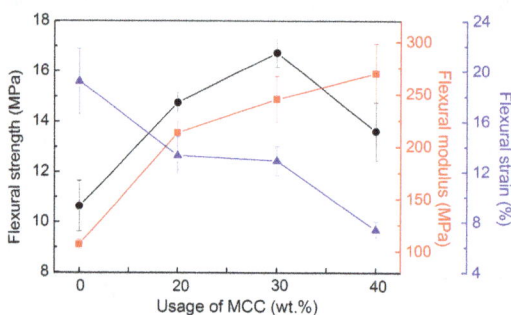

Figure 6.3: Effect of MCC loading on the flexural properties of MCC/AESO composites (● Flexural strength; ■ Flexural modulus; ▲ Flexural strain). Reproduced with permission from [33].

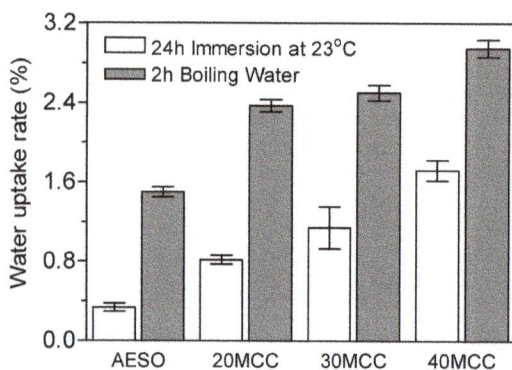

Figure 6.4: Effect of MCC loading on the water absorption of MCC/AESO composites. Reproduced with permission from [33].

[52]. The temperature induced shape-memory effect, thermal properties, blend misci- bility, and hydrolytic degradation were investigated as a function of blend composi- tion. The shape memory effect was found to be strongly influenced by the SP content. In addition, a porous structure or foam was created by high-pressure supercritical carbon dioxide technique (scCO$_2$). The PUDs used in this work is a PCL-based PU. Therefore, it is a biodegradable PU because the soft segment PCL is well known biodegradable and biocompatible thermoplastic material. Blending SP with PCL-based PUDs was considered in this work mainly to improve the mechanical properties and reduce the waster sensitivity of SP. The PUDs used in this work has 70 wt.% PCL-soft segment. The elementary steps for synthesis of PCL-based PUDs are shown in Scheme 6.1. The SP formulation was contained 10 wt.% glycerol as a plasticizer and many other fillers, such as 0.3 wt.% potassium sorbate and sodium sulfite, as well as 0.5 wt.% sodium tripolyphosphate processing aids. The SP was dispersed in water and then mixed with different concentrations of PUDs. The blend mixtures were then casted onto petri- dishes. The casted mixtures were dried at room temperature for three days and under vacuum for another three days at 50 °C [52]. The obtained solid films were then used for mechanical and morphological characterization.

Only one T_g was observed for each SP/PU blend, indicating that the two polymer components are miscible over the entire range of composition as clearly seen 5a. The

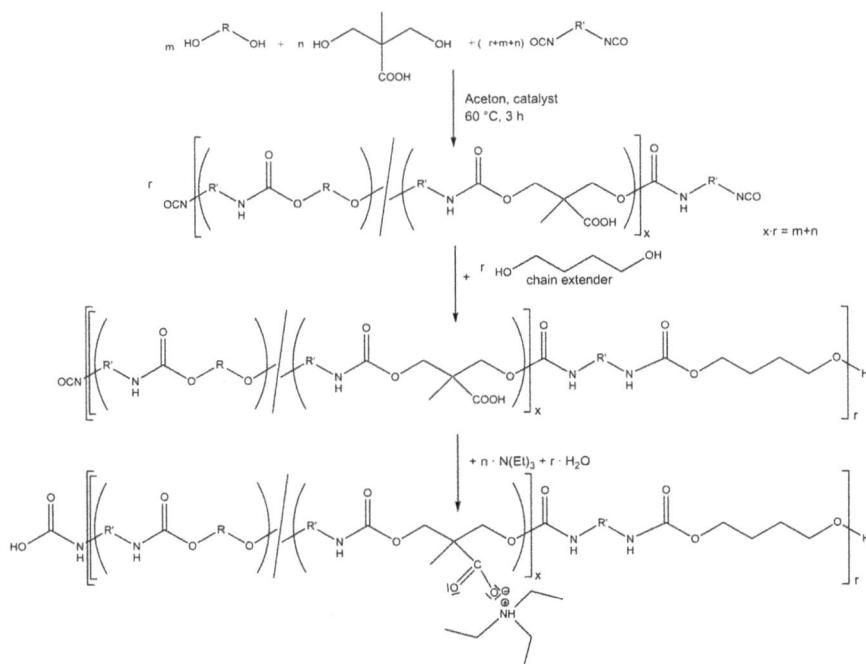

Scheme 6.1: Elementary steps for synthesis PCL-based aqueous PUDs. Reproduced with permission from [52].

T_{gs} of the blends were measured by DSC and dynamic mechanical thermal analysis (DMTA). A very good agreement between the T_{gs} values obtained from DSC and DMTA was observed as can be seen in Figure 6.5a. The SP was also influenced strongly on the crystallization behavior of the PCL soft segments of PU as clearly seen in Figure 6.5b. The magnitude of the melting peak of PCL decreased and shifted to lower temperature with increasing the SP content. The melting temperature and the heat of fusion decreased lineally with increasing the SP content as demonstrated in the inset plot of Figure 6.5b

The solid SP/PU blends were also foamed using the scCO$_2$ technique. Highly porous structure can be obtained using the scCO$_2$ technique in a relatively short time at specific critical temperature and pressure. The solid polymer will be saturated with scCO$_2$ under certain temperature and pressure. The pressure will be then released under a certain depressurization rate. The expanded CO$_2$ can generate the porous structure. The porous morphologies for SP/PU blends with different SP contents are shown in Figure 6.6. Obviously, the porous structure strongly changed with the SP content. Large pore size up to 70 µm was observed with 0.2 wt fraction of SP. No stable porous structure could be obtained at high concentration of SP and no foam can be prepared with 0.6 wt fraction of SP as clearly seen in Figure 6.6. The degree of porosity was also found to be strongly dependent on SP content. For pure PU, the degree of porosity reached up to 80 ± 5% and considerably decreased by adding SP in the blend.

Previously we blended a plasticized SP with PLA to improve the mechanical properties and decrease the water affinity of SP [53]. The PLA/SP blends were fabricated to study the feasibility of using these blends instead of petroleum-based plastics in horticulture crop containers. The biodegradation of PLA/SP blends was studied in soil for up to 30 weeks. The effect of biodegradation on the surface morphology, thermomechanical properties, and thermal stability was investigated. Figure 6.7 a shows the biodegradation

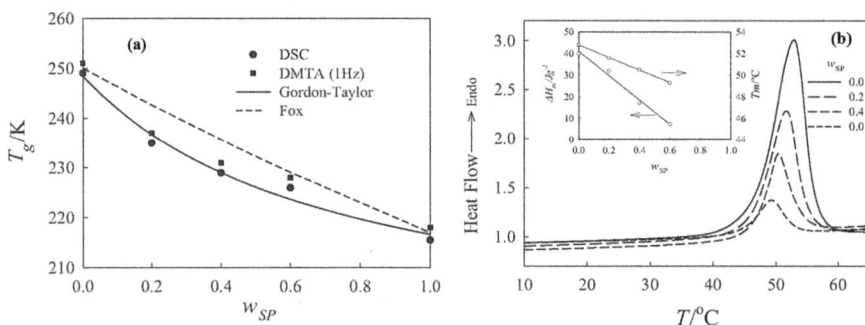

Figure 6.5: (a) SP composition dependent of T_g measured by DSC and DMTA. The solid line was calculated based on Gordon-Taylor equation while the dashed line calculated from Fox equation. (b) Effect of SP on the melting temperature of PCL soft segment of PU. The inset plot demonstrates the melting temperature and heat of fusion of PCL as a function of SP concentration. Adapted with permission from [52].

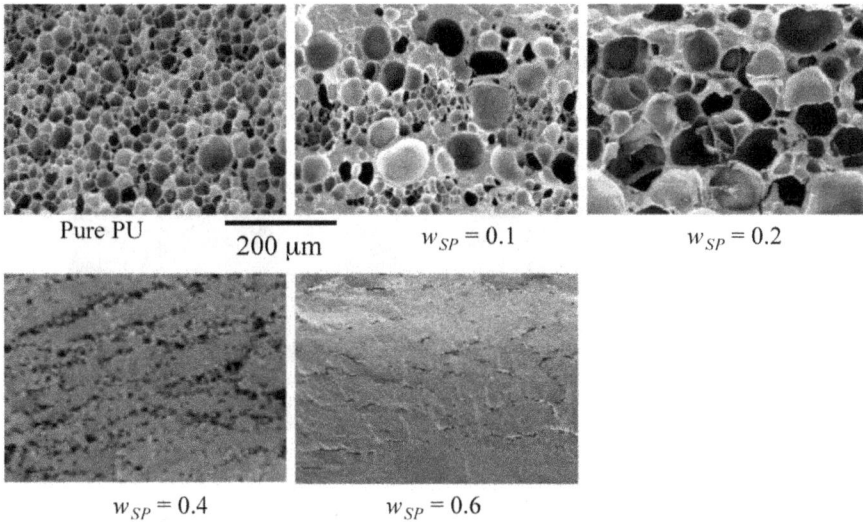

Pure PU $\overline{200\ \mu m}$ $w_{SP} = 0.1$ $w_{SP} = 0.2$

$w_{SP} = 0.4$ $w_{SP} = 0.6$

Figure 6.6: SEM porous structure for PU/SP blends with different SP contents. The porous structure created using scCO$_2$ technique at 100 bar, 35 °C and, 30 min. Reproduced with permission from [52].

time dependent on the weight loss of pure PLA, pure SP.A, and two different blends of SP.A/PLA. The SP.A has 4 wt.% adipic anhydride compatibilizer. Clearly, the SP.A biodegraded totally in less than four weeks in soil, while no weight loss was observed for pure PLA even after 24 weeks. Biodegradation time indicates that SP has extremely fast biodegradation process compare to PLA in soil under identical condition. The SP.A/PLA 50/50 had a biodegradation weight loss of approximately 47% compare to 30% for SP.A/PLA 33/67 blend. Based on this experimental fact the SP did not induce any weight loss of PLA in the blend but the PLA is almost not degradable in soil during the time period of this work (up to 24 weeks). It must be mentioned here that, although there is no weight loss after 24 weeks but the blends lost their integrity and became so brittle and can be converted into power easily under low force. We also tested the blends in a trial for horticulture crop containers and the obtained result confirmed that the blends are qualified for manufacturing horticulture crop containers and provided fertilizer effect to the plants. Figure 6.7b shows a comparison of plants grown in different three containers for six weeks. The plants grown in the SP.A/PLA (33/67) container has better structure integrity and appearance than SP.A/PLA (50/50) container. On the other hand the SP.A/PLA (50/50) container provided higher fertilizer rate, plant and root quality than the SP.A/PLA (33/67) container.

The surface morphology of PLA after the biodegradation process was investigated by SEM. Smooth surface was observed for PLA even after 24 weeks biodegradation time [53]. The effect of biodegradation process on the SEM surface morphologies of SP.A/PLA 33/67

Figure 6.7: (a) Weight loss as a function of biodegradation time in soil for different biodegradable materials. (b) Plants (Marigold) grown in different containers in a greenhouse environment for 6 weeks. Marigold plants after six weeks of growth in a greenhouse environment. Adapted with permission from [53].

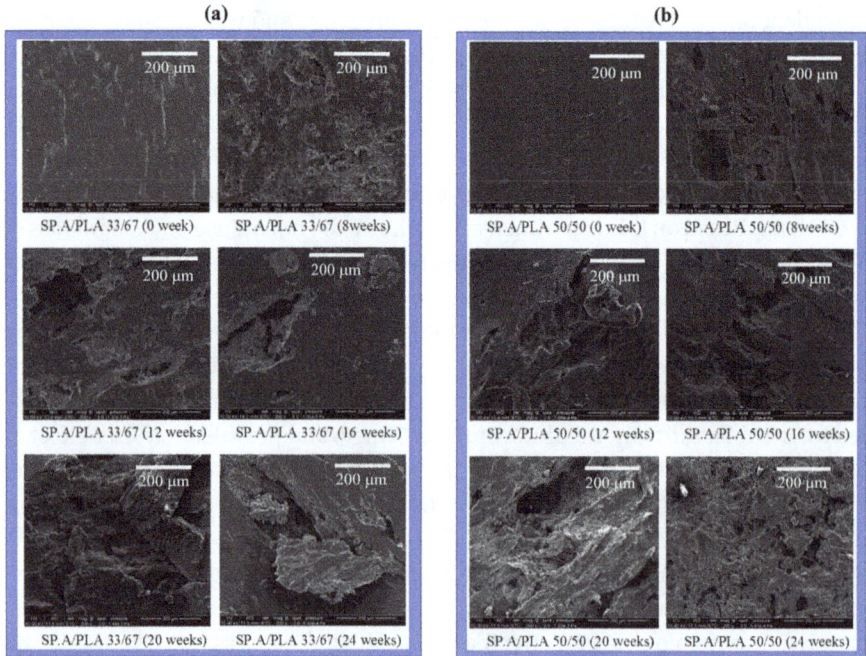

Figure 6.8: (a) SEM surface morphology for SP.A/PLA (33/67) blends after different biodegradation times in soil. (b) SEM surface morphology for SP.A/PLA (50/50) blends after different biodegradation times in soil. Adapted with permission from [53].

and SP.A/PLA 50/50 blends is shown in Figure 6.8a and b, respectively. Clearly the two blends showed good strong interfacial adhesions between the blend components before the biodegradation process (zero week). At the early stage of the biodegradation process both blends showed many cracks and tiny corrosive holes. With increasing biodegradation time, the cracks became wider and larger and the hopes became deeper as seen in Figure 6.8a and b. Aggregated small eroded regions and irregular pits with very rough surface were developed in the two blends after 24 weeks biodegradation time. It is also clear that the SEM micrographs of SP.A/PLA (50/50) have more significant corrosive holes and disorderly erosive aggregates than that of SP.A/PLA (33/67) as expected due to the higher biodegradation rate as mentioned above (see Figure 6.7a).

6.4 Conclusion

Soy protein is a natural material that can be isolated from soybean, which is a major agricultural crop in the U.S. The viability of soybean-based polymers and composites is questioned due to their high-water absorption and poor mechanical properties. To make soybean-based polymers commercially viable, the water sensitivity issue needs to be addressed. There have been many environmentally friendly attempts to improve the properties of soybean polymers, such as blending with biodegradable polymers including PLA, PCL, polyurethane, etc. To be able to turn soybeans and soybean waste into a usable material for polymer production, the raw beans must go through several processing steps to facilitate the removal of soluble non-protein matter from the soybean and increasing the overall protein percentage in the soybean. Abundant and inexpensive soybean hulls can be modified with dilute acid hydrolysis to be used in natural fiber reinforced composites. Environmentally friendly PUD was mixed with soybean protein to develop shape-memory degradable polyurethane/SP blends. Only one glass transition temperature was observed for each SP/PU blend, indicating that the two polymer components are miscible over the entire range of composition. The magnitude of the melting peak of PCL decreased and shifted to lower temperature with increasing the SP content. The melting temperature and the heat of fusion decreased lineally with increasing the SP content. The solid SP/PU blends were also foamed using high-pressure supercritical carbon dioxide $scCO_2$ technique. Highly porous structure was obtained using the $scCO_2$ technique in a relatively short time at specific critical temperature and pressure. The PLA/SP blends were fabricated to study the feasibility of using these blends instead of petroleum-based plastics in horticulture crop containers. The biodegradation of PLA/SP blends was studied in soil for up to 24 weeks. The effect of biodegradation on the surface morphology, thermomechanical properties, and thermal stability was investigated. A trial for horticulture crop containers made from SP.A/PLA blends was investigated. The obtained result confirmed that the blends are qualified for manufacturing horticulture crop containers and provided fertilizer effect to the plants. The plants grown in the SP.A/PLA (33/67) container has better structure integrity and appearance than SP.A/PLA (50/50) container. On the other hand the SP.A/PLA (50/50) container provided higher fertilizer rate, plant and root quality than the SP.A/PLA (33/67) container. At the early stage of the biodegradation process the SP.A/PLA blends showed many cracks and tiny corrosive holes. With increasing biodegradation time, the cracks became wider and larger and the hopes became deeper. Aggregated small eroded regions and irregular pits with very rough surface were developed in after 24 weeks biodegradation time.

Author contributions: All the authors have accepted responsibility for the entire content of this submitted manuscript and approved submission.
Research funding: None declared.
Conflict of interest statement: The authors declare no conflicts of interest regarding this article.

References

1. Tiseo Pby I, 27 J. Global plastic production 1950-2018 [Internet]. Available from: https://www.statista.com/statistics/282732/global-production-of-plastics-since-1950/ [Accessed 15 Feb 2021].
2. SoyInfo Center. Henry Ford and his employees [Internet]. Available from: https://www.soyinfocenter.com/HSS/henry_ford_and_employees.php [Accessed 15 Feb 2021].
3. Sue H-J, Wang S, Jane J-L. Morphology and mechanical behaviour of engineering soy plastics. Polymer 1997;38:5035–40.
4. Visakh PM, Nazarenko O, editors. Soy protein-based blends, composites and nanocomposites. John Wiley & Sons; 2017.
5. Srinivasan G. Soy protein polymers: enhancing the water stability property [Master thesis]. Ames Iowa: Iowa State University; 2010.
6. Baschek G, Hartwig G, Zahradnik F. Effect of water absorption in polymers at low and high temperatures. Polymer 1999;40:3433–41.
7. Labet M, Thielemans W. Synthesis of polycaprolactone: a review. Chem Soc Rev 2009;38: 3484–504.
8. Kumar R, Choudhary V, Mishra S, Varma IK, Mattiason B. Adhesives and plastics based on soy protein products. Ind Crop Prod 2002;16:155–72.
9. Berk Z. Technology of production of edible flours and protein products from soybeans. Viale delle Terme di Caracalla, Rome, Italy: FAO; 1992.
10. Fernández-Espada L, Bengoechea C, Cordobés F, Guerrero A. Thermomechanical properties and water uptake capacity of soy protein-based bioplastics processed by injection molding. J Appl Polym Sci 2016;133. https://doi.org/10.1002/app.43524.
11. Adamy M, Verbeek CJ. Injection-molding performance and mechanical properties of blood meal-based thermoplastics. Adv Polym Technol 2013;32:1–7.
12. Mo X, Sun XS, Wang Y. Effects of molding temperature and pressure on properties of soy protein polymers. J Appl Polym Sci 1999;73:2595–602.
13. Huang M-C, Tai C-C. The effective factors in the warpage problem of an injection-molded part with a thin shell feature. J Mater Process Technol 2001;110:1–9.

14. Mohanty AK, Tummala P, Liu W, Misra M, Mulukutla PV, Drzal LT. Injection molded biocomposites from soy protein based bioplastic and short industrial hemp fiber. J Polym Environ 2005;13: 279–85.
15. Félix M, Martín-Alfonso JE, Romero A, Guerrero A. Development of albumen/soy biobased plastic materials processed by injection molding. J Food Eng 2014;125:7–16.
16. Paetau I, Chen C-Z, Jane J. Biodegradable plastic made from soybean products. II. Effects of cross-linking and cellulose incorporation on mechanical properties and water absorption. J Environ Polym Degrad 1994;2:211–7.
17. McCrum NG, Buckley CP, Bucknell CB. Principles of polymer engineering, 2nd ed. New York: Oxford University Press; 1997:7 p.
18. Balla VK, Tadimeti JG, Kate KH, Satyavolu J. 3D printing of modified soybean hull fiber/polymer composites. Mater Chem Phys 2020;254:123452.
19. Joshi SV, Drzal LT, Mohanty AK, Arora S. Are natural fiber composites environmentally superior to glass fiber reinforced composites? Compos Appl Sci Manuf 2004;35:371–6.
20. Pandey JK, Nagarajan V, Mohanty AK, Misra M. Commercial potential and competitiveness of natural fiber composites. In: Biocomposites. Woodhead Publishing; 2015:1–15 pp.
21. Merlini C, Soldi V, Barra GMO. Influence of fiber surface treatment and length on physico-chemical properties of short random banana fiber-reinforced castor oil polyurethane composites. Polym Test 2011;30:833–40.
22. Jawaid M, Khalil HPSA, Bakar AA, Hassan A, Dungani R. Effect of jute fibre loading on the mechanical and thermal properties of oil palm–epoxy composites. J Compos Mater 2012;47: 1633–41.
23. Cao XV, Ismail H, Rashid AA, Takeichi T, Vo-Huu T. Maleated natural rubber as a coupling agent for recycled high density polyethylene/natural rubber/kenaf powder biocomposites. Polym Plast Technol Eng 2012;51:904–10.
24. Majid RA, Ismail H, Taib RM. Effects of polyethylene-g-maleic anhydride on properties of low density polyethylene/thermoplastic sago starch reinforced kenaf fibre composites, Iran. Polym J 2010;19:501–10.
25. Jacob M, Francis B, Thomas S, Varughese KT. Dynamical mechanical analysis of sisal/oil palm hybrid fiber-reinforced natural rubber composites. Polym Compos 2006;27:671–80.
26. Balla VK, Kate KH, Satyavolu J, Singh P, Tadimeti JG. Additive manufacturing of natural fiber reinforced polymer composites: processing and prospects. Compos B Eng 2019;174:106956.
27. Mauck SC, Wang S, Ding W, Rohde BJ, Fortune CK, Yang G, et al. Biorenewable tough blends of polylactide and acrylated epoxidized soybean oil compatibilized by a polylactide star polymer. Macromolecules 2016;49:1605–15.
28. Inkinen S, Hakkarainen M, Albertsson A-C, Södergård A. From lactic acid to poly(lactic acid) (PLA): characterization and analysis of PLA and its precursors. Biomacromolecules 2011;12:523–32.
29. Bhardwaj R, Mohanty AK. Advances in the properties of polylactides based materials: a review. J Biobased Mater Bioenergy 2007;1:191–209.
30. Garlotta D. A literature review of poly(lactic acid). J Polym Environ 2001;9:63–84.
31. Pelletier H, Gandini A. Preparation of acrylated and urethanated triacylglycerols. Eur J Lipid Sci Technol 2006;108:411–20.
32. Chang K, Robertson ML, Hillmyer MA. Phase inversion in polylactide/soybean oil blends compatibilized by poly(isoprene-b-lactide) block copolymers. ACS Appl Mater Interfaces 2009;1: 2390–9.
33. Liu W, Fei M-E, Ban Y, Jia A, Qiu R. Preparation and evaluation of green composites from microcrystalline cellulose and a soybean-oil derivative. Polymers 2017;9:541.
34. Izzati Zulkifli N, Samat N, Anuar H, Zainuddin N. Mechanical properties and failure modes of recycled polypropylene/microcrystalline cellulose composites. Mater Des 2015;69:114–23.

35. Janardhnan S, Sain MM. Isolation of cellulose microfibrils - an enzymatic approach. Bioresources 2006;1:176–88.
36. Pinkl S, Veigel S, Colson J, Gindl-Altmutter W. Nanopaper properties and adhesive performance of microfibrillated cellulose from different (ligno-)cellulosic raw materials. Polymers 2017;9:326.
37. Hua S, Chen F, Liu Z-Y, Yang W, M-bo Y. Preparation of cellulose-graft-polylactic acid via melt copolycondensation for use in polylactic acid based composites: synthesis, characterization and properties. RSC Adv 2016;6:1973–83.
38. Pan Y, Pan Y, Cheng Q, Liu Y, Essien C, Via B, et al. Characterization of epoxy composites reinforced with wax encapsulated microcrystalline cellulose. Polymers 2016;8:415.
39. Ummartyotin S, Pechyen C. Microcrystalline-cellulose and polypropylene based composite: a simple, selective and effective material for microwavable packaging. Carbohydr Polym 2016;142: 133–40.
40. Spoljaric S, Genovese A, Shanks RA. Polypropylene–microcrystalline cellulose composites with enhanced compatibility and properties. Compos Appl Sci Manuf 2009;40:791–9.
41. Deng F, Zhang Y, Ge X, Li MC, Li X, Cho UR. Graft copolymers of microcrystalline cellulose as reinforcing agent for elastomers based on natural rubber. J Appl Polym Sci 2015;133:1–11.
42. Deng F, Ge X, Zhang Y, Li M-C, Cho UR. Synthesis and characterization of microcrystalline cellulose-graft-poly(methyl methacrylate) copolymers and their application as rubber reinforcements. J Appl Polym Sci 2015;132:1–10.
43. Yakubu A, Umar T, Mohammed S. Chemical modification of microcrystalline cellulose: improvement of barrier surface properties to enhance surface interactions with some synthetic polymers for biodegradable packaging material processing and applications in textile, food and pharmaceutical industry. Adv Appl Sci Res 2011;2:532–40.
44. Çetin NS, Özmen Çetin N, Harper DP. Vinyl acetate-modified microcrystalline cellulose-reinforced HDPE by twin-screw extrusion. Turk J Agric For 2015;39:39–47.
45. Xiao L, Mai Y, He F, Yu L, Zhang L, Tang H, et al. Bio-based green composites with high performance from poly(lactic acid) and surface-modified microcrystalline cellulose. J Mater Chem 2012;22: 15732.
46. Lithner D, Larsson Å, Dave G, et al. Environmental and health hazard ranking and assessment of plastic polymers based on chemical composition. Sci Total Environ 2011;409:3309–24.
47. He M, Zhou J, Zhang H, Luo Z, Yao J. Microcrystalline cellulose as reactive reinforcing fillers for epoxidized soybean oil polymer composites. J Appl Polym Sci 2015;132. https://doi.org/10.1002/app.42488.
48. Ashori A, Nourbakhsh A. Performance properties of microcrystalline cellulose as a reinforcing agent in wood plastic composites. Compos B Eng 2010;41:578–81.
49. Liu W, Xie T, Qiu R. Improvement of properties for biobased composites from modified soybean oil and hemp fibers: dual role of diisocyanate. Compos Appl Sci Manuf 2016;90:278–85.
50. Chen T, Liu W, Qiu R. Mechanical properties and water absorption of hemp fibers–reinforced unsaturated polyester composites: effect of fiber surface treatment with a heterofunctional monomer. Bioresources 2013;8:2780–91.
51. Dhakal H, Zhang Z, RichardsoN M. Effect of water absorption on the mechanical properties of hemp fibre reinforced unsaturated polyester composites. Compos Sci Technol 2007;67:1674–83.
52. Madbouly SA, Lendlein A. Degradable polyurethane/soy protein shape-memory polymer blends prepared via environmentally-friendly aqueous dispersions. Macromol Mater Eng 2012;297: 1213–24.
53. Yang S, Madbouly SA, Schrader JA, Srinivasan G, Grewell D, McCabe KG, et al. Characterization and biodegradation behavior of bio-based poly (lactic acid) and soy protein blends for sustainable horticultural applications. Green Chem 2015;17:380–93.

James Goodsel* and Samy Madbouly

7 Biodegradable polylactic acid (PLA)

Abstract: Polylactic acid (PLA) is a biodegradable material that can be processed using the common processing techniques, such as injection molding, extrusion, and blow molding. PLA has widely been researched and tested due to its biodegradable nature. As a biodegradable material, PLA can be subject to some inherently poor qualities, such as its brittleness, weak mechanical properties, small processing windows, or poor electrical and thermal properties. In order to nullify some of these issues, nanofiller composites have been added to the polymer matrix, such as nanocellulose, nanoclays, carbon nanotubes, and graphene. Dye-clay hybrid nanopigments (DCNP) have been used to explore potential applications in the food packaging industry with promising results. Several different compatibilizers have been studied as well, with the goal of increasing the mechanical properties of blends. A key application for PLA is in wound healing and surgical work, with a few studies described in the present chapter. Finally, the superwettability of dopamine modified PLA is examined, with promising results for separation of oily wastewater.

Keywords: biodegradable PLA; nanocomposites; polymer blends.

7.1 Introduction

Polylactic acid (PLA) is an aliphatic polyester derived from lactic acid. PLA is biodegradable and made from renewable resources, such as corn starch or sugarcane [1]. PLA has mechanical properties comparable to that of polyethylene terephthalate (PET), but PLA has a drastically lower maximum continuous use temperature [1]. Poly(L-lactide) (PLLA) is the most important polymer in the PLA family. The semicrystalline polymer has crystallinity around 37%, a glass transition temperature (T_g) of 67 °C and a melting temperature (T_m) of approximately 180 °C. PLA is processable using extrusion, injection molding, blow molding, or fiber spinning techniques to create various products. The resultant products can be recycled after use by remelting and processing the material a second time or by hydrolysis to lactic acid, which is the basic chemical. It is also possible to compost the PLA in order to introduce it into the natural life cycle of all biomass, where it will degrade into CO_2 and water. PLA is a desirable polymer because it can be recycled traditionally, composted like other organic matter,

*Corresponding author: James Goodsel, Behrend College, School of Engineering, Pennsylvania State University, Erie, PA 16563, USA, E-mail: jpg5738@psu.edu
Samy Madbouly, Behrend College, School of Engineering, Pennsylvania State University, Erie, PA 16563, USA

This article has previously been published in the journal Physical Sciences Reviews. Please cite as: J. Goodsel and S. Madbouly "Biodegradable polylactic acid (PLA)" *Physical Sciences Reviews* [Online] 2021, 7. DOI: 10.1515/psr-2020-0072 | https://doi.org/10.1515/9781501521942-007

incinerated in an incineration plant, which is a process that is not harmful to the environment, or be introduced into a traditional waste management system [1].

PLA has high strength and high modulus on top of its good appearance. It is often compared to PS at room temperature due to its high stiffness and strength. However, there are lots of drawbacks limited the applications of PLA, such as low T_g, poor ductility, low impact strength, and is very brittle. PLA does not have a high crystallization rate and under certain processing condition, the crystallinity tends to reduce significantly. PLA is much more susceptible than PET to chemical and biological hydrolysis. PLA has a slow degradation rate and is relatively hydrophobic. The material is also thermally unstable and has poor gas barrier performance. PLA is not a very flexible material and has long molding cycles [2].

Polymer composites are becoming increasingly prevalent in PLA in order to improve the material properties. The next section will discuss several of these composites and their usefulness for PLA.

7.2 Polymer composites

Governments around the world have been pushing for environmental sustainability, which has resulted in accelerated development of biodegradable polymers. The development of these unique polymers has provided nanomaterial science with new challenges and possibilities for improvement. Biodegradable materials, including nanofillers, are becoming increasingly important due to their increasing use in greener markets [3]. Naturally degrading polymers are important in solving common pitfalls of polymer materials and helps to maintain nature's homeostasis. Intrinsic drawbacks of some of these materials include poor physical properties, tight processing conditions, and weak electrical and thermal properties, which can be corrected using nanofiller fortified composites [4]. There are a wide range of applications in different areas for these biodegradable polymer composites due to the characteristic large surface area and larger aspect ratio. Additionally, composites can utilize the symbiotic results from the nanofiller and biodegradable polymer matrix, which results in strengthened properties while remaining environmentally friendly [5]. The main nanofillers incorporated in nanocomposites are nanoclays, carbon nanotubes and some organic nanofillers. Nanocellulose has attracted the most attention worldwide [5].

7.2.1 Nanocellulose

Cellulose is an extract of natural cellulose made out from minuscule structural substances. The cellulose family can be divided into three types. Cellulose nanocrystals, known as nanocrystalline cellulose, cellulose nanofibers, referred to as nanofibrillated cellulose (NFC), and bacterial cellulose (BC), known as microbial cellulose, are the three major types of cellulose [6]. The hierarchical structure of cellulose is shown in Figure 7.1. Cellulose is extracted from flour, wood, beets, ramie, potato tubers, algae and other plants. Bacterial nanocellulose has strong physical properties, biocompatibility, an ultrafine fiber network, as well as high porosity as a nanofiller [5].

7.2.2 Nanoclays

The most commonly used nanolayered silicate nanocomposite is layered silicate, or clay. There are widespread uses in the creation of clay-based composites. Different clay

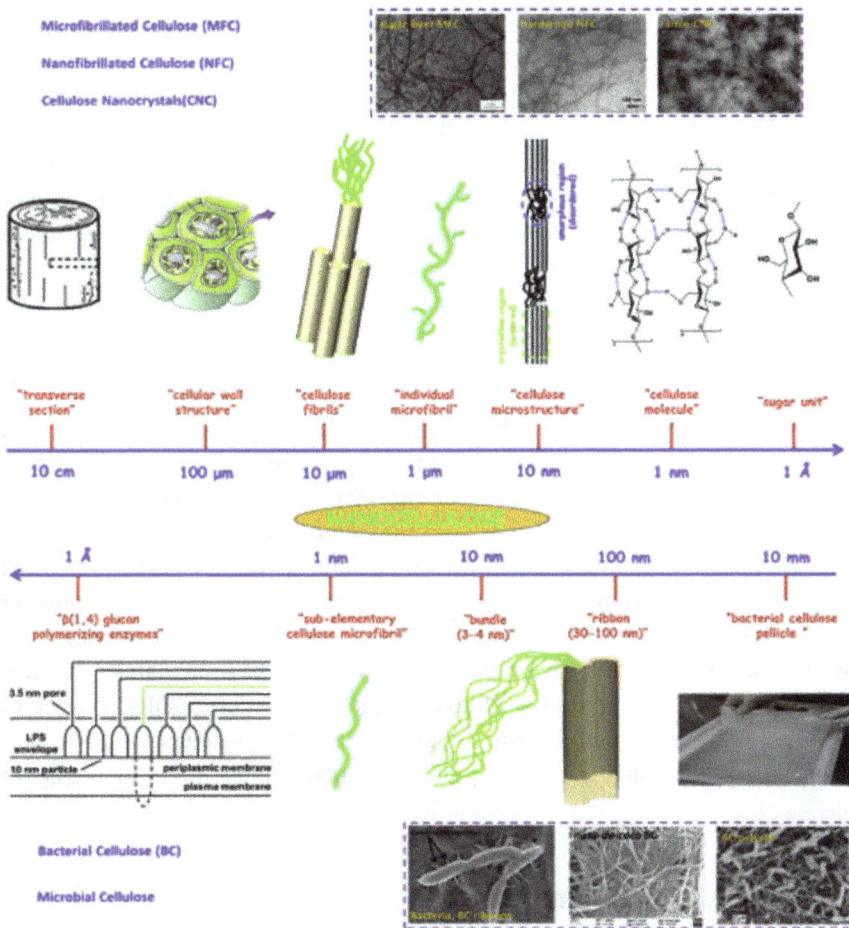

Figure 7.1: Hierarchical structure of cellulose. The top half of the figure shows (from large unit to small unit): cellulose nanocrystals (CNC), micro/nanofibrillated cellulose (MFC and NFC); bottom image (from tiny unit to small unit): bacterial cellulose (BC) [5].

sources, the silicate particles and production technology affect the size of the silicate layers [7]. Biodegradable plastic clay nanocomposites have gained lots of publicity due to improved mechanical and obstruction properties [8]. The composites also have lower burnable points on each native polymer. Nanoclays have been included in many recent, newly advanced biodegradable polymer materials. They are being used in an expanding range of applications due to their strong practicality, low price point, machinability, and thermosetting properties [9–12].

7.2.3 Carbon nanotubes

Carbon nanotubes demonstrate great electrical, magnetic, and mechanical properties. Carbon nanotubes are ideal material nanofiller for creating outstandingly strong polymer composites. However, due to van der Waals interactions, carbon nanotubes tend to form stable bundles. These bundles are extremely hard to disperse and align in the polymer. There are many situations where carbon nanotubes can be utilized, such as polymer composites, electrochemical energy storage and conversion, and hydrogen storage to name a few. Polymer composites are used as functional fillers to both increase the thermal and mechanical properties, as well as to provide additional functions like flame resistance and barrier properties [5].

7.2.4 Graphene

Graphene is comprised of one layer of hybridized carbon atoms ordered in a two-dimensional lattice. Graphene is created through the process of peeling off graphite nanosheets. Graphene is a very special structure which gives it outstanding thermal and mechanical properties. These properties come from the theoretical specific surface area of a graphene sheet, which is 2630–2956 m^2, and its high aspect ratio, which exceeds 2000. One of graphene's most unique applications is nanocomposite polymer fillers. With excellent electrical, mechanical, optical, and transport properties, there have been many applications for graphene. Graphene reinforced nanocomposites display high strength and hardness, although, good dispersion results has been a pressing issue. Another challenge is getting graphene to completely separate to form one single layer or a couple layers of material while maintaining a reasonable lateral size [13].

7.2.5 Other functional nanofillers

Layered double hydroxide (LDH) is a functional nanofiller. LDH has been receiving a lot of focus recently because of its adaptability for the development of different biomedical applications. Bio-based polymer composites with LDH are widely utilized in gene therapy, drug delivery, and tissue engineering, due to their compatibility and noncytotoxic and nonirritating biological systems [14, 15]. LDH can be mainly prepared by monomer exchange and *in situ* polymerization, coprecipitation or polymer displacement, and polymer recombination. Each preparation method is shown in Figure 7.2. *In-situ* polymerization is conducted by filling the interlayer of nanolayers with reactive monomers, causing the monomers to polymerize between layers. The coprecipitation method is named as such because there are two or more cations contained in a solution. The cations are readily available in a homogeneous solution and then the precipitant is introduced. Uniform precipitation of different segments can be obtained after the precipitation reaction. Polymer recombination deals with reactions that result

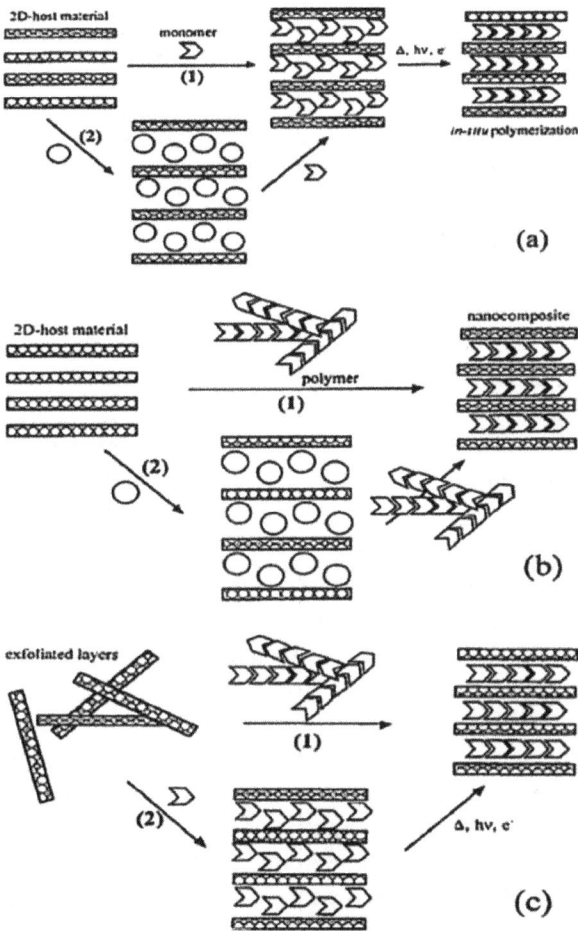

Figure 7.2: Layered double hydroxide (LDH) nanocomposites preparation using different methods, (a) monomer exchange and *in situ* polymerization, (b) direct exchange, and (c) exfoliated layers restacking [5].

in different orders of polymer monomer arrangements. The monomers then react to form new polymers [5].

Biodegradable films are a huge potential market for biodegradable PLA, but coloring biodegradable films can be difficult. This issue can be overcome using dye-clay hybrid nanopigments in the biodegradable PLA. PLA can be filled with a dye-clay hybrid nanopigment (DCNP) in order to create a colored biodegradable/biocompatible film. This film was investigated as a high thermomechanical resistant barrier that has exceptional mass transport and light properties to be used in packaging applications for different foods [16]. A DCNP was manufactured using a wet chemical process yielded 76%. It was then integrated into a PLA matrix, at different concentration levels, using a solution casting method. After analysis, it was found that, "The morphological characterizations revealed partially intercalated/exfoliated structure for PLA–DCNP films," [17]. DCNP filled specimens displayed a positive correlation for both modulus and T_g, showing improvements of up to 20% and 12 °C when compared to neat PLA, respectively. PLA's gas permeability and water vapor were reduced by 54 and 36%, respectively, with small amounts of DCNP integrated into the polymer. DCNP filled PLA showed exceptional light protection performance and color characteristics. Better performance was observed from the PLA–DCNP film when compared to a PLA–Cloisite 20A film with identical filler loading. The DCNP–PLA film displayed desirable properties for use in food packaging materials [17].

With growing concerns about long-lasting plastic pollution around the world, biopolymers have gained enormous attention in literature and industry. A potential candidate to be substitute for petroleum-based polymers in a diverse series of applications is PLA [18]. Still, PLA's application for food packaging has been severely restricted by numerous high-impact deficiencies, such as poor mechanical and thermal properties. Additionally, PLA has high light, gas, and water permeability. Using different nanofillers in polymer matrices has been widely regarded as one of the, "most straightforward, efficient, and cost-effective strategies," to circumvent the aforementioned issues [19]. Organically modified montmorillonite, a two-dimensional nanofiller consisting stacks of 1 nm thick clay platelets, drastically increases the thermal, mechanical, and barrier properties of PLA, including at low content when the platelets are intercalated or exfoliated during the dispersion process. Additionally, uniformly dispersed clay layers within PLA can effectively behave as visible light and ultraviolet shielding agents to preserve food containers and the food within from destructive light radiation. Food packaging materials often include colorants for coloration and for preserving foodstuffs against photodegradation. Colorants tend to be sorted on being organic or inorganic. Organic pigments tend to have high tinting strength. However, they struggle with poor heat and light stability as well as a high likelihood of aggregate formation while compounding [20]. Even with comparably better dispersibility in a polymer matrix, good thermal degradation and photodegradation resistance, inorganic pigments retain heavy metals, which can be toxic, and display dull shades and poor color strength. Using a DCNP that has exceptional tinting strength requires a significantly smaller amount of colorants to reach each individual shade in contrast with that of conventional organic and inorganic pigments. Additionally, easy dispersion of DCNP into polymer matrices, superior colorimetric properties, and superior color fastness allow DCNP to stand out compared to those of its counterparts [17].

After adding 1 and 3% DCNP by weight into the PLA matrix, an increase of 19 and 20%, respectively, was observed in the storage modulus of PLA. This finding displays the fortifying effects of DCNP incorporated the polymer matrix. The resultant intercalated/exfoliated morphology enhanced the interfacial adhesion among the PLA molecular chains and the levels of DCNP. By increasing the content of DCNP to 5% by weight in PLA, a deterioration of the storage modulus was observed. The results can be explained by the formation of aggregates along with a reduction in the amount of intercalation/exfoliation of DCNP at high filler levels. The accompanying temperature for the maximum value of tan δ denotes the sample's T_g. The tan δ peak was observed at 64 °C for the neat PLA and increased by 1–12 °C after the addition of the fillers. A possible explanation is the fact that the dispersed clay layers in the PLA matrix cause a reduction in the free volume, which hinders the segmental movements of PLA molecular chains at the interaction point, resulting in an increased T_g [21]. Incorporation of fillers in the PLA matrix caused a decrease in the tan δ peak intensity, and this drop became more intense at high filler levels. Typically, low intensity of the tan δ peak is caused by a low degree of polymer chain mobility [22]. Incorporating high aspect ratio DCNP layers in the PLA molecular chains results in a strong interaction between the two materials, even at lower concentrations, which restricts polymer chain motions. The PLA–DCNP films showed better storage modulus (E') and T_g results when compared to unfilled PLA and PLA filled with C20A, a commercial organoclay, which showed exceptional performance of DCNP to improve the dynamic mechanical properties. Also, the raised values of T_g for PLA–DCNP films versus those of the unfilled PLA film displayed a positive effect of DCNP on PLA thermal stability.

The T_g of PLA (55–65 °C) is lower than common polymers used in food packaging, such as PET with (T_g ranging from 67 to 81 °C) as well as polystyrene with a T_g of approximately 100 °C, which causes PLA to be technically limited in thermally processed packages [23]. A low T_g for food packaging, roughly around the application temperature, is unfavorable because the packaging may be in or near its rubbery state, causing it to display poor mechanical strength. This causes vulnerability in the material, and it will readily lose its dimensional stability. Additionally, with the high mobility of the polymer molecules in the rubbery state, it is more probable for the packaging materials to emigrate from the matrix and contaminate the foodstuffs. In order to avoid these problems, the material should have a T_g sufficiently above the expected use temperature. One notable instance where the application temperature is below the T_g of PLA–DCNP films (up to 76 °C) is mild-temperature pasteurization of water, milk, juice, and beer, which is typically conducted in temperatures ranging from 60–75 °C [17].

The oxygen permeability (OP) and water vapor permeability (WVP) of PLA drop by 36 and 54% after the incorporation of 1 and 3% DCNP by weight. The resulting improved barrier properties allow for the material to pass permeability requirements and also allow for expansion in food packaging material applications for PLA. Using the Nielson tortuous path model, the improvement of PLA's barrier properties after incorporation of DCNP can be explained [24]. Interestingly, well-dispersed DCNP layers act as barrier shields, which elongate the diffusion pathway of water vapor and oxygen within the molecular matrix, while also decreasing the water vapor and gas permeability of PLA [25]. The nucleating effect of clay could also be the cause of this phenomenon. The clay platelets act as a nucleating agent in PLA crystallization, which can cause more crystalline structures to form with increased barrier properties [26]. Increased DCNP levels caused the PLA film to endure 17 and 7% deterioration for WVP and OP, respectively. This was probably due to some DCNP aggregates forming, increasing the number of preferential pathways for permeant species. When examining PLA films filled with DCNP and C20A with the same amount of loading, it was observed that DCNP has higher effectiveness for enhancement of barrier properties of PLA compared to that of C20A. This could be due to DCNP dispersing better in the PLA matrix [17].

PLA/PHB/cellulose biodegradable nanocomposites have been studied to investigate the thermal and mechanical properties. Experiments were conducted using different processing methods and the results are discussed below.

Biodegradable blends and nanocomposites are developed out of PL, PHB, and cellulose nanocrystals (NC) using a single step reactive blending process that required the use of dicumyl peroxide (DCP) as a cross-linking agent. In order to gain a better understanding of the effect of processing techniques on the morphological, thermal, and mechanical properties of these nanocomposites, compression molding, extrusion, and 3D printing processing methods were utilized. The incorporation of DCP enhanced interfacial adhesion and dispersion of NC in nanocomposites, which was shown using scanning electron microscopy and atomic force microscopy. The carbonyl index calculated from Fourier transform infrared spectroscopy displayed higher levels of crystallinity post DCP incorporation in PLA/PHB and PLA/PHB/NC, which was verified using differential scanning calorimetry analyses. NC and DCP displayed nucleating activity and increased crystallization of PLA, improving crystallinity from 16% in PLA/PHB to 38% in DCP crosslinked blend and to 43% in crosslinked PLA/PHB/NC nanocomposite [27].

The thermo-mechanical characterization of uncross-linked and crosslinked PLA/PHB blends and nanocomposites displayed the significance of the processing method. Larger storage modulus measurements resulted from filaments gathered using extrusion and 3D printed meshes rather than compression molded films. Following a similar pattern, the thermogravimetric analysis identified a rise of the onset degradation temperature, with more than 10 °C for PLA/PHB blends and nanocomposites after extrusion and 3D-printing, as compared with compression molding. The research [8] displays that

PLA/PHB products with improved interfacial adhesion, enhanced thermal stability, and mechanical properties are obtainable by choosing the proper processing technique and settings using NC and DCP in order to balance the properties [27].

Figure 7.3a and 7.3b displays the first heating and cooling curves of neat PLA and PHB as well as the first heating curves of PLA/PHB blends and nanocomposites processed as films. Pristine PLA and PHB samples show separate T_{gs}, with the first at 56.7 °C and PHB at a significantly lower temperature of −10.0 °C, which allows for good flexibility at room temperature. While PHB has a higher melting peak than PLA, 164.4 versus 149.5 °C, respectively, the range of melt temperatures is similar for the two materials, ranging from 130 to 160 °C for PLA and from 130 to 180 °C for PHB. The closeness is vital, as it allows the materials to blend in the melt state. PLA displays a small exothermic crystallization peak during the first heating cycle at 120.8 °C as shown in Figure 7.3b. PHB, on the other hand, crystallizes in the processing step. These results were similar to previous studies [28–30].

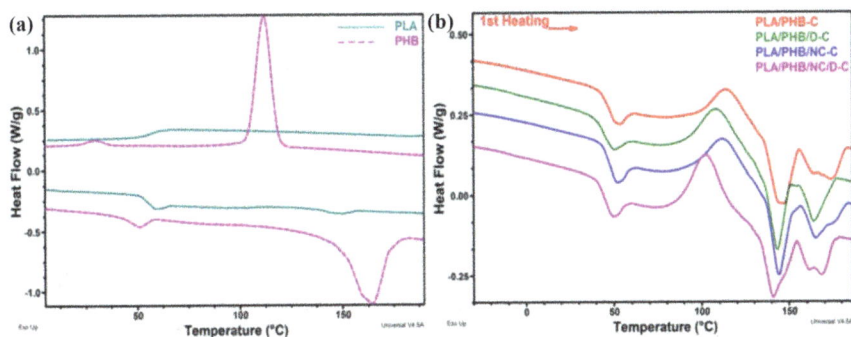

Figure 7.3: Neat PLA and PHB DSC curves—first heating and cooling scans (a) PLA/PHB blends and nanocomposites—first heating scan (b) [27].

Figure 7.4 displays both materials' thermal behaviors. The first degradation peak shown in the derivative thermogravimetric (DTG) curves is generated from the decomposition of the PHB segment. The second curve is caused by PLA's sensitivity to oxidative chain scission and hydrolysis, which is conducive to the literature and matches when compared with the pure components DTG results [31, 32]. The small shoulder observed in the DTG curves around 195 °C is likely caused by the TBC plasticizer being released out of the PHB matrix [27]. Polymer blends are another common way to improve the PLA's properties. Polymer blends will be discussed in the next section.

7.3 Polymer blends

As mentioned above, PLA displays several strengths, but a pressing weakness is its brittleness. An efficient way to toughen PLA is by blending it with biodegradable poly

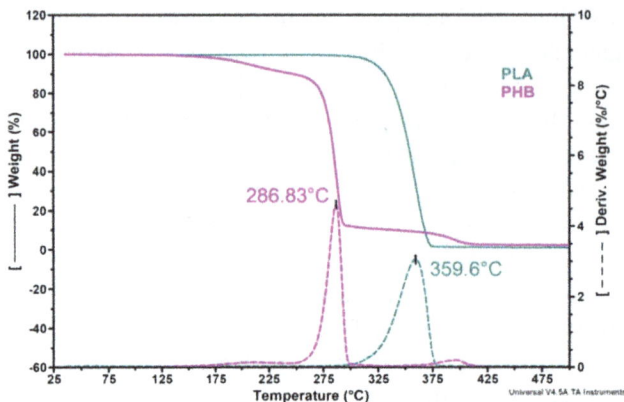

Figure 7.4: Thermogravimetric curves (TG) and their derivatives (DTG) for PLA and PHB [27].

(butylene-adipate-co-terephthalate) (PBAT). An effective compatibilizer of PLA-g-GMA (PLA grafted with glycidyl methacrylate) was produced in a study conducted by melt grafting [33]. It was then introduced as a compatibilizer for the PLA and PBAT phases of the blend. The dynamic rheological responses were examined to analyze the influence of PLA-g-GMA on the rheological responses of PLA/PBAT blends. Additionally, the mechanical properties of 3D printed PLA/PBAT/PLA-g-GMA samples in longitudinal and transverse deposition were identified to analyze against injection molding to expose the effects of compatibilizer in 3D printing. Addition of the compatibilizer drastically increased the blend's compatibility, and the interface binding capacity of PLA and PBAT was positively correlated to the amount of compatibilizer in the blend. At 10% compatibilizer by weight, dynamic rheological analysis displayed that blends had the greatest amount of molecular chain entanglement, complex viscosity and melt strength. Additionally, the compatibilizer strengthened the interfacial adhesion within layers of the 3D printed samples. Orientation of the filaments with microchannels played a large role in increasing the resultant sample behavior of the 3D printed samples too. It was also discovered that the tensile strength of a PLA/PBAT = 70/30 blend with 10% compatibilizer was 42.6 MPa and elongation at break was greater than 200%. These values are both larger than the injection sample (33.7 MPa) but are also higher than the maximum value reported in the literatures [33].

Figure 7.5a showed T_{gs}, centered around −32.8 and 60.4 °C, one broad exothermic peak at 110.5–140.6 °C which is where cold crystallization occurred, and a broad endothermic peak at 150.0–169.8 °C which shows the material's melting phase. After examining the thermal behavior of the blended material with unfilled PLA and PBAT using identical conditions, the resulting T_g measurements shown were accredited to the PBAT and PLA. The exothermic peak was attributed to the crystallization of PLA blocks. The existence of two clear T_g values similar to thee unfilled materials showed the PLA and PBAT phases were immiscible in all cases. Since the compatibilizer had a

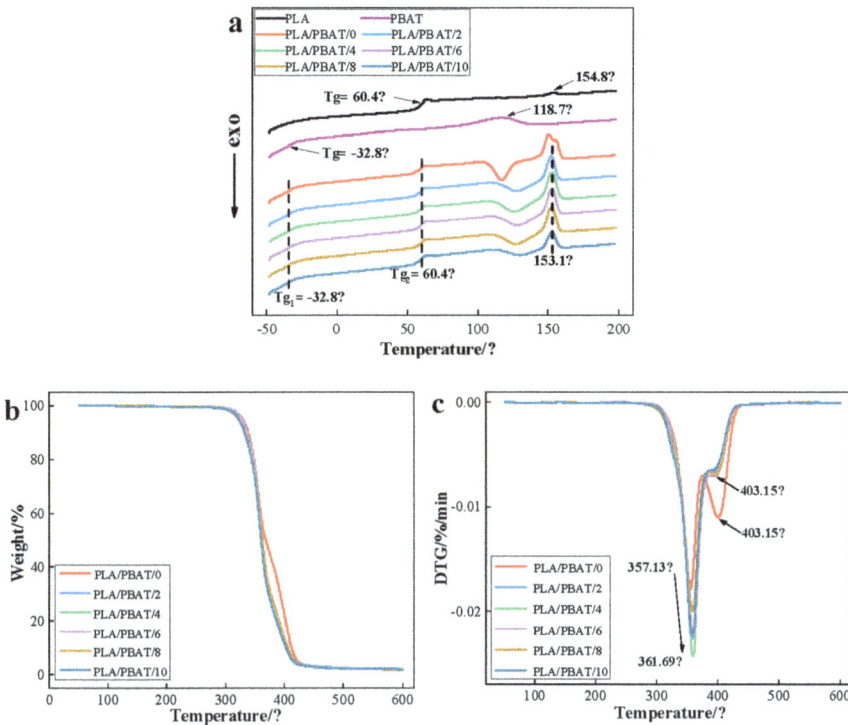

Figure 7.5: Characterization of PLA/PBAT/PLA-g-GMA blends: (a) DSC curves (second heating scan, 10 K/min), (b) TGA curves, and (c) DTG curves. Reproduced with permission from Lyu et al. [33].

high molecular weight, it did not have a significant impact on the free volume, demonstrating the blends' T_g were not changed using varying amounts of compatibilizer. Increasing the compatibilizer caused cold crystallization to be observed at drastically higher temperatures. Figure 7.5b shows the TGA thermograms of PLA/PBAT/PLA-g-GMA blends. All samples were heated to the same starting temperature of 328 °C. The DTG shown in Figure 7.5c clearly shows the PLA/PBAT blend two step degradation, with a first maximum decomposition peak at 357–362 °C and second maximum decomposition peak at 403 °C. Increasing compatibilizer content brought the peaks closer to each other, displaying enhanced compatibility among the two materials [34].

Figure 7.6a shows that incorporating compatibilizer has the ability greatly enhance the tensile strength of injection samples in contrast with PLA/PBAT blends. Increasing the compatibilizer caused the tensile strength and elongation at break to increase, which was attributed to the close combination of PBAT and PLA. The tensile strength and elongation at break improved to 33.7 MPa and 237% respectively with 10% compatibilizer, which was greater than the results reported by other researchers using the same ratio. This result indicates that PLA-g-GMA is an effective compatibilizer for PLA/PBAT blends. Figure 7.6b and 7.6c shows that PLA/PBAT with 10 wt% PLA-g-GMA

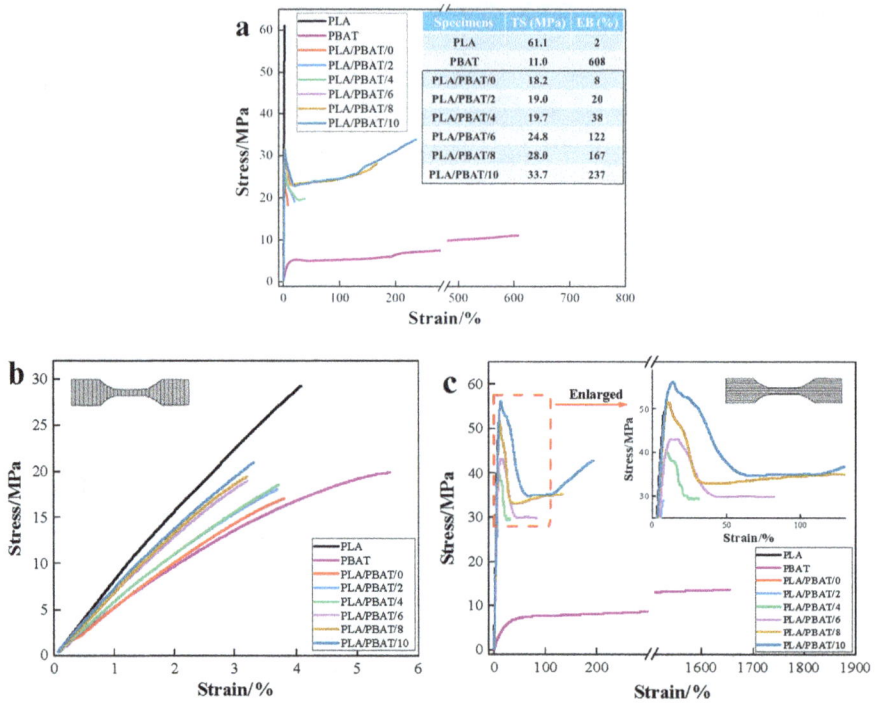

Figure 7.6: PLA, PBAT, and PLA/PBAT/g blends stress-strain curves by injection molding. (a) 3D printing in transverse direction, and (b) 3D printing in longitudinal direction. Reproduced with permission from Lyu et al. [33].

displayed the greatest tensile strength in each direction for the 3D printed tests. The tensile strength of the compatibilized blend in the longitudinal direction was greater than the injection sample and retained an elongation at break of over 200% [33].

PLA-based biodegradable polyester blends are immiscible. Using an effective compatibilizer, compatibility and mechanical properties improved, and these results will be discussed in the following section. In order to improve the mechanical properties of PLA-based biodegradable polyester blends, block copolymer can be used as an effective compatibilizer. However, finding an "A–C" type of di-block copolymer that is used as an effective compatibilizer to improve the miscibility of A and B blends is difficult. Monomethoxy poly(ethylene glycol)-polylactide (MPEG-PLA) di-block copolymers were orchestrated to compatibilize PLA-based polyester blends, like PLA with PBAT, PBS, or PBSA. Examination displayed that the di-block compatibilizer favorably lowered the interfacial tension and enhanced the interfacial adhesion, which reduced the dispersed phase size from 2 to 0.3 µm when MPEG-PLA copolymer was added using a segment ratio of 2.5 EG/LA in the PLA/PBAT blends. The elongation at break for the blends had a maximum value 296%, which was 10 times that of the virgin blends. Additionally, the MPEG-PLA di-block copolymer with a specific structure enhanced the mechanical properties and altered the phase morphology of PLA/PBS and PLA/PBSA blends. The PEG segment in the "A–C" di-block copolymer behaved as B blocks in "A–B" di-block copolymers, resulting in improved compatibility in the PLA-based biodegradable polymer blends [35].

DSC and DMA analysis were used to characterize the thermal properties of the virgin blends as well as the blends with compatibilizer. There were two distinct peaks of tan δ shown on the DMA curves shown in Figure 7.7a. The two T_{gs} of the blends were 60 °C for the PLA matrix and another –30 °C for the other component [36]. The T_g of PLA decreased when the di-block copolymer was added. This drop was due to two factors. The first was the compatibilization of PLA with the supplemental segment caused the T_g to shift toward each other. The other cause of PLA's T_g drop was that the di-block copolymers have a lower melting point and lower molecular weight than the component in the blends, causing it to act as a plasticizer, accelerating the movement of the chains, which resulted in plasticization. The degree of crystallinity greatly influences the mechanical properties of polymer materials and polymer blends [37]. Figure 7.7b shows the two distinct T_g of the blends, which demonstrates that the PLA-based blends are immiscible. This result also confirmed the results of the DMA test. Incorporating di-block copolymers caused PLA's crystallinity to increase. The χ_c of the PLA/PBAT blends with ME45L18 compatibilizer was 26.8% compared to 9.0% for the blends without compatibilizer. This result was caused by the di-block copolymer acting like a nucleating agent, which enhanced PLA's crystallinity, and the PBAT phase of the blend affected PLA's crystallization. Figure 7.7 shows the PBAT phase spread throughout the polymer matrix as nanomicelles. The nanomicelles effectively performed as a successful nucleating agent to promote the crystallization of the PLA/PBAT blend. PBAT maintains a faster crystallization rate, which speeds up the crystallization of PLA [35].

Efficient compatibilizers both reduce particle dispersion and enhance the blends' physical properties. "Brittle fractures, spinning phenomena, and skin-core," were present in the tensile test for the immiscible blends due to the incompatibility between

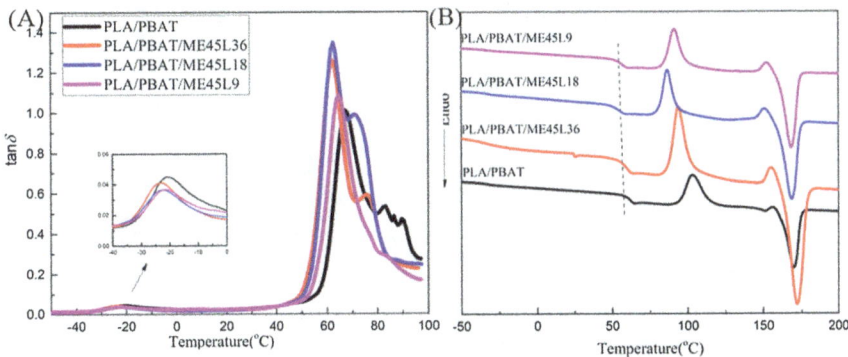

Figure 7.7: The thermo properties of the PLA/PBAT blends: (a) tan δ and (b) DSC curves [35].

the two phases [35]. Figure 7.8a and 7.8b shows an elongation at break of 30% for the neat blends. When 3% by weight ME45L18 di-block copolymers was incorporated, an elongation at break of 296% was observed, which is nearly 10 times higher than the specimen without compatibilizer. Stress whitening, necking and stress hardening phenomena occurred during the tensile tests for the specimens with ME45L18 di-block copolymer. With that result, it was clear that the MPEG-PLA di-block copolymers efficiently enhanced the blends' ductility [35].

One of biodegradable PLA's inherent weaknesses is its impact toughness. As discussed in the following section, blending PLA with PHA can enhance PLA's impact strength. The focus of this [38] was to find a way to enhance the impact toughness of PLA without a negative impact on the bio-based carbon content and compostability of PLA. Low-crystallinity and amorphous polyhydroxyalkanoates (PHA) copolymers have proven effective in enhancing PLA's toughness at modest loading levels of 10–20% by weight. The performed study included the preparing and characterizing PLA/PHA blends using two variants of amorphous PHA copolymers. Applying an annealing process post molding was also examined to potentially improve the manufactured PLA/PHA blends' impact performance. X-ray diffraction measurements were taken in order to study the material morphology and crystallization differences pre and post annealing of the prepared PLA/PHA blends as well as the virgin PLA. Excellent impact behavior was observed in the PLA/PHA2 blends [38].

Semicrystalline polymers' material properties are very reliant on the degree of crystallinity, crystalline form and crystalline morphology. The formation of oriented shish-kebab structures can enhance the mechanical properties of polymers [39–41]. It has already been found in the literature that unfilled PLA displays a small shoulder peak prior its principal melting peak [42]. This shoulder peak may be related to the phase transition or recrystallization of α'-crystal form, which has loose and disarranged

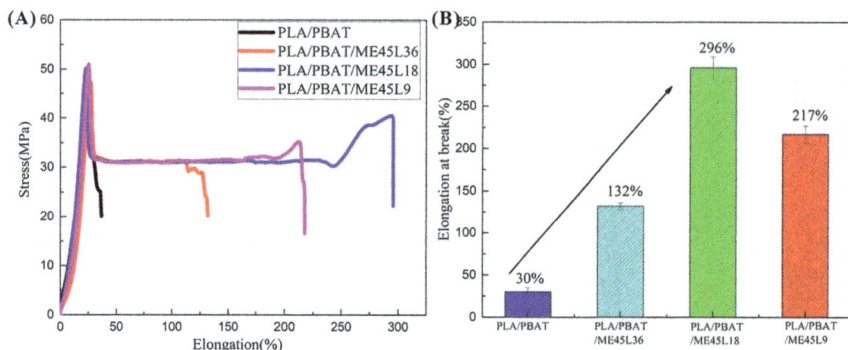

Figure 7.8: (a) Compatibilized blends stress–strain curves and (b) compatibilized blends elongation at break [35].

chain packing, to an arranged α-crystal form while being heated in the DSC. During the study, an almost identical peak shoulder appeared immediately before PLA's melting peak, shown by the black line in Figure 7.9. After including the PHA2, the small shoulder peak was removed for the PLA/PHA2 blends. The two blend ratios displayed a similar melting peak at approximately 178 °C. Increasing the melt temperature and the disappearance of the shoulder peak indicates the selected PHA types could even be effective for nucleating PLA. However, the PHA2 type appears to be only option able to aid in α-crystal formation when processed in injection molding and blended with PLA. Typically, PLA's crystallization process is improved incorporating only 10% of PHA by weight [43]. PLA/PHA3 blends have a similar effect, except the small shoulder peak did not entirely go away, which is shown by the red line in Figure 7.9. The cold crystallization peak observed in the first heating scan was repositioned to lower temperatures for both studied PLA/PHA blends [38].

On average, the tensile strength and tensile modulus of PLA/PHA blends were not as strong as the neat PLA, as shown in Figure 7.10. This was caused by the rubber-like behavior displayed by the PHAs in the formulated blends. The blends were annealed at 100 °C for 1 h. After the blends were annealed, the tensile modulus improved for the PLA and PLA/PHA 80/20 blend. The enhanced tensile modulus may be caused due to a pairing of different crystal morphology and increased overall crystallinity, as displayed by the DSC. The maximum tensile strength result, post annealing, was found in the PLA/PHA3 blend, with a 90/10 blending ratio, reaching almost 50 MPa. This value was similar to that of virgin PLA after annealing [38].

PLA has found many new applications in recent years. Some of these applications are discussed in the next section.

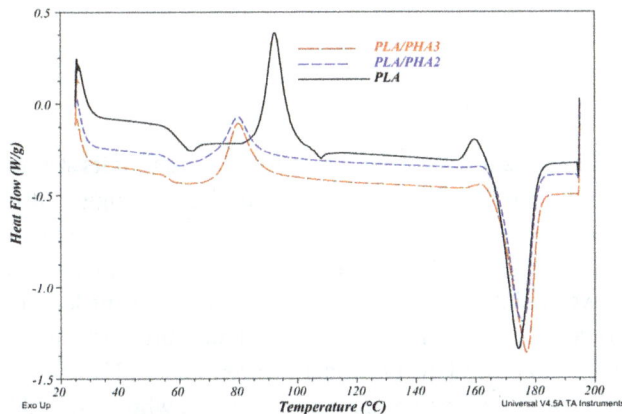

Figure 7.9: Injection molding specimens – first DSC heating scan for PLA and PLA/PHA blends. Reproduced with permission from Burzic et al. [38].

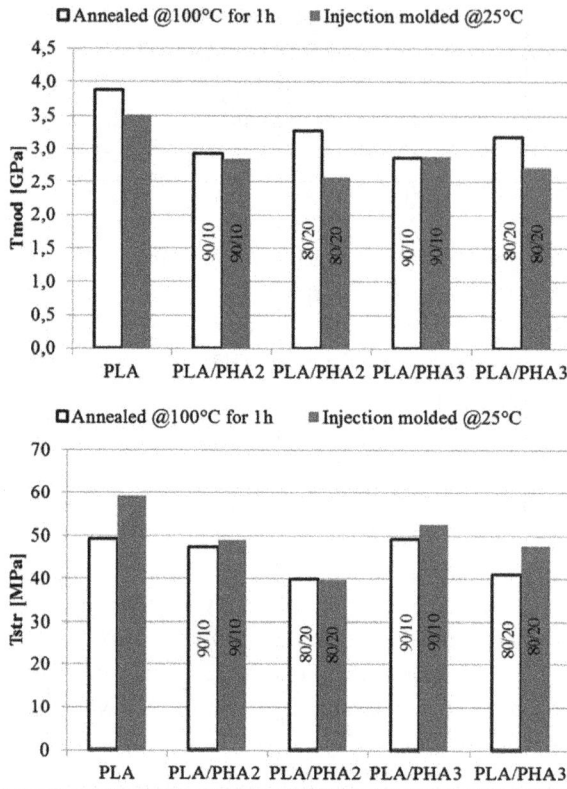

Figure 7.10: Tensile modulus and tensile strength of PLA and PLA/PHA blends before and after annealing. Reproduced with permission from Burzic et al. [38].

7.4 Applications

PLA can be used in a variety of implants in orthopedics surgeries because of its strong mechanical characterization and biomedical properties. These implants struggle with slow degradation rates when they are used in practical scenarios. PLA was developed using additive manufacturing techniques to fabricate and further assess the material's mechanical characterization as well as its degradation tendencies using varying factors [44]. These scaffolds can be shown in Figure 7.11. A digital weight measure was used to find the change in weight of the scaffolds and the pH was measured with a pH meter. A universal testing machine was used to test the compressive strength, while and SEM and EDS were utilized to characterize the morphology and elemental composition, respectively. The biocompatible nature and apatite formation of fabricated scaffolds were investigated using an *in vitro* simulated body fluid study. The results showed that the scaffold with 60% infill displayed the greatest porosity, and this is considered an advantage for apatite formation and osseointegration. Average compressive strength change was measured after 14 and 28 days, measuring 49.79 and 46.11 MPa, respectively. The average change in pH was 5.67 and 5.27 after 14 and 28 days of incubation, respectively. Specimen 1 had a 27.92% higher degradation rate than specimen 3, specimen 5 was 35.69% higher than that of specimen 3, and specimen 9 was 87.98% higher than specimen 3. The study concludes that process parameters have a positive effect on the degradation rate and biocompatible behavior of PLA implants [44].

It has been found that each specimen slightly increased in weight after two weeks which can be seen in Figure 7.12 [44]. This result could be caused by the apatite formation on the specimens' surface at the start, which would cause the rise in weight that was seen. After 28 days, each specimen's weight dropped, which could be expected as the degradation rate would be the dominant factor. The degradation process relies on the diffusion of water into the polymer chain. It begins on the outside and then moves inward eventually leading to the creation of ester bonds with hydrolytic chain reaction [45]. PLA has a higher deterioration rate than other polymers occurring *in vivo* operation at implantation sites because of its dependence on hydrolytic cleavage. This could be possible because of the degradation rate and is usually remedied by enzymes whose behavior can differ between subjects. The release of oligomers caused the molecular weight of the PLA scaffolds to lessen, which was then followed by the acidic environment formation [46]. These low weight oligomers were chiefly responsible for the solubilization of biospecimens in the medium which caused weight loss to begin. The data collected from the degradation study showed that the weight loss percentage of specimen 3 was the highest among all nine specimens after the 28 days incubation

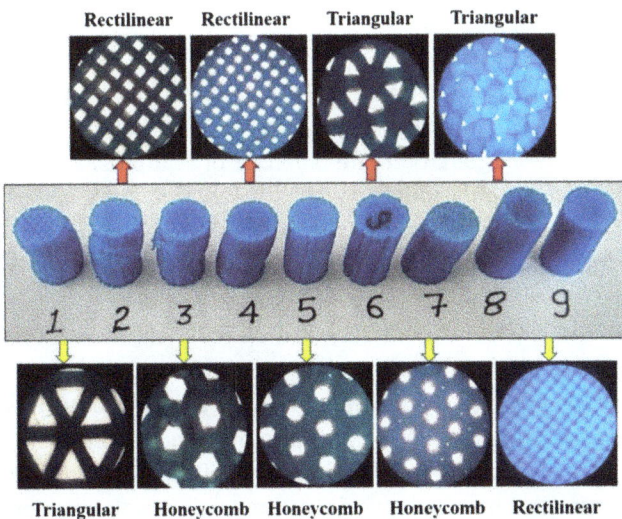

Figure 7.11: Different specimens examined in the study. Reproduced with permission from Singh et al. [44].

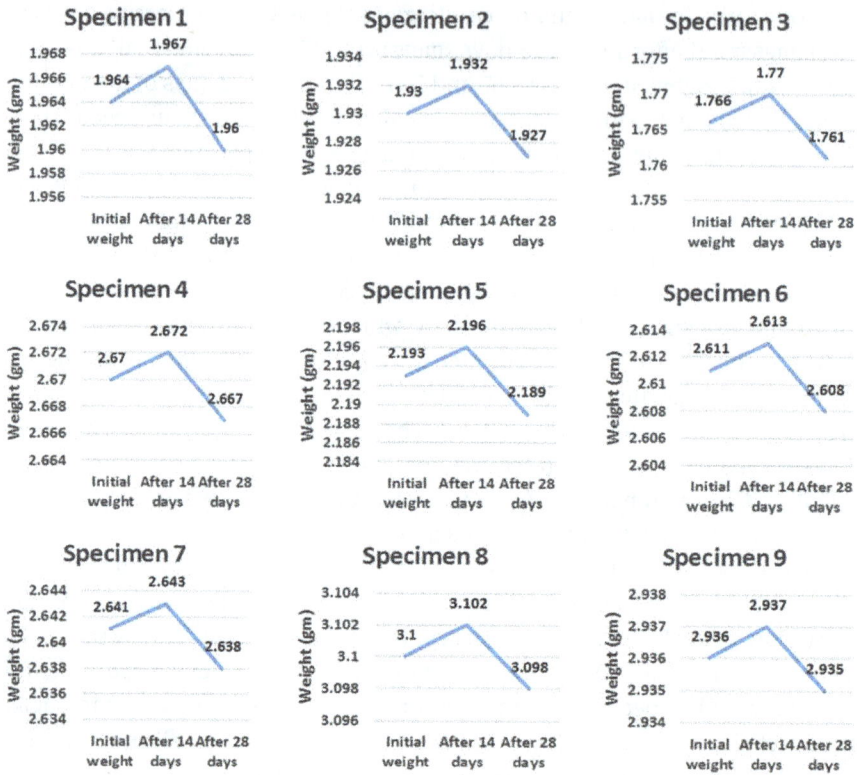

Figure 7.12: Scaffold weight measurements for all nine specimens. Reproduced with permission from Singh et al. [44].

period. The results showed that the dispersion status of PLA scaffolds fabricated with varying infill pattern, infill percentage, and layer thickness in simulated body fluids had a large impact in promoting the degradation rate behavior of scaffolds. For this reason, the degradation effect had a higher weight for the longer incubation period, which correlated with the change in pH of the simulated body solution [44].

One of the most challenging applications is wound repair. PLA's usefulness in the wound healing industry is discussed in the following section.

The repair process for wound healing is necessary and difficult, challenging researchers. Tissue engineering scaffolds demonstrate a viable method for wound repair. In this study [47], nanofibers with Poly(y-glutamic acid) (y-PGA) as the core with a PLA shell material were formulated using a coaxial electrospinning process. Flow rate effects were examined on the structure and diameter of the core-shell nanofibers. Using SEM and TEM, the microstructure and morphology of the core-shell nanofibers were investigated. In laser scanning confocal microscopy experiments, core-shell nanofibers loaded with rhodamine and coumarin-6 were excited using red and green light respectively in order to further evaluate the structure of the core-shell nanofibers. The hydrophobicity and degradation of core-shell nanofibers were also studied by performing both water contact angle tests and degradation performance test experiments. The *in vitro* cell culture results proved the favorable biocompatibility of nanofibers. The *in vivo* animal experiment displayed that PLA/γ-PGA core-shell nanofiber membrane was helpful in wound repair, where more than 90% reepithelialization was reported. This result suggests that the material is promising and dependable for tissue engineering and wound healing [47].

In the procedure, 2.80 g of PLA and 3.60 g of γ-PGA were broken down in 20 mL of CH_2CL_2 and CH_3COOH, respectively, at 40 °C for 10 h using magnetic stirring in order to formulate 10 and 15% by weight of the spinning solution. A stainless steel coaxial spinneret was selected to perform the electrospinning. The aforementioned γ-PGA and PLA solution were filled into 20 mL plastic syringes and then ejected through the coaxial spinneret with a shell flow rate of 1.0 mL h^{-1} and core flow rate of 0.2, 0.4, and 0.6 mL h^{-1}, respectively. Figure 7.13 shows a scheme diagram of the electrospinning process for γ-PGA/PLA core-shell nanofibers [47].

The therapeutic success related to wound healing was explored through the use of animal models and the wounds were photographed. As shown in Figure 7.14a, specimens from treated and untreated mice were examined after 1, 4, 8, and 14 days from when the wounds were created. The normal dermal and epidermal tissues in the afflicted wound area was restored and noted in the samples of the nanofiber membrane treated mouse within two weeks being treated as compared to the control sample. The variation curve for the wound size of two groups of mice was labeled and it is displayed in Figure 7.14b. The A group wound size shrank from 1.2 to 0.3 cm and the B group's wound size shrank from 1.0 to 0.5 cm in 14 days. The stained specimen of treated mice

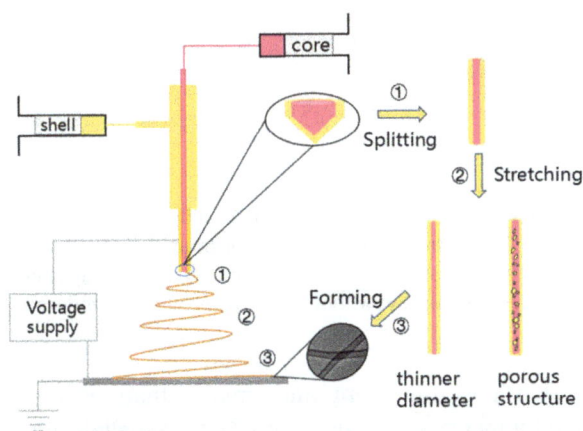

Figure 7.13: Scheme of electrospinning process of γ-PGA/PLA core-shell nanofibers. Reproduced with permission from Fang et al. [47].

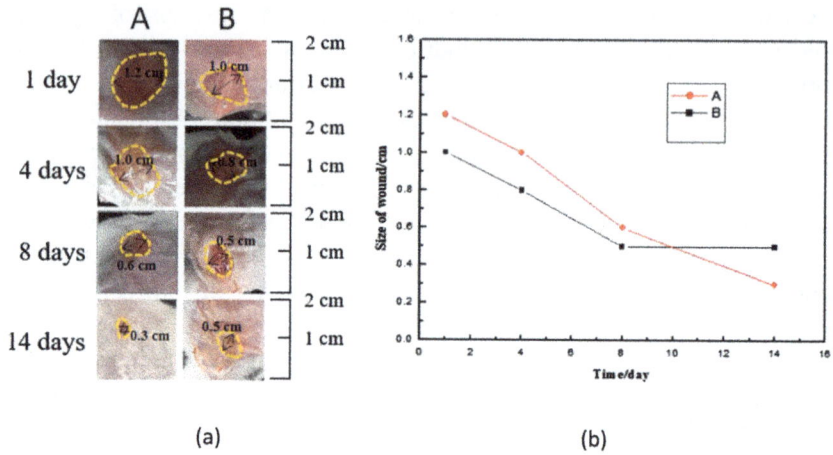

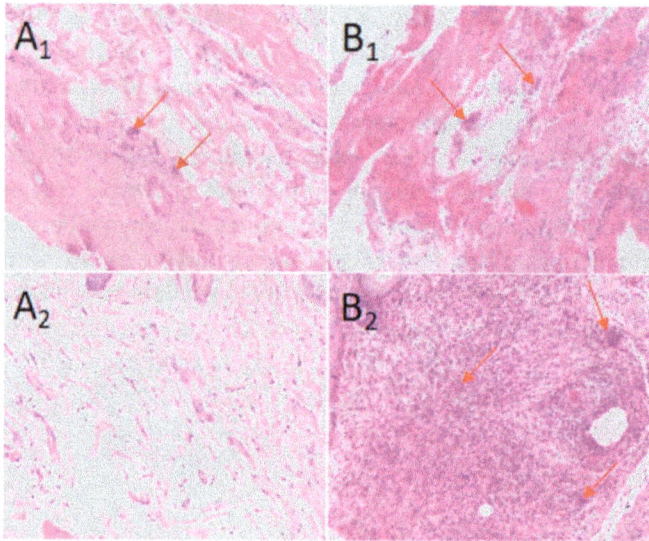

(a)

(b)

(c)

Figure 7.14: "(a) Photographic images of the extent of wound healing: graphical illustration of the changes in wound size on 1, 4, 8, and 14 days: (A) treatment group; (B) control group. (b) wound size variation curve (c) H&E staining of the tissues from the wound of mice: (A1) and (B1) on first day after infection, 20×; (A2) 14 days with membrane, 20×; (B2) 14 days without membrane, 20×". Reproduced with permission from Fang et al. [47].

was similar to that of normal skin post treatment and greater than 90% re-epithelialization was also significantly increased, while only 25% reepithelialization was observed to have improved in the control group. Prior to being treated, numerous dead cell debris, loose connective tissue and neutrophils and the collagen nanofibers were disordered as shown in Figure 7.14c (A1) and (B1). Two weeks after the initial wound creation, the stained specimens of the treated mice was similar to normal skin and reepithelialization also drastically improved in as shown in Figure 7.14c (A2), as opposed to the mice in the control group (B2). Obvious dermal fibroblasts as well as neovascularization filled with numerous red blood cells were noticed thereafter being treated. The gathered data suggests that the nanofiber membrane could efficiently and demonstratively enhance the wound healing process. Proliferation and migration of dermal fibroblast and keratinocytes improved, resulting in faster wound healing within 2 weeks of treatment. This showed that the core-shell nanofibrous scaffolds hold strong potential for future wound repair applications [47].

PLA-b-PEG-b-PLA copolymers have the potential to be used in biliary reconstruction in liver transplantation. The possibility of using the copolymer in a resorbable internal biliary stent will be explored in the following section. Internal biliary stenting in the course of biliary reconstruction in liver transplantation decreases anastomotic biliary complications. Implanting a resorbable internal biliary stent (RIBS) has potential because it can circumvent an ablation gesture. This study [48] focused on evaluating the sufficiency of certain PLA-b-PEG-b-PLA copolymers in resorbable internal biliary stents focused on securing biliary anastomose in the healing process and to stop potential problems, such as bile leak and stricture. The degradation kinetics and physical properties of a RIBS prototype were investigated, focusing on the main bile duct stenting requirements for liver transplants. The degradation of RIBS under biliary mimicking solution compared to the standard phosphate buffer control solution was communicated. "Morphological changes, mass loss, water uptake, molecular weight, permeability, pH variations, and mechanical properties," were analyzed [48]. The permeability and mechanical properties were examined using simulated biliary conditions to determine how useful the PLA-b-PEG-b-PLA RIBS in securing biliary anastomosis. The resulting analysis displayed that there was not any pH effects on the degradation kinetics, as the degradable RIBS remained impermeable for at least 8 weeks and retained their physical properties for 10 weeks. The RIBS totally degraded after 6 months. PLA-b-PEG-b-PLA RIBS demonstrate the necessary *in vitro* degradation behaviors to secure biliary anastomosis in liver transplantation and envision *in vivo* applications [48].

The mass loss and the water uptake are displayed in Figure 7.15. An exponential function was detected for the polymer water uptake over the period of the study. Water uptake was 25, 50, 100%, and more than 1000% at 2, 4, 6, and 12 weeks, respectively. There are clearly two distinctive periods. A lower mass loss rate was observed during the three first weeks compared to the following weeks. The decrease of molecular weight was quasi linear following this period. After the 12th week, weight loss measurements could not be performed due to their friable consistency [48].

In order to quantify the impact of the hysteresis of the polymer, the amount of work needed to load and unload the material at 10, 20, and 50% was calculated. The resultant values are displayed in Figure 7.16. In regards to the stiffness, the total work

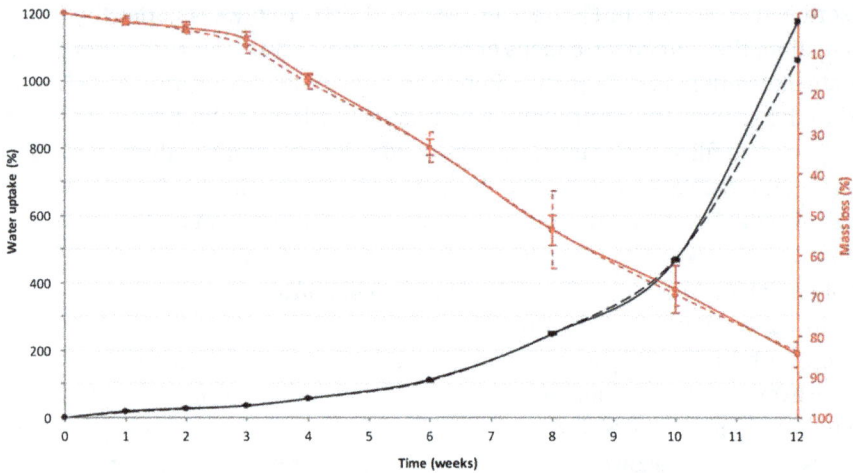

Figure 7.15: "Weight loss and water uptake of PLA-b-PEG-b-PLA triblock copolymer during 12 weeks of degradation at 37 °C. Solid line: "control group" with PBS (pH 7.2) as degradation solution, dotted line: "biliary group" with bile like serum (pH 8.4). (Data are represented as means ± SD with n = 3). PBS, phosphate-buffered saline; PEG, poly(ethylene glycol); PLA, poly(lactide)" [48].

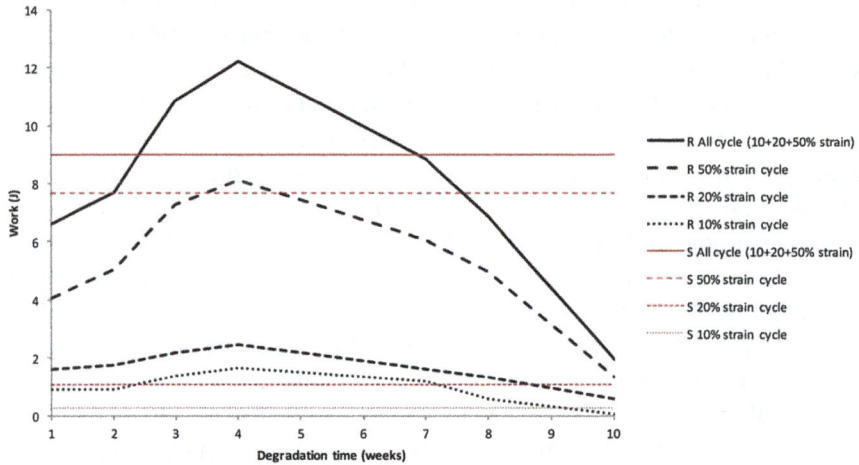

Figure 7.16: "RIBS compressive mechanical tests. Work evolution during degradation for RIBS (R) in black, compare to work for silicone stent (S) in red. RIBS, resorbable internal biliary stent" [48].

increases for a few weeks before dramatically lowering, which illustrate the deteriorating mechanical properties. The work curves drop was more salient compared with the stiffness [48].

Biodegradable PLA is being considered for oil and water separation. Its potential use is discussed in the following paragraphs. The construction of superwettability materials for oil and water separation has been developed rapidly, but the postprocess of the used separation materials is a problem due to their nondegradation in a natural environment. The functionalization of PLA nonwoven fabric as superoleophilic and superhydrophobic material for efficient treatment of oily wastewater with ecofriendly post-treatment due to the well-known biodegradable nature of PLA matrix was examined and deemed as a viable candidate to be used for oil-polluted water treatments [49].

Dopamine was used first to modify PLA nonwoven fabric [50]. In order to provide the PLA nonwoven fabric with superoleophilic and superhydrophobic properties, hierarchical micro/nanoparticles comprised of hydrophobic PS microspheres and silica (SiO_2) nanoparticles were arranged densely on the polydopamine (PDA)-modified PLA fabric to increase its surface roughness. The resulting SiO_2/PS/PLA hybrid nonwoven fabric with hierarchical porous structures displayed high oil-absorption capacity and selectivity separation for oil/water mixtures. The fabric even showed stable wettability after withstanding scrupulous friction and stretching. The above mentioned advantages allow this novel PLA-based hybrid fabric to separate oily wastewater effectively and be conveniently disposed of due to its biodegradability [49].

Oil/water separation are heavily dependent on the surface superhydrophobicity. The surface wettability of PLA nonwoven fabric after modification was analyzed using water contact angle (WCA) measurement. The WCA of the PLA nonwoven fabric studied was 117 ± 3.0°, showing simultaneously oleophilic and hydrophilic surface, which is not able to properly separate oil/water mixtures. Modified with PDA, PLA nonwoven fabric becomes more hydrophilic with a WCA of 23 ± 2.3°. This improvement is caused by the introduction of abundant hydroxyl and amino groups on the surface of PLA fiber. Different sizes of PS microspheres and SiO_2 nanoparticles were introduced simultaneously on the surface of PLA to obtain an improved hydrophobic PLA nonwoven fabric. When the mass proportion of SiO_2 nanoparticles and PS microspheres gets close to 18:1, superhydrophobic PLA fabric with a WCA of 152.0 ± 2.1° can be accomplished. In this state, a water droplet will roll away quickly once in contact with the slightly tilted superhydrophobic SiO_2/PS/PLA nonwoven fabric [49].

7.5 Conclusion

As the push for biodegradable materials and sustainability increases, PLA will have rapidly advancing applications. While PLA can be subject to some inherently poor qualities, such as brittleness, there are many techniques explored to nullify some of the poor properties. Composites of PLA with nanofillers, such as nanocellulose, nanoclays, carbon nanotubes, and graphene, are promising options for improving material properties of PLA. PLA nanocomposites have been used to explore potential applications in the food packaging industry and shown positive results. Several different compatibilizers have been studied as well, which have demonstrated improved mechanical properties of PLA blends. In the case of key application of wound healing and surgical work, possible solutions have shown strong results. Finally, the super-wettability of dopamine modified PLA was examined, with promising results for separation of oily wastewater. It is expected that the PLA, blends, and composites will continue grow and will be used in wide range of industrial applications, which will have strong positive impact on the plastic industry and the environment.

Author contributions: All the authors have accepted responsibility for the entire content of this submitted manuscript and approved submission.
Research funding: None declared.
Conflict of interest statement: The authors declare no conflicts of interest regarding this article.

References

1. Pang X, Zhuang X, Tang Z, Chen X. Polylactic acid (PLA): research, development and industrialization. Biotechnol J 2010;5:1125–36.
2. Omnexus. Polylactide (PLA): complete guide to accelerate your 'green' approach. Available from: https://omnexus.specialchem.com/selection-guide/polylactide-pla-bioplastic [Accessed 3 Sep 2021].
3. Sridhar V, Lee I, Chun HH, Park H. Graphene reinforced biodegradable poly(3-hydroxybutyrate-co-4-hydroxybutyrate) nano-composites. Express Polym Lett 2013;7:320–8.
4. La Mantia FP, Morreale M. Green composites: a brief review. Compos Appl Sci Manuf 2011;42:579–88.
5. Sun J, Shen J, Chen S, Cooper MA, Fu H, Wu D, et al. Nanofiller reinforced biodegradable PLA/PHA composites: current status and future trends. Polym J 2018;10:505.
6. Lin N, Dufresne A. Nanocellulose in biomedicine: current status and future prospect. Eur Polym J 2014;59:302–25.
7. Chivrac F, Pollet E, Avérous L. Progress in nano-biocomposites based on polysaccharides and nanoclays. Mater Sci Eng R Rep 2009;67:1–17.
8. Majeed K, Jawaid M, Hassan A, Abu Bakar A, Abdul Khalil HPS, Salema AA, et al. Potential materials for food packaging from nanoclay/natural fibres filled hybrid composites. Mater Des 2013;46:391–410.
9. Malin F, Znoj B, Šegedin U, Skale S, Golob J, Venturini P. Polyacryl–nanoclay composite for anticorrosion application. Prog Org Coating 2013;76:1471–6.
10. Hakamy A, Shaikh FUA, Low IM. Characteristics of hemp fabric reinforced nanoclay–cement nanocomposites. Cement Concr Compos 2014;50:27–35.
11. Felbeck T, Bonk A, Kaup G, Mundinger S, Grethe T, Rabe M, et al. Porous nanoclay polysulfone composites: a backbone with high pore accessibility for functional modifications. Microporous Mesoporous Mater 2016;234:107–12.
12. Shettar M, Achutha Kini U, Sharma SS, Hiremath P. Study on mechanical characteristics of nanoclay reinforced polymer composites. Mater Today Proc 2017;4:11158–62.
13. Young RJ, Kinloch IA, Gong L, Novoselov KS. The mechanics of graphene nanocomposites: a review. Compos Sci Technol 2012;72:1459–76.
14. Hule RA, Pochan DJ. Polymer nanocomposites for biomedical applications. MRS Bull 2011;32:354–8.

15. Millon LE, Wan WK. The polyvinyl alcohol–bacterial cellulose system as a new nanocomposite for biomedical applications. J Biomed Mater Res B Appl Biomater 2006;79B:245–53.
16. Mahmoodi A, Ebrahimi M, Khosravi A, Mohammadloo HE. A hybrid dye-clay nano-pigment: synthesis, characterization and application in organic coatings. Dyes Pigments 2017;147:234–40.
17. Mahmoodi A, Ghodrati S, Khorasani M. High-strength, low-permeable, and light-protective nanocomposite films based on a hybrid nanopigment and biodegradable PLA for food packaging applications. ACS Omega 2019;4:14947–54.
18. Raquez J-M, Habibi Y, Murariu M, Dubois P. Polylactide (PLA)-based nanocomposites. Prog Polym Sci 2013;38:1504–42.
19. Dai X, Li X, Zhang M, Xie J, Wang X. Zeolitic imidazole framework/graphene oxide hybrid functionalized poly(lactic acid) electrospun membranes: a promising environmentally friendly water treatment material. ACS Omega 2018;3:6860–6.
20. Herbst W, Hunger K. Industrial organic pigments: production, properties, applications. Hoboken, New Jersey: John Wiley & Sons; 2006.
21. Darie RN, Paslaru E, Sdrobis A, Pricope GM, Hitruc GE, Poiata A, et al. Effect of nanoclay hydrophilicity on the poly(lactic acid)/clay nanocomposites properties. Ind Eng Chem Res 2014;53: 7877–90.
22. Nair SS, Chen H, Peng Y, Huang Y, Yan N. Polylactic acid biocomposites reinforced with nanocellulose fibrils with high lignin content for improved mechanical, thermal, and barrier properties. ACS Sustainable Chem Eng 2018;6:10058–68.
23. Bastarrachea L, Dhawan S, Sablani SS. Engineering properties of polymeric-based antimicrobial films for food packaging: a review. Food Eng Rev 2011;3:79–93.
24. Nielsen LE. Models for the permeability of filled polymer systems. J Macromol Sci, Chem 1967;1: 929–42.
25. Duan Z, Thomas NL, Huang W. Water vapour permeability of poly (lactic acid) nanocomposites. J Membr Sci 2013;445:112–8.
26. Chen W, Chen H, Yuan Y, Peng S, Zhao X. Synergistic effects of polyethylene glycol and organic montmorillonite on the plasticization and enhancement of poly(lactic acid). J Appl Polym Sci 2019; 136:47576.
27. Frone AN, Batalu D, Chiulan I, Oprea M, Gabor AR, Nicolae C-A, et al. Morpho-structural, thermal and mechanical properties of PLA/PHB/cellulose biodegradable nanocomposites obtained by compression molding, extrusion, and 3D printing. Nanomaterials 2020;10:51.
28. Panaitescu DM, Nicolae CA, Frone AN, Chiulan I, Stanescu PO, Draghici C, et al. Plasticized poly(3-hydroxybutyrate) with improved melt processing and balanced properties. J Appl Polym Sci 2017;134:44810.
29. Frone AN, Panaitescu DM, Chiulan I, Nicolae CA, Vuluga Z, Vitelaru C, et al. The effect of cellulose nanofibers on the crystallinity and nanostructure of poly (lactic acid) composites. J Mater Sci 2016; 51:9771–91.
30. Armentano I, Fortunati E, Burgos N, Dominici F, Luzi F, Fiori S, et al. Processing and characterization of plasticized PLA/PHB blends for biodegradable multiphase systems. Express Polym Lett 2015;9: 583–96.
31. Arrieta MP, Fortunati E, Dominci F, Rayon E, Lopez J, Kenny JM. Multifunctional PLA-PHB/cellulose nanocrystal films: processing, structural and thermal properties. Carbohydr Polym 2014;107: 16–24.
32. Dhar P, Tarafder D, Kumar A, Katiyar V. Thermally recyclable polylactic acid/cellulose nanocrystal films through reactive extrusion process. Polymer 2016;87:268–82.
33. Lyu Y, Chen Y, Lin Z, Zhang J, Shi X. Manipulating phase structure of biodegradable PLA/PBAT system: effects on dynamic rheological responses and 3D printing. Compos Sci Technol 2020;200: 108399.

34. Mukesh K, Mohanty S, Nayak SK, Parvaiz MR. Effect of glycidyl methacrylate (GMA) on the thermal, mechanical and morphological property of biodegradable PLA/PBAT blend and its nanocomposites. Bioresour Technol 2010;101:8406–15.

35. Ding Y, Feng W, Huang D, Lu B, Wang P, Wang G, et al. Compatibilization of immiscible PLA-based biodegradable polymer blends using amphiphilic di-block copolymers. Eur Polym J 2019;118:45–52.

36. Fu Z, Wang H, Zhao X, Horiuchi S, Li Y. Immiscible polymer blends compatibilized with reactive hybrid nanoparticles: morphologies and properties. Polymer 2017;132:353–61.

37. Zhou SY, Huang HD, Ji X, Yan DX, Zhong GJ, Hsiao BS, et al. Super-robust polylactide barrier films by building densely oriented lamellae incorporated with ductile in situ nanofibrils of poly(butylene adipate-co-terephthalate). ACS Appl Mater Interfaces 2016;8:8096–109.

38. Burzic I, Pretschuh C, Kaineder D, Eder G, Smilek J, Másilko J, et al. Impact modification of PLA using biobased biodegradable PHA biopolymers. Eur Polym J 2019;114:32–8.

39. Liang S, Wang K, Yang H, Zhang Q, Du RN, Fu Q. Preparation and characterization of poly (urea-formaldehyde) microcapsules filled with epoxy resins. Polymer 2006;47:7115–22.

40. Su R, Zhang ZQ, Gao X, Ge YA, Wang K, Fu Q. Polypropylene injection molded part with novel macroscopic bamboo-like bionic structure. J Phys Chem B 2010;114:9994–10001.

41. Kalay G, Bevis MJ. Processing and physical property relationships in injection-molded isotactic polypropylene. 2. Morphology and crystallinity. J Polym Sci, Part B: Polym Phys 1997;35:265–91.

42. Zhang J, Tashiro K, Tsuji H, Domb AJ. Disorder-to-order phase transition and multiple melting behavior of poly (L-lactide) investigated by simultaneous measurements of WAXD and DSC. Macromolecules 2008;41:1352–7.

43. Bai H, Huang C, Xiu H, Zhang Q, Fu Q. Enhancing mechanical performance of polylactide by tailoring crystal morphology and lamellae orientation with the aid of nucleating agent. Polymer 2014;55:6924–34.

44. Singh D, Babbar A, Jain V, Gupta D, Saxena S, Dwibedi V. Synthesis, characterization, and bioactivity investigation of biomimetic biodegradable PLA scaffold fabricated by fused filament fabrication process. J Braz Soc Mech Sci Eng 2019;41:121.

45. Liu H, Webster TJ. Less harmful acidic degradation of poly (lactic-co-glycolic acid) bone tissue engineering scafolds through titania nanoparticle addition. Int J Nanomed 2006;1:541–5.

46. Zhang JF, Sun X. Mechanical properties of poly(lactic acid)/starch composites compatibilized by maleic anhydride. Biomacromolecules 2004;5:1446–51.

47. Fang Y, Zhu X, Wang N, Zhang X, Yang D, Nie J, et al. Biodegradable core-shell electrospun nanofibers based on PLA and γ-PGA for wound healing. Eur Polym J 2019;116:30–7.

48. Girard E, Chagnon G, Moreau-Gaudry A, Letoublon C, Favier D, Dejean S, et al. Evaluation of a biodegradable PLA–PEG–PLA internal biliary stent for liver transplantation: in vitro degradation and mechanical properties. J Biomed Mater Res 2021;109B:410–9.

49. Gu J, Xiao P, Chen P, Zhang L, Wang H, Dai L, et al. Functionalization of biodegradable PLA nonwoven fabric as superoleophilic and superhydrophobic material for efficient oil absorption and oil/water separation. ACS Appl Mater Interfaces 2017;9:5968–73.

50. Liu YL, Ai KL, Lu LH. Polydopamine and its derivative materials: synthesis and promising applications in energy, environmental, and biomedical fields. Chem Rev 2014;114:5057–115.

Samy A. Madbouly*

8 Bio-based polyhydroxyalkanoates blends and composites

Abstract: Polyhydroxyalkanoates (PHAs) are linear semicrystalline polyesters produced naturally by a wide range of microorganisms for carbon and energy storage. PHAs can be used as replacements for petroleum-based polyethylene (PE) and polypropylene (PP) in many industrial applications due to their biodegradability, excellent barrier, mechanical, and thermal properties. The overall industrial applications of PHAs are still very limited due to the high production cost and high stiffness and brittleness. Therefore, new novel cost-effective production method must be considered for the new generation of PHAs. One approach is based on using different type feedstocks and biowastes including food byproducts and industrial and manufacturing wastes, can lead to more competitive and cost-effective PHAs products. Modification of PHAs with different function groups such as carboxylic, hydroxyl, amine, epoxy, etc. is also a relatively new approach to create new functional materials with different industrial applications. In addition, blending PHA with biodegradable materials such as polylactide (PLA), poly(ε-caprolactone) (PCL), starch, and distiller's dried grains with solubles (DDGS) is another approach to address the drawbacks of PHAs and will be summarized in this chapter. A series of compatibilizers with different architectures were successfully synthesized and used to improve the compatibility and interfacial adhesion between PHAs and PCL. Finer morphology and significantly improvement in the mechanical properties of PHA/PCL blends were observed with a certain type of block compatibilizer. In addition, the improvement in the blend morphology and mechanical properties were found to be strongly influenced by the compatibilizer architecture.

Keywords: biodegradation, compatibilizers, DDGS, modification, polyhydroxyalkanoates, starch

8.1 Introduction

Plants carbohydrates (wheat, corn, potato, sugar cane, etc.), natural oils (castor, pine, soybean, rapeseed, etc.), and proteins (zein, chitin, collagen, soybean, etc.) are commonly used to produce environmentally friendly bio-based polymers. The family of polyhydroxyalkanoates (PHAs) including polyhydroxybutyrate (PHB), polyhydroxyvalerate (PHV), and their copolymers are semicrystalline aliphatic

*Corresponding author: Samy A. Madbouly, School of Engineering, Behrend College, Pennsylvania State University, Erie, PA 16563, USA, E-mail: sum1541@psu.edu

This article has previously been published in the journal Physical Sciences Reviews. Please cite as:
S. A. Madbouly "Bio-based polyhydroxyalkanoates blends and composites" *Physical Sciences Reviews* [Online]
2021, 7. DOI: 10.1515/psr-2020-0073 | https://doi.org/10.1515/9781501521942-008

thermoplastics that can be synthesized by microorganisms through metabolic pathways. PHAs are linear polyesters produced naturally by a wide range of microorganisms for carbon and energy storage [1, 2]. PHAs can be used as replacements for petroleum-based polyethylene (PE) and polypropylene (PP) in many industrial applications due to their biodegradability, excellent barrier, mechanical, and thermal properties. PHAs can be processed using the common convention processing techniques, such as extrusion, injection molding, compression molding, thermoforming, and blow molding. In addition, PHAs are biodegradable, biocompatible, and its degraded fragments are nontoxic. Therefore, the PHAs are suitable for wide range of biomedical applications [3, 4]. The PHAs can be created with up to 90 different types of monomers into the polymer backbone via microorganisms [5]. Therefore, the chemical and physical properties can be tailored for specific applications based on the different types of monomers.

The PHB is the first member of PHAs to be discovered in microorganisms and it has a potential to be a high-volume commercial product. The PHB can be obtained in a large production capacity from a readily convertible polymeric chemical precursor in biological systems as an inert granular material in a cell [8]. The PHB, is a highly crystalline, rigid, and brittle homopolymer, which lacks the superior mechanical properties required for different industrial application, particularly in packaging industry. On the other hand, PHAs contain medium chain length monomer of 6–14 carbons, such as carboxymethylchitosan-g-polyhydroxyalkanoates (PHAMCL), which has elastomeric behavior, high flexibility, and low crystallinity and mechanical strength. The chemical structure of PHAs can be seen below with an attached R-group, which can be varied to obtain different biodegradable materials with overall different mechanical and thermal properties (see Figure 8.1). PHAs are accumulated in granules as amorphous mobile polymers of different sizes as clearly seen in Figure 8.1. They are surrounded by a phospholipid monolayer. They can be observed intracellularly as light-refracting granules or as electron-lucent bodies that, in overproducing mutants, cause a striking alteration of the bacterial shape (Figure 8.1). All PHAs are biodegradable and can be totally decomposed to CO_2 and H_2O via a natural microbiological mineralization [9]. The biodegradation rate of PHAs is relatively fast compared to that of other biodegradable polymers, such as polylactide (PLA) and bio-based polyamide (PA) due to the easy hydrolysable ester bonds in the chemical structure of PHAs (see Figure 8.1). The biodegradation or depolymerization process can be enhanced by microorganisms and commonly produces low molecular weight products that can be absorbed as nutrients by living cells. The degradation behavior of PHAs and their copolymers has been investigated in different environments, such as soil, buffer solutions in laboratory atmosphere, activated sludge, sea water, and lake water [10]. Water and moisture content, as well as temperature, pH, and nutrient supply are important factors that influence the degradation rate of PHAs. In addition, the chemical structure, percentage of crystallinity, surface area, and nature and concentration of additives play an important role in the degradation kinetics of PHAs. PHAs can be used in wide industrial applications, such as 3D printing materials, textile, medical

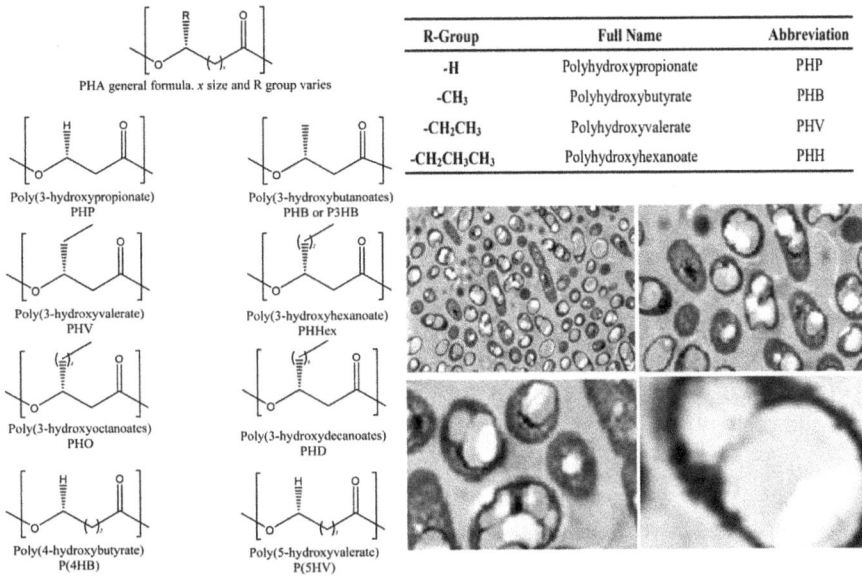

PHA general formula. x size and R group varies

R-Group	Full Name	Abbreviation
-H	Polyhydroxypropionate	PHP
-CH₃	Polyhydroxybutyrate	PHB
-CH₂CH₃	Polyhydroxyvalerate	PHV
-CH₂CH₃CH₃	Polyhydroxyhexanoate	PHH

Poly(3-hydroxypropionate)
PHP

Poly(3-hydroxybutanoates)
PHB or P3HB

Poly(3-hydroxyvalerate)
PHV

Poly(3-hydroxyhexanoate)
PHHex

Poly(3-hydroxyoctanoates)
PHO

Poly(3-hydroxydecanoates)
PHD

Poly(4-hydroxybutyrate)
P(4HB)

Poly(5-hydroxyvalerate)
P(5HV)

Figure 8.1: Chemical structure of PAHs with an attached R group. Reproduced with permission from Hawkins et al. [6]. The table shows the different types of PHAs obtained with different R groups. SEM photograph for PHAs. Adapted with permission from Sharma et al. [7].

implants, drug delivery carriers, packaging materials, agriculture mulching films, animal nutritional supplements, drugs, as well as new types of biofuels [11–13].

8.2 Synthesis of PHAs

The PLAs can be manufacturing from large number of biorenewable feedstocks, such as lignin, cellulose, plant oils, fatty acids, etc. The PHAs can also be synthesized from solid wastes such as municipal and organic solid wastes [14, 15]. Synthesis of PHAs from glucose was also reported in industry [16]. There are many factors play very important roles in the creation process of PHA by microorganisms. Lignin-based pentoses were previously used as precursor for synthesis of PHAs [16–18]. Figure 8.2a shows the microbial production process of PHA and its major or important factors [19, 20]. Figure 8.2b demonstrates the two main routes commonly used for synthesis of PHA from biomass [21]. The first method is the synthesis of the bio-based monomers obtained from the pretreatment process of lignocellulose. These bio-base low molecular weight monomers will be chemically polymerized to produce PHA. The second route which is the most common for synthesis of PHA is the biomass fermentation by microorganisms. The choice of the renewable resource to be used for the production of PHA relies heavily on local biomass availability. Diversification of feedstock is beneficial because it allows multiple farming activities (including aquaculture) to contribute to PHA production, making the process more versatile, scalable, and sustainable. Many combinations have been employed in an attempt to balance costs and the desired degree of purity. Figure 8.2b shows also the different applications of the bio-based PHAs including packing materials, agriculture applications, foam and insulating materials for construction as well as medical applications (wound healing, drug delivery, and biomedical devices) [21]. The mechanical, thermal, biocompatibility, and biodegradability of PHA are the main important properties required for specific applications. The PHAs are very pure semicrystalline polymers with T_{gs} between −40 and 5 °C and melting temperature between 50 and 180 °C. The overall mechanical and physical properties can be tailored by blending with other bio-based polymers or fillers as well described in detail later in this review article.

Synthetic biology and metabolic engineering approaches have proved to be relatively new strategies aimed to improve the PHA-producing microbial systems. Kumarapillai et al. reviewed the applications of synthetic biology and metabolic engineering approaches toward the enhancement of material properties and production efficiency of PHAs from microbial systems and the development of novel, eco-friendly, and economical polymer recovery options [22]. The authors reported that, the modifications in cellular volume and morphology based on the synthetic biology have dramatically improved the PHA-accumulating property of the bacterial cells. The susceptible of cells to gravity separation after the fermentation process was modified. Channeling the metabolic flux to change the PHAs biosynthetic pathways toward the PHAs metabolism was found to be greatly improved the PHAs accumulation efficiency up to 120% [22]. Promoter engineering is an essential aspect of synthetic biology, has a great influenced on the improvement of PHA yield. The chemical lysis and the solvent extraction methods (conventional polymer extraction) are reported to deteriorate the polymer properties and cause environmental pollution. Therefore, new novel methods for extraction of PHA are required. The cell lysis systems (selfdisruptive) based on the

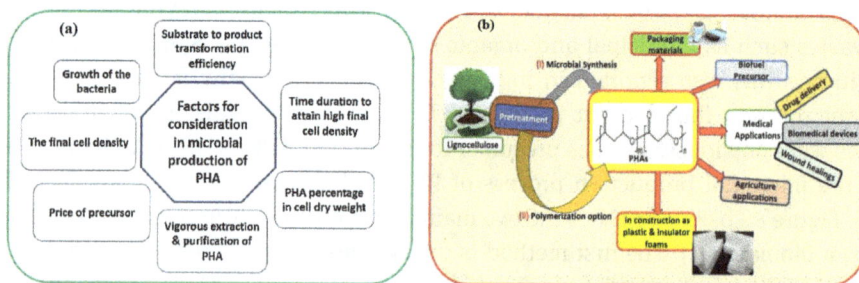

Figure 8.2: (a) Production of PHA through microbial fermentation process and the major factors affecting on it. Reproduced with permission from Adeleye et al. [21]. (b) Different routes for synthesis of PHAs from pretreatment of lignocellulose and their varied industrial applications. Reproduced with permission from Adeleye et al. [21].

holin–endolysin and lysozyme mechanisms are promising, but with low recovery efficiency [22]. Based on the above it appears that the use of synthetic biology and metabolic engineering can contribute to the properties, manufacturing cost, and performance of PHAs.

The bacterial fermentation manufacturing of PHAs include three processes, namely fermentation, isolation, and purification [23, 24]. Reducing the high cost associated with the fermentation and downstream processes [25] (about 15 times higher than that for conventional polyolefins) is very crucial for commercial production of PHAs [26]. One approach is based on using different types of feedstock and biowaste including food byproducts and industrial and manufacturing wastes, can lead to more competitive products [27]. Different organic wastes, such as molasses [28], olive and palm oil mill effluent [29, 30], fermented fruit waste [31], and cheese whey (CW) [32] have been used as substrates to produce PHAs [28]. Cost effective production of PHAs is necessary for wide applications in different industries. The high manufacturing costs of PHAs are the main reasons for not finding PHAs in many applications, although their properties are very similar to many petroleum-based polymers. Efforts to synthesize cost effective PHAs with outstanding properties have been investigated in the last decade. Very recently Quirino et al. summarizes the basic aspects of biofloc technology, and the production of PHAs from bioflocs [6]. Moreover, other types of wastes, such as organic fraction of municipal solid waste (OFMSW), and wastewaters municipal are additional resources for the production of PHAs. So far there are very limited studies have been used the OFMSW for making PHAs [33]. Similar study showed that the production of PHAs can greatly improve by using OFMSW during the accumulation stage [34].

8.3 Chemical modification of PHA

PHAs can be chemically modified into multifunctional materials by adding a chemical group in the structure of PHA. For instance, the hydrolytic degradation rate can be accelerated and improved by incorporating hydrolysis-prone chemical groups. As clearly illustrated in Figure 8.3, the chemical modification of PHAs can be performed using different methods. Addition of carboxylic function group to the chemical structure of PHAs can increase the hydrophilicity, increase the hydrolytic degradation, and increase the biocompatibility [35, 36]. Halogenation is also an excellent modification for PHA to improve the overall properties, functions, and applications of PHA. For example, chlorination of PHA can improve the mechanical properties (increase the stress at break) and hardness [37]. Modification of PHA with other function groups, such as epoxy or hydroxyl groups can improve the reactivity to create multiblock polymers, such as polyurethane or three-dimensional network through crosslink reaction of the epoxy groups. A decrease in the melting temperature and melting enthalpy of PHA can be expected by replacing the olefinic bonds by epoxy bonds [38]. Grafting of PHA with another polymer can be used for modification the crystallization

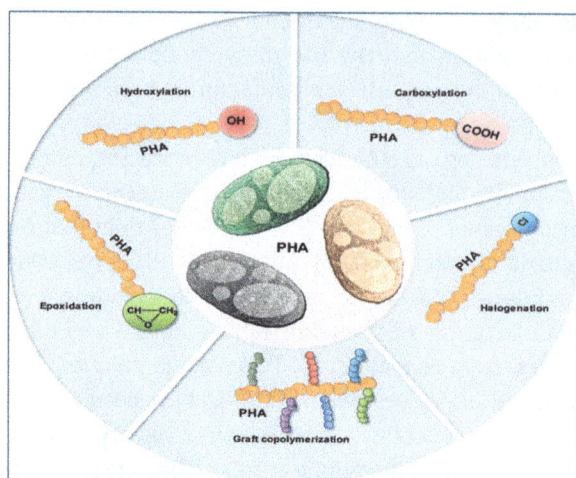

Figure 8.3: Different chemical modification methods of PHA. Reproduced with permission from Sharma et al. [7].

behavior, mechanical properties, optical properties, processibility and production cost of PHA. Grafting process can be performed through radiation [39], free radical [40], ionic [41], and enzymatic [42] methods.

8.4 PHAs blends and composites

Blending of PHAs with other biodegradable polymers or fillers is an effective method to improve the mechanical and thermal properties as well as enhancing the processability. Copolymerization is also a well-established approach to enhance the overall properties of PHAs. PHAs as bioplastic materials have very similar properties like petroleum-based thermoplastics, such as PP and PE. Therefore, PHAs can be used as alternative to the PP and PE in many industrial applications which might be more environmentally friendly products and help the conservation of ecosystem [43, 44]. There are a wide range of applications where PHAs act as biomaterials for different uses. Synthesis of 4-hydroxybutyrate (4-PHB) and 3-hydroxybutyrate (3-PHB) co-polymers with different ratios can be produced bio-based polymers with different overall mechanical and thermal properties. The 3-PHB is a brittle semicrystalline material with approximately 50% crystallinity and T_g of 4 °C. However, 4-PHB is more elastic with $T_g = -48$ °C and very low crystallinity percentage. Most of the efforts aimed at improving the toughness of PHA have focused on blending PHA with other polymers such as PLA or thermoplastic starch [45–48].

Bio-based polyamide (PA) was also blended with PHA to improve its flexibility and impact resistance for increased industrial application [49]. The PA is a resin containing monomers of amides joined by peptide bonds, and can be manufactured through solid-phase synthesis or step-growth polymerization. The bio-based PA resin can be manufactured from tall oil and processed through dimerization and polycondensation. Tall oil is a viscous dark yellow liquid that can be obtained from pine tree and is a byproduct of the paper industry and the pulp. This bio-based PA has high flexibility, low softening temperature, high viscosity, and low water content. The bio-based PA is widely used in industrial applications such as, adhesives, ink binders, and coatings. Blending of PHA with bio-based PA has been investigated to mainly improve the toughness of PHA.

A series of bio-based PA and PHB blends were mechanically melt mixed with different concentrations using a twin-screw extruder [49]. Small-amplitude oscillatory shear flow experiments were used to study the effects of blending on viscoelastic properties as a function of blend composition and angular frequency. The mechanical, thermal, and morphological properties of the blends were investigated using dynamic mechanical analysis, differential scanning calorimetry, thermogravimetric analysis, scanning electron microscopy, and tensile tests. The complex viscosity of the blends increased significantly with increasing the concentration of PHA and reached a maximum value for 80 wt.% PHA blend. In addition, the tensile strength of the blends increased considerably as the content of PHA increased. For blends containing PA >50 wt.%, samples failed only after a very large elongation (up to 465%) without significant decrease in tensile strength. The particle size of the dispersed phase increased greatly, and the blends became more brittle with increasing concentration of PHA. In addition, the concentration of the PA had a substantial effect on the glass relaxation temperature of the resulting blends. Our results demonstrate that the thermomechanical and rheological properties of PHA/PA blends can be tailored for specific applications, and that blends of PHA/PA can fulfill the mechanical properties required for flexible, impact-resistant bio-based polymers [49].

Ausejo et al. mixed PHA and PLA to improve the mechanical properties and tailor the biodegradation behavior [50]. The authors demonstrated the 3D printing of the PHA/PLA blend as a biocompatible material for potential biomedical applications. The PHA was found to be a mixture of mainly PHB with small amount of PHV. The filament for 3D printing consisted of PHA, which contains predominantly 3-hydroxybutyrate units and a small amount of 3-hydroxyvalerate units, as revealed by multistage mass spectrometry (ESI-MS^n). This research found that the properties of 3D printed species before and during abiotic degradation are dependent on the printing orientation. Furthermore, the 3D printed specimens exhibited a good biocompatibility with HEK293 cells, indicating real promise as biological scaffolds for tissue engineering applications. The processing conditions were found to have a considerable influence on the thermal and mechanical properties as well as the degradation profile of the investigated PLA/PHA specimens. The vertical specimens have improved mechanical properties compared to horizontal specimens; due to the more homogenous mixture and deformation induces crystallization effect.

Hydrolytically degradation of PHA/PLA blends was investigated at 50 and 70 °C in demineralized water (pH = 5.6) over a period of 70 days according to the ISO norm [50]. The thermal and mechanical properties as well as failure analysis have been investigated after the degradation process. Visual evaluation after hydrolytical degradation at different temperatures for different time intervals for pure PLA and PLA/PHA blend (dumbbell-shaped) showed erosion through the breaking of the specimens, which began at day three for all specimens degraded at 70 °C and at day 42 for all specimens degraded at 50 °C. Disintegration of the specimens began at day seven for all specimens degraded at 70 °C. Erosion due to water absorption during the degradation of the

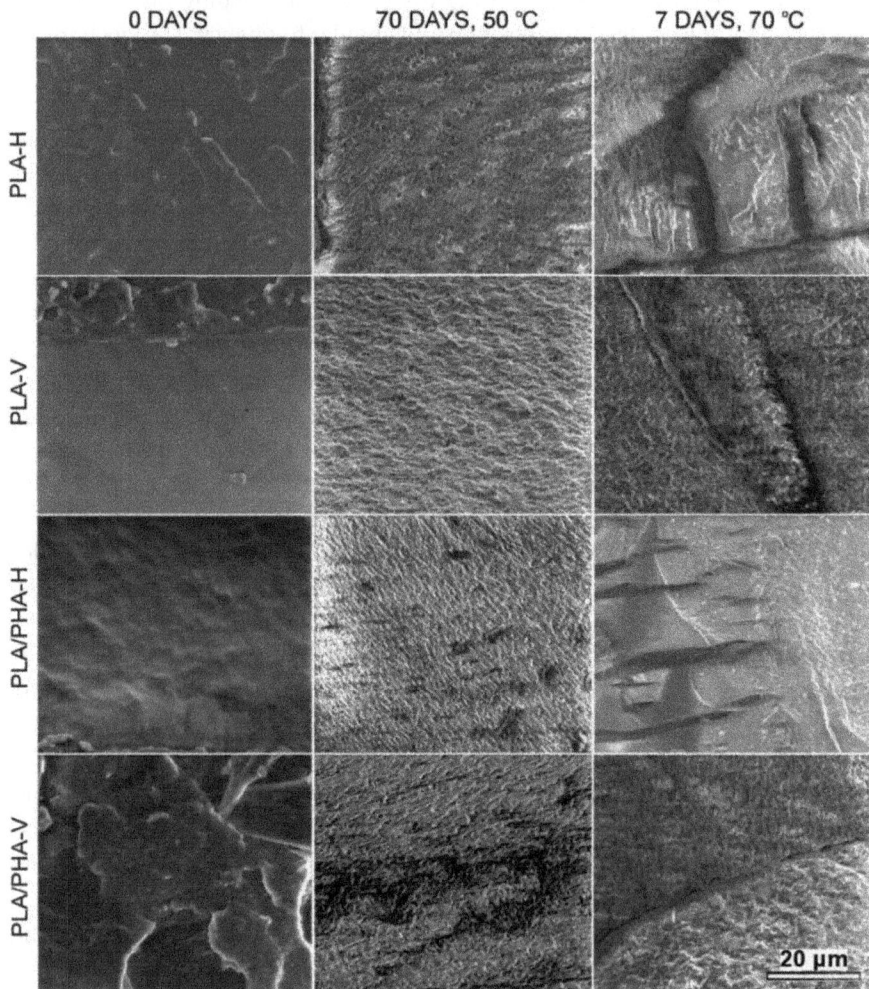

Figure 8.4: SEM micrographs for PLA and PLA/PHA dumbbell-shaped specimens obtained by 3D printing in horizontal (H) and vertical (V) directions before and after 70 days of hydrolytic degradation test at 50 °C and after seven days of hydrolytic degradation test at 70 °C. Reproduced with permission from Ausejo et al. [50].

dumbbell-shaped specimens at 50 and 70 °C was observed in the surface morphology of the materials using SEM as demonstrated in Figure 8.4. The degradation at a temperature above T_g of PLA, leads to a larger surface cracking as clearly seen in Figure 8.4.

Ecker et al. mixed PLA with an amorphous PHA copolymer to improve its impact strength [51]. The mechanical properties of PLA and PLA/PHA blends prepared by 3D printing were compared with that obtained for the same materials prepared by injection molding. It has been found that, the PLA/PHA parts produced by 3D printing showed higher impact strength and notched impact strength values than that obtained for the parts prepared using injection molding. The SEM micrographs for the fractured surface of PLA/PHA blend processed by 3D printed and injection molded revealed an even rougher surface compared to the pure PLA processed by injection molding technique. Good dispersion of the amorphous PHA was observed in all PLA/PHA blends, with PHA amorphous regions being 4–5 μm in size (see Figure 8.5a, b, and c) [51].

The mechanical properties of blends of poly(ε-caprolactone) (PCL) and PHA including poly(butylene succinate) (PBS) with and without plasterers were investigated by Nishida et al. [52]. Two-phase morphology with a sea-island structure was developed to improve the mechanical properties of PHA particularly the elongation at break. On the other hand, the impact strength of PHA did not show any significant improvement by adding the PCL due to the dynamical constrain by blending which suppresses the polymer chains mobility. In addition, proxy crosslinker was also employed to change the sea-island structure into a homogenous structure with a significant improvement in the impact strength and a sharp decrease in the elongation at break. Furthermore, the author used a graft copolymer with styrene-acrylonitrile side chains as a compatibilizer. The compatibilizer was found to have a considerable improvement on both impact strength and elongation without any observed change in the sea-island structure.

Figure 8.6 shows schematic diagram for the X-ray CT microfocus morphology for 50/50 PCL/PHA blends. The effect of the compatibilizer on the morphology of the blend after the mechanical tests is demonstrated in this figure. Figure 8.6a shows the schematic morphology for PHA dispersed phase in PCL matrix or what is called sea-island

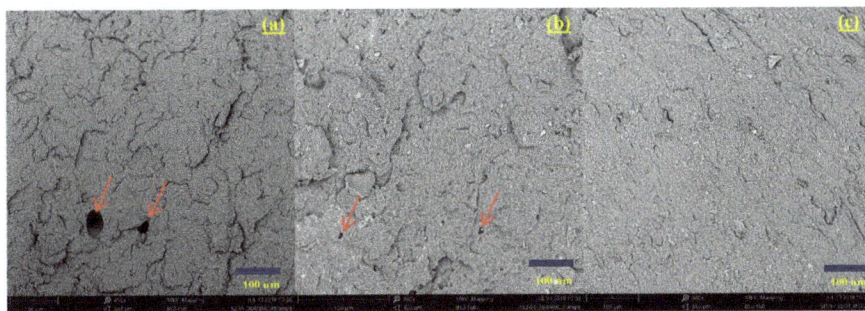

Figure 8.5: SEM micrographs from fracture surfaces of PLA/PHA_3D blends at different angles (a) layer at 45° (b) layer at 0°, and (c) PLA/PHA_IM. Reproduced with permission from Ecker et al. [51].

morphology. The obtained morphology indicates that the interaction between PHA and PCL is very weak due to the heterogenous structure before the mechanical test. Regardless the similar chemical structure of PBS and PCL, the interaction between the PHA and PCL is not strong enough to produce better morphology, such as co continuous structure or finer morphology. The PHA phase was not easily deformed, while PCL was readily deformed in the quasi static condition as clearly seen in Figure 8.6a. The blend morphology shows an obvious broken structure at small scale after the dynamic tensile test. No morphology was observed in the crosslinked blend with a large broken structure after both quasi static and dynamic tests. For the graft copolymer, the compatibilizer located on the interface between PHA and PCL to improve the interfacial adhesion. Still the PHA cannot easily deform in after both quasi static and dynamic tests. In the quasi static condition, smaller broken structure was observed compared to the structure after dynamic test.

Satoh et al. improved the toughness behavior of PHA by blending with biodegradable PCL [53]. The miscibility of PHA and PCL was enhanced greatly by using different types of block copolymer compatibilizers, such as the biodegradable poly(3-hydroxybutyrate-co-3-hydroxyhexanoate) (PHBH) as block A and PCL as block B, or simply AB block copolymer or generally BCPs. The authors used large number of copolymers architecturally varied including AB, ABA, AB_2, A_2B, and A_2BA_2. These architecturally different block copolymer compatibilizers were synthesized from the reaction of PCLs (azido-functionalized) and PHBH (propargyl functionalized) via click reactions [53]. The type and architecture of the compatibilizers were strongly influenced on the morphology and mechanical properties of the blends of the two immiscible PHA and PCL components. It has been found that a 10 wt.% of PHBH-b-PCL

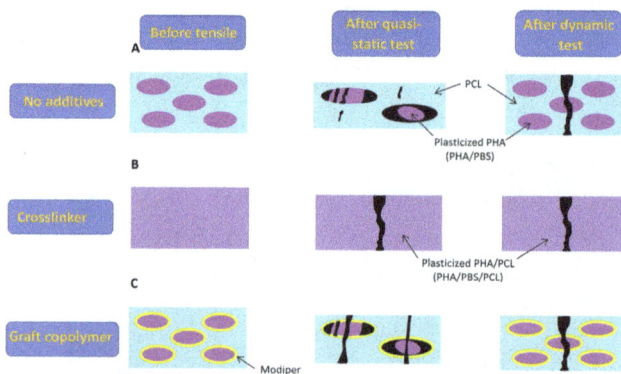

Figure 8.6: Schematic nanostructural morphology for PHA/PCL 50/50 blends before tensile, after quasi-static, and after dynamic tests for PHA/PCL blends with no additives, crosslinked, and grafted blends. Adapted with permission from Nishida et al. [52].

compatibilizer increased the elongation at break by 5–10 folds compared to the results for same blend composition with no compatibilizer.

The effect of the compatibilizer architecture on the morphology of PHA/PCL of different compositions is shown in Figure 8.7 as high resolution TEM micrographs. The different blend samples with and without compatibilizers were prepared in a compression molding at 150 °C and then all the samples were thermally aged for seven days. After that the samples microtomed into thin sections and then stained with RuO_4. The dark domains in Figure 8.7 are the dispersed PCL phase, while the bright matrix is the PHA phase. The average particle size of the dispersed phase was calculated as a function of blend composition and different compatibilizer architecture. The average particle size of PCL-phase (D_{PCL}) was approximately of 4.9 μm for PHA/PCL 75/25 blend with no compatibilizer. Finer morphologies were obtained by adding 10 wt.% compatibilizer to the PHA/PCL 76/23 blend as clearly seen in Figure 8.7. The most observed finer morphology was for A_2B type compatibilizer and the D_{PCL} was 0.51 μm. Based on this experimental fact, it appears that the authors successfully synthesized a series of very effect compatibilizers with different architectures that strongly enhance the interfacial adhesion between PHA and PCL, reduce the particle size and might improve the mechanical properties. To confirm that, the effect of different compatibilizers on the mechanical properties of PHA/PCL blends will be evaluated in the section below.

The effect of different compatibilizers on the stress–strain curves of PHA/PCL blends is shown in Figure 8.8. The pure PHA (PHBH) and the PHA/PCL = 75/25 blend without compatibilizer are very brittle and have very small strain at break (e.g., ~10% strain) as

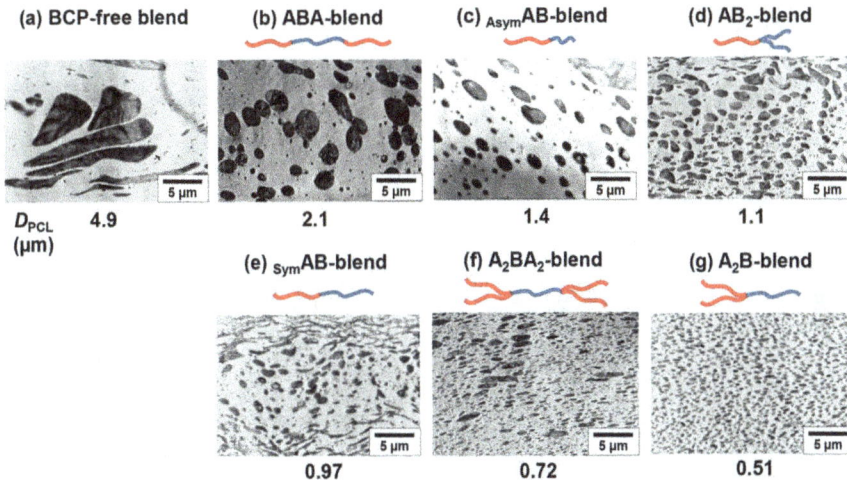

Figure 8.7: TEM micrographs for different blend compositions of PHA/PCL with and without compatibilizers: (a) PHA/PCL = 75/25 (b) PHA/PCL/ABA = 67/23/10, (c) PHA/PCL/AB$_{asym}$ 67/23/10, (d) PHA/PCL/AB2 67/23/10, (e) PHA/PCL/AB$_{sym}$ 67/23/10, (f) PHA/PCL/A$_2$BA$_2$ 67/23/10, and (g) PHA/PCL/A2B 67/23/10. Reproduced with permission from Oyama et al. [53].

clearly seen in Figure 8.8. Secondary crystallization process might be induced during stretching the samples and might forming voids or cracks at low strain for pure PHA and unplasticized blend. Adding 25 wt.% PCL to the PHA have not shown any improvement in the toughness of PHA due to the phase separated structure and poor interfacial adhesion between the two phases without compatibilizer. Adding 10 wt.% of different compatibilizers to the PHA/PCL 67/23 blend was found to have a dramatic improvement (except for AB_{asym} compatibilizer) on the mechanical properties which depends on the architecture of the compatibilizer. For example, a huge improvement in the toughness of the blend was observed for ABA-type compatibilizer, while a considerable decrease in the mechanical properties of the blend was seen for AB_{asym} compatibilizer which might be related to the interfacial instability. In general, the data in Figure 8.8 confirmed that the elongation of the blend improved significantly with certain type of compatibilizer, such as ABA and A_2BA_2. Less improvement was observed with AB_2, A_2B, and AB_{Sym}.

Chen et al. improved the interfacial adhesion of PHA with starch through *in-situ* grafted chemical reaction using a free radical initiator (dicumyl peroxide, DCP) [54]. The effect of DCP content on the PHA/starch gel yield was evaluated. The gel yield was found to be increased with increasing the DCP concentration up to 2 wt.%. The graft reaction between PHA and starch was investigated using FTIR. Blends of PHA/starch with DCP showed clear plastic deformation or stretched fibrils structure at the interface between the two polymer components compared to the same blend concentration with no crosslink agent. The adhesion factor obtained from DMA measurements, mechanical strength, and crystallization behavior were found to be strongly influenced by the concentration of starch in the blend. The thermal stability of the PHA/starch blend was significantly improved by adding DCP. In addition, the overall mechanical properties of the PHA/ starch blends enhanced with the increase of the gel yield. Grafting PHA with starch was

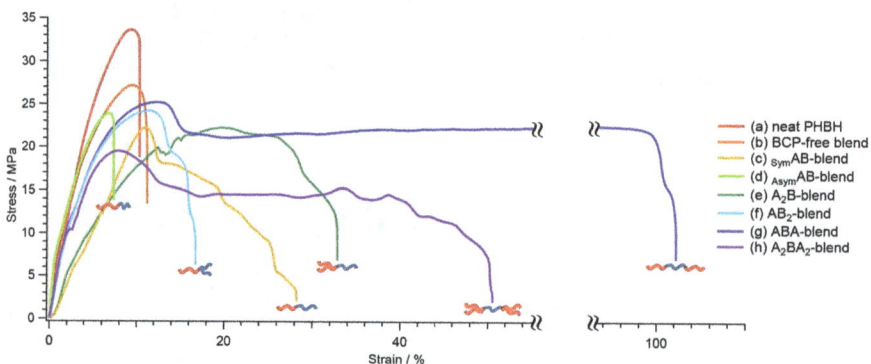

Figure 8.8: Stress–strain curves for pure PHA (PHBH) and PHA/PCL blends with and without compatibilizers: (a) PHA 100 wt.%, (b) PHA/PCL 75/25, (c) PHA/PCL/$_{sym}$AB 67/23/10, (d) PHA/ PCL/$_{Asym}$AB 67/23/10, (e) PHA/PCL/A_2B 67/23/10, (f) PHA/PCL/AB_2 67/23/10, (g) PHA/PCL/ABA 67/ 23/10, and (h) PHA/PCL/A_2BA_2 67/23/10. Reproduced with permission from Oyama et al. [53].

carried out *in-situ* at 170 °C with different DCP concentrations. The mechanism of grafting PHA and starch can be explained based on the high affinity of the free radical obtained from the DCP initiator at a high temperature to abstract the hydrogen atoms from starch and PHA. Stark et al. revealed that the DCP-free radicals are inclined to abstract H, particularly from tertiary –CH groups commonly found in many polymers [55].

The mechanism of the *in-situ* grafting process of PHA/starch with DCP initiator is shown in Figure 8.9a. The authors used FTIR to evaluate the crystallization behavior of PHA after grafting with starch. The FTIR was also used to confirm the grafting reaction between PHA and starch. The crystallization of PHA can be seen in the C–C stretching bands at absorbance peak of 980 cm^{-1} and the C–O–C stretching bands at 1228 cm^{-1} absorbance peak [56, 57]. As seen in Figure 8.9b, the intensity of both PHA crystallization bands just mentioned significantly decreased indicating that the crystallization behavior of PHA strongly inhabited by adding starch. However, on the other hand, the grafting blends showed some increase in the intensity of the FTIR crystallization bands of PHA indicating that the crystallization behavior of PHA in grafted blends with starch improved. Furthermore, the observation of -OH stretching bands at ~3300 cm^{-1} for the PHA/starch gel confirmed the grafting reaction between PHA and starch.

Composite of PHA and DDGS (distiller's dried grains with solubles) has been developed to control the mechanical, crystallization, and degradation behaviors of PHA [58]. The composite was developed as an alternative to the petroleum-based plastics particularly for bio-based and biodegradable container. The DDGS was found to have a significant impact on the biodegradation behavior of PHA. The PHA/DDGS 90/10 composite degraded in soil to about 45 wt.% of its initial weight after about 30 weeks compared to about 7 wt.% for pure PHA under same biodegradation condition. The surface morphology of the samples after the biodegradation process was investigated using SEM. The effect of biodegradation process on the crystallization behavior of PHA was investigated by DSC. In addition, the melt viscosity of the composites was measured using dynamic rheology to understand the relationship between

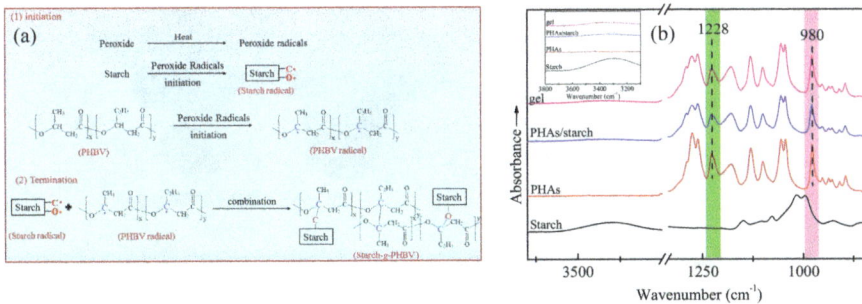

Figure 8.9: (a) Elementary steps for free radial grafting process of PHA/starch blend wing DCP free radical imitator. (b) FTIR spectrum for the pure PHAs, starch, their grafted PLA/starch blend, and gel. Adapted with permission from Xu et al. [54].

viscoelastic properties and biodegradation. The cold crystallization of PHA and the zero-shear viscosity of the composite decrease with increasing the biodegradation time. The weight loss as a function of biodegradation time for pure PHA and PHA/DDGS 90/10 composite in soil under landscape condition is shown in Figure 8.10a. Clearly small concentration of DDGS (e.g., 10 wt.%) can accelerate the biodegradation process of PHA dramatically. The biodegradation process of PHA can be accelerated by many time (approximately six times) by adding just 10 wt.% DDGS. Therefore, the DDGS can simply increase the biodegradation rate, improve the mechanical properties and crystallization behavior and very importantly reduce the cost of PHA.

The main reason for the biodegradation effect of DDGS is based on its structure. The DDGS has a large concentration of fiber, amino acids, protein, and other materials as well as nutrients that can accelerate the biodegradation process in soil. The excellent compatibility of PHA and DDGS plays a crucial role for the accelerated biodegradation process. The PHA/DDGS 90/10 composite was also manufactured as a bio-pot and its effect on the health and growth rate of tomato plant was investigated in comparison with a commercially available PP-pot. It has been found that, the PHA/DDGS bio-pot has no visual or evidence of phytotoxicity and both bio-pot and PP-based pot have identical results related to the plant health, color, height, or growth rate as shown in Figure 8.10b. Therefore, the PHA/DDGS could be an excellent biodegradable candidate suitable for manufacturing of high-quality horticultural crop containers.

The surface morphology can help us to understand the mechanism of biodegradation process of both pure PHA and its composite with DDGS. Figure 8.11 shows the SEM surface morphology for PHA/DDGS 90/10 composite after different biodegradation times up to 24 weeks. Smooth surface of the composite can be seen before the biodegradation process in Figure 8.11 (0 week). Numerous cracks on the surface of the composites were detected after eight weeks degradation time. Areolate surface

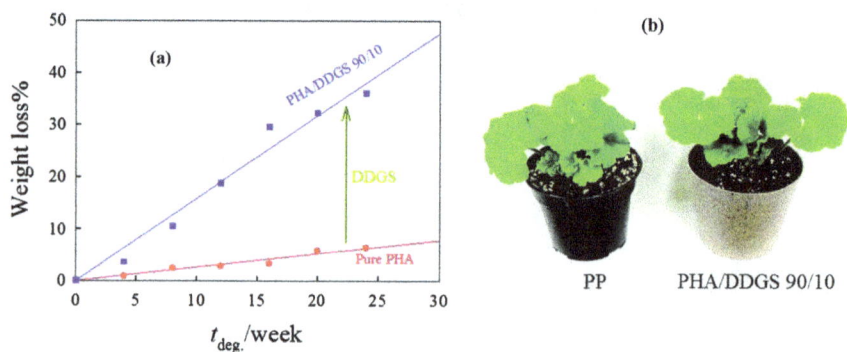

Figure 8.10: (a) Weight loss as a function of biodegradation time in soil for pure PHA and PHA/DDGS 90/10 blend. (b) Photograph for geranium plant growing in PP-container compare to sample plant growing in PLA/DDGS container under identical biodegradation condition. Reproduced with permission from Madbouly et al. [58].

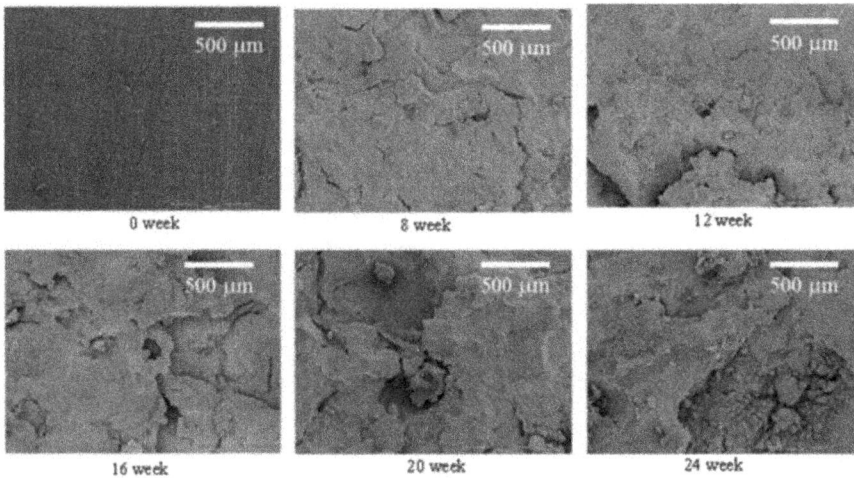

Figure 8.11: Surface morphologies SEM micrographs for PHA/DDG composite after different biodegradation time intervals of PHA/DDGS 90/10 composites after different durations of biodegradation in soil. Reproduced with permission from Madbouly et al. [58].

appearance with pits and deeper cracks obtained with increasing the biodegradation time as clearly seen in Figure 8.11. Based on this SEM micrograph, one can say that the deeper cracks and erosion pits can be used to explain the biodegradation behavior and weight loss in Figure 8.10a.

8.5 Conclusion

The PHAs and their copolymers are semicrystalline aliphatic thermoplastics that can be synthesized by microorganisms through metabolic pathways. PHAs can be processed using the common convention processing techniques, such as extrusion, injection molding, compression molding, thermoforming, and blow molding. In addition, PHAs are biodegradable, biocompatible, and its degraded fragments are nontoxic. The biodegradation rate of PHAs is relatively fast compared to that of other biodegradable polymers, such as PLA and bio-based PA due to the easy hydrolysable ester bonds in the chemical structure of PHAs. PHAs can be used in wide industrial applications, such as 3D printing materials, textile, medical implants, drug delivery carriers, packaging materials, agriculture mulching films, animal nutritional supplements, as well as new types of biofuels. Water and moisture content, as well as temperature, pH, and nutrient supply are important factors that influence the degradation rate of PHAs. In addition, the chemical structure, percentage of crystallinity, surface area, and nature and concentration of additives play an important role in the degradation kinetics of PHAs. The PHAs can be manufacturing from large number of biorenewable feedstocks, such as lignin, cellulose, plant oils, fatty acids, etc. The PHAs can also be synthesized from solid wastes such as municipal and organic solid wastes. The high manufacturing costs particularly fermentation and the purification processes of PHAs are the main reasons for not finding PHAs in many industrial applications. Cost effective production of PHAs is necessary for wide applications in different industries. Modification of PHA with other function groups, such as epoxy or hydroxyl groups can improve the reactivity to create multiblock polymers, such as polyurethane or three-dimensional network through crosslink reaction of the epoxy groups. Blending of PHAs with other biodegradable polymers or fillers is an effective method to improve the mechanical and thermal properties as well as enhancing the processability. Significant improvement in the biodegradation behavior and mechanical properties in blends with PA, PLA, PCL, and DDGS have been observed. In addition, a series of compatibilizers with different architectures were successfully used to improve the compatibility and interfacial adhesion of PHAs and PCL. More research and industrial efforts need to be continued to reduce the manufacturing cost and improve the toughness of PHAs to be suitable for replacing petroleum-based plastics in many industrial applications.

Author contributions: All the authors have accepted responsibility for the entire content of this submitted manuscript and approved submission.
Research funding: None declared.
Conflict of interest statement: The authors declare no conflicts of interest regarding this article.

References

1. Verlinden RA, Hill DJ, Kenward MA, Williams CD, Radecka I. Bacterial synthesis of biodegradable polyhydroxyalkanoates. J Appl Microbiol 2007;102:1437–49.
2. Braunegg G, Lefebvre G, Genser KF. Polyhydroxyalkanoates, biopolyesters from renewable resources: physiological and engineering aspects. J Biotechnol 1998;65:127–61.
3. Hazer DB, Kılıçay E, Hazer B. Poly (3-hydroxyalkanoate) s: diversification and biomedical applications: a state of the art review. Mater Sci Eng C 2012;32:637–47.
4. Hazer B, Steinbüchel A. Increased diversification of polyhydroxyalkanoates by modification reactions for industrial and medical applications. Appl Microbiol Biotechnol 2007;74:1–2.
5. Yang S, Madbouly SA, Schrader JA, Grewell D, Kessler MR, Graves WR. Processing and characterization of bio-based poly (hydroxyalkanoate)/poly (amide) blends: improved flexibility and impact resistance of PHA-based plastics. J Appl Polym Sci 2015;132. https://doi.org/10.1002/app.42209.
6. Hawkins S, Fonseca IB, Lima da Silva R, Quirino RL. Aquaculture waste: potential synthesis of polyhydroxyalkanoates. ACS Omega 2021;6:2434–42.
7. Sharma V, Sehgal R, Gupta R. Polyhydroxyalkanoate (PHA): properties and modifications. Polymer 2021;212:123161.
8. Holmes PA. Applications of PHB-a microbially produced biodegradable thermoplastic. Phys Technol 1985;16:32.

9. Al-Ashraf A, Ramachandran H, Huong KH, Kannusamy S. Microbial-based polyhydroxyalkanoates: upstream and downstream processing. Shrewsbury, Shropshire, UK: Smithers Rapra; 2015.
10. Verma ML, Kumar S, Jeslin J, Dubey NK. Microbial production of biopolymers with potential biotechnological applications. In: Biopolymer-based formulations. Amsterdam, Netherlands: Elsevier; 2020:105–37 pp.
11. Wang S, Chen W, Xiang H, Yang J, Zhou Z, Zhu M. Modification and potential application of short-chain-length polyhydroxyalkanoate (SCL-PHA). Polymers 2016;8:273.
12. Gao X, Chen JC, Wu Q, Chen GQ. Polyhydroxyalkanoates as a source of chemicals, polymers, and biofuels. Curr Opin Biotechnol 2011;22:768–74.
13. Muneer F, Rasul I, Azeem F, Siddique MH, Zubair M, Nadeem H. Microbial polyhydroxyalkanoates (PHAs): efficient replacement of synthetic polymers. J Polym Environ 2020;28:2301–23.
14. Ivanov V, Stabnikov V, Ahmed Z, Dobrenko S, Saliuk A. Production and applications of crude polyhydroxyalkanoate-containing bioplastic from the organic fraction of municipal solid waste. Int J Environ Sci Technol 2015;12:725–38.
15. Babu RP, O'connor K, Seeram R. Current progress on bio-based polymers and their future trends. Prog Biomater 2013;2:1–6.
16. Wang Q, Tappel RC, Zhu C, Nomura CT. Development of a new strategy for production of medium-chain-length polyhydroxyalkanoates by recombinant *Escherichia coli* via inexpensive non-fatty acid feedstocks. Appl Environ Microbiol 2012;78:519–27.
17. Yang ST, Yu M. Integrated biorefinery for sustainable production of fuels, chemicals, and polymers. Bioprocessing technologies in biorefinery for sustainable production of fuels, chemicals, and polymers 2013;1:1–26.
18. Koller M, Atlić A, Dias M, Reiterer A, Braunegg G. Microbial PHA production from waste raw materials. In: Plastics from bacteria. Berlin, Heidelberg: Springer; 2010:85–119 pp.
19. Tan D, Wu Q, Chen JC, Chen GQ. Engineering Halomonas TD01 for the low-cost production of polyhydroxyalkanoates. Metab Eng 2014;26:34–47.
20. Chen GQ, Patel MK. Plastics derived from biological sources: present and future: a technical and environmental review. Chem Rev 2012;112:2082–99.
21. Adeleye AT, Odoh CK, Enudi OC, Banjoko OO, Osigbeminiyi OO, Toluwalope OE, et al. Sustainable synthesis and applications of polyhydroxyalkanoates (PHAs) from biomass. Process Biochem 2020;96:174–93.
22. Pillai AB, Kumar AJ, Kumarapillai H. Synthetic biology and metabolic engineering approaches for improved production and recovery of bacterial polyhydroxyalkanoates. Next Gener Biomanuf Technol 2019;9:181–207.
23. Acevedo F, Villegas P, Urtuvia V, Hermosilla J, Navia R, Seeger M. Bacterial polyhydroxybutyrate for electrospun fiber production. Int J Biol Macromol 2018;106:692–7.
24. Fabra MJ, Lopez-Rubio A, Lagaron JM. Nanostructured interlayers of zein to improve the barrier properties of high barrier polyhydroxyalkanoates and other polyesters. J Food Eng 2014;127:1–9.
25. Laycock B, Halley P, Pratt S, Werker A, Lant P. The chemomechanical properties of microbial polyhydroxyalkanoates. Prog Polym Sci 2013;38:536–83.
26. Choi JI, Lee SY. Process analysis and economic evaluation for poly (3-hydroxybutyrate) production by fermentation. Bioprocess Eng 1997;17:335–42.
27. Reis M, Albuquerque M, Villano M, Majone M. 6.51 - mixed culture processes for polyhydroxyalkanoate production from agro-industrial surplus/wastes as feedstocks. In: Moo-Young M, editor. Comprehensive biotechnology, 2nd ed. Burlington: Academic Press; 2011: 669–83 pp.
28. Fernández-Dacosta C, Posada JA, Kleerebezem R, Cuellar MC, Ramirez A. Microbial community-based polyhydroxyalkanoates (PHAs) production from wastewater: techno-economic analysis and ex-ante environmental assessment. Bioresour Technol 2015;185:368–77.

29. Dionisi D, Carucci G, Papini MP, Riccardi C, Majone M, Carrasco F. Olive oil mill effluents as a feedstock for production of biodegradable polymers. Water Res 2005;39:2076–84.
30. Hassan MA, Shirai Y, Kusubayashi N, Karim MIA, Nakanishi K, Hashimoto K. The production of polyhydroxyalkanoate from anaerobically treated palm oil mill effluent by *Rhodobacter sphaeroides*. J Ferment Bioeng 1997;83:485–8.
31. Melendez-Rodriguez B, Castro-Mayorga JL, Reis MAM, Sammon C, Cabedo L, Torres-Giner S, et al. Preparation and characterization of electrospun food biopackaging films of poly(3-hydroxybutyrate-co-3-hydroxyvalerate) derived from fruit pulp biowaste. Front Sustain Food Syst 2018;2:38.
32. Colombo B, Pepè Sciarria T, Reis M, Scaglia B, Adani F. Polyhydroxyalkanoates (PHAs) production from fermented cheese whey by using a mixed microbial culture. Bioresour Technol 2016;218: 692–9.
33. Morgan-Sagastume F, Hjort M, Cirne D, Gérardin F, Lacroix S, Gaval G, et al. Integrated production of polyhydroxyalkanoates (PHAs) with municipal wastewater and sludge treatment at pilot scale. Bioresour Technol 2015;181:78–89.
34. Korkakaki E, Mulders M, Veeken A, Rozendal R, van Loosdrecht MC, Kleerebezem R. PHA production from the organic fraction of municipal solid waste (OFMSW): overcoming the inhibitory matrix. Water Res 2016;96:74–83.
35. Roy I, Visakh PM, editors. Polyhydroxyalkanoate (PHA) based blends, composites and nanocomposites. Cambridge, UK: Royal Society of Chemistry; 2015:141–82 pp.
36. Kurth N, Renard E, Brachet F, Robic D, Guerin P, Bourbouze R. Poly (3-hydroxyoctanoate) containing pendant carboxylic groups for the preparation of nanoparticles aimed at drug transport and release. Polymer 2002;43:1095–101.
37. Arkin AH, Hazer B, Borcakli M. Chlorination of poly (3-hydroxy alkanoates) containing unsaturated side chains. Macromolecules 2000;33:3219–23.
38. Park WH, Lenz RW, Goodwin S. Epoxidation of bacterial polyesters with unsaturated side chains. II. Rate of epoxidation and polymer properties. J Polym Sci, Part A: Polym Chem 1998;36:2381–7.
39. Torres MG, Muñoz SV, Rosales SG, del Pilar Carreón-Castro M, Muñoz RA, González RO, et al. Radiation-induced graft polymerization of chitosan onto poly (3-hydroxybutyrate). Carbohydr Polym 2015;133:482–92.
40. Nguyen S, Marchessault RH. Graft copolymers of methyl methacrylate and poly ([R]-3-hydroxybutyrate) macromonomers as candidates for inclusion in acrylic bone cement formulations: compression testing. J Biomed Mater Res B Appl Biomater, An Official Journal of The Society for Biomaterials. The Japanese Society for Biomaterials, and The Australian Society for Biomaterials and the Korean Society for Biomaterials 2006;77:5–12.
41. Macit H, Hazer B, Arslan H, Noda I. The synthesis of PHA-g-(PTHF-b-PMMA) multiblock/graft copolymers by combination of cationic and radical polymerization. J Appl Polym Sci 2009;111: 2308–17.
42. Iqbal HM, Kyazze G, Tron T, Keshavarz T. A preliminary study on the development and characterisation of enzymatically grafted P (3HB)-ethyl cellulose based novel composites. Cellulose 2014;21:3613–21.
43. Sidek IS, Draman SF, Abdullah SR, Anuar N. Current development on bioplastics and its future prospects: an introductory review. INWASCON Technol Mag 2019;1:03–8.
44. Li T, Elhadi D, Chen GQ. Co-production of microbial polyhydroxyalkanoates with other chemicals. Metab Eng 2017;43:29–36.
45. Gerard T, Budtova T. Morphology and molten-state rheology of polylactide and polyhydroxyalkanoate blends. Eur Polym J 2012;48:1110–7.

46. Parulekar Y, Mohanty AK. Extruded biodegradable cast films from polyhydroxyalkanoate and thermoplastic starch blends: fabrication and characterization. Macromol Mater Eng 2007;292: 1218–28.

47. Shamala TR, Divyashree MS, Davis R, Kumari KL, Vijayendra SV, Raj B. Production and characterization of bacterial polyhydroxyalkanoate copolymers and evaluation of their blends by fourier transform infrared spectroscopy and scanning electron microscopy. Indian J Microbiol 2009;49:251–8.

48. Zheng Z, Deng Y, Lin XS, Zhang LX, Chen GQ. Induced production of rabbit articular cartilage-derived chondrocyte collagen II on polyhydroxyalkanoate blends. J Biomater Sci Polym Ed 2003; 14:615–24.

49. Yang S, Madbouly SA, Schrader JA, Grewell D, Kessler MR, Graves WR. Processing and characterization of bio-based poly (hydroxyalkanoate)/poly(amide) blends: improved flexibility and impact resistance of PHA-based plastics. J Appl Polym Sci 2015;27:132.

50. Ausejo JG, Rydz J, Musioł M, Sikorska W, Sobota M, Włodarczyk J, et al. A comparative study of three-dimensional printing directions: the degradation and toxicological profile of a PLA/PHA blend. Polym Degrad Stabil 2018;152:191–207.

51. Ecker JV, Burzic I, Haider A, Hild S, Rennhofer H. Improving the impact strength of PLA and its blends with PHA in fused layer modelling. Polym Test 2019;78:105929.

52. Nishida M, Tanaka T, Hayakawa Y, Ogura T, Ito Y, Nishida M. Multi-scale instrumental analyses of plasticized polyhydroxyalkanoates (PHA) blended with polycaprolactone (PCL) and the effects of crosslinkers and graft copolymers. RSC Adv 2019;9:1551–61.

53. Oyama T, Kobayashi S, Okura T, Sato S, Tajima K, Isono T, et al. Biodegradable compatibilizers for poly (hydroxyalkanoate)/poly (ε-caprolactone) blends through click reactions with end-functionalized microbial poly (hydroxyalkanoate) s. ACS Sustainable Chem Eng 2019;7:7969–78.

54. Xu P, Zeng Q, Cao Y, Ma P, Dong W, Chen M. Interfacial modification on polyhydroxyalkanoates/starch blend by grafting in-situ. Carbohydr Polym 2017;174:716–22.

55. Wei L, McDonald AG, Stark NM. Grafting of bacterial polyhydroxybutyrate (PHB) onto cellulose via in situ reactive extrusion with dicumyl peroxide. Biomacromolecules 2015;16:1040–9.

56. Bloembergen S, Holden DA, Hamer GK, Bluhm TL, Marchessault RH. Studies of composition and crystallinity of bacterial poly (β-hydroxybutyrate-co-β-hydroxyvalerate). Macromolecules 1986;19: 2865–71.

57. Hong SG, Chen WM. The attenuated total reflection infrared analysis of surface crystallinity of polyhydroxyalkanoates. E-Polymers 2006;6. https://doi.org/10.1515/epoly.2006.6.1.310.

58. Madbouly SA, Schrader JA, Srinivasan G, Liu K, McCabe KG, Grewell D, et al. Biodegradation behavior of bacterial-based polyhydroxyalkanoate (PHA) and DDGS composites. Green Chem 2014;16:1911–20.

Emily Archer, Marissa Torretti* and Samy Madbouly

9 Biodegradable polycaprolactone (PCL) based polymer and composites

Abstract: Polycaprolactone (PCL) is a biodegradable polyester that has advantages over other biopolymers, making it an extensively researched polymer. PCL is a hydrophobic, slow-degrading, synthetic polymer making it particularly interesting for the preparation of long-term implantable devices and a variety of drug delivery systems. Recently, PCL has been used for additional applications including food packaging and tissue engineering. In this chapter, the processing methods and characterization of PCL will be discussed. The chapter will summarize the synthesis of poly(α-hydroxy acid) and the ring-opening polymerization of PCL. Discussion on the biodegradability of PCL will be reviewed. The biomedical applications of PCL, such as, drug-delivery systems, medical devices, and tissue engineering will be also summarized. Finally, the chapter will conclude with a characterization section outlining recent studies focusing on PCL based composites and films.

Keywords: biodegradable; crystallization; hydrophobic; morphology; poly-caprolactone; processability; synthetic; tissue engineering.

9.1 Introduction

Polycaprolactone (PCL) is a hydrophobic, slow-degrading, biocompatible, synthetic thermoplastic polymer that is often used in biomedical applications ranging from controlled drug delivery systems to implants for orthopedic surgery. Other applications include food packaging and tissue engineering. The wide range of applications and its potential to be degraded by microorganisms make PCL one of the most appealing biodegradable polymers on the commercial market [1–32, 34–41, 43–50]. Drug-delivery devices and suture materials are common applications of PCL because of its slow degradation rate. Some disadvantages that make PCL less likely to be chosen are low melting temperature and low mechanical properties which limit its industrial uses [1–7]. Overall, PCL has advantages over other biopolymers, that cause this polymer to

*Corresponding author: Marissa Torretti, Penn State Behrend-Plastics Engineering Technology, 4701 College Drive, Erie, PA 16563, USA, E-mail: mlt37@psu.edu
Emily Archer and Samy Madbouly, Penn State Behrend-Plastics Engineering Technology, 4701 College Drive, Erie, PA 16563, USA

This article has previously been published in the journal Physical Sciences Reviews. Please cite as: E. Archer, M. Torretti and S. Madbouly "Biodegradable polycaprolactone (PCL) based polymer and composites" *Physical Sciences Reviews* [Online] 2021, 7. DOI: 10.1515/psr-2020-0074 | https://doi.org/10.1515/9781501521942-009

be extensively researched, low-cost and is approved by the U.S. Food and Drug Administration (FDA) for biomedical applications [1, 4, 8–12, 42, 47, 49]. Studies have shown that PCL is degradable in many natural environments, such as soil, seawater, and active sludge [40].

PCL is a semicrystalline polymer with a low melting point range of 58–64 °C depending on the degree of crystallization and molecular weight and a glass transition temperature of –60 °C [3, 4, 6, 7, 21, 33]. The average molecular weight of PCL samples may vary from 3000– 80,000 g/mol [10]. PCL is soluble in a wide range of organic solvents, such as chloroform, dichloromethane, carbon tetrachloride, benzene, toluene, cyclohexanone, and 2-nitropropane at room temperature [1, 45]. PCL has low solubility in acetone, 2-butanone, ethyl acetate, dimethylformamide, and acetonitrile. It is insoluble in alcohol, petroleum either and diethyl either. PCL has tailorable degradation kinetics and mechanical properties, along with ease of processing and manufacturing [16, 45]. Its structure can enable favorable cell responses with the addition of function groups, making the polymer more hydrophilic, adhesive, or biocompatible [1, 4]. Inorganic particles are used to improve physical properties of the polymer [2]. Biomedical research emerged because of the enticing properties of PCL. However, PCL lacked the mechanical properties to launch its long-term use in medical devices and drug-delivery systems [1]. Some other physical and mechanical properties of PCL can be seen in Table 9.1 [45].

In the 1990s and 2000s, PCL made its way into potential application in tissue engineering [1]. The low degradation rate (up to two years *in vivo*) of PCL allows for tissue growth where there is bone present, in other words where long-term remediation (4–8 weeks) may be required. Toxicity of initiators is a concern when it deals with the synthesis of PCL [1, 45]. Owing to this concern, more biocompatible initiators have been studied. Two ways to enhance bioactivity, which is desirable for various applications, are copolymerization or blending with many other polymers. Crystallinity, solubility, and degradation patterns can indirectly be affected by the

Table 9.1: Physical and mechanical properties of PCL.

Property	Range	Unit
Melting temperature, T_m	56–65	°C
Number average molecular weight, M_n	42.5–64	kDa
Weight average molecular weight, M_w	50.4–124	kDa
Tensile elastic modulus, E_t	251.9–440	MPa
Yield stress, σ_y	8.2–17.8	MPa
Tensile strength, σ_m	10.5–27.3	MPa
Strain at yield, ε_y	2.4–7	%
Strain at break, ε_b	80–800	%

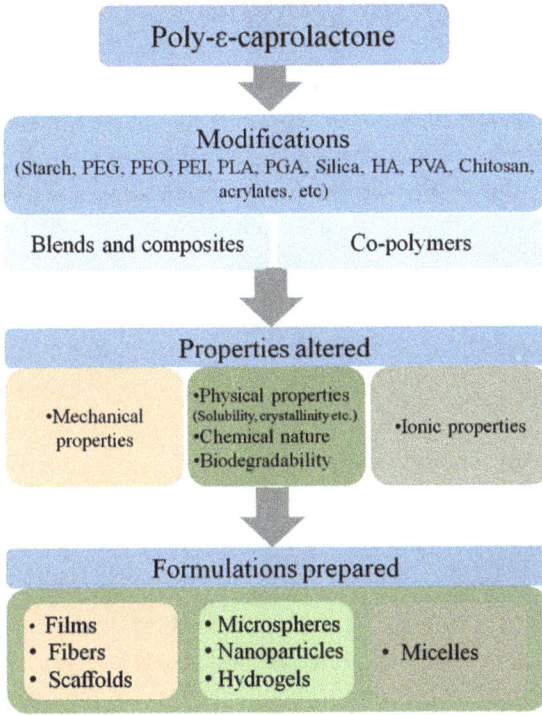

Figure 9.1: Modifications of PCL and their preferences for different formulations [39].

chemical property during copolymerization, which is why PCL possesses potential properties for drug delivery. Some materials that are prepared after copolymerization include PCL micelles, hydrogels, and dendrimers. The blending of PCL is what leads to altered physical properties and biodegradation. In regard to changing mechanical properties, PLC is often used in formulations of tissue engineering such as scaffolds, fibers, and films [10, 39, 44]. Numerous polymers have been analyzed for their compatibility to modify the thermal, rheological as well as biophysical properties of PCL based on application. Some polymers that are compatible with PCL include, starch, chitosan, and synthetic polymers such as polyethylene oxide (PEO) and polyvinyl alcohol (PVA). With the addition of synthetic polymers to PCL, the modifications satisfy the required biophysical properties for most of the formulation that is currently used in drug delivery. Figure 9.1 demonstrates the different modification approaches to create PCL of improved properties. Research has found that PCL is relatively inexpensive and easily synthesized in comparison with other aliphatic polyesters. PCL has become very valuable to the driven polymer markets. The synthesis of poly (α-hydroxy acid) will be examined in more detail to gain a better understanding.

9.2 Synthesis of poly(α-hydroxy acid)s

The Carothers group first synthesized PCL in the early 1930s. PCL is prepared by ring-opening polymerization of the cyclic monomer ε-caprolactone (CL) with free radical ring-opening polymerization catalyst as shown in Figure 9.2 [1, 3, 10, 20, 27]. Ring-opening polymerization from the respective lactone is used to synthesize poly(α-hydroxy acid)s. A key factor in control of the polymerization process is the initiator or catalyst chosen.

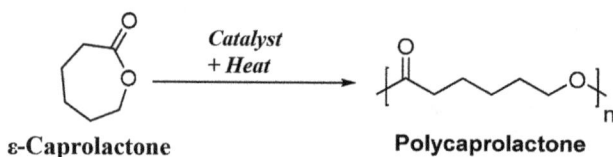

ε-Caprolactone   **Polycaprolactone**

Figure 9.2: Ring-opening polymerization of the cyclic monomer ε-CL to PCL [10].

PCL can be synthesized from the cyclic ester ε-CL through ring-opening polymerization. A simple schematic is shown in Figure 9.3. Metal alkoxides, metal carboxylates or ionic initiators act as catalysts at elevated temperatures [48]. A disadvantage of the ring-opening method is the catalyst requirement. Since the catalysts are metal based, they are often toxic and can be difficult to remove during purification. The calcium and magnesium-based catalysts have been investigated because of their low toxicity and high molecular weights with low dispersity. The result of this process is a semicrystalline polymer with crystallinity dependent on molecular weight, cooling rate, and presence of matrix impurities. At room temperature, PCL is soluble in solvents including toluene, benzene, chloroform, cyclohexanone, carbon tetrachloride, tetrahydrofuran (THF), dimethyl carbonate (DMC), dioxane, 2-nitropropane, and dichloromethane (DCM). PCL is partially soluble in acetone, ethyl acetate, dimethyl formamide (DMF), 2-butanone, and acetonitrile. PCL is insoluble in water, alcohols, diethyl ether, and petroleum ether [1, 10, 45].

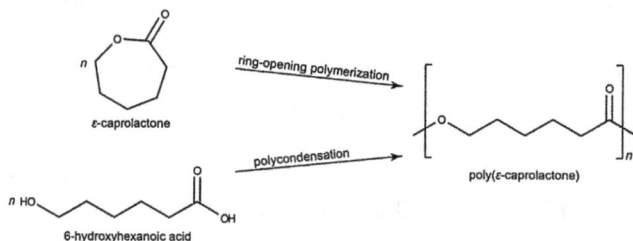

Figure 9.3: Synthetization of PCL from cyclic ε-CL through ring-opening polymerization [45].

9.2.1 Initiators/catalysts for the synthesis of poly(α-hydroxy acid)s

The process of choosing an initiator, or catalystis important in the control of polymer synthesis because it can impact the properties of a polymer, in regards to their physical and chemical states. Some impacts include the crystallinity, melt and glass transition temperatures, molecular weight, molecular weight distribution, end groups, sequence distribution, and the presence of residual monomers [1]. Low molecular weight alcohols can be used to control the molecular weight of the polymer [10]. Tin- and aluminum-based initiators are the most common uses when it comes to ring-opening polymerization of (di)lactones, such as CL. Comparing the two tin- and aluminum-based initiators, tin compounds are better transesterification catalysts. The development of tin-based initiators provides greater control over macromolecular architecture, along with having a greater hydrolytic stability and low cost.

The most common tin-based initiator for the synthesis of poly(α-hydroxy acid)s is tin (II) 2-ethylhexanoate, or stannous octoate (Sn(Oct)$_2$) [1, 10, 20]. Usually an alcohol compound is needed, as a co-initiator, because ring-opening polymerization using Sn(Oct)$_2$ requires an active hydrogen compound. The coexistence of processes such as transesterification, are why there are quasi living systems observed during the synthesis of poly(α-hydroxy acid)s. The polymerization can be affected by different mechanisms which include anionic, cationic, coordination, and radical. These methods may affect the molecular weight, molecular weight distribution, end group composition, and chemical structure of the copolymers [10]. A key factor for the determination of the rate of intra- and intermolecular transesterification in the ring-opening polymerization is the ionic nature of the initiator, or propagating species.

9.3 Biodegradability of PCL

The mechanism of biodegradation is influenced by the chemical and physical properties of plastics. Some other parameters including surface area, hydrophilic, and hydrophobic properties affect the degradation process of PCL as well. The first order structures such as chemical structure, molecular weight, and molecular weight distribution along with the high order structures, such as glass transition temperature, melting temperature, modulus of elasticity, high hydrophobicity, crystallinity, and crystal structure play a role in biodegradation [26, 38].

It has been found that increasing molecular weight of the polymer decreases its degradability. In addition, the morphology of a polymer can strongly affect the rate of biodegradation. Enzymes mainly attack amorphous domains of polymers so the degree of crystallinity will affect biodegradability. The amorphous regions are more susceptible to degradation because they are loosely packed unlike crystalline regions that are tightly packed. Other studies show that, the higher the melting temperature, the lower the biodegradation of the polymer [38]. All these factors are important to consider when examining the biodegradability of PCL which will be further discussed.

It is important to understand something that is biodegradable is not always bio-resorbable. This means it is not necessarily removed from the body as it degrades and moves away from their site of action *in vivo*. Bioresorbable is much different in the sense where there are no residual side effects with the total elimination of the initial foreign materials and bulk degradation products-by-products. Understanding the difference between biodegradable, bioresorbable, bioabsorbable, and bio erodible are important in the discussion of polymer-based materials involved in biomedical applications [1].

PCL is not biodegradable in animal and human bodies because of the lack of suitable enzymes; however, they can be degraded by outdoor living organisms such as bacteria and fungi [38, 45, 48]. PCL could be bioresorbable but the process of propagation via hydrolytic degradation takes a long time. Poly(α-hydroxy) ester can hydrolytically degrade through surface or bulk degradation pathways. The pathway is determined by the diffusion–reaction phenomenon. Surface degradation involves the hydrolytic breakdown of the polymer backbone at the surface. For this to happen, the rate of hydrolytic chain scission and production of oligomers and monomers is faster than the rate of water intrusion into the polymer bulk. The result is some polymer thinning without affecting the molecular weight or bulk of the polymer. This type of surface degradation is predictable, making rates determinable for drugs [1].

When water penetrates the entire polymer bulk, bulk degradation occurs. Hydrolysis occurs throughout the entire polymer matrix. Molecular weight would reduce from the random hydrolytic chain scission. Equilibrium for the diffusion-reaction phenomenon can be reached when water molecules diffuse into the polymer bulk and hydrolysis within the chains allows the monomers or oligomers to diffuse out. If equilibrium is disturbed, the degradation could cause internal autocatalysis via the carboxyl and hydroxyl end group by-products. In surface degradation, the oligomers and carboxyl groups can diffuse freely into the surroundings [1]. Whereas in bulk degradation, the internal concentration of autocatalysis products can form an acidic gradient. This leaves a lower molecular weight and degraded interior with a higher molecular weight skin. Bimodal molecular weight distribution thus defines the degradation mechanism. A higher molecular weight structure is created when tiny thinner oligomers rapidly diffuse through the outer layer. In combination with the oligomers, a decreased chain scission rate and weightless result a hollowed-out structure. Inflammatory reactions *in vivo* can occur from the rapid release of these oligomers and acid by-products [1].

Depending on the starting molecular weight of the device or implant, the homopolymer PCL has 2–4 years of degradation time. Copolymerization with other lactones or glycolides/lactides can change the rate of hydrolysis. Studies have shown both hydrolytic degradation rates were similar and the enzymatic involvement in the first stage of degradation did not impact the degradation process [45]. PCL experiences a two-stage degradation process. The first stage involves the non enzymatic hydrolytic breakdown of ester groups. The second stage consists of the intracellular degradation when the polymer is more highly crystalline and has a low molecular weight. Within the first stage, the degradation rate of PCL is very similar to the surface hydrolysis at 40 °C. The study concluded that the degradation of PCL was because of random hydrolytic chain scission of ester linkages, resulting in decreased molecular weight [1].

9.3.1 Other microorganisms that degrade PCL

Aerobic and anaerobic microorganisms found in various ecosystems are shown to degrade PCL. The *Penicillium* sp. Strain 26-1 (ATCC 36507) was used to investigate the degradation rate of higher molecular weight PCL. Within 12 days, PCL was almost completely degraded. This strain can take in unsaturated aliphatic and alicyclic polyesters but not aromatic polyesters. Another microorganism, *Aspergillus* sp. strain ST-01, was used to investigate degradation. PCL was completely degraded after 6 days incubation at 50 °C. Under anaerobic conditions, it was found that microorganisms belonging to the genus *Clostridium* also degraded PCL [38].

PCL can be degraded by lipases and esterases. Lipase is an enzyme that catalyzes the hydrolysis of fats; while esterases is an enzyme that creates hydrolysis. The molecular weight and degree of crystallinity of PCL determine its degradation rate [4, 38, 45]. Within amorphous regions, faster degradation occurred with the use of *Aspergillus flavus and Penicillium funiculosum*. The copolymerization with aliphatic polyesters can increase the biodegradability of PCL. Copolymers have lower crystallinity and lower melting temperatures than homopolymers making it more susceptible to degradation [38]. Table 9.2 summarizes the different degradation mechanisms of PCL.

A study showed that composites used in tissue engineering are prepared by sol–gel method. Hybrid materials with varying compositions are synthesized with the sol–gel technique because of the low processing temperature. Thermolabile molecules in an inorganic glassy matrix were then able to be trapped. Hydrolysis and polycondensation of the metal alkoxides led to gelation. This consisted of titanium dioxide (TiO_2)/PCL or zirconia oxide (ZrO_2)/PCL hybrid fillers synthesized with a PCL matrix. The hybrid filler ZrO_2/PCL consists of two materials that are both classified as biocompatible whether the form is bulk or as coating. Organic–inorganic ZrO_2/PCL hybrid materials can be synthesized by the sol–gel method and result with an amorphous Class I material [14].

The organic component of the hybrid organic–inorganic ZrO_2/PCL was varied [6, 12, 24] and synthesized for the sol–gel method. First, a solution of PCL in chloroform was added to a solution of zirconium prop oxide ($Zr (OC_3H_7)_4$) in an ethanol acetylacetone–water mixture. The purpose for the acetylacetone mixture was to reduce the hydrolytic activity of zirconium alkoxide. A magnetic stirrer was used after each reactant was added, which allowed for uniform and homogenous sols. Shown in Table 9.3, the gelation process lasted from 8 to 18 days depending on the material [14].

Table 9.2: Extended list of degradation types including contributing factors and their mechanisms.

Degradation Type	Contributing factors	Mechanism
Radical-mediated Generally neutral – Advantageous for medical devices – Ubiquitous across environments – Rarely a dominant mechanism	End groups – Hydroxy – accelerates in air Presence of reactive species – Superoxide radical anion (O_2^-) – Hydroxyl radical (–OH) – Singlet oxygen (O_2) – Hydrogen peroxide- derived radicals	Generally – Hydrogen abstraction – Cascade of scission events – Termination via recombination with other radicals or through additional hydrogen abstraction
Thermal Generally undesired – Occurs during processing	End groups – Hydroxy – accelerates in air – Carboxyl – autocatalysis and reactivity – Ester – most stable General reaction acceleration catalyst residues	Combination of – Random chain scission – Radical-medicated cleavage driven by hydroxy end group
pH-mediated Generally desired – Used for accelerated degradation or surface modification	Base catalysis – Limited swelling – Surface erosion Acid catalysis – Pronounced swelling – Bulk degradation	Ester hydrolysis catalyzed by acid/base
Enzymatic Generally desired – Advantageous for medical devices – Useful for environmental biodegradation	Highly dependent on environmental conditions and presence of specific enzymes	Four step sequence 1. Enzyme adsorbs onto surface 2. Transition complex is formed 3. Specific chain scission occurs – Polymer–enzyme complex further interactions with other portions of the polymer
Intracellular Generally desired – Advantageous for medical devices	Proceeds once molecular weight has sufficiently decreased (10 kDa)	Involves – Surface adhesion of macrophages, giant cells, and cells from the local tissue – Complete fragmentation within macrophages and cells from tissue – All products cleared after digestion

Table 9.3: Gelation process with varying PCL weight content and gelification time results [14].

Simple name	Systems composition	Gelification time
Zr(0)	ZrO_2	18 days
Zr(1)	ZrO_2 + PCL 6 wt.%	15 days
Zr(2)	ZrO_2 + PCL 12 wt.%	13 days
Zr(3)	ZrO_2 + PCL 24 wt.%	9 days
Zr(4)	ZrO_2 + PCL 50 wt.%	8 days

Figure 9.4 provides a more in-depth flow-chart of the hybrid synthesis process by the sol–gel method. A dip coating technique was used with the materials from the sol phase to coat glass slides and resulted with a transparent and uniform film after drying at 45 °C for 24 h. This technique modified the surface properties of a substrate at a low cost and therefore proved to be promising for the deposition of functional coatings [14].

The addition of PCL to ZrO_2 allowed for a more porous and crack free coating surface. This could be due to the higher elasticity of the polymer. Shown in Figure 9.5a, the ZrO_2 obtained a severe number of cracks compared to Figure 9.5b, ZrO_2/PCL obtained no visible cracks. These cracks are caused from the coating's ease of gas removal and the thermal gradient within the coating was small. The content of PCL changed the film properties, therefore the higher the PCL content, the presence of cracks on coating decreased and porosity increased. This resulted in the improvement of film bioactivity. These coatings can then be used to coat metal implants to improve the bioactivity, or bone bonding ability [14].

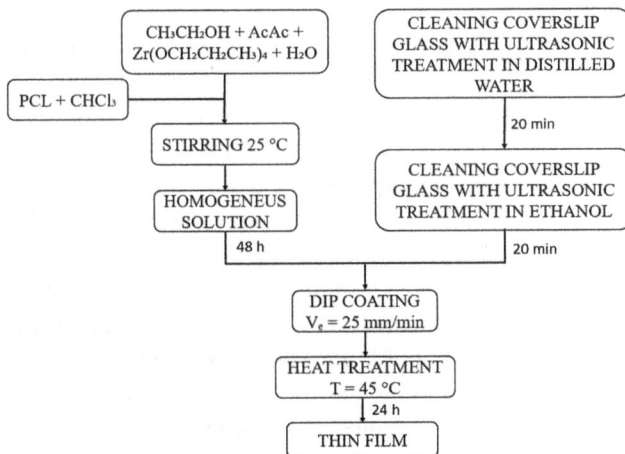

Figure 9.4: Hybrid synthesis of ZrO_2/PCL gel by the sol–gel method [14].

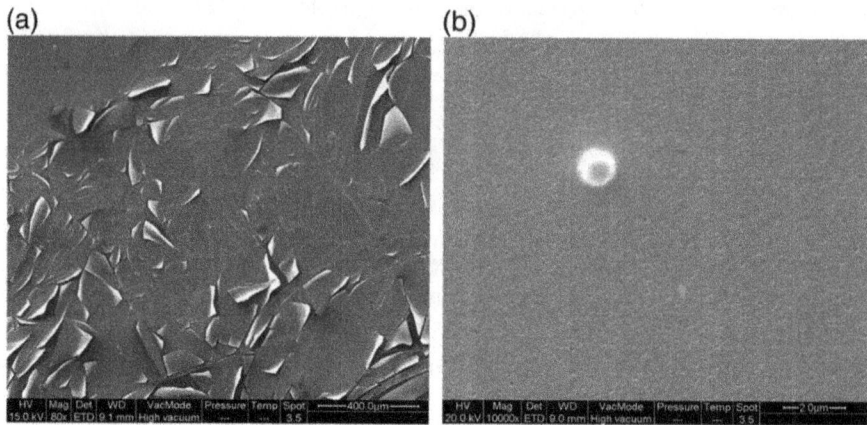

Figure 9.5: Scanning electron microscope (SEM) results of (a) ZrO_2 and (b) ZrO_2/PCL coatings [14].

Scaffolds are used for tissue engineering with a pre specified shape and are biodegradable polymeric porous structures. *In vivo* systems are where these scaffolds are commonly used; this is where damaged tissue is repaired or replaced in the body and providing mechanical support. Scaffolds can be used to accelerate tissue growth or healing or to prevent infections by being used as a carrier for growth factors or antibiotics. PCL and its copolymers are widely studied with clinical establishment of biodegradability and biocompatibility while preferably used in 3D cell culture [39].

Nanofibrous scaffolds are an application that is arising in the medical field and has significant applications in regenerative medicine [45]. The enhancement of stem cell applications relies on a key factor of the appropriate cell growth on synthetic scaffolds. Scaffolds obtain the ability to mimic an extracellular matrix (ECM) because of the high porosity and high volume to surface ratio. ECM is an essential characteristic of the regeneration of injured tissues and other potential applications regarding tissue engineering [19].

Both PCL and PLA are commonly used in the process of engineering scaffolds for skin and internal bone fracture. The low cost, slow degradation rate, flexible, hydrophobic, distinct viscoelastic and rheological properties, and low immunogenicity characteristics of PCL made the material an exceptional choice for the application [8, 19]. PCL is outstanding in tissue engineering because the material naturally resorbs within months or years, allowing for long term implantation. Solvent casting/particulate leaching, spin coating, and separation phase techniques are used to process a PCL-based scaffold. Nanofibrous scaffolds from PCL have been prepared onto which mesenchymal stem cells (MSC) derived from bone marrow and from adipose tissue were seeded. This allowed for the determination of the scaffold support ability in tissue engineering applications. Possible applications were then determined by comparing the impact of PCL and PLA scaffolds on MSC proliferation and cytotoxicity [19].

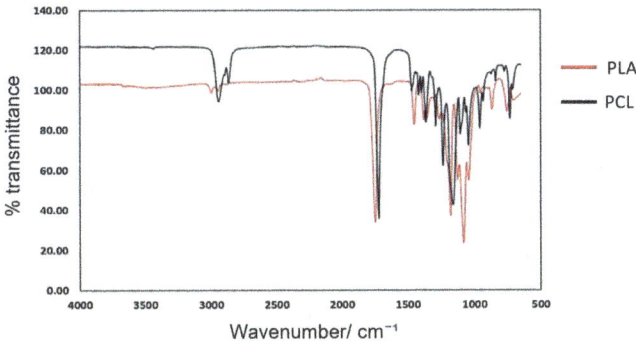

Figure 9.6: FTIR results of PCL and PLA nanofibrous scaffolds with a spectral region between 4000 and 500 cm^{-1} and a resolution of 2 cm^{-1} within 20 scans [19].

Rather than the sol–gel method, the scaffolds are processed using an electrospinning method, which is one of the most effective techniques regarding the product [4, 19, 22, 41]. A total of 22-gauge syringes were used to insert the PCL solution into the electrospinning machine. The electric generator's anode and cathode had the syringe tops and collectors connected, respectively [19]. Chloroform is commonly used as a solvent for electrospinning PCL but then produces microfibers rather than nanofibers [22]. PCL-based nanofibrous scaffolds possess highly interconnected, meaning the porosity is greater than 96%, and hierarchically structured pores, meaning the size ranges from sub microns to hundreds of microns. Therefore, making the scaffold 3D, compared to a substrate that is 2D [8].

Figure 9.6 provided results regarding the Fourier transform infrared (FTIR) spectroscopy of PCL and PLA nanofibrous scaffolds. FTIR was used to confirm the structure of the material and be certain of the absence of any chemical contamination during manufacturing. The typical peaks of both PCL and PLA were pictured in Figure 9.6. With a focus on the PCL curve, there are eight peaks that draw importance and are known as the asymmetric aliphatic CH_2 stretching at 2949 cm^{-1}, symmetric CH_2 stretching at 2865 cm^{-1}, C=O stretching 1727 cm^{-1}, C–O and C–C stretching in the crystalline phase at 1293 cm^{-1}, asymmetric C–O–C stretching at 1240 cm^{-1}, O–C–C stretching at 1190 cm^{-1}, symmetric C–O–C stretching at 1170 cm^{-1}, and the C–O and C–C stretching vibrations in the amorphous phase at 1157 cm^{-1} [19].

To enhance the cell adherence within the PCL structure, a coating to the surface using collagen or gelatin must be added because of the high hydrophobicity of the material [50]. A study proved that PCL nanofibers were able to successfully work with bone marrow derived mesenchymal stem cells (BMSC) with an increase in feasibility ensuring the biocompatibility. Adipose tissue stem cells (ASC) are another type of cell that properly works with PCL nanofibers, because of the high surface area to volume ratio of the nanofibers. A result of PCL, there was a large quantity of cells within the fibers because of the porosity of the material [19].

One common application in the medical field is the use of sutures, which allows for wound closure and has been around for many centuries. Fibers of synthetic polymers are typically what make up the suture material. These fibers could be absorbable or nonabsorbable. The key advantage of using a synthetic absorbable suture is the reproducible degradability inside a biological environment. Over the past four decades, aliphatic polyesters that are used to make biocompatible sutures have been studied. In Europe, PCL is a common material used as a biodegradable suture because of the material being regarded as tissue compatible. Hydrolytic degradation occurs due to the presence of the hydrolytically labile aliphatic ester linkages in the human body which is considered a physiological condition. As mentioned, PCL possesses a slow degradation time of two years, which means copolymers have been synthesized to accelerate the rate of biosorption. The blend of ε-CL and glycolides (GL) improve the disadvantages of sutures which include tissue drag and trauma and potential cause of infection through the interstices of the braid structure. GL offers the property of reduced stiffness to the suture. Studies show the blend material for the suture displayed excellent handling properties, minimal resistance during passage through tissue, and excellent tensile properties. The absorption of these sutures is complete between the 91st and 119th days after implantation, with slight tissue reaction [10].

9.4 Characterization behavior of PCL

A previous study showed to improve mechanical properties of PCL, PCL/calcium sulfate (CS) whisker composites that contained different CS contents (0, 5, 10, 15, and 20 wt. %) were prepared by melting and co-precipitation blending methods [2]. Figure 9.7 provides X-ray powder diffraction (XRD) spectra of CS whisker, pure PCL, and the composites prepared by both blending methods. Shown by the graph on the left, curve A represents the pure PCL that contained three reflections at the angles (2θ) of about 21.4°, 22.0°, and 23.7°. Comparing the results of the pure PCL to the PCL/CS whiskers, the same strong reflections are present in both curves. Therefore, the crystal structure of the PCL matrix in composites obtained an orthorhombic crystal structure. For the whisker composites, results show that the PCL reflection at 21.4° moved more towards the larger angles as the CS content increased. This means that the d-spacing between PCL in the whisker composites decreased. The cell could have been compressed due to the when whisker impacted the growth of the crystal grain. However, during the melting process, some whiskers, and a part of the molecular chain of PCL were broken down by shear stress resulting in a higher number of whiskers. The crystallization rate could be enhanced because of this change causing an increase in the nucleation rate. The shorter molecular chain of PCL also had a nucleation effect, causing an increase in the nucleation rate of composites. Composites prepared by the melting method obtained a slightly higher crystallinity value compared to the coprecipitation method [2].

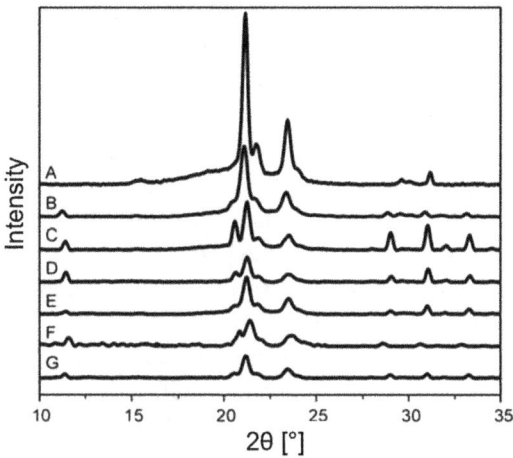

Figure 9.7: X-ray spectra for (A) pure PCL and composites prepared by melt blending method (B–D) 5 wt. %, 15 wt. %, and 20 wt. %, and prepared by co-precipitation method (E–G) 5 wt. %, 15 wt.%, and 20 wt.% whisker [2].

Table 9.4 provided information determined from differential scanning calorimetry (DSC) regarding the thermal and crystallization properties parameters of pure PCL compared to PCL/CS whisker composites. The first parameter listed is the crystallization temperature (T_c), as the content of CS whisker increased the Tc also increased. Therefore, the whisker allowed for a nucleating effect on the crystallization of PCL. Table 9.4 shows that the T_c is slightly higher for the composites prepared by the melt

Table 9.4: PCL/CS whisker composites prepared by melting and coprecipitation methods with crystallization and thermal data including, crystallization temperature, melting temperature, glass transition temperature, undercooling, half peak width, and crystallinity [2].

Crystallization and thermal parameters	Fabrication method	Content of CS whisker [%]			
		0	5	15	20
Crystallization temperature (T_c [°C])	Coprecipitation	27.6	31.5	30.7	30.5
	Melting		30.4	33.0	33.1
Melting temperature (T_m [°C])	Coprecipitation	57.9	58.7	58.3	58.5
	Melting		58	59	58.4
Glass transition temperature (T_g [°C])	Coprecipitation	−62.0	−62.4	−61.5	−61.1
	Melting		−62.5	−63.2	−62.8
Undercooling (ΔT [°C])	Coprecipitation	30.3	27.2	27.6	27.0
	Melting		27.6	26.0	25.3
Half peak width {$(\Delta T)_p$ [°C]}	Coprecipitation	12.0	8.0	7.0	7.0
	Melting		5.0	5.5	5.0
Crystallinity (Cr [%])	Coprecipitation	54.6	46.1	47.4	54.6
	Melting		52.7	54.4	54.9

blending method compared to the coprecipitation. This was caused by the shear stress of the screw during processing for melt blending which caused the number of whiskers to increase. The increase in nucleation rate and enhancement of the crystallization rate was caused by the increase in the number of whiskers [2].

The undercooling (ΔT) in Table 9.4 is calculated by taking the difference of the melting temperature (T_m) and the crystallinity temperature (T_c); both were determined by the DSC. Undercooling is inversely proportional to the crystallization rate. Table 9.4 shows that the pure PCL obtained a 30.3 °C change and as the CS percentage increased the ΔT decreased to 25.3 °C. The composite that was prepared by the melt blending obtained a lower ΔT than the coprecipitation value. The whisker size affected the nucleation effect to PCL [2]. The half peak width (ΔT) p of the crystallization curve can be referred to as the crystallization perfection. Table 9.4 shows that the pure PCL obtained a value of 12.0 °C, and the composite prepared by the melt blending had a value of 5 °C, which made it about 7 °C narrower than the pure PCL. Comparing the coprecipitation and melt blending methods, the melt blended composite obtained a (ΔT) p that was 2 °C narrower. The filler shape influenced why the melt blending composites were superior to the pure PCL and coprecipitation method. Gain growth can be inhibited by the fillers; therefore fillers with a high aspect ratio obtain a higher chance of causing a restraining effect [2].

A previous study provided information regarding the spherulite morphology of pure PCL prepared by both the melting and coprecipitation blending method. Figure 9.8 showed the spherulite structure of pure PCL along with 5, 15, and 20 weight percentages of whisker content, respectively. Shown in Figure 9.8, there are clear grain boundaries within the spherulites; therefore the morphology of pure PCL was imperfect. The CS whisker content has a direct impact on the grain refining effect to PCL. This is seen by the difference between the pure PCL and all composites reveal a change in spherulite size. As the content of CS whiskers increased, the spherulite size significantly decreased [2].

Crystallization perfection of the composites that were prepared by the melt blending method obtained better results compared to the coprecipitation method samples. Figure 9.8e–g showed how the coprecipitation blending method samples resulted in much smaller spherulites compared to the melt blending method samples shown in Figure 9.8b–d. This information provided that the size of the whisker impacted the grain refining effect, therefore the long whiskers are stronger than the short whiskers [2].

9.5 Development of a biodegradable PCL film

Common polymers such as polyethylene terephthalate (PET), PP, and polystyrene (PS) are widely found in the packaging industry because of their low cost and high processability [21, 45]. However, rising environmental issues make these products

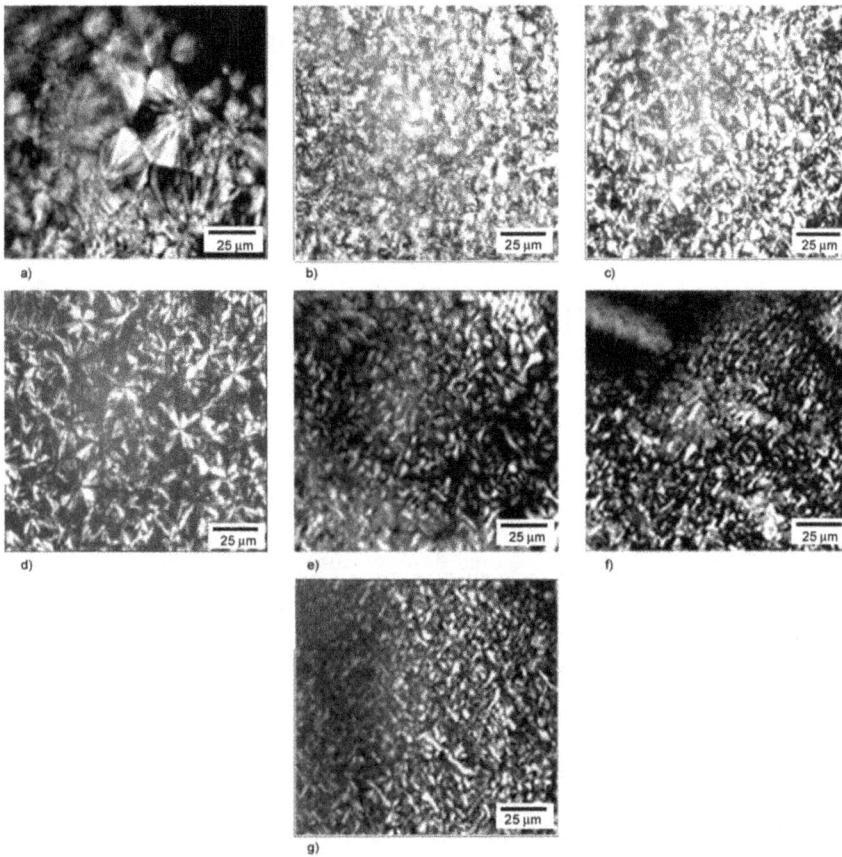

Figure 9.8: Polarized light microscope (POM) spherulite morphology results with a magnification of 400×: (a) pure PCL prepared by the melting method containing, (b) 5 wt.%, (c) 15 wt.%, and (d) 20 wt.% whisker content, and by coprecipitation method containing (e) 5 wt.%, (f) 15 wt.%, and (g) 20 wt.% whisker content [2].

undesirable because they are not biodegradable. Non biodegradable plastic products are becoming an issue at the forefront of global warming. An effort to combat these biodegradable problems, a new film is being developed [21].

A new antimicrobial packaging film using PCL and grapefruit seed extract (GSE) are being developed. They are used as a polymeric material and antimicrobial packaging film, respectively [21, 36]. PCL is easy to process because it is a hydrophobic semi crystalline linear aliphatic polyester with a low melting point and glass transition temperature [45]. PCL has not been explored regularly in the food packing department because of its low thermal stability. Though the low melting point allows easy processing for PCL, it limits its application. PCL would be appropriate for refrigerated foods such as fresh salads and cheeses [21].

Food deterioration is caused by microbial growth on the surface. Antimicrobial agents are incorporated into the polymeric materials or blended during the processing stage to create an antimicrobial packaging film [21]. GSE, a natural antimicrobial substance, has been used to retard or reduce bacterial growth. Bioactive flavonoids such as naringin, naringenin, hesperidin, contribute to the antimicrobial activity of GSE. Antimicrobial activity of GSE also involves organic acids such as ascorbic acid and citric acid. Naringin, one of the components of GSE, is a major constituent found in the seed of ripe grapefruits. A wide range of applications opens because this product is water-soluble and nontoxic even when consumed in excessive amounts. GSE is heat-stable up to 120 °C which means it can be used in polymeric packaging materials without thermal decomposition under manufacturing conditions [21].

The ease and simplicity of solution casting technique has made it widely popular in film manufacturing preparation. However, solution casting is not practicable for commercial film production because it is hard to scale-up and requires long processing time. The extrusion technique is a better option for preparation of packaging materials and is preferred in industrial application. Extrusion technique can be performed as a continuous unit operation, controlling time, temperature, size, moisture, and shape. The extrusion technique provides more structured film and better dispersion of active compounds in the polymers. There are very few studies that show the preparation of biodegradable antimicrobial films using extrusion [21]. The results of the second objective regarding the structural, colorimetric, thermal, and mechanical properties of the PCL films will be discussed in this section.

Biodegradable PCL/GSE composite film was prepared with a co rotating twin-screw extruder. The extruder had a screw diameter of 11 mm and length of 440 mm. PCL pellets and GSE at different concentrations (1, 3, and 5 wt.%) were mixed and fed into the feeder. Pellets were extruded and pelletized three times to ensure homogeneous distribution. The speed of the feeder was 5 rpm as the screw speed was 50 rpm. Using a

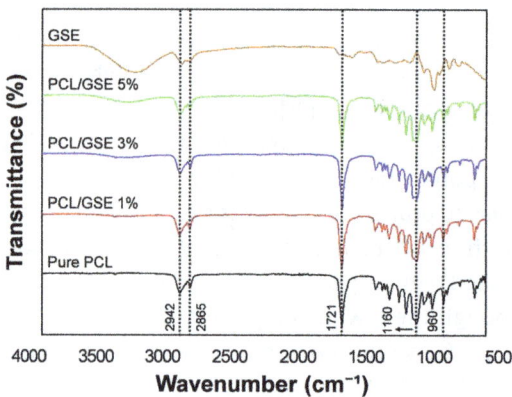

Figure 9.9: FTIR spectra of pure PCL, GSE, and PCL/GSE composite films [5].

heat press machine, the extruded sheets were thinned to films. The films were obtained with a thickness of 0.12 ± 0.01 mm and kept in a thermo–hygrostat.

The chemical interaction between PCL and GSE were investigated with the FTIR spectroscopy. Figure 9.9 shows the FTIR spectra of pure PCL and PCL/GSE composite films. In the pure PCL film spectrum, weak peaks observed at 2942 cm^{-1} and 2865 cm^{-1} correspond to asymmetric elongation of the methylene–oxygen (CH_2–O) and symmetric methylene groups (CH_2–). The vibration of –C=O bonds is represented by the sharp and strong peaks at 1721 cm^{-1}. In addition, the stretching of the oxime bond (960 cm^{-1}) as well as the C–O–C bond (1166 cm^{-1}) was observed. In the GSE spectrum, the typical OH vibration of phenolic/aromatic compounds is observed over the band at 3600–3100 cm^{-1}. The peaks of PCL/GSS were like those of pure PCL film however, the broad peak at 3300 cm^{-1} was present in the PCL/GSE composite film spectra. The peak appeared gradually as the GSE content increased which may be due to the flavonoid group in GSE. The FTIR graph showed that there were no significant structural changes due to the addition of GSE in PCL into the polymer matrix [14].

In Figure 9.10 below, the thermogravimetric analysis results of the PCL films incorporated with various concentrations of GSE are shown. A simple decomposition profile of a single transition temperature was shown for pure PCL. The thermal stability of pure PCL film was slightly affected by the addition of GSE. A three-step decomposition was shown for all composites. The first step at 90 °C was due to water evaporation inside of the PCL/GSE composite film. Decomposition of glycerol and GSE in PCL polymer was observed in the second and third steps of degradation. The second stage was at approximately 120–200 °C which is attributed to the decomposition of glycerol. The main degradation, the degradation of PCL, was observed at approximately 400 °C. Incorporation of GSE into the PCL matrix appeared to decrease thermal stability but the decrease was not significant. From the TGA results, it could be concluded that the

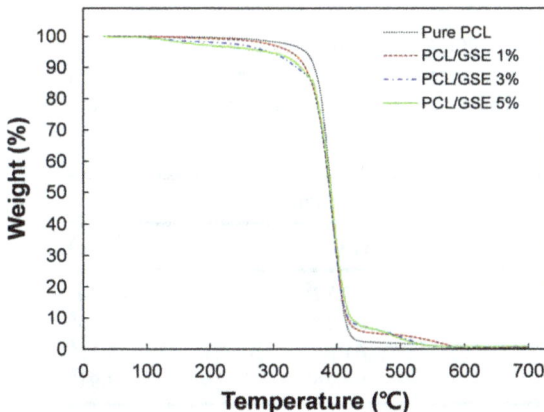

Figure 9.10: TGA graphs of pure PCL and PCL/GSE composite films [14].

Table 9.5: Color parameters of the pure PCL and PCL/GSE composite films. Data are expressed as the mean ± standard deviation of five replicates. L*, lightness; a*, redness; b*, yellowness; delta; and E*, total difference. The different uppercase letters A–D in the same column indicate a significant difference ($p < 0.05$), as assessed by Duncan's multiple range test [14].

Composite	Color parameters			
	L^*	a^*	B^*	ΔE^*
Pure PCL	94.40 ± 0.06^D	3.86 ± 0.04^D	1.16 ± 0.06^A	2.72 ± 0.06^A
PCL/GSE 1%	93.59 ± 0.16^C	3.78 ± 0.04^C	2.76 ± 0.16^B	4.07 ± 0.18^B
PCL/GSE 3%	91.08 ± 0.06^B	3.70 ± 0.02^B	6.83 ± 0.24^C	8.61 ± 0.21^C
PCL/GSE 5%	90.24 ± 0.17^A	3.33 ± 0.02^A	9.65 ± 0.14^D	11.32 ± 0.09^D

thermal stability of PCL-based films is not significantly affected by the addition of GSE into the polymer matrix [14].

The color parameters of the pure PCL and PCL/GSE composite films are presented in Table 9.5 below. The L* and a* values were significantly decreased while the b* were significantly increased with the addition of GSE. The variation in color suggests that the PCL/GSE composite film had more dark, green, and yellow tones compared to the pure PCL film. The GSE has an opacity and yellowish tint [14].

The thermal properties of the pure PCL and PCL/GSE compost film were investigated using the DSC analysis. The main thermal parameters, including the melting temperature, the enthalpy of melting, and the degree of crystallinity are shown in Table 9.6. There was a clear melting peak at 60 °C for each film. The pure PCL film had a melting temperature of 62.93 °C and enthalpy of melting of 67.72 ± 0.25 J/g which was close to reported values. These values were slightly decreased for the PCL/GSE composite films, though their thermal behaviors were very similar. The crystallinity

Table 9.6: Thermal properties of the pure PCL and PCL/GSE composite films. Data are expressed as the mean of three replicates which include melting temperature, the enthalpy of melting, and the degree of crystallinity. The different uppercase letters (A-C) in the same column indicate a significant difference [14].

Composite	Thermal properties		
	T_m (°C)	ΔH_m (J/g)	X_c (%)
Pure PCL	62.93^C	67.72^C	49.76^C
PC L/C SE 1%	61.77^B	64.58^B	47.45^B
PCL/GSE 3%	61.33^{AB}	61.66^A	45.31^A
PCL/CSE 5%	60.91^A	59.76^A	43.91^A

Table 9.7: Mechanical properties of the pure PCL and PCL/GSE composite films. Data are expressed as the mean ± standard deviation of five replicates. The different uppercase letters (A and B) in the same column indicate a significant difference [14].

Composite	Mechanical properties	
	Tensile strength (MPa)	Elongation at break (%)
Pure PCL	29.59 ± 1.22[B]	302.96 ± 46.54[A]
PCL/GSE 1%	28.31 ± 1.08[AB]	334.87 ± 20.71[AB]
PCL/GSE 3%	27.99 ± 1 0.73[AB]	360.96 ± 25.62[AB]
PCL/GSE 5%	27.31 ± 2.27[A]	360.53 ± 15.13[B]

percentage of the PCL/GSE composite films was gradually decreased as the GSE content increased [14].

The mechanical properties including tensile strength and elongation at break of the pure PCL and PCL/GSE composite films were investigated. The mechanical properties were necessary to determine the effect of the GSE in the PCL/GSE composite films; results are shown below in Table 9.7. An extremely high elongation at break and average tensile strength of 20–30 MPa characterize pure PCL film. As shown in Table 9.7, the pure PCL film developed in this study showed an average tensile strength of 29.59 ± 1.22 MPa and elongation at break of 302.96 ± 46.54%. No significant change was observed in tensile strength until the addition of 3% GSE. There was a significant decrease with the PCL/GSE 5% film. The decrease may be accounted for from the addition of GSE contents in the PCL matrix. It is known that typical additives except for cross linking agents decrease tensile strength and increase elongation of the film. It was observed that the elongation at break gradually improved as the amount of GSE decreased. Each percentage of GSE addition increased its elongation value. These results match previous reports of improvement in elongation at break due to addition of GSE into the polymer matrix. The increase in elongation is due to the glycerol components in GSE that usually act as a plasticizer. PCL/GSE composite film becomes a potential use in flexible packaging materials because of its advanced flexibility [14].

9.6 Biomedical applications of PCL

There are three main characteristics that make PCL in fiber form suitable for controlled drug delivery: high permeability, excellent biocompatibility, and the ability to be fully excreted from the body once bioresorbable [1, 16, 17, 45]. PCL has slow biodegradation compared to other polymers making it most suitable for long-term delivery. PCL can form compatible blends with other natural polymers such as starch, chitosan and synthetic polymers to construct desired release profiles [10]. Blended PCL can improve stress, crack resistance, dyeability, and control over release rate of drugs. The type of formulation, method of preparation, content, and size will determine the drug release rate of PCL [1].

Factors that lead to why PCL is commonly used in the medical field are biodegradability, biocompatibility, pliability, good solubility, low melting point, and exceptional blend-compatibility. PCL is applied to many different spectrums of the medical device sector including sutures, three dimensional scaffolds in tissue engineering and cartilage applications, wound dressing, fixation devices, contraceptive devices, dentistry, etc. [1–3, 10, 18, 45–48]. Owing to the slow degradation rate and biocompatibility, PCL is one of the most common biopolymers in the medical field [6, 23]. Metal items such as plates, screws, and nails are being replaced by biodegradable implants. PCL lacks mechanical properties to be applied for these high load bearing applications [1, 17, 37]. Numerous aliphatic polyesters such as, PCL, poly(lactide)s, poly(glycolide)s, and their copolymers are commonly investigated for biomedical and pharmaceutical applications [32]. The repair of orbital fracture is possible using three-dimensional and directional porous structures, a 3D-printed PCL mesh was developed, which will provide the characteristic of biocompatibility. Therefore, cell ingrowth and regeneration of the surrounding host tissue is allowed [35].

Tissue engineering is a part of the medical field where alternatives are found regarding harvested tissues and organs for transplantation. There are many different approaches when it comes to tissue engineering. One of the most common approaches is when cells are seeded and cultured to develop into tissues on a biodegradable scaffold. PCL is a three-dimensional porous substrate capable of supporting a cellular structure [47]. Tissue engineering scaffolds such as fixation of facial fractures and chemotherapeutic implants, along with applications related to shape memory have drawn a great deal of attention using a PCL/PLA blend. A disadvantage of this blend is poor mechanical properties [24]. Some of the most commonly used synthetic polymers in tissue engineering are PCL, PLA, poly(glycolic acid) (PGA), poly lactic-*co*-glycolic acid (PLGA), polyurethane (PU), polypropylene (PP), etc [7, 10].

9.7 Conclusion

The purpose of this chapter was to discuss PCL-based polymer and composites regarding processing and characterization. The reader is first introduced to PCL as a synthetic, commercially available polymer that is often characterized for its large set of biodegradation and mechanical properties. As shown in this chapter, PCL has been widely investigated for particle, industrial and medical applications. In comparison with other polyesters, this polymer is easy to manufacture and manipulate making it appealing to a multitude of applications. To understand the formulation of PCL, the synthesis of poly(α-hydroxy acid)s was discussed. PCL is prepared by ring-opening polymerization of the cyclic monomer ε-CL catalyst or with free radical ring-opening polymerization. The process of choosing an initiator or catalyst is important in the control of polymer synthesis because it can impact properties of the polymer, regarding the physical and chemical states. As the chapter title entails, it was important to discuss the biodegradability of PCL. Regarding the biodegradability, the degradation mechanism, and other microorganisms that degrade PCL were highlighted. PCL has been widely researched and tested for mechanical applications. This chapter discussed how PCL can be applied in drug-delivery systems, medical devices, and tissue engineering. One of the biggest applications, tissue engineering, was further discussed with the use of a study. The study outlined the sol–gel process, scaffolds, and sutures. Finally, the chapter was complete with a section on characterization. Characterization of PCL was examined through a PCL/calcium sulfate composite and a biodegradable PCL film. Non biodegradable plastics are becoming an issue at the forefront of global warming. An effort to combat these biodegradable problems has led to this chapter's discussion on PCL today.

Author contributions: All the authors have accepted responsibility for the entire content of this submitted manuscript and approved submission.
Research funding: None declared.
Conflict of interest statement: The authors declare no conflicts of interest regarding this article.

References

1. Woodruff MA, Hutmacher DW. The return of a forgotten polymer—polycaprolactone in the 21st century. Prog Polym Sci 2010;35:1217–56.
2. Liu JY, Reni L, Wei Q, Wu JL, Liu S, Wang YJ, et al. Fabrication and characterization of polycaprolactone/calcium sulfate whisker composites. Express Polym Lett 2011;5. https://doi.org/10.3144/expresspolymlett.2011.72.
3. Kakroodi AR, Kazemi Y, Rodrigue D, Park CB. Facile production of biodegradable PCL/PLA in situ nanofibrillar composites with unprecedented compatibility between the blend components. Chem Eng J 2018;351:976–84.
4. Cipitria A, Skelton A, Dargaville TR, Dalton PD, Hutmacher DW. Design, fabrication and characterization of PCL electrospun scaffolds: a review. J Mater Chem 2011;21:9419–53.
5. Wei J, Chen F, Shin JW, Hong H, Dai C, Su J, et al. Preparation and characterization of bioactive mesoporous wollastonite–polycaprolactone composite scaffold. Biomaterials 2009;30:1080–8.
6. de Menezes BRC, Montanheiro TLDA, Sampaio ADG, Koga, Ito CY, Thim GP, et al. PCL/βAgVO3 nanocomposites obtained by solvent casting as potential antimicrobial biomaterials. J Appl Polym Sci 2020;50130. https://doi.org/10.1002/app.50130.
7. Bhattarai DP, Aguilar LE, Park CH, Kim CS. A review on properties of natural and synthetic based electrospun fibrous materials for bone tissue engineering. Membranes 2018;8:62.
8. Zhu B, Bai T, Wang P, Wang Y, Liu C, Shen C. Selective dispersion of carbon nanotubes and nanoclay in biodegradable poly (ε-caprolactone)/poly (lactic acid) blends with improved toughness, strength and thermal stability. Int J Biol Macromol 2020;153:1272–80.

9. Permyakova ES, Kiryukhantsev-Korneev PV, Gudz KY, Konopatsky AS, Polčak J, Zhitnyak IY, et al. Comparison of different approaches to surface functionalization of biodegradable polycaprolactone scaffolds. Nanomaterials 2019;9:1769.

10. Azimi B, Nourpanah P, Rabiee M, Arbab S. Poly (-caprolactone) fiber: an overview. J Eng Fibers Fabrics 2014;9. https://doi.org/10.1177/155892501400900309.

11. Gong C, Shi S, Dong P, Kan B, Gou M, Wang X, et al. Synthesis and characterization of PEG-PCL-PEG thermosensitive hydrogel. Int J Pharm 2009;365:89–99.

12. Park JH, Kim DI, Hong SG, Seo H, Kim J, Moon GD, et al. Poly (ε-caprolactone)(PCL) hollow nanoparticles with surface sealability and on-demand pore generability for easy loading and NIR light-triggered release of drug. Pharmaceutics 2019;11:528.

13. del Ángel-Sánchez K, Borbolla-Torres CI, Palacios-Pineda LM, Ulloa-Castillo NA, Elías-Zúñiga A. Development, fabrication, and characterization of composite polycaprolactone membranes reinforced with TiO2 nanoparticles. Polymers 2019;11:1955.

14. Catauro M, Bollino F, Veronesi P, Lamanna G. Influence of PCL on mechanical properties and bioactivity of ZrO2-based hybrid coatings synthesized by sol–gel dip coating technique. Mater Sci Eng C 2014;39:344–51.

15. Lu F, Lei L, Shen YY, Hou JW, Chen WL, Li YG, et al. Effects of amphiphilic PCL–PEG–PCL copolymer addition on 5-fluorouracil release from biodegradable PCL films for stent application. Int J Pharm 2011;419:77–84.

16. Dubey N, Varshney R, Shukla J, Ganeshpurkar A, Hazari PP, Bandopadhaya GP, et al. Synthesis and evaluation of biodegradable PCL/PEG nanoparticles for neuroendocrine tumor targeted delivery of somatostatin analog. Drug Deliv 2012;19:132–42.

17. Iqbal M, Valour JP, Fessi H, Elaissari A. Preparation of biodegradable PCL particles via double emulsion evaporation method using ultrasound technique. Colloid Polym Sci 2015;293:861–73.

18. Grossen P, Witzigmann D, Sieber S, Huwyler J. PEG-PCL-based nanomedicines: a biodegradable drug delivery system and its application. J Contr Release 2017;260:46–60.

19. Marei NH, El-Sherbiny IM, Lotfy A, El-Badawy A, El-Badri N. Mesenchymal stem cells growth and proliferation enhancement using PLA vs PCL based nanofibrous scaffolds. Int J Biol Macromol 2016;93:9–19.

20. Mellinas-Ciller AC, Ramos M, Grau-Atienza A, Jordá Sánchez A, Burgos N, Jiménez A, et al. *Biodegradable poly (ε-Caprolactone) active films loaded with MSU-X mesoporous silica for the release of α-tocopherol*; 2020.

21. Lyu JS, Lee JS, Han J. Development of a biodegradable polycaprolactone film incorporated with an antimicrobial agent via an extrusion process. Sci Rep 2019;9:1–11.

22. Mochane MJ, Motsoeneng TS, Sadiku ER, Mokhena TC, Sefadi JS. Morphology and properties of electrospun PCL and its composites for medical applications: a mini review. Appl Sci 2019;9:2205.

23. Van der Schueren L, De Schoenmaker B, Kalaoglu Öl, De Clerck K. An alternative solvent system for the steady state electrospinning of polycaprolactone. Eur Polym J 2011;47:1256–63.

24. Zhu B, Bai T, Wang P, Wang Y, Liu C, Shen C. Selective dispersion of carbon nanotubes and nanoclay in biodegradable poly (ε-caprolactone)/poly (lactic acid) blends with improved toughness, strength and thermal stability. Int J Biol Macromol 2020;153:1272–80.

25. Kumar A, Mir SM, Aldulijan I, Mahajan A, Anwar A, Leon CH, et al. Loadbearing biodegradable PCLPGAbeta TCP scaffolds for bone tissue regeneration. J Biomed Mater Res B Appl Biomater 2020.

26. Rogina A, Antunović M, Milovac D. Biomimetic design of bone substitutes based on cuttlefish bonederived hydroxyapatite and biodegradable polymers. J Biomed Mater Res B Appl Biomater 2019;107:197–204.

27. Malinowski R. Mechanical properties of PLA/PCL blends crosslinked by electron beam and TAIC additive. Chem Phys Lett 2016;662:91–6.
28. Scaffaro R, Lopresti F, Botta L, Maio A. Mechanical behavior of polylactic acid/polycaprolactone porous layered functional composites. Compos B Eng 2016;98:70–7.
29. Scaffaro R, Lopresti F, Sutera A, Botta L, Fontana RM, Puglia AM, et al. Effect of PCL/PEGbased membranes on Actinorhodin production in streptomyces coelicolor cultivations. Macromol Biosci 2016;16:686–93.
30. Scaffaro R, Botta L, Gallo G, Puglia AM. Influence of drawing on the antimicrobial and physical properties of chlorhexidinecompounded poly (caprolactone) monofilaments. Macromol Mater Eng 2015;300:1268–77.
31. Choong C, Yuan S, Thian ES, Oyane A, Triffitt J. Optimization of poly (εcaprolactone) surface properties for apatite formation and improved osteogenic stimulation. J Biomed Mater Res 2012;100:353–61.
32. Croisier F, Duwez AS, Jérôme C, Léonard AF, Van Der Werf KO, Dijkstra PJ, et al. Mechanical testing of electrospun PCL fibers. Acta Biomater 2012;8:218–24.
33. Douglas P, Andrews G, Jones D, Walker G. Analysis of in vitro drug dissolution from PCL melt extrusion. Chem Eng J 2010;164:359–70.
34. Teo EY, Ong SY, Chong MSK, Zhang Z, Lu J, Moochhala S, et al. Polycaprolactone-based fused deposition modeled mesh for delivery of antibacterial agents to infected wounds. Biomaterials 2011;32:279–87.
35. Kim SY. Application of the three-dimensionally printed biodegradable polycaprolactone (PCL) mesh in repair of orbital wall fractures. J Cranio-Maxillofacial Surg 2019;47:1065–71.
36. Olewnik-Kruszkowska E, Burkowska-But A, Tarach I, Walczak M, Jakubowska E. Biodegradation of polylactide-based composites with an addition of a compatibilizing agent in different environments. Int Biodeterior Biodegrad 2020;147. https://doi.org/10.1016/j.ibiod.2019.104840.
37. Eskitoros-Togay ŞM, Bulbul YE, Tort S, Korkmaz FD, Acartürk F, Dilsiz N. Fabrication of doxycycline-loaded electrospun PCL/PEO membranes for a potential drug delivery system. Int J Pharm 2019;565:83–94.
38. Tokiwa Y, Calabia BP, Ugwu CU, Aiba S. Biodegradability of plastics. Int J Mol Sci 2009;10:3722–42.
39. Dash TK, Konkimalla VB. Poly-ε-caprolactone based formulations for drug delivery and tissue engineering: a review. J Contr Release 2012;158:15–33.
40. Al Hosni AS, Pittman JK, Robson GD. Microbial degradation of four biodegradable polymers in soil and compost demonstrating polycaprolactone as an ideal compostable plastic. Waste Manag 2019;97:105–14.
41. Hodge J, Quint C. The improvement of cell infiltration in an electrospun scaffold with multiple synthetic biodegradable polymers using sacrificial PEO microparticles. J Biomed Mater Res 2019;107:1954–64.
42. Haryńska A, Kucinska-Lipka J, Sulowska A, Gubanska I, Kostrzewa M, Janik H. Medical-grade PCL based polyurethane system for FDM 3D printing—characterization and fabrication. Materials 2019;12:887.
43. Scaffaro R, Maio A, Sutera F, Gulino EF, Morreale M. Degradation and recycling of films based on biodegradable polymers: a short review. Polymers 2019;11:651.
44. Zhu B, Wang X, Zeng Q, Wang P, Wang Y, Liu C, et al. Enhanced mechanical properties of biodegradable poly (ε-caprolactone)/cellulose acetate butyrate nanocomposites filled with organoclay. Composit Commun 2019;13:70–4.
45. Bartnikowski M, Dargaville TR, Ivanovski S, Hutmacher DW. Degradation mechanisms of polycaprolactone in the context of chemistry, geometry and environment. Prog Polym Sci 2019;96:1–20.

46. Izquierdo R, Garcia-Giralt N, Rodriguez MT, Caceres E, Garcia SJ, Gómez Ribelles JL, et al. (2008). Biodegradable PCL scaffolds with an interconnected spherical pore network for tissue engineering. J Biomed Mater Res Part A 85, 25–35.
47. Augustine R, Hasan A, Patan NK, Augustine A, Dalvi YB, Varghese R, et al. Titanium nanorods loaded PCL meshes with enhanced blood vessel formation and cell migration for wound dressing applications. Macromol Biosci 2019;19.
48. Bartnikowski M, Dargaville TR, Ivanovski S, Hutmacher DW. Degradation mechanisms of polycaprolactone in the context of chemistry, geometry and environment. Prog Polym Sci 2019;96: 1–20.
49. Liu C, Gong C, Pan Y, Zhang Y, Wang J, Huang M, et al. Synthesis and characterization of a thermosensitive hydrogel based on biodegradable amphiphilic PCL-Pluronic (L35)-PCL block copolymers. Colloid Surface Physicochem Eng Aspect 2007;302:430–8.
50. Sheng D, Li J, Ai C, Feng S, Ying T, Liu X, et al. Electrospun PCL/Gel-aligned scaffolds enhance the biomechanical strength in tendon repair. J Mater Chem B 2019;7:4801–10.

Ty Burford*, William Rieg and Samy Madbouly

10 Biodegradable poly(butylene adipate-*co*-terephthalate) (PBAT)

Abstract: Poly(butylene adipate-*co*-terephthalate), PBAT, is a synthetic and 100% biodegradable polymer based on fossil resources. Most conventional plastics utilized today are produced from petroleum-based products, making them nondecomposable. With polymer manufacturing companies under constant scrutiny due to the effect nondegradable plastics have on the environment, biodegradable polymer production is growing at an exponential rate. However, developing new biodegradable polymers that can maintain the requirements of expected material properties has been a challenge for material manufacturers. When compared to other polymers, PBAT is classified as polyester. Aliphatic polyesters biodegrade efficiently because of ester bonds in the soft chain portion of the polymer. These ester bonds are broken down through hydrolysis, making the polymer degradable in almost any environment. In this chapter, key components of biodegradable PBAT and prominent blends of PBAT will be reviewed and analyzed for suitable end-use applications. This chapter will also provide a general understanding of the chemical composition of PBAT and how the addition of components effect the properties of the material.

Keywords: aliphatic; aromatic; biodegradation; morphology; polybutylene terephthalate; polyethylene terephthalate.

10.1 Introduction

Poly(butylene adipate-*co*-terephthalate), PBAT, has been the most promising and popular biodegradable polymer that has been developed thus far. PBAT is a random copolymer that is obtained by polycondensation of butanediol (BDO), adipic acid (AA), and terephthalic acid (Figure 10.1). This polymer recipe has proven to be the best combination of mechanical properties and biodegradability currently on the market.

The formation of polyesters generally is synthesized by polycondensation of diols and dicarboxylic acids. The specific diols and acids used to produce are shown in Figure 10.2. Common catalysts for this polycondensation reaction include compounds based on zinc, tin, and titanium. Proper preparation of PBAT requires the use of a

*Corresponding author: **Ty Burford**, School of Engineering, Behrend College, Pennsylvania State University, 4701 College Drive, Erie, PA 16563, USA, E-mail: ty.burford98@gmail.com
William Rieg and Samy Madbouly, School of Engineering, Behrend College, Pennsylvania State University, 4701 College Drive, Erie, PA 16563, USA, E-mail: wsr5062@psu.edu (W. Rieg), sum1541@psu.edu (S. Madbouly)

This article has previously been published in the journal Physical Sciences Reviews. Please cite as: T. Burford, W. Rieg and S. Madbouly "Biodegradable poly(butylene adipate-*co*-terephthalate) (PBAT)" *Physical Sciences Reviews* [Online] 2021, 7. DOI: 10.1515/psr-2020-0078 | https://doi.org/10.1515/9781501521942-010

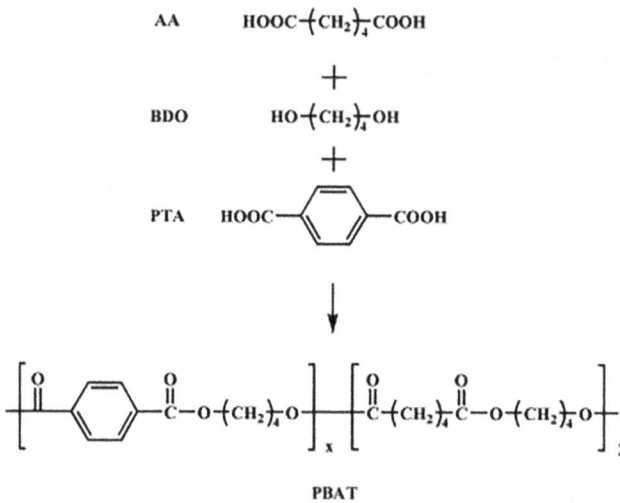

Figure 10.1: Synthesis of PBAT [1].

strong vacuum with a temperature of at least 190 °C. This is required in order to remove the lighter, less important molecules such as water. Nucleating agents are often used when producing PBAT to improve crystallization ability. In order to obtain PBAT with good overall mechanical and physical properties, the use of compounds like talc, chalk, mica, or silicon oxides are used as nucleating agents. Phosphoric acids can also be included in the creation of PBAT as a color stabilizer but reduces condensation rate.

While standalone PBAT does not provide enough properties (Table 10.1) for consumer acceptance, the addition of low-cost materials such as starch or materials with strong mechanical properties such as polylactic acid (PLA) can improve PBAT consumer standards. The combination of these materials decreases the final price when compared to PBAT while increasing mechanical properties and maintaining the biodegradability of PBAT. However, the most important advantage of blending PBAT with PLA and starch is the material's ability to be processed using modern plastic processing equipment. This leads to PBAT-based products being used widely in film and packaging applications [1].

With a growing interest in biodegradable polymers like PBAT, research is being done to find ways to lower its high production costs and improve multiple thermomechanical properties. Limitations of some properties restrict the use of sustainable materials; otherwise, biodegradable polymers would be used more often to help the environment. Adding different compositions of monomers and the selective addition of natural fillers can act as alternatives to develop better PBAT polymers with a performance that could match or exceed common commodity plastics.

Biodegradable polymers can fall under two categories: natural and synthetic. Natural polymers are the most abundant on earth and are found in nature. Examples are cellulose, chitin/chitosan, and other essential macromolecules like proteins.

Table 10.1: Mechanical and thermal properties of PBAT [1].

Properties	Test method	PBAT
Tensile strength (MPa)	ASTM D638	21–36
Elongation at break (%)	ASTM D638	650–700
Flexural strength (MPa)	ASTM D790	7.5
Young's modulus (MPa)	ASTM D790	20–35
Flexural modulus (MPa)	ASTM D790	126
Melting point (°C)	DSC	115–125
Crystallization point (°C)	DSC	60
Heat distortion temperature (°C)	DSC	55
Glass transition temperature (°C)	DSC	−30
Melt flow index	ASTM D1238	4.0
Specific gravity	ASTM D792	1.22

PBAT is a synthetic thermoplastic polymer, based on fossil resources, and is known for being completely biodegradable with high elongation at break and high flexibility. Specifically, it can be produced by a polycondensation reaction of 1,4-butanediol with both adipic and terephthalic acids or butylene adipate. Zinc acetate can be used as a catalyst. Preparation of PBAT requires a long reaction time, a high vacuum, and a temperature typically higher than 190 °C. These conditions are required for condensation reactions to remove the lighter molecules, like water, to produce consistent properties [4].

Some of these properties are comparable to low-density polyethylene, making PBAT a very promising biodegradable material that could replace it in some industrial applications. However, its lower mechanical properties have limited its application range.

PBAT gets its biodegradability from the aliphatic section of the molecule chain. Mechanical properties come from the aromatic section. Compared to most biodegradable polyesters like PLA and poly(butylene-*co*-succinate), the mechanical properties of PBAT make it more flexible [4].

10.2 Biodegradation

Recently, the biodegradation behavior of PBAT with thermoplastic starch (TPS) of different compositions has been investigated [2]. This blend is commonly used in the food packaging industry. The addition of grafted-malleated PBAT (PBAT-g-MA) and maleic anhydride (MA) was investigated as coupling agents for mechanical properties, morphology, melt rheology, and biodegradability of PBAT/TPS blends. This work will focus primarily on the biodegradation of the blends [2].

Starch alone is a biopolymer obtained from renewable resources. In order to produce TPS, disruption of starch granules must occur as well as mixing water and any other plasticizer. In order to remain relevant in the manufacturing industry, TPS is commonly blended with hydrophobic polymers, such as polyesters. This is because TPS alone can rarely meet application specifications in terms of processing, mechanical properties, and durability due to high moisture sensitivity and high biodegradation tendencies. However, with PBAT being relatively hydrophobic and TPS being hydrophilic, this leads to poor interfacial adhesion between the two phases. The addition of a coupling agent (PBAT-g-MA or MA for this study) improves compatibility between the two phases, while also maintaining a good balance in mechanical properties.

In this study, different content levels of PBAT and TPS were examined. The TPS was firstly prepared by mixing natural starch with glycerol at a ratio of 80/20 wt% with additional 10 wt% of water added. This was extruded through a twin-screw extruder to create TPS [3]. The same twin-screw extruder was then used to produce all PBAT/TPS blends. The TPS/PBAT blends were made with TPS contents of 40, 50, and 60 wt%.

The biodegradation test was taken according to ISO 14855-2:2018 standards. A series of vessels containing compost and the polymeric material was prepared, along with a vessel containing only compost to be used as a baseline. The amount of CO_2 produced was measured using absorption columns that were measured periodically for the weight of carbon dioxide and water. The test was performed at 58 °C under a 10 mL/min air flow rate. Air was used to bubble through distilled water to maintain moisture content of greater than 50%. The following equation was used to calculate biodegradation as a percentage of polymer carbon mineralized [2, 3].

$$\text{Biodegradation (\%)} = \frac{(CO_2)_s - (CO_2)_c}{\text{ThCO}_2} * 100$$

Where, $(CO_2)_s$ is the amount of CO_2 produced in the sample and $(CO_2)_c$ is the amount of CO_2 produced in the control group. The ThCO_2 is known as the theoretical carbon dioxide, calculated with the following formula.

$$\text{ThCO}_2 = 44 * \frac{C_{\text{total}} M_{\text{total}}}{12}$$

Where, 44, C_{total}, M_{total}, and 12 are the amount of total carbon per total weight, total weight, the molar mass of CO_2, and the molar mass of carbon, respectively.

The results of the study are explained in terms of biodegradation percentage over a given number of days. For all samples, CO_2 production levels steadily increase, which was expected with the biodegradability properties of TPS and PBAT blends. Standalone TPS, PBAT, and cellulose (reference value) all reached a degree of biodegradability of at least 90%, indicating that all three materials are fully biodegradable. Examining the coupling agents, the addition of PBAT-g-MA caused a mineralization decrease of approximately 12 wt%, suggesting an overall decrease in carbon dioxide production in blends with TPS content between 40 and 60 wt%. This change in carbon dioxide production is occurring because the addition of PBAT-g-MA allows the TPS and PBAT to produce a more homogenous blend, generating an interfacial attraction between the two components that is more difficult to degrade by bacteria. When MA was used as a coupling agent, a degradation profile similar to that of cellulose and PBAT was observed along with a higher hydrophilic property. This indicates that the PBAT was more encapsulated in the TPS phase when compatibilized with MA rather than PBAT-*g*-MA [2].

10.3 Mechanical, thermal, and rheological properties of PBAT

Some strategies have been implemented to produce PBAT with more improved mechanical properties. The mechanical properties of this polymer strongly depend on monomer composition and molecular weight, specifically the amount of terephthalate content. The Young's modulus will increase with more terephthalate, while elongation at break will decrease with less terephthalate. The tensile strength increases with molecular weight while elongation at break decreases. The mechanical properties of PBAT can also be linked to process variables such as pressure and temperature since both shift the reaction toward the products that affect the molecular weight.

The addition of fillers is a helpful way to improve the properties of PBAT while reducing the overall material and production costs and keeping overall biodegradability. Composites are usually added by three different methods: *in-situ* polymerization, melt mixing, and solvent casting. Each of these methods has major advantages and disadvantages.

In-situ polymerization takes natural fillers and homogeneously disperses them into the solution containing the polymer monomers, leading to an efficient load transfer and ultimately improving the mechanical properties of the composite. However, the polymerization mechanism is complex with major factors involved such as pressure, temperature, monomers, etc. These can degrade the fillers during manufacturing. Fillers dispersed in the monomer can prevent the diffusion of smaller molecules, leading to a lower degree of polymerization [4].

Melt mixing, by extrusion or injection molding, is the most common method of preparing PBAT-based composites. As a high shear force method, dispersion and distribution are promoted on a large-scale production. It ultimately improves tensile strength and the modulus of elasticity. However, the hydrophobic matrix of PBAT may cause a hydrophilic effect which will negatively change the dispersion of fillers. With a high viscosity and a nonpolar chemical structure, the mechanical properties of the composite will decrease. To avoid these problems, composites using other biodegradable polymers can blend with PBAT [4].

Solvent casting shows material properties optimized since the long reaction time gives the particles enough time to interface and organize. This process generates a rigid 3D network. Unfortunately due to economic and environmental reasons, solvent casting is not an efficient way to prepare composites, especially compared to the other processing techniques [4].

PBAT-based composites can be combined with many different fillers like cellulose nanocrystals (CNC), montmorillonites (MMT), natural fibers, nanofibrillated cellulose (CNF), red mud, distillers dried grains with solubles, and even coffee grounds. However, the fillers have a weak compatibility with PBAT, due to it being a low polarity polymer, is a major issue. Surface modification of fillers is an effective way to improve wetting while providing better compatibility. Most of these fillers improve the mechanical and rheological properties of the composites by melt mixing [5].

An *in-situ* grafting polymerization of CNC into PBAT is an excellent method of reinforcing the material. This material is a more ideal choice as a filler because of its good biodegradability, light weight, large specific surface area, low cost, and abundance in nature. In particular, CNC possesses an almost perfect crystal structure with a high elastic modulus, making it a good mechanical reinforcement for many biodegradable polymers, including PBAT. However, the hydrophilic nature of CNC limits the dispersibility into the hydrophobic PBAT. Surface modification of CNC has been used to enhance the inter-compatibility and make dispersion more uniform. A good dispersion of nanomaterials in the matrix and a well-bonded interface between the inclusions and polymer chains are necessary to facilitate the stress transfer and to achieve high reinforcement [5].

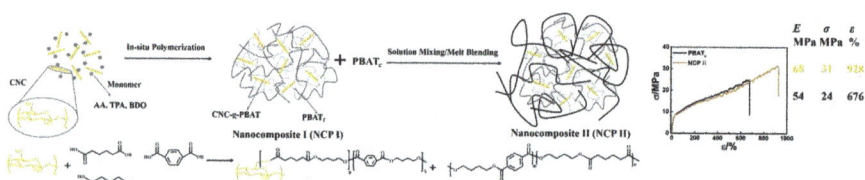

Figure 10.2: Preparation of several CNC/PBAT nanocomposites. AA is adipic acid, TPA is terephthalate, and BDO is 1,4-butanediol. PBAT$_c$ is commercial PBAT, PBAT$_f$ is ungrafted (free), and CNC-*g*-PBAT is CNC grafted by PBAT [5].

With a low usage of *in-situ* polymerized CNC content, Young's modulus, tensile strength, elongation at break, and toughness are all simultaneously enhanced in PBAT [5]. The efficiency of this reinforcement is one of the highest among similar biodegradable polymer nanocomposites. The use of CNC within the PBAT matrix increases the crystallinity of the polymer matrix, leading to its strengthening and toughening effect. This filler is greatly beneficial to producing high-performance biodegradable polymer nanocomposites, allowing for more uses in applications.

4-phenylbutyl isocyanate-modified CNC was used in the solvent casting method and was reported to increase the elastic modulus and the tensile strength of the nanocomposites when compared to the base composites [5]. Similar results were observed for nanocomposite materials using the melt mixing method of PBAT and modified CNC by acetic anhydride. This improves the thermal stability and the mechanical properties of the nanocomposite with a higher melt elasticity, complex viscosity, and storage modulus after the addition of bio-nanofillers into the PBAT matrix.

PBAT-based nanocomposites mixed with CNF exhibited improvements in the storage modulus (G') and the dynamic viscosity (η^*). Thermal and mechanical properties of clay nanocomposites containing unmodified and organically modified MMT through melt blending with PBAT also were greatly improved. Unmodified and modified clay nanoparticles such as sepiolite, MMTs, and fluorohectorites in the PBAT matrix increased the thermal stability, whereas the higher elastic modulus and hardness were related to the reinforcing effect of nanomaterials. Sepiolite particles were able to promote polymer crystallization of PBAT, improving crystallinity while the layered silicates slightly hindered the crystallization to a small degree [5].

Since PBAT is a semicrystalline material, developing semicrystalline polymers requires knowledge of the crystallization behaviors and thermal properties of PBAT. Moderate crystallinity and good thermal stability can make processing the semicrystalline polymers much easier.

The crystallization and melting behaviors of PBAT were investigated through a differential scanning calorimeter (DSC) with a cooling and heating rate of 10 °C/min. The thermal stability of PBAT was analyzed by thermogravimetric analysis (TGA) with a heating rate of 20 °C/min under nitrogen conditions. The thermal data results are summarized in Table 10.1 above. PBAT melts with a broad peak at about 123 °C and crystallizes with a hollow peak at about 60 °C. PBAT has a good crystallization and sensitive thermal stability. It still has great processing stability, enough to be used alone or blended with other materials through conventional manufacturing processes such as extrusion, injection molding, and blow molding.

Naturally biodegradable polymers or nanocomposites can also reinforce the mechanical and thermal properties of PBAT with fillers. Examples include layered silicates, layered double hydroxides, and cellulose nanocrystals. A more prominent example would be sustainable nanochitin, which can be introduced into PBAT without a compatibilizer through melt blending and compression molding. Chitin is most commonly used to enhance the physical properties of polymers due to its low-cost, biocompatibility, biodegradability, and antibacterial properties [6].

Traditional approaches for the preparation of chitin in polymer composites are solvent casting and melt processing. Solvent casting achieves a uniform dispersion of chitin nanocrystals in the PBAT matrix. Melt mixing techniques offer a more economically viable synthetic strategy for large-scale composite manufacturing. The shear forces from processing assist with the dispersion, giving nanochitin particles a stronger adhesion to PBAT polymer chains. This adhesion restricts movement for the polymer chains but also prevents crack propagation.

Adding a small amount of nanochitin had a heterogeneous nucleation effect on the PBAT, promoting the formation of crystallites in the polymer matrix during the cooling of the melt. This improves thermal stability and increases the glass transition temperature (T_g). If too much nanochitin content is added, the particles tend to aggregate. This is because chitin is hydrophilic while PBAT is mildly hydrophobic [6].

Remarkable increases in the tensile strength and the elongation at break were found when PBAT was melt blended with nanochitin. Massive improvements in mechanical properties were made when compared to pure PBAT. However, this improvement to rigidity is only seen in low concentrations of nanochitin. At higher nanochitin concentrations, nanochitin agglomeration induces stress cracking in the PBAT composite, decreasing the elongation at break and overall toughness.

A nanochitin composite with a high degree of crystallinity and excellent toughness is possible for PBAT. The nanochitin seems to be more efficient at improving PBAT with lower content added. Any higher concentrations were detrimental to the mechanical and thermal properties of the composite [6].

PBAT is known to have a more sensitive thermal stability, which can lead to degradation during processing. To help compensate for this, a chain extender called Joncryl can be incorporated into the reactive processing of PBAT. The epoxide groups of Joncryl can react with both the hydroxyl and the carboxyl groups of polyesters, resulting in the extending and balancing of their molecular chains.

Adding Joncryl during PBAT processing can significantly increase the molecular weight and intrinsic viscosity. This proves that the chains extend which can be considered recuperation for the decrease of the molecular weight during melt extrusion. The polymer-solvent interactions are decreasing, but the polymer-polymer interactions are increasing in strength [7].

As for PBAT's rheological behavior, viscosity decreases over time, indicating that thermal degradation does occur. This phenomenon is related to a decrease in the molecular weight and intrinsic viscosity. The complex viscosity gradually decreases with the angular frequency which is typical shear-thinning behavior. So adding small amounts of Joncryl can enhance the shear-thinning behavior of PBAT and consequently shift the Newtonian plateau to lower angular frequencies. The introduction of chain branches due to chain-extending reactions improves the melt elasticity. Storage modulus becomes less shear sensitive when chain extension/branching agent content is increased, showing a more cohesive network viscosity [7].

The use of Joncryl increases viscosity and storage modulus, and it requires more processing since high melt viscosity and high elasticity are required in the processes such as thermoforming and foaming. The activation energy in the modified PBAT is higher than pure PBAT. This can be attributed to the polar interactions (hydrogen bonding) between Joncryl and the polymer matrix which impedes polymer chain mobility. This clearly indicates that the linear structure of PBAT has not only extended but also branched [7].

Incorporating a chain extension/branching agent, such as Joncryl, by reaction extrusion into PBAT shows an improvement to its thermal stability. This property is strongly dependent on the reaction time because of the competing branching and degradation reactions. The epoxy reactive functions have been successfully used to increase the molar mass, intrinsic viscosity, shear-thinning, and elasticity during melt processing due to forming branching chains [7].

Overall, improving the mechanical and thermal properties of PBAT-based composites is best done by introducing fillers, chain reinforcement agents, or biodegradable polymers to blend with. Fillers can act as nucleating agents, affecting the crystallinity of the PBAT. The uniform filler dispersion in PBAT and the improved interaction between the filler and polymer matrix will allow better dissipation of energy throughout the polymer matrix. Reinforcing agents can extend or branch the polymer chains of PBAT, enhancing its weak thermal stability and increasing various properties. Biodegradable polymers tend to improve both the mechanical and thermal properties if fillers are added in low concentrations, retaining full biodegradability. More details about improving the thermal properties of PBAT will be discussed in the next section.

10.4 PBAT thermal degradation

PBAT can be processed at high temperatures and high shear rates. However, PBAT is highly susceptible to thermal degradation during processing, making it costly to produce and hard to compete with conventional polymers. In this specific study, the thermostabilization of PBAT is conducted with two different types of stabilizers using a torque rheometer at 60 rpm at two temperature levels. The stabilizers were used as masterbatches with a weight percentage of 10% of the additives in the PBAT. The molecular weight, torque values after 10 min of mixing, and the absorbance at 400 nm were used to understand the stabilization process. The primary and secondary antioxidants used both had a positive effect at the processing temperatures of 180 °C and 200 °C. These results indicate that the antioxidants can be used to protect PBAT against thermal degradation reactions, eliminating one of the main issues associated with this material.

The highest concern of biodegradable polymers is the amount of time required for biodegradation and bioassimilation during the degradation process. This process is defined by the attack of the microorganisms at the ester links that enable fast fragmentation. These links are sensitive to the degradation process caused from high temperatures and shear rates, as well as hydrolysis due to the large amount of moisture present. This degradation occurs mainly during processing since PBAT is worked at high temperatures and shear rates. The degradation of PBAT during processing is based on the hydrolysis of ester linkage, main-chain scissions, and β-C-H hydrogen transfer. Stabilizers are very important to keep the physical properties of PBAT post processing. The objective is to evaluate the thermomechanical stabilization of PBAT through the use of antioxidants [8].

Commercial PBAT was used as an experiment. The primary stabilizer used was Irganox 1010 (*P*, 0.4 w/w%) and the secondary was Irgafos 168 (*S*, 0.5 w/w%). Since the concentration of additives is very low, masterbatches were prepared with a 10 wt% of additives in the PBAT matrix. Then, a mixture of virgin PBAT and a masterbatch was made to get the desired concentrations. The concentrates, then, were prepared by being carried out in a mechanical mixer and then cut to obtain small sizes. The thermomechanical degradation was performed on a torque rheometer at 60 rpm for 10 min. Two levels of the temperature inside the rheometer were used, 180 °C and 200 °C. PBAT and PBAT/masterbatch were dried in an oven for 1 h at 70 °C before processing. The evaluation of the stabilization process was done by analyzing the torque value after 10 min with size exclusion chromatography (SEC) and UV-visible spectroscopy. The SEC tests were run in a Viscoteck with a series of columns at 40 °C and a refractive index detector.

Table 10.2 below presents the average values of M_n and M_w, polydispersity (PD), torque after 10 min, and the moisture absorbance at 400 nm. Virgin PBAT was used as a control group for comparison with the PBAT and the additives processed on the rheometer for 10 min under 180 °C and 200 °C. The molecular weight data in Table 10.2 indicates that a higher temperature has a larger influence on the thermal degradation of PBAT as there is a higher drop in molecular weight values for samples without stabilizers processed at 200 °C when compared to samples processed at 180 °C.

At the same time, the value of PD had increased. This shows that the process involved in the thermal degradation of PBAT is controlled by the scission chain reactions. This is accurate because the rheometric data since the value of the torque at 10 min has decreased greatly at 200 °C. For additives at 180 °C, there is no major

Table 10.2: Molecular weight and torque data of the PBAT samples processed with and without stabilizers at 180 °C and 200 °C [8].

Sample/Temperature	$\overline{M}_n$ (g/mol)	$\overline{M}_w$ (g/mol)	PD	T_{10min} (N.m)	Abs$_{400nm}$ (u.a.)
Neat PBAT	40.600	84.400	2.08	–	0.00829
PBAT/180 °C	36.450	76.950	2.11	1.8	0.03297
PBAT/200 °C	33.250	70.200	2.11	1.5	0.03928
PBAT + P/180 °C	40.500	80.000	1.97	4.5	0.02516
PBAT + S/180 °C	39.750	80.900	2.03	4.1	0.01699
PBAT + PS/180 °C	39.700	79.100	1.99	4.0	0.02539
PBAT + P/200 °C	40.100	78.300	1.95	2.6	0.02216
PBAT + S/200 °C	36.850	74.000	2.01	2.5	0.02516
PBAT + PS/200 °C	38.300	74.700	1.95	2.0	0.02321

difference between the primary and secondary antioxidants when taking into account only molecular weight and torque data. Molecular weight values remained practically unchanged for 180 °C when compared with virgin PBAT. Torque values at 10 min were similar for all three compositions [8].

For adding both the primary and secondary antioxidants, the molecular weight data for the sample additives processed at 200 °C showed a positive effect. There is no trend with PD because the scissions and crosslinking reactions competed during the thermal degradation process of PBAT. Torque values at 10 min are lower than samples processed at 180 °C and higher than samples without additives. Higher temperatures had stronger thermal degradation effects in the PBAT even with additives.

Absorbance of the polymer solutions at 400 nm is used to describe two types of phenomena, an increase of chromophore groups or a high dispersion of the light from insoluble fragments in polymers during crosslinking reactions. At both temperatures, the presence of additives decreased the overall absorbance, showing the positive results of adding the stabilizers that help prevent thermal degradation. The highest decrease of Abs_{400nm} is for the sample processing at 180 °C, in the presence of the secondary stabilizer. The additives could prevent the crosslinking reactions at this temperature but not the chain scissions since the drop in the molecular weight is the same for the processing of the samples with the primary stabilizer, alone or in combination with the secondary one.

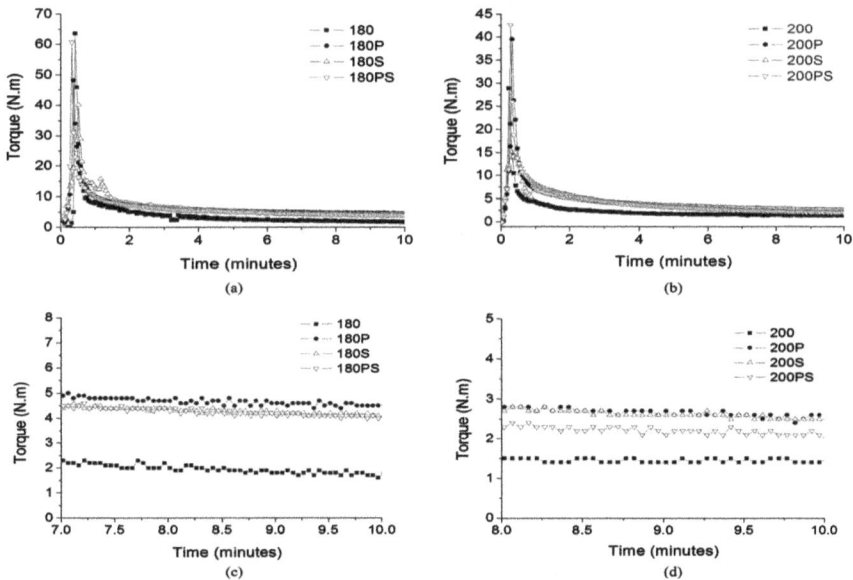

Figure 10.3: The rheometer torque curves of PBAT processed at 180 °C (a) and 200 °C (b) with and without the stabilizers, P = primary antioxidant, S = secondary oxidant, and PS = both oxidants. Detailed curves are around 10 min for 180 °C (c) and 200 °C (d) [8].

Figure 10.3 below shows the rheometer torque curves of the PBAT processed at 180 °C and 200 °C with and without the stabilizers (Figure 10.3). As shown, all compositions with the stabilizers at both temperatures led to torque values above PBAT without the stabilizers. The stabilizers were stronger for the lower temperature of 180 °C, especially the primary oxidant. The primary stabilizers act when the free radicals deactivate. The secondary stabilizers act on hydroperoxide decomposition. It is indicated that the thermostabilization of PBAT is more driven by the deactivation of the free radicals than the decomposition of hydroperoxides [8].

Figure 10.4 displays the molecular weight distribution curves of virgin PBAT, PBAT processed at 180 °C, and PBAT processed at and 200 °C, all with and without the stabilizers. At both temperatures, it is proven that the molecular weight curve is more displaced by a lower molecular weight when the polymer is processed with no stabilizers. The best stabilizer effect is obtained by the presence of the primary antioxidant at both temperature conditions. These results are in agreement with the previous results.

In conclusion, the thermostabilization of PBAT was experimented on with a torque rheometer and a primary or secondary antioxidant. The molecular weight, torque values after 10 min of mixing, and the absorbance at 400 nm were used to evaluate the stabilization process. The results showed that these two types of stabilizers worked well together or alone. The concentration used with the primary antioxidant is the best choice to stabilize PBAT during processing. But better control of the drying process of the samples must be performed to avoid any hydrolysis effects, which cannot be prevented by the stabilizers. The torque rheometer technique proved to be able to generate results in a simple way, with indicators for the best stabilizer choice.

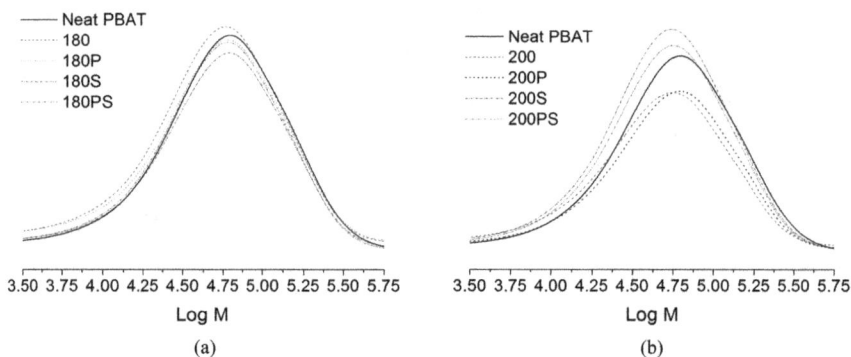

Figure 10.4: The molecular weight distribution curves of virgin (neat) PBAT and PBAT processed at 180 °C (a) and 200 °C (b) with and without the stabilizers. P = primary antioxidant, S = secondary oxidant, and PS = both oxidants [8].

10.4.1 PBAT blends (PLA)

PLA is aliphatic polyester synthesized by the ring-opening polymerization of lactides which are the cyclic dimers of lactic acids and are derived from corn starch fermentation. PLA is considered environmentally biodegradable because of its polyester chains having a high molecular weight and can hydrolyze to a lower molecular weight under certain temperature and moisture environments. Microorganisms in these environments convert these lower molecular weight components into natural resources like carbon dioxide, water, and humus soil. Standard PLA has a high modulus of 3 GPa and a strength of 50–70 MPa. However, its low toughness and aging issues prevent its application in the medical and consumer fields. It is also very brittle with strains breaking at 3.8%. Blending PLA with another material is a more practical and economical approach. This can substantially modify the mechanical and thermal properties, degradation rate, and permeability. Many of these blends are immiscible or only partially miscible and may need compatibilizers to increase their compatibility [9]. Commercial PLA (Natureworks PLA 4032D, Natureworks) has a density of 1.25 g/cm^3, a weight-average molecular weight of 207 kDa, a PD of 1.74 (gel permeation chromatography [GPC] analysis), a T_g of 66 °C, and a T_m of 160 °C [9].

Commercial PBAT (Ecoflex F BX 7011, BASF Corp.) has a density of 1.26 g/cm^3, a weight-average molecular weight of 145 kDa, a PD of 2.40 (GPC analysis), a T_g of –29 °C, and a T_m of 115 °C through a DSC analysis and a dynamic mechanical analysis (DMA). With its high toughness and biodegradability (strain at break ~710%), PBAT is a good material choice to toughen PLA. The idea is to increase the weaker properties while retaining overall biodegradability [9]. PLA and PBAT are both biodegradable thermoplastics that are commonly blended together. PLA has a high strength and modulus but relatively brittle, while PBAT is more flexible and tough. Blending these materials together can enhance both their properties without compromising the biodegradability of the blend. This section will focus on the extrusion and injection molding process and will discuss each material's basic properties as well as the resulting properties of the blends.

In this study, PLA and PBAT were melt-blended together using a twin-screw extruder. Melt elasticity and viscosity of the blends increased with more PBAT in the blend [9]. Crystallization of the PLA component, phase morphology of the blend, mechanical properties, and toughening were also investigated. The blend was an immiscible, two-phase system with the PBAT evenly dispersed within the PLA matrix. The PBAT component accelerated the crystallization rate of PLA but had little effect on its final degree of crystallinity. With the increase in PBAT content (5–20 wt%), the blend showed a small decrease in tensile strength and modulus; however, elongation and toughness were significantly increased in comparison. With the addition of PBAT, the failure mode changed from a brittle fracture of pure PLA to a ductile fracture of the blend as demonstrated by the tensile tests and (*scanning electron microscope* SEM) micrographs. Debonding between the PLA and PBAT domains induces large plastic deformation in PLA matrix ligaments [9].

A co-rotating twin-screw extruder (Leistritz ZSE-18) was used to blend the base materials. The PLA/PBAT blends studied contained 5, 10, 15, and 20 wt% PBAT and were compared to the base materials as a control. The thermal properties were studied by a DMA and a DSC. The cold crystallization morphology of PLA and PLA/PBAT blends was studied using a polarized optical microscope. The standard tensile (Pleas D638, Type III) and Izod impact (American Society for Testing Materials [ASTM] D256) test samples were prepared by injection molding. Tensile testing was performed using a universal testing machine (Instron 4466) equipped with a 10 kN electronic load cell. The Izod impact test was performed using a standard impact tester (TMI 43-1). All the tests were carried out according to the ASTM standard, and five replicates were tested for each SEM. The dynamic rheological properties of the PLA/PBAT blends were assessed using a strain-controlled rheometer. A strain sweep test was initially conducted to determine the linear viscoelastic region of the materials. A dynamic frequency sweep test was subsequently performed to determine the dynamic properties of the blends [9].

PBAT has a higher steady shear viscosity than PLA. Adding PBAT resulted in a steady increase in the viscosity of the blends. PLA has a longer Newtonian region than PBAT, so adding more PBAT reduced the Newtonian region of the blends. The steady shear viscosities of the PLA/PBAT blends are shown in Figure 10.5. Although the rheology data indicated a higher viscosity in PBAT than in PLA at 180 °C, the addition of PBAT increased the processability of PLA in extrusion. This is because PBAT has a lower melting point, which allows it to melt first in the barrel during extrusion and act as a lubricant for the

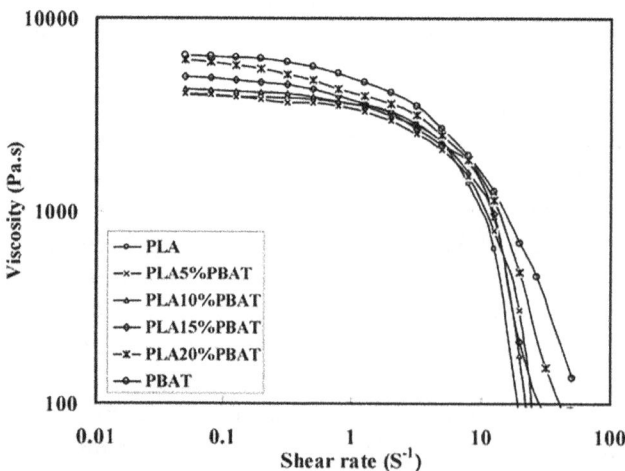

Figure 10.5: Steady shear viscosities of PLA, PBAT, and their blends [9].

PLA's rigid pellets. This reduced overall extrusion pressure, but a high injection pressure would be needed if using injection molding since PBAT has a higher viscosity.

DSC results show that each polymer on its own had one T_g while the blends showed two T_{gs}, one for PBAT and one for PLA. The T_{gs} stayed unchanged with varying PBAT concentrations, indicating a lack of significant molecular interactions between PLA and PBAT.

PLA has a cold crystallization temperature, adding PBAT decreased this temperature and narrowed the peak width, indicating an enhanced crystallinity in the PLA but still staying amorphous. The blend started to crystallize at a lower temperature than the virgin PLA.

In the DSC data obtained at 5 °C/min, virgin PLA had a melting peak with a shoulder. Adding PBAT clearly separated the melting peak and the shoulder of virgin PLA into two individual peaks, suggesting the presence of a new crystalline structure induced by the PBAT.

When the 10 °C/min heating rate was used, the DSC thermogram was different. The melting shoulder from the virgin PLA at 5 °C/min was gone. This difference could be because of the shorter time required for PLA crystals to reorganize at 10 °C/min. But like 5 °C/min, adding PBAT enhanced the cold crystallization of virgin PLA, proven by the narrow cold crystallization peaks and lower peak temperatures.

The fracture behavior of the specimen in the tensile tests changed from a brittle fracture of the PLA to a more ductile fracture with the blends (Figure 10.6). This is shown in the tensile stress and extension curves below in Figure 10.7. The PLA had a distinct yield point with neck instability failure, and its strain at break was only 3.7%. However, all the blends showed distinct yielding and stable neck growth through cold drawing. Even at 5 wt% of PBAT, the elongation of the blend was significantly increased by more than 200%, and elongation still continued to increase with the PBAT content. The tensile strength and the modulus of the PLA/PBAT blends decreased with more PBAT content. The tensile strength decreased by 25% (63 MPa on virgin PLA to 47 MPa on a

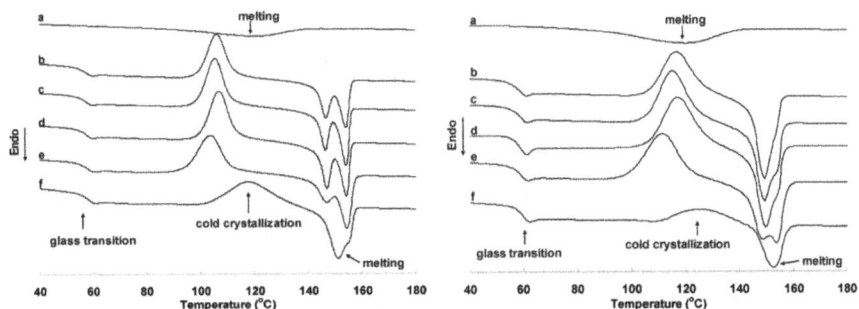

Figure 10.6: The DSC heating curves of PLA, PBAT, and the PLA/PBAT blends after crystallization from the melt. The top graph is for a heating rate of 5 °C/min and the bottom is for 10 °C/min. They are both ordered as PBAT (a); 20% PBAT (b); 15% PBAT (c); 10% PBAT (d); 5% PBAT (e); and PLA (f) [9].

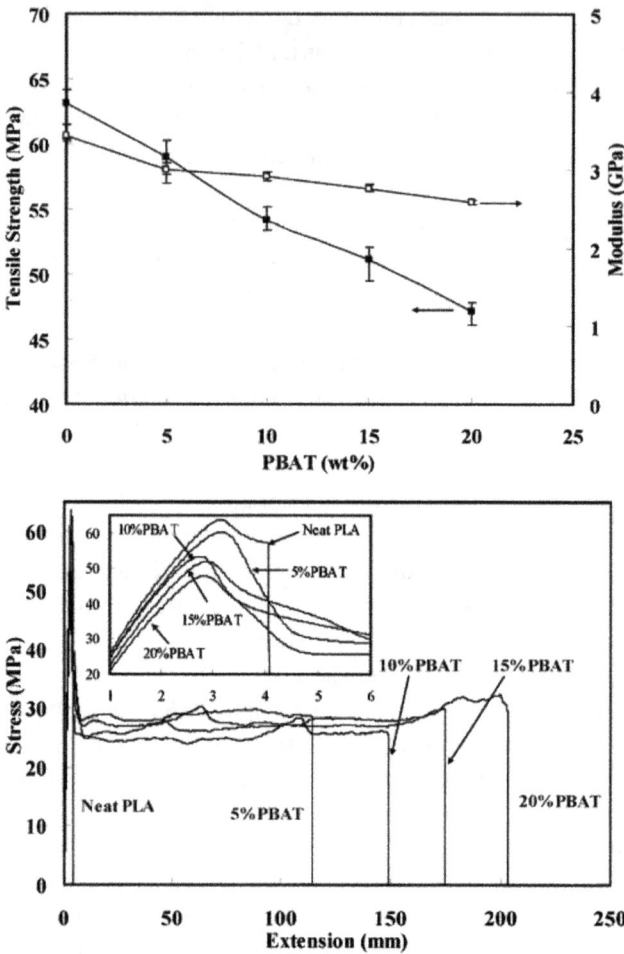

Figure 10.7: The tensile stress (top) and extension (bottom) curves [9].

20% PBAT composite) and the modulus decreased by 24% (3.4 GPa on virgin PLA to 2.6 GPa on a 20% PBAT composite). This outcome was expected as PBAT has a lower modulus and tensile strength than PLA [9].

Impact strength was also increased from 2.6 GPa on virgin PLA to 4.4 GPa on a 20% PBAT composite, as shown in Figure 10.8 below. SEM scans on the specimen surfaces show evidence of ductile fractures since longer fibers can be seen as the PBAT content increases. Crazing, cracks, and shear yielding have been identified as how energy dissipates in the impact fracture of the toughened polymer.

Previous studies have determined that the viscosity ratio of the dispersed polymer to the matrix polymer is a critical variable when producing a desired blend's properties. In the study observed, PBAT materials with varying melt viscosities were prepared.

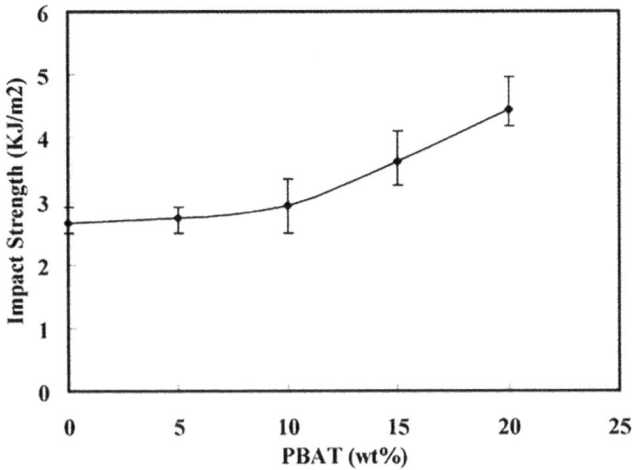

Figure 10.8: Impact strength as more PBAT is added to PLA in the blends [9].

Table 10.3: Flow testing of PBAT/PLA blends.

Samples	MFI (g/10 min)	η^* (10 rad/s, Pa.s)	Viscosity ratio of PBAT to PLA
PBAT-1	90.8	112.6	0.23
PBAT-2	49.3	194.3	0.40
PBAT-3	27.4	373.5	0.77
PBAT-4	7.68	947.7	1.95
PLA	28.8	485.9	–

Also, PLA/PBAT blends (70/30 wt) with different viscosity ratios were compounded in a counter-rotating twin-screw extruder under constant processing conditions. The study investigated the morphology, mechanical, thermal, and rheological properties of the blends. Properties of each blend analyzed are shown in Table 10.3.

The screw torque was evaluated during the compounding of the polymers. As shown in Figure 10.9, the torque peaked in the initial 40 s that the material was introduced to the screw. This was attributed to the melting of the PBAT pellets. DCP (dicumyl peroxide) was also added during extrusion to initiate chain extension of the PBAT. A dramatic increase in the torque was also shown with an increase in contents of DCP added, indicating the production of the molecular weight or long-chain branching. The torque values decreased for both blends that incorporated DCP after reaching a mixture time of 150 s, speculating thermal degradation of the polymer blend.

Viscosity and melt flow index (MFI) are common parameters used to aid in the understanding of how the polymer can be processed. These values depend greatly on the molecular weight and chain structure of the polymer. Figure 10.10 shows the viscosity and MFI for different blends. As indicated by the viscosity plot, with an increase in DCP

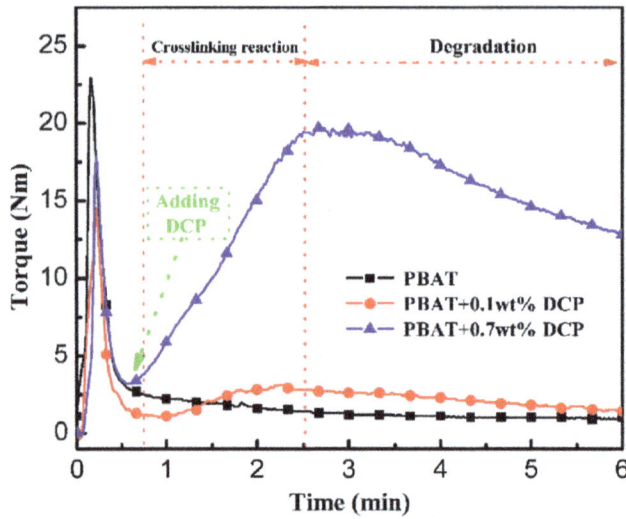

Figure 10.9: Torque evaluation of PBAT blends under constant processing conditions.

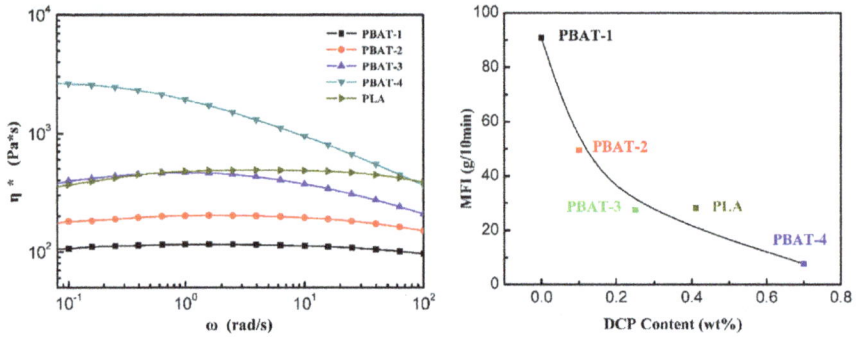

Figure 10.10: (a) Complex viscosity plot and (b) Melt flow index plot of PBAT blends.

content, the viscosity of the PBAT will increase. In accordance with the entanglement theory, the DCP intensified the polymer's ability to produce long-chain branching or introduce a cross-linking network of polymer chains. This result indicates that DCP produces favorable cross-linking properties when used as a chain-extender in PBAT blends. The MFI plot shows an agreement with the concept of DCP being a strong cross-linking agent [13].

In another study, PLA and PBAT were melt mixed at a ratio of 40:60 and extruded and cast into a film. Then, these film samples were buried in soil. The degraded samples were taken regularly from the soil and analyzed through a DSC over 4 months as shown in Figure 10.11 below [10].

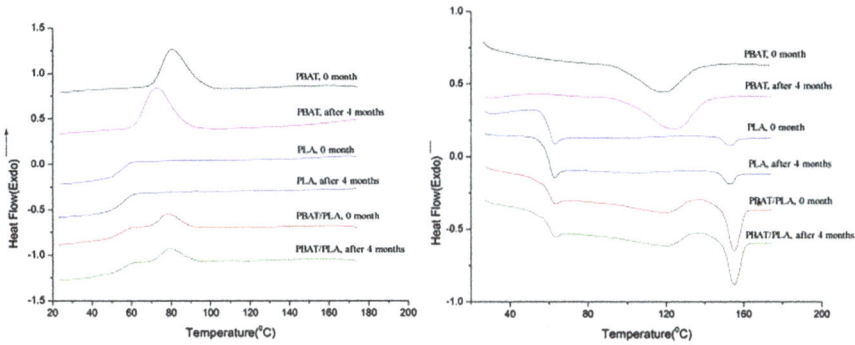

Figure 10.11: DSC results of PLA, PBAT, and the 40:60 blend after 4 months [10].

The analysis showed that the PLA and the PBAT in the blends had different biodegradation mechanisms. After biodegradation, the carbon atom content in the molecular structure of the samples decreased, while the oxygen atom content increased, indicating that all the samples did still degrade. The biodegradation rate of the PLA/PBAT blends was not the same as were for the materials on their own [10].

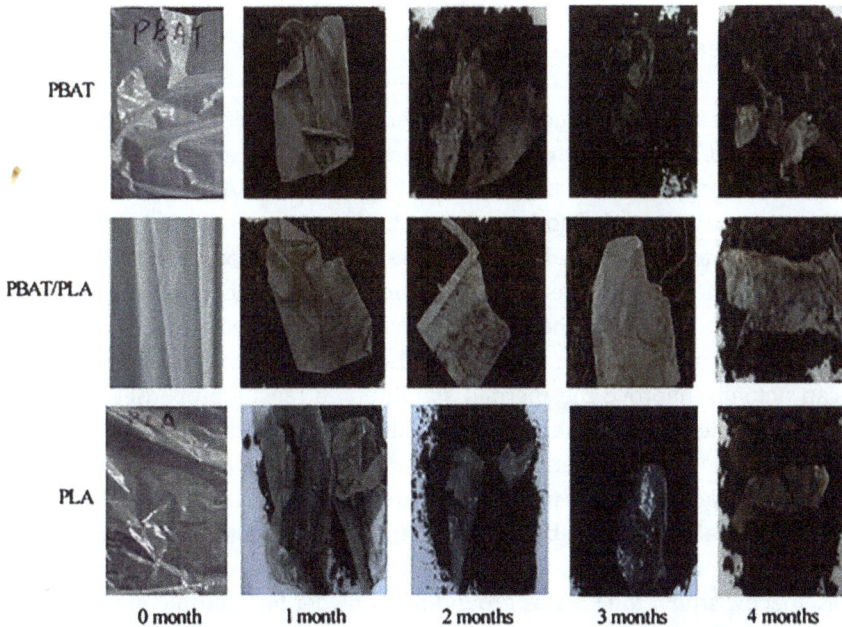

Figure 10.12: Degradation rates after being buried in soil [10].

There are different degradation rates for PLA, PBAT, and the blends as shown below in Figure 10.12. We can see from the DSC curves above that the melt temperature of PBAT after degradation slightly decreased, while the melt temperature of PLA after degradation slightly increased. The melt temperature and melt point changes of the various components in the blend basically followed the changing process of the respective single polymers [10].

In conclusion, PLA and PBAT were blended together using a twin-screw extruder and turned into pellets meant for injection molding. Rheological data revealed that PBAT has a higher melt elasticity and a higher viscosity than PLA. These same properties increased when more PBAT was introduced into the blends. PBAT also increased the processability of PLA by acting as a lubricant during extrusion since the materials were dispersed in each other evenly. DMA and DSC results show that this blend is immiscible. Adding PBAT accelerated the crystallization rate of PLA but had little effect on the degree of crystallinity. The semi-crystalline PLA experienced extensive cold crystallization, making it amorphous in the molded products after processing, this is common in blends. Even with just 5 wt% of PBAT, the toughness of the PLA was greatly increased without any major losses in the tensile strength or modulus. The impact strength of the blend was also greatly improved, changing from a brittle failure to a ductile failure after adding PBAT. The biodegradation rates of the blends are slightly longer than the materials on their own, but the overall biodegradation process has not been compromised.

10.4.2 PBAT blends (corn stovers)

PBAT can also be blended with more natural materials such as corn stovers (CS). These stovers are the by-product of corn grown on farms, and they are typically burned. Factors such as the particle size and amount of CS modify PBAT in multiple ways. Adding CS improves the modulus and accelerates the crystallization process, but it also decreases the thermal stability and increases the moisture absorption of PBAT. The strength and toughness also decrease as the morphology of the blend is negatively affected. This can be avoided if the particle size of the CS is reduced by grinding it down, allowing for more uniform dispersion of the CS throughout the PBAT. This method can actually improve the thermal stability and tensile properties of PBAT. As already mentioned above, the PBAT is flexible, biodegradable, and has good mechanical properties, but its thermal stability is low and that can restrict its application uses. CS can act as organic filler that improves the functionality of PBAT while retaining biodegradability. CS is abundant, cheap, and is completely biodegradable, making it a solid choice to blend with PBAT as it can also reduce agricultural waste.

In this specific study, the CS was ground into powder for multiple meshes to be dispersed through the PBAT. Both materials were dried for 12 h at 80 °C and then melt-blended together to avoid chemical treatment and reducing production costs. The machines used for this study began with an S3500 laser particle size analyzer to make the mesh sizes of 25, 100, and 200. The composites were then blended by a 50EHT 3Z "Plastograph" Mixer at 160 °C for 6 min with a screw speed of 100 rpm. The blends were then made by compression molding with an LP-S-50 Vulcanizing Press at 130 °C, making 2 mm thick sheets. The number of parts per hundred of resin by weight (phr) measured the amount of CS in each composite tested which were 1, 5, 15, and 30 [11].

For a moisture absorption test, the blend samples were 10 × 10 mm × 0.7 mm. They, then, were dried in an oven at 50 °C for 24 h to get a constant weight (W_1) and then submerged into purified water at 25 °C for 24 h and then weighed again (W_2). Five specimen of each sample were tested to get the averages. The water absorption (W_w) can be calculated according to the following equation:

$$W_w\% = \frac{W_2 - W_1}{W_1} \, x \, 100$$

The morphology of the PBAT/CS blends was analyzed by SEM scans on the cryo-fractured surface with the results shown below in Figure 10.13. PBAT exhibited great toughness with only a few cracks from ductile fractures on the surface, observed in the PBAT control group (a). With CS added in, the fracture surface became rougher and showed clear cracks. In the second image, there is a honeycomb structure with a few voids between the PBAT and the CS, showing poor adhesion of the polymer matrix to the larger particles (b). These large cavities will act like defects and cause stress concentrations, decreasing mechanical performance for the tensile test. However, the smaller CS particles in the remaining images (c–g) disperse well into the PBAT without many particles clumping together. These agglomerations happen because CS is hydrophilic, so it is easy to aggregate water. The voids between phases disappear, and the particles are embedded into the polymer matrix since they are smaller and more uniform [11].

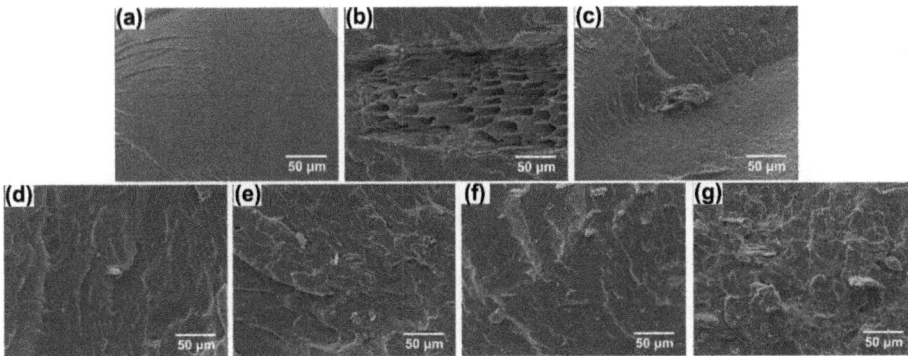

Figure 10.13: (a) PBAT; (b) 25M-5CS/PBAT; (c) 100M-5CS/PBAT; (d) 200M-1CS/PBAT; (e) 200M-5CS/PBAT; (f) 200M-15CS/PBAT; and (g) 200M-30CS/PBAT [11].

In the above images (d–g), adding more CS content made the surface textures rough. This indicates that the compatibility of PBAT and CS worsens with more CS content. This is because, PBAT is hydrophobic and cannot completely mix with CS particles since they are hydrophilic. Still, no major particle clumps show up on the surface of the PBAT/CS blends. The uniform dispersion of the CS into the PBAT can be attributed to the fact that the aromatic structure of the lignin in the CS has more complex forms. The trend seems to be that increasing the amount of CS content and increasing the particle size has the same negative impact on blend morphology.

The thermal stability of the PBAT/CS blends was investigated with TGA in a nitrogen atmosphere. PBAT starts to degrade at 358.3 °C and at 416.3 °C the rate of degradation is maximum. At 430 °C, the degradation rate slows down, and only 10% of the PBAT is unchanged. But for CS, moisture content is lost between 50 °C and 250 °C. 40% of CS is still remaining at 600 °C due to the slow decomposition of the lignin in the CS [11].

Two-step degradation does not occur during the decomposition process of the blends even though the temperature maximum of PBAT is almost 100 °C higher than CS, proving there is an even dispersion of CS particles into PBAT's matrix. By introducing the CS particles, the temperature of the PBAT material decreases because of the lower thermal stability, decomposing at nearly 200 °C. The thermal stability of the PBAT/CS blends is improved as the particle size decreases. This is also due to the lignin

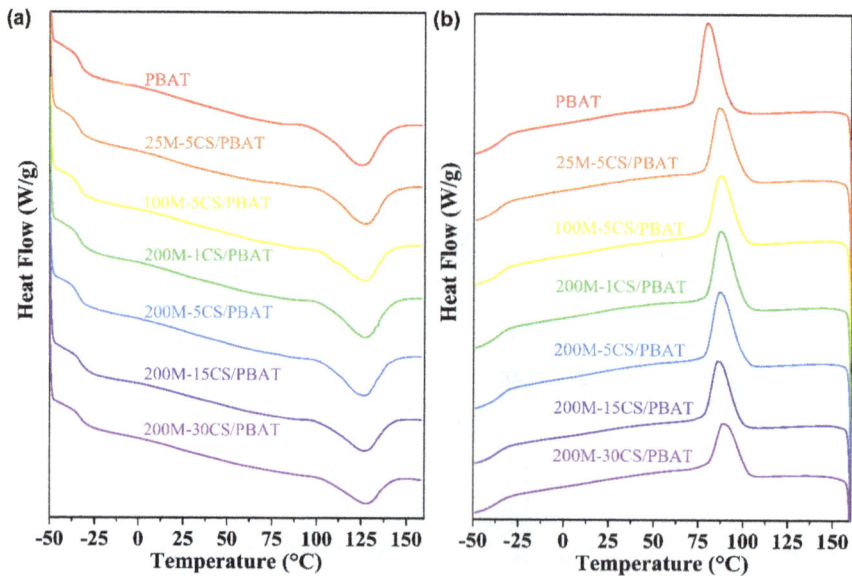

Figure 10.14: DSC heating (a) and cooling (b) curves for PBAT/CP blends [11].

in the CS, but the enhancement decreases as the lignin loading increases, so it can only be helpful in small concentrations.

The DSC heating and cooling curves of the PBAT/CS blends are shown below in Figure 10.14. Semi-crystalline PBAT has both amorphous and crystalline phases, so 2 phase transition regions will be found. Pure PBAT has a lower T_g, providing a strong impact resistance. The molecular chains of PBAT have both aliphatic and aromatic units, but only the aromaticunits have a melting process in the heating curve.

As CS is introduced, T_g is not super affected, but T_c increases while ΔH_m and ΔH_c decrease. Changing the particle size and CS content does not strongly alter T_c across the blends. The decrease of the ΔH_m and ΔH_c shows that the CS particles not only accelerate the crystallization process of PBAT, but they also prevent crystallite growth and decrease overall crystallinity. Negative effects from the CS particles on PBAT increase with more CS content. CS particles restrict the packing and reorganization of the molecular chains in PBAT. But with more CS, the slightly increased T_m of the PBAT/CS blends indicates that more uniform grains are produced than that of virgin PBAT.

Virgin PBAT is soft polyester with a high elongation at break and a low modulus. By adding a small amount of CS content, the modulus increases. This shows that the CS makes the PBAT stiffer, operating as rigid filler for the polymer. Making the CS particles smaller will further enhance the modulus with a more efficient stress transfer process. The tensile properties of the blends also improve with smaller CS particles. The mechanical properties of the PBAT/CS blends are shown to be very sensitive to CS particle size and the amount of CS content in the blend, even at low loads. Large particles become stress concentrators in the blend and can cause early failures. But the rigid CS fillers could restrict the chain mobility of PBAT's polymer matrix during the tensile process, causing a decreasing trend of plastic deformation. However, the CS particles still cause a negative effect on the tensile strength and the elongation at break, implying how sensitive PBAT is to impurities. Poor adhesion from the high amount of CS content leads to an undesirable stress transfer process. Although adding CS particles causes some decrease in PBAT's toughness, the CS does still give it a higher tensile modulus and yield strength, which will lower the production cost and give a wider use of applications.

Since PBAT is hydrophobic, it absorbs water of 0.69% its weight after being submerged in clean water for 24 h. The amount of moisture absorption increases to more than 1.00% in the blends due to the hydrophilic cellulose in the CS. Pulverization of it into a powder form reduces the CS size and makes the cellulose more exposed to water, but the size of the CS particles show no significant effect on the moisture absorption of the blends.

The CS content influences the moisture absorption of the blends by increasing gradually with the growth as more CS is added. Because of the weak interaction between the CS fillers and the PBAT polymer matrix, increasing CS particle sizes will directly lead to an increase in the amount of water absorption channels. Clearly, the hydrophilic traits of the cellulose in the CS dominate the water absorption of the composites instead of the molecular interactions between the cellulose and PBAT.

In conclusion of this section, the effects of CS particle size, and content affected the structure and properties of the PBAT in the blends of PBAT/CS. Larger CS particles have poor adhesion to PBAT's polymer matrix, but compatibility improves as the size and amount of content of CS decreases. Adding CS improves the crystallization of PBAT but restricts its growth with more CS content. Due to the uniform dispersion of CS, the blends with smaller and fewer CS particles have better thermal stability and better mechanical properties. Acting as rigid and hydrophilic filler, CS remarkably improves the stiffness of PBAT, but also slightly increases the amount of moisture absorption. Compared to virgin PBAT, the PBAT/CS blends still have the advantage of waste reduction, lower production costs, faster crystallization, and higher modulus. This increases the range of applications to single-use items like plastic shopping bags or mulch bags.

10.4.3 PBAT/TPS blends

Another way to help prevent plastic waste with PBAT is to blend it with a thermoplastic starch (TPS). All of these materials are eco-friendly, so the biodegradability of PBAT will not be negatively affected. However, an issue with PBAT is its weak thermal stability, meaning it can easily degrade under the wrong processing conditions. But with the addition of a 1% of chain extender additive, called Joncryl, blends with up to 30% of TPS can have this degradation process reverted to a degree, even when processing at higher temperatures.

PBAT is a synthetic semi-crystalline thermoplastic with aromatic-aliphatic copolyester. It has mechanical and thermal properties similar to certain polyethylene grades. It is biodegradable and compostable and can be processed using the conventional methods and conditions. But it can degrade between processing temperatures of 140–230 °C by hydrolysis of the ester bond due to humidity. As a result, the average molar mass decreases, limiting application uses.

The TPS is composed of carbon, hydrogen, and oxygen, and is obtained by the plasticization of starch, with water or glycerol heating up. Because of its limited application use on its own, it is often used in blends. Adding TPS to another biodegradable polymer is a common method to make a compostable product, thus making TPS a good choice to blend with PBAT. The issues with this kind of blend are the cost, compatibility, and water sensitivity with increasing TPS content. This can prevent common application uses such as packaging. TPS is a complex material composed of linear amylose (20–25%) and branched amylopectin (75–80%), and it has low molecular weight plasticizers. Crystallinity is between 20 and 40%, so TPS is considered a semi-crystalline material with amorphous regions [12].

The studies done on different PBAT/TPS blends indicate that the use of compatibilizers, particularly at high TPS contents, is necessary to achieve desirable properties. Several compounds such as soybean oil, citric, and tartaric acids, maleic anhydride, and glycidyl methacrylate can be used as compatibilizers to improve the thermal, mechanical, and biodegradable properties. Transesterification reactions, the transfer of the organic group R″ of an ester with the organic group R′ of alcohol, also occur during the melting process of the ternary PBAT/TPS blends.

In the experiment conducted, PBAT/TPS blends were processed at different temperatures in an internal mixer for 15 min. After 10 min of processing, the chain extender Joncryl PR010, also acting as a compatibilizer, was mixed into the blends during the last 5 min. It can extend the molecular chain because it is an oligomer with epoxy and methacrylate residues, which is compatible with the PBAT/TPS composite. The torque and temperature was monitored since the data gathered revolves around the torque recovery and polymer degradation.

The blends made were primarily composed of PBAT with TPS at 10, 20, and 30% of starch by weight, and they were made in a Haake Rheomix 3000 internal mixer at a speed of 60 rpm, kept at a constant temperature of 140, 170, and 200 °C. The total processing time was 15 min. PBAT and TPS were separately subjected to the same treatment to have a control group to compare to. 1% of Joncryl was added to all the materials after 10 min [12]. The results are shown in the following bar figures.

Figure 10.15 shows the adjusted torque of the PBAT and the blends in terms of processing temperature and composition, reflecting the viscosity of the melts. PBAT is

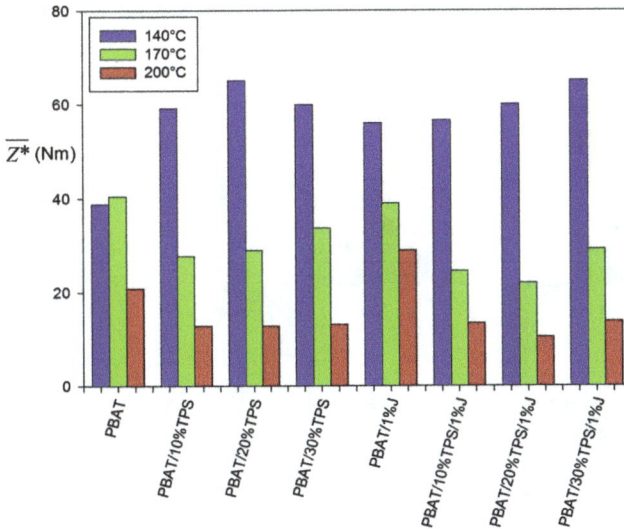

Figure 10.15: PBAT and all the blends with and without Joncryl, at three different temperatures [12].

shown to be more viscous than TPS. But the viscosity of the blends was higher than the viscosity of the components at low processing temperatures. The viscosity of the blends was intermediate between the components at higher temperatures. This effect is due to the ternary blend's viscosity on the individual component's viscosities, coupled with the significant difference in their temperature dependence. The viscosity of the blends is virtually independent of the composition and the additive Joncryl. Viscosity comparisons are not used to study the degradation process.

Figure 10.16 displays the higher rate of degradation in TPS as compared to PBAT, revealing that the degradation is independent of temperature at the lower processing temperatures of 140 °C and 170 °C. The chain extending additive, Joncryl, had a striking effect on virgin PBAT. The additive not only recovered the losses due to degradation but also extended the molecular chains of the PBAT. The degradation rate of the blends without the additive was higher than the rate of degradation of the components and increased with more TPS content especially at the higher processing temperature of 200 °C. Introducing Joncryl resulted in a partial temperature-dependent recovery. The effect of the additive is null and void at the lower processing temperatures but is very significant at the higher processing temperatures, where the recovery is significantly dependent on the amount of TPS content.

Joncryl is effective at increasing the molar mass of polyesters like PBAT, but not for polysaccharides like TPS. The addition of more TPS content also increased the degradation rate of PBAT. The combined effect is responsible for the uncommon pattern of recovery seen.

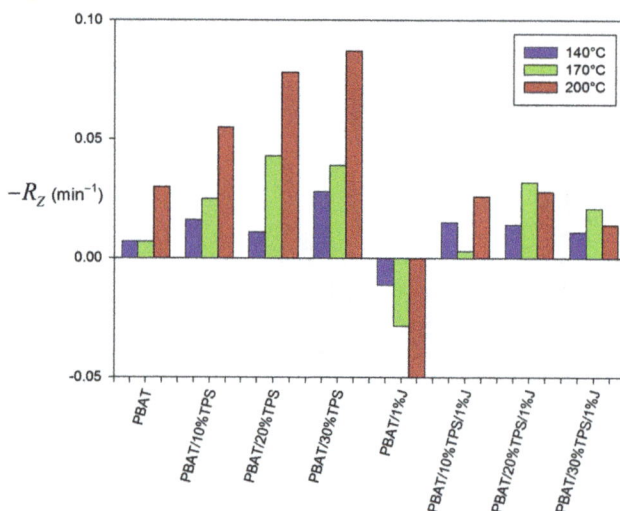

Figure 10.16: Rate of change of the PBAT and the blends with and without Joncryl, at three different temperatures. Negative values indicate that the molar mass increased due to chain extension [12].

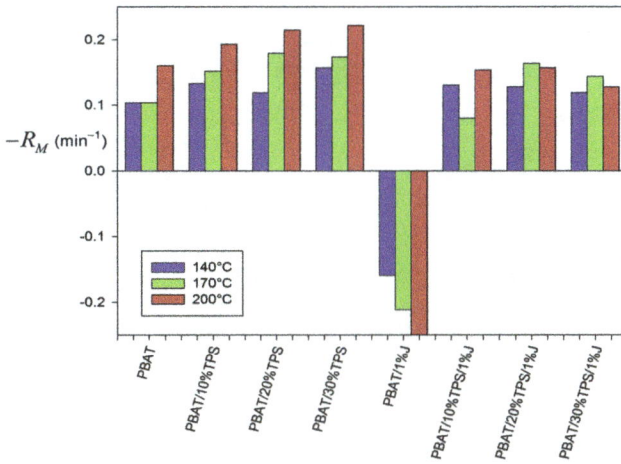

Figure 10.17: Rate of change of the molar mass during the processing of the PBAT and the blends with and without Joncryl, at three different temperatures [12].

Figure 10.17 shows how the molar mass decreases for the blends in terms of processing temperature and composition, in order to observe degradation and recovery. As expected, the estimated rate of change of molar mass ($-R_M$) trends follow the adjusted torque molar mass ($-R_Z$) trends exactly. However, the differences between the composites and the temperatures are flat due to the influence of the viscosity dependence on molar mass.

In this study, polymer degradation and recovery under processing was investigated in fully biodegradable blends. Viscosity comparisons were unable to reveal the temperature and composition dependence of polymer degradation under processing conditions. But it had a superior performance of relative rate of torque (viscosity) changes in the terminal stage of processing. TPS had a negative effect on the degradation of the PBAT/TPS blends and it has a significant temperature dependence. Not only did TPS degrade more than PBAT during processing but it also increased the degradation of PBAT. The chain extender additive Joncryl, at a 1% of concentration, was helpful for the degradation recovery in PBAT/TPS blends, particularly at higher processing temperatures. Joncryl was effective at increasing the molar mass of PBAT, but not as much for TPS. Though TPS increases degradation, it has a low processing cost, and including the right additives can make it an excellent material to blend PBAT with, especially since it retains biodegradability. PBAT can in turn make TPS tougher, so these two materials can improve upon each other in a blend.

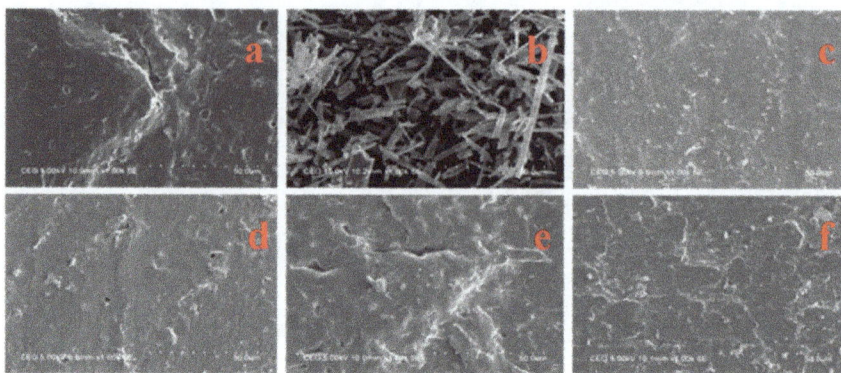

Figure 10.18: SEM micrographs of a) Neat PBAT, b) Wollastonite powder, c) PBAT/W 1% wt, d) PBAT/W 3% wt, e) PBAT/W 5% wt, and f) PBAT,W 7% wt.

10.5 Processing of PBAT/wollastonite biocomposites used in medical applications

Polymer morphology, mechanical properties, and dispersion quality of PBAT composites with 0–7 wt% wollastonite (W) have been investigated. W is an inorganic, calcium-silicate-based ceramic that is known to be highly biodegradable and is a material that is being strongly considered for use in bone tissue regeneration. When used as filler in polymers, W will enhance the bioactivity and mechanical properties and is known for its excellent biocompatibility in medically approved materials. This section will only consider the morphological studies of the PBAT when W is introduced.

Figure 10.18 displayed the SEM images of PBAT biocomposites with different W loading contents. In Figure 10.18c, e, it was observed that W contents of 1–5 wt% were uniformly dispersed throughout the PBAT polymer matrix with minimal agglomeration, indicating a strong interfacial adhesion. For samples with a W content of 7 wt%, dispersion and distribution of the filler were adequate, but W particles aggregated into small portions on the surface of the polymer matrix, as shown in Figure 10.18f. Generally, this uniform distribution and dispersion of the filler into the polymer matrix will promote the crystallinity of the polymer composite [14].

10.6 Conclusion

PBAT used as a blend component is continuing to grow in the polymer processing industry. This is mostly due to increased awareness of nonbiodegradable negative effects on the environment. Aliphatic groups within PBAT's chemical structure provide degradative balance in a wide variety of environments while aromatic groups provide strong compatibility with other materials, allowing the enhanced mechanical properties of the blend. With environmentally friendly materials being of utmost importance, PBAT is the ideal replacement for applications that require strong mechanical properties, while also exhibiting ease of biodegradability in a wide variety of environments. Good mechanical properties are acquired by blending biodegradable PBAT with materials such as PLA, starches, and other natural resources that can degrade naturally in a variety of environments. PBAT has been studied thoroughly in order to optimize the mechanical properties without sacrificing its ability to naturally degrade. This chapter explains the effect of biodegradability and property changes of PBAT when blended with different materials. The structure and mechanical properties have been explained and summarized in this book chapter.

Author contributions: All the authors have accepted responsibility for the entire content of this submitted manuscript and approved submission.
Research funding: None declared.
Conflict of interest statement: The authors declare no conflicts of interest regarding this article.

References

1. Jian J, Xiangbin Z, Xianbo H. An overview on synthesis, properties and applications of poly(butylene-adipate-co-terephthalate)-PBAT. Adv Ind Eng Polymer Res 2020;3:19–26.
2. Herrera R, Franco L, Rodriguez-Galan A, Puiggali J. Characterization and degradation behavior of poly(butylene adipate-co-terephthalate)s. J Polym Sci 2002;40:4141–57.
3. Kijchavengkul T, Auras R, Rubino M, Selke S, Ngouajio M, Fernandez T. Biodegradation and hydrolysis rate of aliphatic aromatic polyester. Polym Degrad Stabil 2010;95:2641–7.
4. Ferreira FV, Cividanes LS, Gouveia RF, Lona LMF. An overview on properties and applications of poly(butylene adipate-co-terephthalate)-PBAT based composites. Polym Eng Sci 2019;59:E7–15.
5. Lai L, Wang S, Li J, Liu P, Wu L, Wu H, et al. Stiffening, strengthening, and toughening of biodegradable poly(butylene adipate-co-terephthalate) with a low nanoinclusion usage. Carbohydr Polym 2020;247:116687.
6. Meng D, Xie J, Waterhouse GIN, Zhang K, Zhao Q, Wang S, et al. Biodegradable Poly(butylene adipate-co-terephthalate) composites reinforced with bio-based nanochitin: preparation, enhanced mechanical and thermal properties. J Appl Polym Sci 2020;137:48485.
7. Al-Itry R, Lamnawar K, Maazouz A. Improvement of thermal stability, rheological and mechanical properties of PLA, PBAT and their blends by reactive extrusion with functionalized epoxy. Polym Degrad Stabil 2012;97:1898–914.
8. Chaves RP, Fechine GJM. Thermo stabilisation of poly (butylene adipate-co-terephthalate). Polímeros 2016;26:102–5.
9. Jiang L, Wolcott MP, Zhang J. Study of biodegradable polylactide/poly(butylene adipate-co-terephthalate) blends. Biomacromolecules 2006;7:199–207.
10. Weng Y-X, Jin Y-J, Meng Q-Y, Wang L, Zhang M, Wang Y-Z. Biodegradation behavior of poly(butylene adipate-co-terephthalate) (PBAT), poly(lactic acid) (PLA), and their blend under soil conditions. Polym Test 2013;32:918–26.

11. Xu Z, Qiao X, Sun K. Environmental-friendly corn stover/poly(butylene adipate-co-terephthalate) biocomposites. Mater Today Commun 2020;25:101541.
12. Marinho VAD, Pereira CAB, Vitorino MBC, Silva AS, Carvalho LH, Canedo EL. Degradation and recovery in poly(butylene adipate-co-terephthalate)/thermoplastic starch blends. Polym Test 2017;58:166–72.
13. Lu X, Zhao J, Yang X, Xiao P. Morphology and properties of biodegradable poly (lactic acid)/poly (butylene adipate-co-terephthalate) blends with different viscosity ratio. Polym Test 2017;60: 58–67.
14. Bheemaneni G, Saravana S, Kandaswamy R. Processing and characterization of poly(butylene adipate-co-terephthalate)/wollastonite biocomposites for medical applications. Mater Today 2018;5:1807–16.

Medhat S. Farahat Khedr*

11 Bio-based polyamide

Abstract: Biobased polymers are sustainable polymers produced from renewable resources such as biomass feedstocks instead of the industrial fossil resources such as petroleum and natural gases. This trend helps in creating an environmentally friendly chemical processing that is characterized by low carbon footprint emission to the globe which in turn will limit the increase of the atmospheric carbon dioxide concentration even after their incineration. Synthesis of polymeric materials from biobased resources also solves the problem of polymer waste recycling. This chapter covers a basic background on the origin and importance of biobased polyamides, different synthetic routes of their starting monomeric materials obtained from biomass feedstocks, and a brief summary of the physical and chemical properties and applications of some common aliphatic, semiaromatic and fully aromatic polyamides. This chapter ends with a recent published data on the growth of the global market of biobased poly-amides to emphasize on the economic importance of this manufacturing trend.

Keywords: aliphatic polyamides; aramids; aromatic polyamides; biobased polyamides; nylons; phthalamides.

11.1 Introduction

Polymers and plastics are usually utilized in packaging and coating applications. Around 80–85% polymeric materials are generated from the petroleum sector. The growing use of plastics is putting stress on the environment with its growing carbon footprint CFT owing to greenhouse gas production, along with other aspects such as soil and water pollution. The principal feedstock for the high production volume of these polymeric materials depends basically on the petrochemical industries, which in turn elevates the levels of CFT in the globe. This fact motivated many scientists to search for alternative green sources based on biomass feedstocks to supply the desired raw monomeric materials and lower the dependence on the petrochemical sources. The benefits of such trend toward green production of polymeric materials, also known as environmentally friendly materials, or "eco-materials" are multiples from the economic and environmental points. The ease of disposal, recycling and biodegradation of these eco-friendly polymeric materials add more benefits to this industry [1–3]. This growing need for eco-friendly materials is synchronously increasing with the increasing stress towards sustainable bio-based polymers and plastics in the upcoming

*Corresponding author: Medhat S. Farahat Khedr, Arts and Science, University of North Florida, 1 UNF Drive, FL 32224, Jacksonville, FL, USA, E-mail: mfarahat2007@gmail.com

This article has previously been published in the journal Physical Sciences Reviews. Please cite as: M. S. F. Khedr "Bio-based polyamide" *Physical Sciences Reviews* [Online] 2021, 7. DOI: 10.1515/psr-2020-0076 | https://doi.org/10.1515/9781501521942-011

decades. Replacing the conventional polymeric materials by biodegradable polymers will lower CFT levels in the globe and keep the environment greener. The raw materials needed in the production of biopolymers are obtainable from agricultural wastes, and biopolymer manufacturers will play a prime role in the municipal solid waste MSW management. The use of raw materials form biomass sources has a positive impact on the life-cycle assessment LCA of plastic products as well. Biopolymer market can be classified into different biodegradable polymer grades, for instance; biodegradable polyester, regenerated cellulose, bio polyethylene Bio-PE, biodegradable starch blends, polyhydroxyalkanoate PHA, bio-polyethylene terephthalate Bio-PET, poly-lactic acid PLA, polyamides PAs and other material types based on the origin of the polymeric material type. PAs have been widely considered for many decades because they exhibit a good flow behavior with keeping their mechanical properties like tensile strengths at elevated temperature and high resistant to wear and abrasion and dimensional stability, and physical properties like low permeability to gases and electrical insulation with good chemical resistance all at a high level. PAs are of prime importance also due to their wide range of applications in different industries in particular the medical sector [1–3]. Castor oil is considered on top of the main sources for the various raw monomeric materials required for the synthesis of bio-based polyamides bio-PAs including both components of dibasic acids and aliphatic diamine compounds [4].

11.1.1 Global warming potential GWP (kg CO_2 equivalent)

GWPs are relative to the impact of carbon dioxide. GWPs are an index for estimating the relative global warming contribution due to the atmospheric emission of a kg of a particular greenhouse gas compared to the emission of a kg of carbon dioxide. Carbon footprint CFP is a term sometimes used regarding this concept. The carbon footprint is a measure of the exclusive total amount of carbon dioxide emissions that is directly and indirectly caused by an activity or is accumulated over the life stages of a product. Studies that deal with GWP indicators are usually based on kg CO_2 eq [2].

11.2 Polyamides and nylon: background

The amide linkage [–NH–C=O–] is a common and very important functional group which contributes to the special properties of peptides, proteins, beta-lactam antibiotics and numerous synthetic polymers. The discovery of its rigid structure and planarity and high tendency for hydrogen bonding sped Linus Pauling's discovery of the alpha-helix structure which is the template for deoxyribonucleic acid DNA strands and most protein secondary structures. The amide planarity and rigidity were discovered by Pauling and co-workers in the 1930s, which was also supported by resonance theory [5, 6]. Aliphatic PAs, are commonly known as nylons, are synthesized by multiple synthetic routes, via polycondensation reactions of dicarboxylic acids with diamines, or by polycondensation of amino acids or via ring-opening polymerization ROP of cyclic amides (lactams). The resulting PAs might be of AABB-type and AB-type, respectively [5]. The number of carbon atoms between the amide bonds are described by the number(s) after the prefix PA for the AB-type. But on the other hand, the first number(s) after the prefix PA for the AABB-type refers to the number of carbon atoms between the two amine groups and the second number(s) refers to the total number of carbon atoms of the dicarboxylic acid [7]. Figure 11.1 shows structures of some PAs synthesized by ROP of different lactam monomers.

Figure 11.2 shows structures of some PAs polymers synthesized by polycondensation reactions of diamines with dicarboxylic acids along with their name abbreviations.

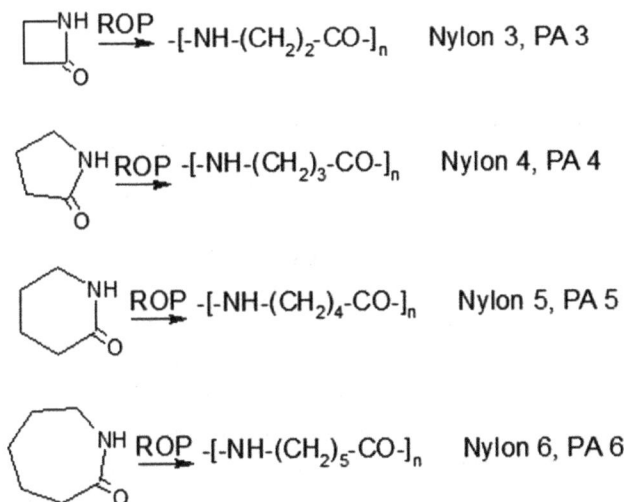

Figure 11.1: Structures of different PAs synthesized by ROP of lactam monomers.

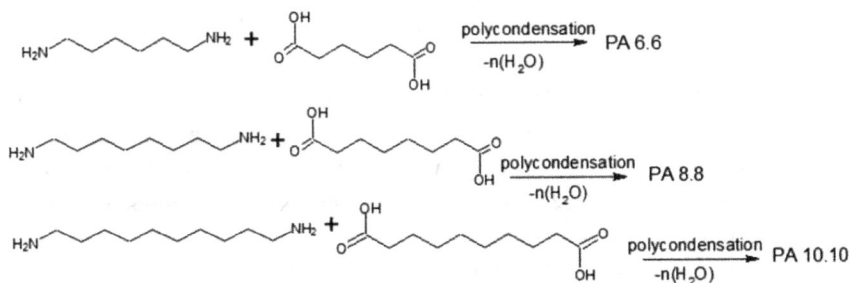

Figure 11.2: Structures of different PAs synthesized by polycondensation reactions.

The existence of the amide linkages along the polyamide chain make it more hydrophilic than most functional group polymers. The reason such high polarity and hydrophilicity can be attributed to the rigidity and planarity of the amide linkage allow resonance structures which places negative and positive charges on O-atom and N-atom respectively as shown by Figure 11.3 below.

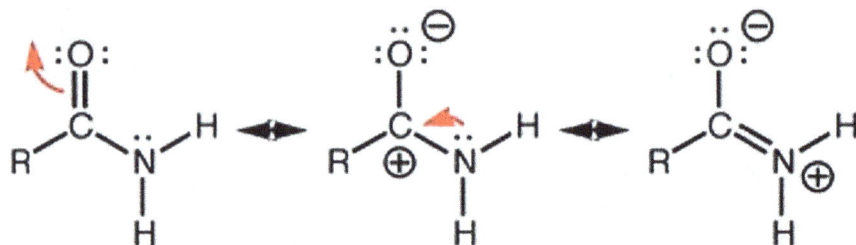

Figure 11.3: Resonance structures of the amide linkage.

Aliphatic PAs (nylons) are produced on a much larger scale than aromatic PAs (aramids), their structures are in general amorphous or semicrystalline due to presence of the aliphatic chains. On contrary, aromatic PAs are more crystalline due to the existence of the aromatic rings and are characterized by their higher mechanical strength, flame retardancy and higher dimensional stability. Resistance to degradation under different acids, bases and organic solvents compared with other functional polymers were studied extensively and reported, and in general PAs exhibit superior properties over other functional polymers including gas permeability as well [5, 6].

The chain length of aliphatic PAs control most of their physical and chemical properties. For instance, long chain aliphatic PAs such as PA11, PA10.10, and PA6.10 are characterized by their lower density, lower moisture absorption, lower strength and stiffness and lower melting temperature (T_m) with good continuous operating temperature in comparison with shorter aliphatic PAs like PA6 [8]. Natural and synthetic fillers can be incorporated into the long chain aliphatic PAs composition for the purpose of compensating the loss in their mechanical properties and thermal stability. For instance, applying fillers like bentonite (sodium montmorillonite) as nanocomposites with PA10.10 prepared by intercalating polymerization, or composites with multiwalled carbon, and composites of PA11 with nanoclay prepared by melt-compounding method showed higher modulus of elasticity and onset temperature of decomposition compared with neat unfilled PA10.10 and PA11. It was also reported that a high increase in hardness, friction and wear mechanical properties of PA11 occurred when filled with 20 wt% of short glass fibers and copper or 6% of bronze powders processed by extrusion followed by injection molding [9–12]. The stiffness properties of glass fibers reinforced PA6, PA4.6 and other aliphatic PAs compete with those of metals which make them considered as replacement parts in many industrial applications. But unfortunately, the high moisture absorption showed by PA6 and PA4.6 which results in deterioration of their mechanical behavior by time limit their application in this field [13].

11.3 Biobased aliphatic bio-PAs

i. Different synthetic routes for the starting monomeric materials.

Biobased aliphatic PAs (bio-PAs) are PAs that were fully or partially synthesized from renewable resources. Renewable natural resources, such as plant oils, fatty acids, cellulose and lignin, have been widely pursued as precursors for manufacturing sustainable PAs. There are many challenges in the production of biobased plastics. Bio-PAs are obtained by several methods: (i) from raw or chemically modified natural polymers; (ii) from reacting a mixture of monomeric raw materials obtained from both biomass and petroleum feedstocks; (iii) through the polymerization of chemically tailored monomers that are fully obtained from biomass feedstocks after complicated chemical transformations to synthesize bio-PAs [14–17]. Poor solubility of these natural polymers limits their processing and applications in addition to the difficulty of removing a diverse of organic impurities and undesirable chemical compounds due to their negative influence on their properties if were not removed. Bio-PAs synthesized from mixtures of petroleum based and biobased resources or exclusively biobased resources are more applicable routes in this field. Plant oils and fats have been used for a long time as main biomass feedstocks for the synthesis of bio-PAs. Castor oil was the oldest known source for the monomeric raw materials obtained from its chemical transformation processes, a fact that it was considered as castor refinery due to the great value of the obtained products [4, 15–17]. In production of commercially available bio-PAs, castor oil from *Ricinus communis* plant is used as the main biomass feedstock. Castor oil has been long considered as feedstock for manufacturing important consumer products as soaps, lubricants and coatings. Today, castor oil appears as a valuable feedstock for biorefineries, nonedible and noncompeting with the food chain, suitable for manufacturing biofuels, biochemicals and biopolymers [4]. The privilege of castor oil is due to its chemical composition which contains 85–90% triglyceride of ricinoleic (12-hydroxy-9-octadecenoic) acid which is characterized by the presence of a double bond between C_9 and C_{10} atoms, and a hydroxyl group on C_{12} atom. These three functional groups in ricinoleic acid structure are crucial for its further chemical derivation which opens the way to a large variety of organic synthesis routes. Further chemical transformations of ricinoleic acid produce many essential monomers for the synthesis of bio-PAs [16, 17]. Essential monomers which can be produced by chemical transformations of ricinoleic acid include sebacic acid, amino undecanoic acid and decamethylene diamine DMDA [18]. Figure 11.4 shows some chemical transformations of castor oil to different starting monomeric materials for the synthesis of bio-PAs.

Monomers used in bio-PAs synthesis can be obtained partially or fully from biomass feedstocks. Currently, biomass-derived monomers with the highest industrial meaning for the synthesis of bio-PAs are adipic acid, sebacic acid, 1,4-butanediamine, 1,4-pentanediamine, 1,10-decanediamine, 11-aminoundecanoic acid and caprolactam.

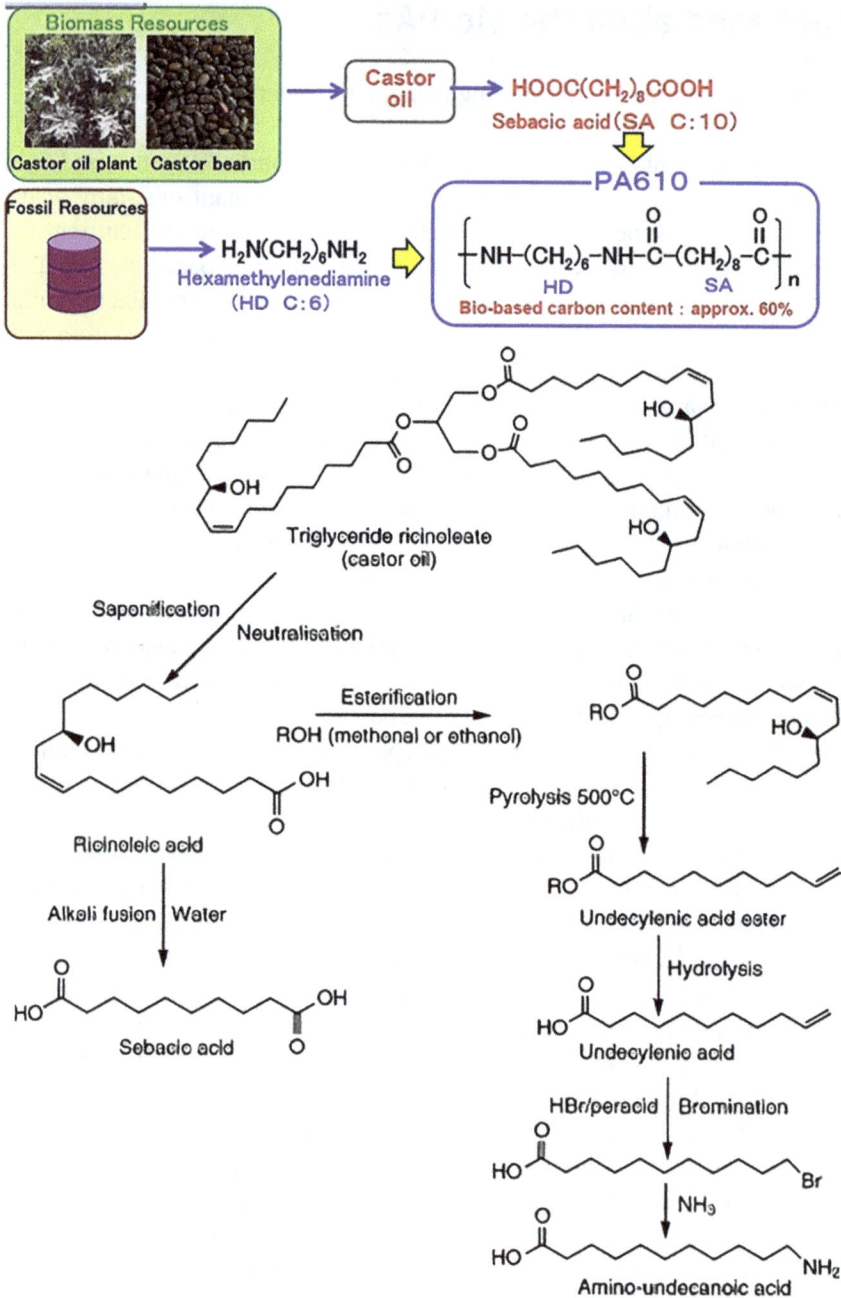

Figure 11.4: Chemical transformations of castor oil into starting monomeric materials for bio-PAs.

The concepts of their synthesis from natural resources were briefly described before [19, 20]. Figure 11.5 shows a schematic diagram for the chemical transformations of these biomass feedstocks into monomeric materials used in the synthesis of bio-PAs.

Carbohydrate biomasses represent an additional valuable feedstock for polyesters and polyamides because they provide access to monomers that are hard to produce from petrochemical feedstocks. Recently, Lankenau and Kanan reported a new chemical route for transformation of lignocellulose to tetrahydro furan bicyclic lactam monomer [21]. Their synthetic route is focused on carbonate-promoted C–H carboxylation chemistry that enables streamlined syntheses of 5-(aminomethyl)-furan-2-carboxylic acid (**1**) and 8-oxa-3-azabicyclo[3.2.1]octan-2-one (**2**) from furfurylamine (**3**). Lignocellulose is a raw nonedible biomass when subjected to acid-catalyzed depolymerization and dehydration yields Furfural which is considered the starting material in their work [22]. Figure 11.6 shows a comparison between their innovative work with the previous traditional synthesis routes to obtain the same monomers (**1** and **2**).

ii. Synthetic methods of some common aliphatic bio-PAs

Aliphatic bio-PAs of the AABB-type can be synthesized by polycondensation of monomeric materials from biomass conversion processes. Synthesis of bio-PAs of AB-type can only be obtained by ROP of cyclic lactams which can be obtained from biomass feedstocks. For instance, PA6 is formed by ROP of ε-caprolactam which can be

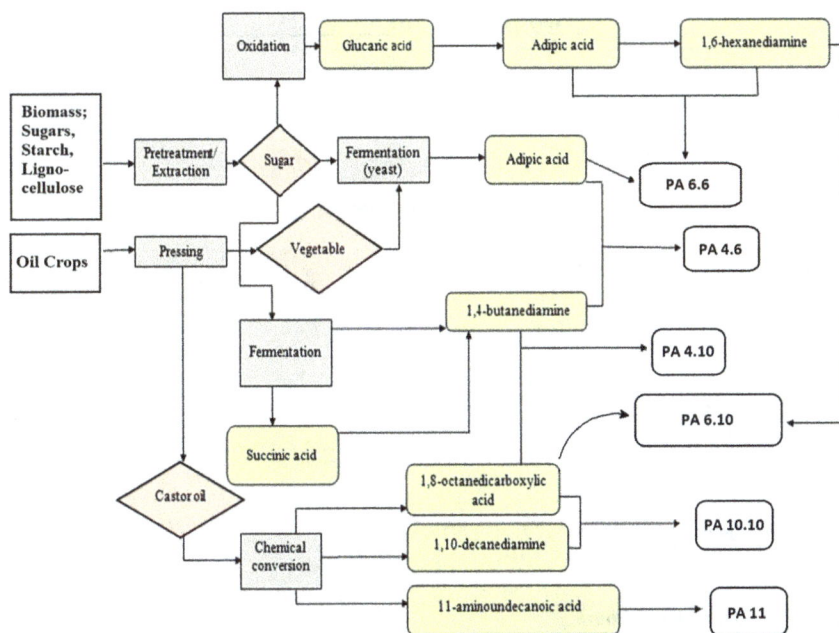

Figure 11.5: Chemical transformations of various biomass feedstocks into monomeric materials.

obtained by glucose fermentation of crops or sugar where hydroxymethyl furfural is the major fermentation product. Hydroxymethyl furfural can be then transformed to ε-caprolactam through different chemical routes [21, 23, 24]. The same chemical procedures can be applied to obtain PA6.6, PA4.6, PA6.10, PA4.10 and PA10.10 (Figure 11.5).

Aliphatic bio-PAs containing longer chain dicarboxylic acid monomers like (PA4.10, PA5.10, PA6.10, PA6.12, PA10.10 and PA10.12) are PAs of AABB-type where the sebacic acid and 1,12-dodecanedioic acid are obtained from biomass feedstocks [16, 17] and the diamine monomers like 1,4-diaminobutane is obtained by fermentation of oil crops, and 1,5-diaminopentane is obtained from lysine through glucose fermentation process [25]. Recent studies enabled the production of 1,5-diaminopentane by metabolic engineering of the soil bacterium "corynebacterium glutamicum" through various fermentation processes [26]. PA6.10 is synthesized from HMDA and sebacic acid and both are produced from castor oil (Figure 11.4). PA6.12 is the polycondensation product of HMDA and 1,12-dodecanedioic acid which are originally obtained from fermentation of plant oil or glucose and its CFT value is 4.6 kg CO_2 eq. Biotechnology

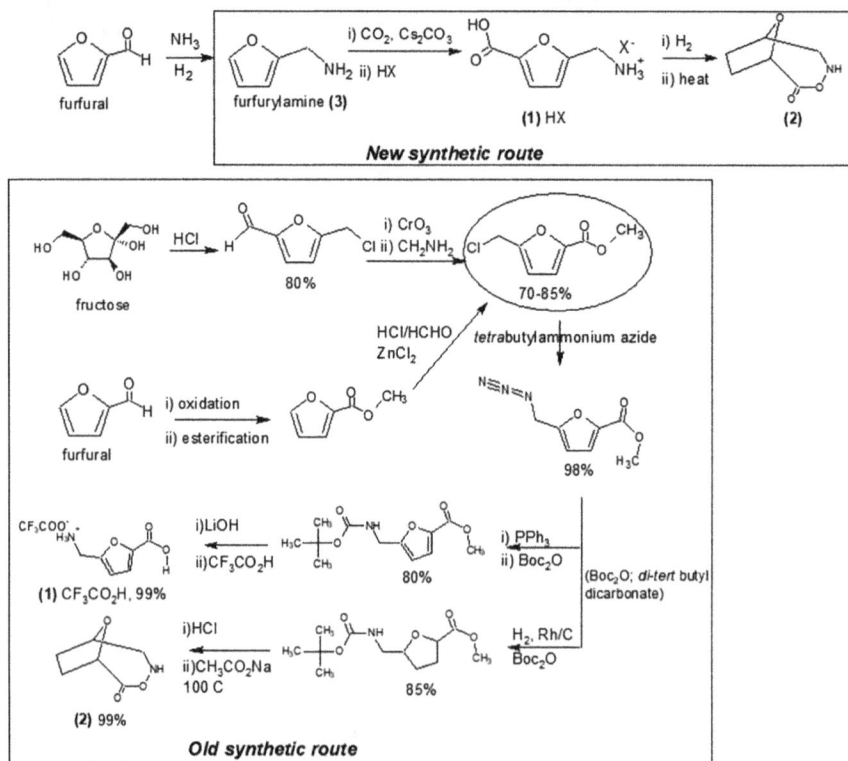

Figure 11.6: Preparation of tetrahydrofuran bicyclic lactam (2) via different chemical routes.

introduced new innovative ways to overcome the limitations and disadvantages coming from the petrochemical processing [27]. PA10.10 is the polycondensation product of DMDA and sebacic acid where both monomers are obtainable from castor oil, thus making PA10.10 100% bio-PA [16, 17]. PA10.12 contains a more longer hydrocarbon segment and both diamine and dibasic acid are obtained from castor oil and it has CFT value of 5.2 kg CO_2 eq. PA10.12 is an alternative to PA12 and PA12.12. PA12.12 which is commercially manufactured by polycondensation reaction of 1,12-dodecanediamine and 1,12-dodecanedioic acid. Synthetic routes and characteristics for PA12.12 are very similar to those of PA11 and PA12. PA13.6 can be synthesized from 1,13-tridecanediamine and adipic acid and both can be obtained from biomass feedstocks.

A new biobased strategy based on catalytic conversion of lactones (cyclic esters) which were subjected to chemical upgrading into lactams (cyclic amides) using cellulose and hemicellulose as biomass feedstock was also developed [18, 24, 28]. A recent economic friendly process for the catalytic conversion of corn stover to γ-valerolactone combined with a catalytic conversion of γ-valerolactone to ε-caprolactam showed good yields from both processes which are suitable for commercial-scale production [29]. The world annual production of PA6.6 is the second largest volume of all aliphatic PAs (3.4 million tons) where about 55% of PA6.6 production is used basically as fibers and 35% is applied in the field of engineering thermoplastics. The production of PA6.6 involves the use of two monomers: adipic acid obtained through fermentation of plant-oil or glucose and hexamethylenediamine HMDA, which are readily obtainable from biomass sources (Figure 11.5) [30]. Commercially biobased synthetic routes from renewable feedstocks were developed to obtain the monomers for PA6.6 production. The production of biobased adipic acid on a large commercial scale is a well-established process but the production of biobased HMDA on a large scale still considered a new development platform process. The large-scale production of biobased HMDA requires unique technologies with significant cost advantages to replace products obtained from petrochemical industries. PA11 was well known for many decades as the first biobased high-performance engineering PA. PA11 is synthesized by fhe polycondensation reaction of 11-aminoundecanoic acid which is a product of chemical transformation of castor oil (Figure 11.4). Star-shaped PA11 structures were obtained via one-pot co-polycondensation of 11-aminoundecanoic acid with a multifunctional agent, either trifunctional *bis*-hexamethylene triamine (BHMTA) or tetrafunctional 2-oxocyclohexane-1,1,3,3-tetrapropionic acid. As can be seen that 11-aminoundecanoic acid is an amino acid where both of amino and carboxylic groups exist in the same monomer and its polycondensation yields PA of AB-type. It can be copolymerized through polycondensation reaction with variable ratios from polyamine and poly carboxylic acid monomers to yields a variety of star-shaped polymers that can be tailored to reach the desired rheological and solid-state properties products [31]. 12-Aminododecanoic acid is an amino acid monomer that has been known for many decades to produce PA12. Bio-PA11 can be synthesized by polycondensation reaction of

12-aminododecanoic acid which can be entirely derived from renewable sources (palm kernel oil) by one-step fermentation process. PA12 can be also synthesized by ROP of laurolactam, a cyclic amide obtained from petrochemical industry, using anionic initiators [25, 32].

11.4 Mechanical characteristics and applications of common aliphatic PAs

Aliphatic Pas, commonly known as (*Nylon*) are produced on a much larger scale than the fully aromatic PAs due to their cost and ease of production, and consequently they represent the more important class of engineering thermoplastics. The aromatic PAs, commonly known as (*Aramids*), are characterized by having higher mechanical strength, flame retardancy, chemical resistance to most organic solvents and higher dimensional stability than all aliphatic PAs, but are a way higher expensive and require sophisticated procedures to be produced. The mechanical properties of PAs are controlled by the nature of hydrocarbon segments between the polyamide linkages which are either aliphatic, semi-aromatic or fully aromatic. The reason of the superior mechanical properties of the fully aromatic PAs is due to the strong hydrogen bonds between the aramid chains resulting in high melting temperature T_m (>475 °C), ultra-high tensile strength at low weight and excellent flame retardancy in addition to good dimensional stability and solvent resistance at moderate and elevated temperatures. PA6 and PA6.6 are the most important and the largest production volume of all aliphatic PAs due to their high mechanical characteristics. High flexibility, good resilience, high tensile and impact strengths (toughness) along with low creep are the main characteristic properties of PA6 and PA6.6. They are easy to dye and exhibit excellent resistance to wear due to their low coefficient of friction (self-lubricating). PA6 and PA6.6 have high T_m (227–267 °C) and T_g which resulted in their high mechanical properties at elevated temperatures. For instance, the heat deflection temperature (HDT) of PA6.6 lies between (180 and 240 °C) which exceeds those of polycarbonates and polyesters. In addition, they exhibit good resistance to oils, bases, fungi, and many organic solvents and they are widely used in textile and automotive industries [33].

11.5 PA6 and PA6.6

The high melting temperatures T_m of PA6 (≈ 220 °C) and PA6.6 (≈ 265 °C) put them on top of production volumes of all aliphatic PAs. The world production of PA6 exceeds four million tons annually due to its highest demands in many engineering thermoplastic applications. PA6 possess high stiffness and strength under dry conditions, excellent heat and chemical resistance. Its relatively high moisture absorption limits its application under humid environment. PA6 is commonly used to replace metal in automotive parts due to its design flexibility and heat and chemical resistance are highly demanded. Examples for PA6 applications in automotive industry include radiator grilles, airbag containers, air intake manifolds, relay boxes, engine covers where it is exposed to high temperature and drastic chemical conditions. PA6.6 shows improved thermal and chemical resistance and its moisture absorption is (8–9 wt.%) which is slightly lower than that for PA6 (10 wt.%) and PA4.6 (13 wt.%) according to ISO 1110 [34]. Aliphatic PAs absorb moisture from air due to their amorphous structures which cause changes in their mechanical characteristics over a time span. Moisture act as a plasticizer which reduces their stiffness properties but increases their elongation and impact strength. The concentration of the amide groups (hydrophilic) versus the length of poly(methylene) segments (hydrophobic) depicts the moisture absorption behavior of the aliphatic PAs such that the longer the poly(methylene) segment the lower the moisture absorption and vice versa. The higher thermal stability of PA6.6 makes it more suitable for industrial yarns, airbags, radial tires, and most of under the hood automotive parts. Comparison between basic features of PA6 and PA6.6 are summarized below in Table 11.1.

11.6 PA11, PA12 and PA12.12

PA11 and PA12 have lower moisture absorption compared with PA6 and PA6.6 due to their increased length of poly(methylene) segment and consequently provide better dimensional stability. In addition, bio sourced PA11 and PA12 possess superior thermal and chemical resistance, durability, ageing and flexibility [35]. PA11 and PA12 are suitable for automotive and electronic industries where precision moldings are highly demanded. Table 11.2 summarizes the main features of PA11.

PA12 performance is like that of PA11 and its mechanical properties lies between that of PA6 and PA6.6. PA12 has a lower amide group concentration and longer poly(methylene) segment than any other commercially available PAs which make it reduces its moisture absorption to a minimum level with excellent chemical resistance to hydraulic fluids, oil, gasoline, grease, salt water and organic solvents. It is characterized by the following: lower impact resistance, good resistance to abrasions and UV light, good dimensional stability, and reasonable electrical properties. Due to these mentioned reasons, PA12 is suitable for making cables covering and insulating material in the field of electronics and oil and gasoline-resistant tubes in the field of automobile industry. PA12 has also a broad spectrum of film applications for food packing materials and sterilized films and bags for use in pharmaceutical and medical applications. PA12 sheets and sintered powder are commonly applied for coating metals in addition to the textile industry and leisure goods as well [29]. Table 11.3 summarizes the

Table 11.1: Comparison between characteristics of PA6 versus PA6.6.

Property	PA6	PA6.6
Dimensional stability		
Coefficient of linear thermal expansion	$5–12 \times 10^{-5}/°C$	$5–14 \times 10^{-5}/°C$
Shrinkage	0.5–1.5%	0.7–3%
Water absorption 24 h	1.6–1.9%	1–3%
Electrical properties		
Arc resistance	118–125 s	130–140 s
Dielectric constant	4–5	4–5
Dielectric strength	10–20 kV/mm	20–30 kV/mm
Dissipation factor	$100–600 \times 10^{-4}$	$100–400 \times 10^{-4}$
Volume resistivity	14×10^{15} Ω cm	14×10^{15} Ω cm
Fire performances		
Fire resistance (LOI)	23–26%	21–27%
Flammability UL94	HB	HB
Mechanical properties		
Elongation at break	200–300%	150–300%
Elongation at yield	3.4–140%	3.4–30%
Flexibility (Flexural modulus)	0.8–2 GPa	0.8–3 GPa
Hardness Rockwell M	30–80	30–80
Hardness Shore D	80–95	80–95
Stiffness (Flexural modulus)	0.8–2 GPa	0.8–3 GPa
Strength at break (Tensile)	50–95 MPa	50–95 MPa
Strength at yield (Tensile)	50–90 MPa	45–85 MPa
Toughness (Notched izod impact at room temperature)	50–160 J/m	50–1150 J/m
Toughness at low temperature (Notched izod impact)	16–210 J/m	27–35 J/m
Young modulus	0.8–2 GPa	1–3.5 GPa
Optical properties		
Gloss	130–145%	65–150%
Physical properties		
Density	1.12–1.14 g/cm³	1.13–1.15 g/cm³
Glass transition temperature	60 °C	55–58 °C
Radiation resistance		
Gamma radiation resistance	Fair	Fair
UV light resistance	Fair	Poor
Service temperature		
HDT at 0.46 Mpa (67 psi)	150–190 °C	180–240 °C
HDT at 1.8 Mpa (264 psi)	60–80 °C	65–105 °C
Max continuous service temperature	80–120 °C	80–140 °C
Min continuous service temperature	−40 to −20 °C	−80 to −65 °C
Others		
Sterilization resistance (Repeated)	Poor	Poor
Thermal insulation (Thermal conductivity)	0.24 W/m K	0.25 W/m K

Table 11.2: Main features of PA11.

Advantages	Limitations
The lowest water absorption of all commercially available PAs	High cost relative to other PAs
Outstanding impact strength, even at temperatures well below the freezing point	Lower stiffness and heat resistance than other polyamides
Resistant to chemicals, particularly against greases, fuels, common solvents and salt solutions	Poor resistance to boiling water and UV
Outstanding resistance to stress cracking, aging and abrasions	Proper drying before processing is needed
Low coefficient of friction	Attacked by strong mineral acids and acetic acid and are dissolved by phenols
Noise and vibration damping properties	Electrical properties highly depend on moisture content
Fatigue resistant under high frequency cyclical loading condition	
Ability to accept high loading of fillers	
Highly resistant to ionization radiation	

basic features of PA12 in addition to limitations of PA12 applications compared with other PAs.

PA12.12 possesses similar properties to those of PA11 and PA12 and exceeds by showing better shape retention at elevated temperatures. PA12.12 find applications in many areas, for example, electric insulating material in electrical and radio engineering, oil extraction and instruments making as a housing material, vehicle and aviation manufacturing, plus medical and other industries [36].

Table 11.3: Main features of PA12.

Advantages	Limitations
Lowest water absorption of all commercially available PAs	Expensive than other PAs
Outstanding impact strength, even at very low temperatures	Lower stiffness and heat resistance than other PAs
Good chemical resistance, in particularly against greases, fuels, common solvents and salt solutions	Low UV resistance
Outstanding resistance to stress cracking	Proper drying before processing is needed
Excellent abrasion resistance	Electrical properties highly depend on moisture content
Low coefficient of friction	
Noise and vibration damping properties	
Good fatigue resistance under high frequency cyclical loading condition	

11.7 PA4.6 and PA4.10

PA4.6 has the highest melting temperature T_m ($\approx$ 295 °C) of all aliphatic PAs. The chain structure symmetry between its monomeric parts of 1,4-diaminobutane and adipic acid led to its high degree of crystallinity, outstanding flow characteristics and convenient processing which leads to reduced cycle time and increased design freedom. These features gave PA4.6 a technical cutting-edge characteristic above other engineering thermoplastics. The only drawback of PA4.6 is its higher moisture absorption which limits it application in the field of electronics and electrical insulation. Table 11.4 shows the main features of PA46.

Table 11.4: Main features of PA4.6.

Property	Value
Dimensional stability	
Shrinkage	1.5–2%
Water absorption 24 h	1.3–3.7%
Electrical properties	
Dielectric constant	3.4–3.8
Dielectric strength	15–25 kV/mm
Dissipation factor	190–600 10^{-4}
Volume resistivity	15 10^{15} Ω cm
Fire performances	
Fire resistance (LOI)	24%
Flammability UL94	HB
Mechanical properties	
Elongation at break	160–300%
Flexibility (Flexural modulus)	1–3.2 GPa
Hardness Rockwell M	92
Stiffness (Flexural modulus)	1–3.2 GEa
Strength at break (Tensile)	65–85 MPa
Strength at yield (Tensile)	
Toughness (Notched izod impact at room temperature)	30–250 J/m
Young modulus	1–3.3 GPa
Physical properties	
Density	1.17–1.19 g/cm^3
Radiation resistance	
Gamma radiation resistance	Fair
UV light resistance	
Service temperature	
HDT at 1.8 MPa (264 psi)	150–155 °C
Max continuous service temperature	110–150 °C
Min continuous service temperature	–40 °C
Others	
Thermal insulation (Thermal conductivity)	0–3 W/m K

PA4.10 is synthesized by polycondensation reactions of 1,4-diaminobutane with sebacic acid. PA4.10 has higher hydrophobicity and its typical applications include bushings, cams, electrical connectors, automotive parts, tubing, rods and sheets and numerous other extrusion molding parts [37].

11.8 PA6.10, PA6.12, PA5.10, PA10.10, PA10.12, PA10.12 and PA13.6

PA6.10 and PA6.12 possess improved mechanical characteristics over that for PA6, PA6.6, PA11 and PA12 regarding excellent bend recovery and higher resistance against deformation under loads in wet environment. PA6.10 is characterized by its exceptional dimensional stability, high quality surface, good abrasion, wear resistance and lowered CFT value. These improved mechanical characteristics plus its good chemical resistance make PA6.10 suitable for manufacturing high precision injection molding parts for the automotive industry (connectors, housings, nonreturn valves, power steering fluid reservoirs) and electronic industries as well [38]. PA6.10 is used also in tools, machines and building constructions (gears, door handles, fittings, office equipment), and sports and leisure because of its high mechanical properties and low density. Table 11.5 shows the mechanical characteristics of PA6.10.

PA6.12 is a stiffer polymer with a higher burst resistance and excellent resistance to greases, oils, fuels, hydraulic fluids, alkalis, and salt solutions, excellent resistance to stress cracking, low sliding friction coefficient, and high abrasion resistance under dry conditions and lower oxygen permeability when compared those properties of PA6.10, PA11 and PA12. PA6.12 is characterized over PA11 and PA12 by its better UV light resistance, higher HDT, higher tensile and flexural strength, and excellent rebound resilience combined with high wet strength. PA5.10 has mechanical features like those of PA6.6, it is dominantly used for injection and extrusion molding applications for making automotive parts. Either in a pure form or as reinforced with glass fibers, the 100% biobased PA5.10 has high mechanical strength that outweighs that of the well-established PAs from petrochemical sources (PA6 and PA6.6) [39].

PA10.10 is an aliphatic long-chain PA that has similar properties to those of PA11, PA12 and PA12.12 and can replace them in most of their applications, for instance, in automotive and industrial pipes, cables, and injection molding parts for sports and electronics applications. PA10.10 can be used as jacketing material for cables and optical fibers and in a thermoplastic powder-coating material for machines industry due to its low moisture uptake and excellent resistance to hydrolysis [40]. PA10.10 exhibits high degree of rigidity in its unfilled structure and even higher rigidity when reinforced with glass fiber, high thermal stability, low permeability to petrol and gases. The CO_2 footprint of PA10.10 is 4.0 kg CO_2 eq.

Table 11.5: Mechanical characteristics of PA6.10.

Property	Value
Dimensional stability	
Coefficient of linear thermal expansion	$6{-}10\ 10^{-5}/°C$
Shrinkage	1–1.3%
Water absorption 24 h	0.4–0.6%
Electrical properties	
Arc resistance	120 s
Dielectric constant	3–4
Dielectric strength	16–26 kV/mm
Dissipation factor	$400\ 10^{-4}$
Volume resistivity	$14\ 10^{15}\ \Omega\ cm$
Fire performances	
Fire resistance (LOI)	23–27%
Flammability UL94	V2
Mechanical properties	
Elongation at break	150–300%
Flexibility (Flexural modulus)	1–2 GPa
Hardness Rockwell M	1–50
Hardness Shore D	60–85
Stiffness (Flexural modulus)	1–2 GPa
Strength at break (Tensile)	50–65 MPa
Strength at yield (Tensile)	
Toughness (Notched izod impact at room temperature)	70–999 J/m
Young modulus	1–2 GPa
Physical properties	
Density	$1.09{-}1.1\ g/cm^3$
Radiation resistance	
Gamma radiation resistance	Fair
UV light resistance	
Service temperature	
HDT at 0.46 MPa (67 psi)	160–175 °C
HDT at 1.8 MPa (264 psi)	80–85 °C
Max continuous service temperature	80–150 °C
Others	
Thermal insulation (Thermal conductivity)	0.21 W/m K

PA10.12 has a longer poly(methylene) segments than PA10.10 which reduces its moisture absorption to the minimum level and make it more appropriate for making many automotive and industrial parts. PA10.12 can also be processed as films with good transparency. It has a characteristic high mechanical and dimensional stability, chemical resistances, flexibility, broad working temperature which exceeds other semicrystalline PAs [41].

PA13.6 is characterized by its fast crystallization kinetics compared to PA6 and PA6.6, it has melting temperature T_m (206 °C) and glass transition temperature

T_g (60 °C) close to those for PA6, but however, its moisture absorption is much lower than both PA6 and PA6.6 which is due to its much lower amide and high poly (methylene) contents [42].

11.9 Fully aromatic (aramids) and semi-aromatic (polyphthalamides) PPAs

Fully aromatic PAs (aramids) are based on aromatic diamine monomers, like *m*-phenylenediamine MPDA and *p*-phenylenediamine PPDA and aromatic dicarboxylic acids, like isophthalic and terephthalic acids. Examples for the most common aramids include poly(*p*-phenylene terephthalamide), commercially known as Kevlar and poly(*m*-phenylene isophthalamide). Aromatic polyamines and polycarboxylic acids can also be incorporated into their chemical structures to get crosslinked and thermosetting resins to cover a broad spectrum of high mechanical properties as demanded by the engineering plastic industry. On the other hand, semi-aromatic polyamides, also knowns as (polyphthalamides) (PPAs) are synthesized by polycondensation reaction of an aliphatic diamine, most commonly HMDA with terephthalic acid and/or isophthalic acid. The weight percentage of the aromatic component is typically slightly above 50% and the aliphatic component is aimed to be less than 50% by weight for achieving better mechanical properties. This blend between aromatic and aliphatic components greatly reduces moisture absorption which in turn results in little dimensional changes and more stable properties [43]. They are melt-processible and thus filling the performance gap between fully aliphatic PAs such as PA6 and PA6.6 and the highly expensive fully aromatic PAs. They possess high crystallinity and high mechanical strength and stiffness at elevated temperatures. However, their high cost and processing difficulty due to their higher melting temperatures T_m confine their high production volumes compared to aliphatic PAs. Aramids are sometimes blended with aliphatic PAs like PA6.6 in specific ratios to improve their processability and lower their cost. Instances for most common semi-aromatic amides comprise poly(hexamethylene terephthalamide) (PA6T) and poly(hexamethylene isophthalamide) (PA6I). PA6T has a high melting temperatures T_m ($\approx$ 322 °C) and glass transition temperature T_g ($\approx$ 137 °C) and well known for its excellent dimensional stability, low creep at elevated temperatures and good chemical resistance compared to many high-performance engineering thermoplastics. PPAs find wide applications in several industries such as electronic devices, packaging and automotive industries. They can be used in composites for high temperature applications. The ultimate mechanical properties and durability of PPAs expected in applications such as automotive industry depend on the properties of strain hardening and impact strength [44]. An interesting application of the semi-aromatic or fully aromatic PAs is exhibited by reverse osmosis RO membranes. Aramid RO membranes are synthesized by interfacial polymerization IP between MPDA and trimesoyl chloride TMC was reported and used for desalination on the lab-scale [45]. Figure 11.7 shows the chemical structures of MPDA and TMC polymerized by IP on a microporous polyether sulfone PES support.

An industrially important application of semi-aromatic PAs comprises shape memory polymers SMPs that can change and recover the shape repeatedly under different stimuli such as light, pH, electric or heat. They have attracted wide attention and interests in the development of smart materials due to their lightweight, transparency, good mechanical and thermal properties, facile process mode, and biocompatibility. Semi-aromatic polyamides have been developed and progressed rapidly in the past 20 years due to their substantial high thermal stability, chemical resistance, good processability and recyclability, which makes them attractive materials for automobile, structure parts, and circuit boards [46].

11.10 Market analysis and insights: global bio-PAs market

The following data was published in the year 2020 annual report by Data Bridge Market Research, a leader in consulting and advanced formative research. Bio-PAs market will

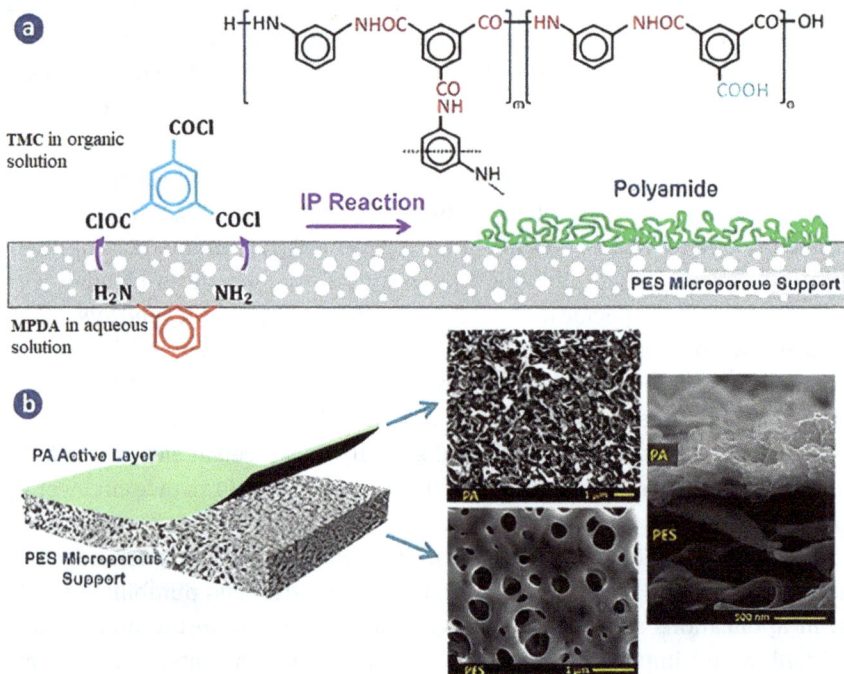

Figure 11.7: Interfacial polymerization IP of fully aromatic polyamide RO membrane.

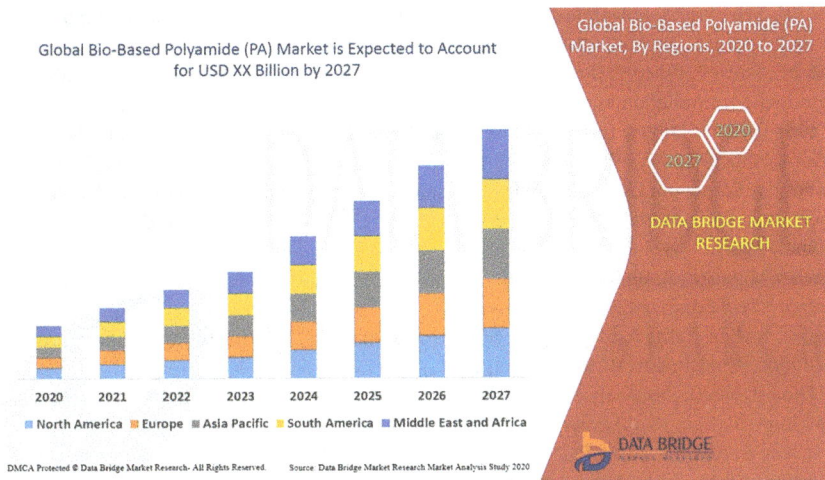

Figure 11.8: Growth of global market of bio-PAs for the period of 2020–2027.

reach an estimated volume of 396.04 thousand tons by 2027, while registering this growth at a rate of 6.20% for the forecast period of 2020–2027. The rising automotive and electrical and electronics industries will act as a driving factor for the bio-PAs market in the above mentioned period. Growing demand for ecofriendly polymers use various end-use industries around the world, increasing automotive and electrical and electronics industries, rising disposable income of the people, growing adoption of bio-based products are some of the factors escalating the market growth in the forecast period of 2020–2027. The fast developing automobile, electrical & electronics and textile sectors will further create several opportunities that will lead to the growth of the bio-PAs market in the above mentioned forecast period. Figure 11.8 shows the trend in the growth of global market of bio-PAs for the period of 2020–2027.

Author contributions: All the authors have accepted responsibility for the entire content of this submitted manuscript and approved submission.
Research funding: None declared.
Conflict of interest statement: The authors declare no conflicts of interest regarding this article.

References

1. Nguyen XH, Honda T, Wang Y, Yamamoto R. Eco-materials. Module-H. Bunkyo City, Tokyo, Japan: University of Tokyo; 2010.
2. Cabeza LF, de Gracia A, Solé A. Sustainable energy technologies. In: Solé A, de Gracia A, Cabeza LF, editors. Chapter: thermal energy storage systems for solar applications. Boca Raton, FL, USA: CRC Press; 2017.

3. Schmitz K, Schepers U. Polyamides as artificial transcription factors: novel tools for molecular medicine? Angew Chem Int Ed 2004;43:2472.
4. Dimian AC, Iancu P, Plesu V, Bonet-Ruiz A-E, Bonet-Ruiz J. Castor oil biorefinery: conceptual process design, simulation and economic analysis. Chem Eng Res Des 2019;141:198–219.
5. Polyamide (nylon) plastics: properties, performance, and military applications, military handbook. MIL-HDBK-797(AR); 1994.
6. Greenberg A, Breneman CM, Liebman JF. The amide linkage: structural significance in chemistry, biochemistry, and materials science, ISBN 13: 9780471358930. Hoboken, NJ, USA: John Wiley & Sons, Inc.; 2000.
7. Winnacker M, Rieger B. Biobased polyamides: recent advances in basic and applied research. Macromol Rapid Commun 2016;37:1391–413.
8. Endres HJ, Siebert-Raths A. Engineering biopolymers markets, manufacturing, properties and applications. Munich: Carl Hanser Verlag; 2011.
9. Liu Z, Zhou P, Yan D. Preparation and properties of nylon-1010/montmorillonite nanocomposites by melt intercalation. J Appl Polym Sci 2004;91:1834–41.
10. Zeng H, Gao C, Wang Y, Watts PCP, Kong H, Cui X, et al. In situ polymerization approach to multiwalled carbon nanotubes-reinforced nylon 1010 composites: mechanical properties and crystallization behavior. Polymer 2006;47:113–22.
11. Liu T, Lim KP, Tjiu WC, Pramoda KP, Chen ZK. Preparation and characterization of nylon 11/ organoclay nanocomposites. Polymer 2003;44:3529–35.
12. Rajesh JJ, Bijwe J. Influence of fillers on the low amplitude oscillating wear behavior of polyamide 11. Wear 2004;256:1–8.
13. Mark HF. Encyclopedia of polymer science and technology, 3rd ed. Weinheim: John Wiley & Sons; 2005.
14. Jiang Y, Loos K. Enzymatic synthesis of biobased polyesters and polyamides. Polymers 2016;8: 243.
15. Karak N. Vegetable oil-based polymers; 2012.
16. Mubofu EB. Castor oil as a potential renewable resource for the production of functional materials. Sustain Chem Process 2016;4. https://doi.org/10.1186/s40508-016-0055-8.
17. Ogunniyi DS. Castor oil: a vital industrial raw material. Bioresour Technol 2006;97:1086–91.
18. Kuciel S, Kuźniar P, Liber-Kneć A. Polyamides from renewable sources as matrices of short fiber reinforced biocomposites. Polimery 2012;57:627–34.
19. Radzik P, Leszczyńska A, Pielichowski K. Modern biopolyamide-based materials: synthesis and modification. Polym Bull 2020;77:501–28.
20. Pagacz J, Raftopoulos KN, Leszczynska A, Pielichowski K. Bio-polyamides based on renewable raw materials, Glass transition and crystallinity studies. J Therm Anal Calorim 2016;123:1225–37.
21. Lankenau AW, Kanan MW. Polyamide monomers via carbonate-promoted C–H carboxylation of furfurylamine. Chem Sci 2020;11:248–52.
22. Lange JP, van der Heide E, van Buijtenen J, Price R. Furfural-a promising platform for lignocellulosic biofuels. ChemSusChem 2012;5:150–66.
23. Jiang Y, Loos K. Enzymatic synthesis of biobased polyesters and polyamides. Polymers 2016;8: 243–53.
24. Buntara T, Noel S, Phua PH, Melián-Cabrera I, de Vries JG, Heeres HJ. Caprolactam from renewable resources: catalytic conversion of 5-hydroxymethylfurfural into caprolactone. Angew Chem Int Ed Engl 2011;50:7083–7.
25. Jiang Y, Loos K. Enzymatic synthesis of biobased polyesters and polyamides. Polymers 2016;8: 243–53.
26. Mimitsuka T, Sawai H, Hatsu M, Yamada K. Metabolic engineering of Corynebacterium glutamicum for cadaverine fermentation. Biosci Biotechnol Biochem 2007;71:2130–5.

27. Mobley DP. Biosynthesis of long-chain dicarboxylic acid monomers from renewable resources. New York, US: GE Corporate Research and Development; 1999, Final Technical Report No. DE-FC36-95G01 0099.
28. Saskiawan I. Biosynthesis of polyamide 4, a biobased and biodegradable polymer. Microbiol Indonesia 2008;2:119–23.
29. Han J. Biorenewable strategy for catalytic ε-caprolactam production using cellulose and hemicellulose derived γ-valerolactone. ACS Sustainable Chem Eng 2017;5:1892–8.
30. Dros AB, Larue O, Reimond A, De Campoa F, Pera-Titus M. Hexamethylenediamine (HMDA) from fossil- vs. bio-based routes: an economic and life cycle assessment comparative study. Green Chem 2015;17:4760–72.
31. Martino L, Basilissi L, Farina H, Ortenzi M-A, Zini E, Silvestro G-D, et al. Biobased polyamide 11: synthesis, rheology and solid-state properties of star structures. Eur Polym J 2014;59:69–77.
32. McKeen LW. Polyamides (nylons), chapter 8. In: Andrew W, editors. Fatigue and tribological properties of plastics and elastomers. Oxford/Amsterdam: Elsevier; 2010.
33. Kohan MI. Nylon plastics handbook, ISBN-13:978-1569901892. Liberty Township, OH, USA: Hanser Publishers, LLC; 1995.
34. International Organization for Standardization. ISO 1110: plastics – polyamides – accelerated conditioning of test specimens. Technical committee: ISO/TC 61/SC 9 thermoplastic materials. Geneva, Switzerland: International Organization for Standardization ISO; 1995.
35. Woishnis WA, Ebnesajjad S, editors. Chemical resistance of thermoplastics. Amsterdam, Netherlands: Elsevier Science; 2012.
36. Cai Z, Liu X, Zhou Q, Wang Y, Zhu C, Xiao X, et al. The structure evolution of polyamide 1212 after stretched at different temperatures and its correlation with mechanical properties. Polymer 2017; 117:249–58.
37. Janssen PGA, Ligthart GBWL, Rulkens R. Polyamide containing monomer units of 1,4-butylene diamine. Patent WO2012110413 A1, 2012.
38. Armioun S, Pervaiz M, Sain M. Biopolyamides and high-performance natural fiber reinforced biocomposites biopolyamides and high-performance natural fiber-reinforced biocomposites. Austin, TX, USA: Scrivener Publishing LLC; 2017.
39. Kind S, Neubauer S, Becker J, Yamamoto M, Völkert M, Abendroth GV, et al. From zero to hero production of biobased nylon from renewable resources using engineered Corynebacterium glutamicum. Metab Eng 2014;25:113–23.
40. Quiles-Carrillo L, Boronat T, Montanes N, Balart R, Torres-Giner S. Injection-molded parts of fully biobased polyamide 1010 strengthened with waste derived slate fibers pretreated with glycidyl- and amino-silane coupling agents. Polym Test 2019;77:1–10.
41. Quiles-Carrillo L, Montanes N, Boronat T, Balart R, Torres-Giner S. Evaluation of the engineering performance of different bio-based aliphatic homopolyamide tubes prepared by profile extrusion. Polym Test 2017;61:421–9.
42. He J, Samanta S, Selvakumar S, Lattimer J, Ulven C, Sibi M, et al. Polyamides based on the renewable monomer, 1,13-tridecane diamine I: synthesis and characterization of nylon 13,T. Green Mater 2013;1:114–24.
43. ASTM D5336-15a. Standard classification system and basis for specification for polyphthalamide (PPA) injection molding materials. West Conshohocken, PA, USA: ASTM International.
44. Djukic S, Bocahut A, Bikard J, Long DR. Mechanical properties of amorphous and semi-crystalline semi-aromatic polyamides. Heliyon 2020;6:e03857.
45. Habib S, Weinman ST. A review on the synthesis of fully aromatic polyamide reverse osmosis membranes. Desalination 2021;502:114939.
46. Yan G-M, Wang H, Li D-S, Lu H-R, Liu S-L, Yang J, et al. Design of recyclable, fast-responsive and high temperature shape memory semi-aromatic polyamide. Polymer 2021;216.

Tanner Alauzen, Shaelyn Ross and Samy Madbouly*

12 Biodegradable shape-memory polymers and composites

Abstract: Polymers have recently been making media headlines in various negative ways. To combat the negative view of those with no polymer experience, sustainable and biodegradable materials are constantly being researched. Shape-memory polymers, also known as SMPs, are a type of polymer material that is being extensively researched in the polymer industry. These SMPs can exhibit a change in shape because of an external stimulus. SMPs that are biodegradable or biocompatible are used extensively in medical applications. The use of biodegradable SMPs in the medical field has also led to research of the material in other applications. The following categories used to describe SMPs are discussed: net points, composition, stimulus, and shape-memory function. The addition of fillers or additives to the polymer matrix makes the SMP a polymer composite. Currently, biodegradable fillers are at the forefront of research because of the demand for sustainability. Common biodegradable fillers or fibers used in polymer composites are discussed in this chapter including Cordenka, hemp, and flax. Some other nonbiodegradable fillers commonly used in polymer composites are evaluated including clay, carbon nanotubes, bioactive glass, and Kevlar. The polymer and filler phase differences will be evaluated in this chapter. The recent advances in biodegradable shape-memory polymers and composites will provide a more positive perspective of the polymer industry and help to attain a more sustainable future.

Keywords: biodegradable; biomedical applications; biopolymer; composites; shape-memory; stimulus.

12.1 Introduction

Shape memory polymers, otherwise known as SMPs, are mechanically active materials that can easily exhibit a change in shape. This change in shape is caused by the introduction of an external stimulus. Owing to shape-memory polymers' ability to withstand large deformations, the material is often referred to as a smart material [1, 2]. SMPs can be fabricated to be: temporary shape, permanent shape, or deformed. The geometries are then able to change shape by triggering the shape-memory effect, which is often done through an external stimulus. Some common types of stimuli include heat

---
*Corresponding author: Samy Madbouly,** Plastics Engineering Technology, Penn State Behrend, Erie, USA, E-mail: sum1541@psu.edu
Tanner Alauzen and Shaelyn Ross, Plastics Engineering Technology, Penn State Behrend, Erie, USA

This article has previously been published in the journal Physical Sciences Reviews. Please cite as: T. Alauzen, S. Ross and S. Madbouly "Biodegradable shape-memory polymers and composites" *Physical Sciences Reviews* [Online] 2021, 6. DOI: 10.1515/psr-2020-0077 | https://doi.org/10.1515/9781501521942-012

and light. On the molecular level, SMPs are made up of switching domains and permanent net points. The permanent net point sections are tangible net points linked with large transition temperatures and covalent net points. The domain switch is performed by an accurate melting temperature or glass transition temperature [3].

Like all polymers, shape memory materials have a glass transition temperature (T_g). The glass transition temperature or the melt temperature (T_m), can also represent the transition temperature of an SMP (T_{trans}) that is thermally induced. A transition temperature is the temperature at which above the original shape is restored, and below the shape freezes in its current orientation [4]. Understanding the transition temperature is crucial when understanding the properties and possibilities of SMPs.

Besides being lightweight, easily processed, and cost-effective, the main advantage of SMPs is their ability to recover from higher strain percentages than other materials, including shape memory alloys [5]. The shape memory effects are broken into the strain recovery rate and strain fixity rate, which are a measure of the polymer's ability to maintain its shape and return to its original shape, respectively [4]. When using a biopolymer in the creation of an SMP, there is the added advantage of biodegradability and biocompatibility. This is made possible because biopolymers are made from naturally occurring materials such as glycerol and lactide that do not harm the environment or bodily functions [5, 6]. This makes them popular in medical applications, where negative interaction with the body is not wanted, and break down of the material while in the body is often desired. The main downfall of shape memory polymers is their decreased strength and stiffness. Fortunately, this disadvantage can be remedied to an extent by including fillers or additives in the material [4].

SMPs have been made into fibers, which act as a functional filler. These fillers can allow the material to act like a multifunctional material. Like all polymers, different SMP applications may require different fillers to achieve the product's desired functionality. In SMPs, too much filler will often lead to loss of mechanical properties and recovery rate. Fillers for SMPs must be carefully selected because only a small number of fillers enhance the shape memory effect. Most fillers common in the polymer market will decrease the shape memory effect in terms of recovery rate and fixation rate [1, 7].

SMP composites are typically reinforced with long fibers, short fibers, and nanofibers. SMPs have also been used in accordance with carbon materials to improve their reinforcement, chemical resistance, and physical properties. Some carbon materials included in SMPs are graphene, carbon nanotubes, and carbon fibers. Other fillers often used with SMPs include glass fibers, silica, and spandex fiber. Long fibers are typically most suitable for obtaining good reinforcement properties in SMPs [1, 8].

12.2 Classification of shape memory polymers

As previously stated, SMPs are able to undergo large changes in shape after exposure to an external stimulus. There are many different types of SMPs, which depend on the composition, type of stimulus, and the shape memory function in the polymer [9]. The categories used for the classification and organization of SMPs are shown in Figure 12.1. Each classification is then broken into multiple sections that help define the different types of SMPs available by net points, composition, stimulus, and shape memory function. The different types of SMPs mentioned in this classification will be explained in more detail in the following sections.

12.2.1 Composition and structure

SMPs can be designed based on the chemical structure and composition. For example, SMPs can be created via crosslinked homopolymers, segmented block copolymers, polymer blends, supramolecular polymer networks, and polymer composites. Cross-linked homopolymers can either be chemically or physically crosslinked. In both cases, crosslinking within the hard segments allows the polymer to memorize the original shape it is molded into whereas the switching segments allow the polymer to undergo the shape change. The soft segments not only allow the polymer to change shape but also allow the shape change to be reversible when the polymer reaches the transition temperature [10, 11]. A block copolymer SMP focuses on the fact that each segment or block has a separate transition temperature. In this situation, the monomer with the higher transition temperature will determine the permanent shape returned to whereas the monomer with the lower transition temperature will determine when the SMP

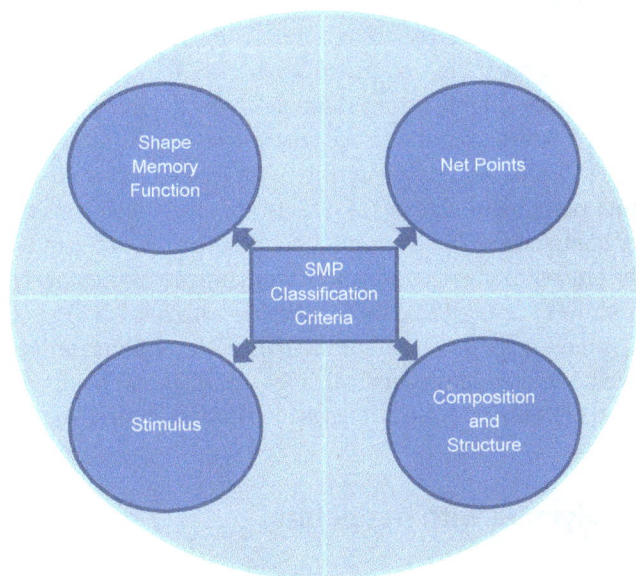

Figure 12.1: Classification of shape memory polymers.

changes shape. A common example of a biodegradable block copolymer SMP is a polylactic acid polycaprolactone-co-lactide (PLA/PCLA) or polylactic acid polylactic-co-glycolic acid (PLA/PLGA) blend, where the addition of the PCLA and PLGA gives the desired permanent phase [9].

SMP blends can be categorized as miscible and immiscible. Similar to the block SMP, one material will determine the permanent phase whereas the other determines the switching phase when the SMP changes shape [10]. Supramolecular polymer networks are different, in that they are made through an *in-situ* polymerization reaction that creates hydrogen or electrostatic bonds, which are classified as supramolecular interactions. These polymers have a high molecular weight, are extremely flexible, and are suitable for a wide range of biomedical applications [12, 13].

Finally, there are SMP composites and nanocomposites. These SMPs include fillers that affect the shape memory ability of the polymer, and the mechanical properties. Because SMPs are typically very flexible, to change shape, small amounts of fillers can be added to improve the mechanical properties of the material. Examples of these reinforcing fillers for a biodegradable polymer are silica, clay, and natural fibers [10]. By including a filler, SMPs are also able to have different stimuli or shape memory effects. This includes, but is not limited to multi, indirect, gradient, and two-way shape memory effects [9]. SMP nanocomposites in particular have such excellent mechanical properties and shape recovery, that they are often included in aerospace applications [10].

When discussing biodegradable SMP composites, these are partially made up of bio-based polymers such as Polylactic Acid (PLA) and Polycaprolactone (PCL) whereas including other materials that allow the material to be classified as a composite. These materials have increased in popularity recently, in an effort to decrease the use of nonrenewable resources and focus on more eco-friendly options. PLA allows the composite to have, along with biodegradability, desirable mechanical strength, and biocompatibility. Unfortunately, it poses limitations because it is very crystalline and difficult to process. When using PLA in an SMP composite, PLA is the hard segment giving the composite its T_g and permanent shape [14]. PCL is another biodegradable polymer used within polymer composites. The material is semicrystalline and can be made into an SMP through the cross-linking or copolymerization processes previously discussed. PLA degrades slower than PCL, and because it has a T_g of approximately −60 °C, the T_m is the trigger for the shape memory effect. Both materials when included in an SMP have flourished in the medical field. Although not nearly as popular as PLA and PCL, starch or chitosan-based SMPs is also valid options [14].

12.2.2 Shape memory polymers with net points

As discussed previously, the composition or structure of an SMP can be broken down into many different types, including crosslinked homopolymers. The cross-linked structure can either be physically or chemically crosslinked. Biodegradable SMPs can be either physically or chemically crosslinked as well. For example, some physically crosslinked SMPs are made of PCL/epoxy blends, which yield more brittle samples because the increase of PCL in the blend interferes with the crosslinking of the epoxy [4].

Chemically cross-linked SMPs are commonly known as thermosets because of the covalent bonds between the polymer chains that do not allow remelting. Once the material is cross-linked during polymerization, the covalent bonds set the initial shape of the SMP and do not allow further change. Cross-linking of the polymer chains also gives the covalent net points needed for the polymer to return to its original shape [9, 15].

Typically, a thermosetting SMP has higher thermal and mechanical properties than a physically cross-linked SMP, so thermoset SMPs are often used for structural support applications. For example, the cross-linking process for polyurethane SMPs has proven to improve the creep resistance of the material and increase the temperature at which the SMP returns to its original shape [16]. This improvement of properties makes the polymer more desirable in applications where strength is crucial. When a chemically cross-linked SMP is not created during polymerization, it is carried out through post-processing methods such as crosslinking agents or electromagnetic radiation [9]. Cross-linking agents such as peroxides are commonly used to start the reaction in materials that are composed of unsaturated carbon bonds to produce the cross-linked polymer. In contrast, when using electromagnetic radiation to create SMPs, photo-initiators and light-sensitive monomers must be included in the SMP for the process to be successful [9].

Physically cross-linked polymers include both amorphous and semi-crystalline regions within the morphology, which allows for the shape change to take place in thermally-induced SMPs. Physically cross-linked SMPs can be remelted and the polymer's original shape that it returns to can be changed. SMPs with amorphous switching segments use the amorphous regions to create weak cross-links from hydrogen bonds, ionic bonds, and van der Waal forces between the molecules [10]. For this type of SMP, the soft segment determines the shape memory effects present, and the transition temperature is the glass transition temperature of this soft segment. The hard segments are typically those with the higher transition temperature so that the material will always return to this initial orientation.

Again, polyurethane SMPs are a good example of how this process works. The addition of monomers allows the T_g and composition of polyurethane SMPs to be changed so that they function as desired. The hard segments use hydrogen bonding or polar interaction to maintain shape whereas the soft segments absorb any stresses by flexing and moving chains [9]. Copolymerization can be used within this process as well, to obtain the desired properties such as toughness and biodegradability.

SMPs with crystalline switching segments use the crystals to temporarily hold the shape of the soft segments in place. Crystalline switching segments are much easier to use because they have a narrower transition temperature region. This makes the right combination of soft segments easily attainable compared to those SMPs with amorphous switching segments so that the shape can be changed easily when below the T_g. Thermoplastic polyurethanes are some of the most commonly used SMPs with semicrystalline segments [9]. Copolymerization can be used within this process as well, to obtain the desired properties such as toughness and biodegradability.

12.2.3 Stimulus method

SMP composites are often categorized based on their stimulus method. This includes heat, electricity, magnetics, light, and water. Thermal, or heat stimulus is the most common way to trigger the shape memory effect of the material. This method uses heat transfer methods such as conduction, convection, and radiation to transfer heat from the environment to the SMP itself and trigger the shape change by exceeding T_{trans} [17]. With this being the most popular stimulus method for SMPs, it is also the basis for many others. An SMP that is not initially stimulated by heat can be changed into a thermally reactive SMP by crosslinking the material [10, 18]. Similarly, SMPs that use heat as a stimulus method can also be made to respond to other stimuli by temporarily creating physical crosslinks.

By incorporating conductive fillers in otherwise thermally resistant materials, researchers have made it substantially easier to heat and cause shape changes in SMPs with an electrical current. As a current is passed through the conductive filler in an SMP, it begins to raise the temperature of the material as a whole. Again, once the material exceeds T_{trans}, the shape begins to revert to its initial shape [17]. Because of the conductive fillers present in these SMPs, they have more uniform heating and revert to their original shape at a faster rate. Conductive fillers such as carbon black nanofibers have been used in the recent creation of biodegradable polycaprolactone polydimethylsiloxane (PCL/PDMS) SMP composites for scaffolds in tissue engineering [19]. The nanofibers were created through electrospinning, and proved to maintain both conductivity and biocompatibility, making the composite a viable option in such medical applications.

Like electrically-induced SMP composites, magnetically activated SMPs use an alternate method of heating, called Joule heating, to raise the temperature of the material until it exceeds T_{trans} [20]. Three ways of heating the SMP composite are through current losses, hysteresis losses, and rotational losses. Within this stimulus method, it is crucial to note that because the heat itself is generated by a magnetic field inside the polymer, the temperature increases at a significantly faster rate. This quick recovery rate makes electrically-induced SMPs very desirable in biomedical applications such as implants and medical instruments, partially accredited to their versatility [17, 21].

Light-induced SMPs are powered by both photochemical reactions and conversion of light to heat to change shape. For photochemical reactions, light-sensitive molecules are included in the polymer matrix so that after so many are included, the switching segments will respond to the wavelength of the light and deform. In contrast, when the functional groups within the material convert light to heat a larger range of light can be absorbed to allow deformation. Light-induced SMP composites are favored in precise applications and are most often made into films [17, 22].

SMPs that are water driven are a recent development and function because water has a plasticizing effect on the material. As water penetrates the polymer it increases the flexibility of the polymer matrix [23]. This drops the T_g of the material and as it reaches the temperature of the surroundings, the SMP will begin to change shape. These SMPs are fairly easy to control and are compatible with multiple materials, so their popularity is growing rapidly [17, 24].

For example, water-induced SMPs can be created with carboxymethyl cellulose sodium (CMC) and polyvinyl alcohol (PVA). The CMC is highly sensitive to water, absorbing it fairly easily. This causes the T_g of the SMP composite to decrease until the shape begins to revert to its original. Along with the shape memory function of this particular composite, the addition of a water stimulus allows the PVA to repair itself while in application [25]. This is shown on a molecular level in Figure 12.2.

As shown on the left, hydrogen bonds within the material are destroyed by swelling when a polymer composite becomes wet. In the same case, the lubricity of the chains is what makes the shape memory function work so well in water-induced SMPs. After reverting to its initial shape, the CMC chains manipulate themselves to entangle once again. Finally, as shown on the right, the PVA and CMC form new hydrogen bonds as the material begins to dry. This self-healing property allows any scratches or tears in the polymer to restore themselves, and any mechanical properties characteristic of the initial dry material and shape [25].

12.2.4 Shape memory function

Understanding shape memory function is crucial to tailoring SMPs properties, abilities, and applications. Each SMP is composed of soft and hard segments, and net points. The

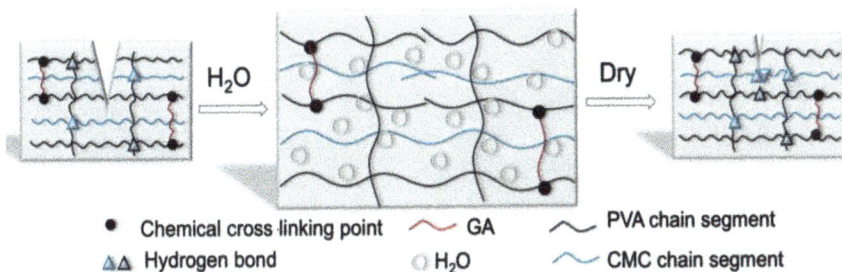

Figure 12.2: Self-healing ability of water-induced CMC/PVA SMP composites on a molecular level. Reproduced with permission from Yang et al. [25].

transition temperature, or T_{trans}, controls the shape of the SMP. When the temperature of a thermally-induced SMP is greater than the determined transition temperature, the soft segment is very flexible and is able to change shape. In contrast, when the temperature of the SMP is below the transition temperature the shape is fixed and has significantly less flexibility.

The shape memory function of a biodegradable SMP composite can be one-way, two-way, multi-shape, triple-way, or multifunctional [4]. Figure 12.3 demonstrates the shape memory effect within a one-way, thermally-induced SMP on a molecular level.

The first step represents the frozen sample in its initial shape. It is then heated above the transition temperature of the material to create flexibility in the soft segments of the polymer chains. The second step shows the newly flexible molecular chains, which then undergo stress-induced deformation to change the shape of the SMP. This begins the programming of the shape memory effect in the polymer. Although still applying the stress to form the secondary shape, the SMP is cooled below the transition temperature to freeze the chains. Steps three and four show the polymer chains as they are being deformed by stress and as their temperature is decreased below the transition temperature to freeze the temporary shape. The SMP is then reheated above the transition temperature in step 5 to return the sample to its initial shape and finally cools off in that initial orientation [10]. This process can then be repeated by introducing heat above the transition temperature of the material, causing the SMP to transition from its temporary shape to its initial shape through heat stimulus rather than stress.

Two-way shape memory effect SMPs have become more popular in recent years. Many two-way SMPs are composed of photo-deformation materials, which have shape

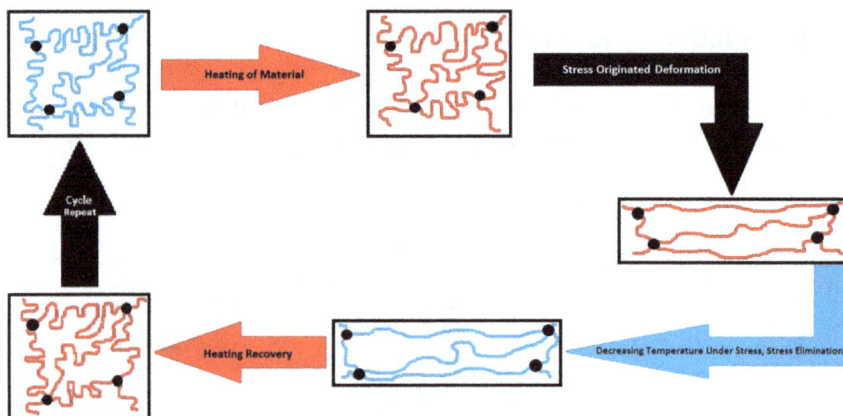

Heating of Material

Stress Originated Deformation

Cycle Repeat

Heating Recovery

Decreasing Temperature Under Stress, Stress Elimination

Figure 12.3: Schematic representation of thermally-induced SMPs. The black dots represent crosslinked net points. The red lines represent the polymer chains when heated whereas the blue lines represent the polymer chains at a cooler temperature.

transitions stimulated by light. The process is similar to that of one-way SMPs, with the exception that a two-way SMP can change shape through both heating and cooling while the stress is applied [4, 26]. If this thermal cycle is completed multiple times to create two programmed shapes before removal of the stress, it is considered a multi-shape SMP rather than a two-way SMP. An example of this would include shape-changing gels, where one-layer changes at a certain transition temperature whereas the other layer changes at another. In this case, after establishing the first programmed shape at an elevated temperature the sample is reloaded to create the second shape. It is then cooled again and finally unloaded to finish programming. To be successful, the temperature for the added shape must be less than the previously programmed permanent shape, and the temperature range for thermal transitioning must be relatively wide [4, 26].

Similarly, triple-way SMPs are different in that they can be programmed to memorize a second temporary shape while remembering the initial and permanent ones as well. This makes a total of three shapes memorized. A triple-way SMP is just the addition of another switch to a one-way SMP, which can be carried out in multiple ways. One way of creating a triple-way SMP is by using the phase change from crystalline to amorphous or crystalline to crystalline within the base polymer as a transition point for programming [27, 28]. Like one-way, the sample would be heated above both transition temperatures and then deformed into the first shape. As the polymer is cooled the stress is held to establish the programmed shape between the initial and permanent stages. This is made possible because within triple-way SMPs, both crosslinking density and T_g are controlled by the types of bonding present in the material: covalent and hydrogen bonds [29–31]. By using a triple-way SMP to establish three different known shapes, more complexity is being added to the system.

Multifunctional SMPs can be in any of the previously discussed forms, but also have added functional properties that are unrelated to the shape memory ability such as optical properties, thermochromism, and biodegradability. With that being said, all biodegradable SMPs are multifunctional because of their degradability [4]. Any other multifunctional properties desired are added through the use of additives and fillers, thus making the SMP more specialized. As a result, multifunctionality makes SMPs more attractive in various applications because of the complex nature of the polymer and the added function from the additives and fillers.

12.3 Shape-memory polymer composites

Polymer composites are materials in which multiple phases exist because of the addition of fillers into the polymer matrix [32]. Fibers or other fillers are introduced to materials to alter the mechanical properties needed for an application. The amount of filler material can also influence the mechanical properties of the polymer. Fillers allow polymers to be modified to fit specific design needs. Biodegradable and natural fillers are at the forefront of research today because of concerns with sustainability in the industry.

One common filler used in SMPs is attapulgite clay. Attapulgite clay is a natural clay which is made up of a three-dimensional network of closely packed rods. The shape memory effect has been investigated for polyurethane and attapulgite clay composites by Xu et al. [33]. The effect of different contents of thermally treated and untreated clay on the shape memory effect of polyurethane was investigated [33]. Four samples at different amounts of clay (30, 20, and 10% clay, pure PU) were analyzed by comparing hardness and load. Every mix that included heat-treated clay showed a hardness value greater than the pure polyurethane material with respect to load. Disproportionately, all the materials with untreated clay showed a hardness value less than pure polyurethane with respect to load. The study also showed that the polyurethane-based SMP's hardness decreased as a function of increasing load. Mechanical mixing of the clay and polyurethane was found to be inefficient and multiple phases. This nonuniform phase prohibited the potential for enhanced mechanical properties for SMPs. TEM images of nontreated and heat-treated clay can be seen in Figure 12.4 [33, 34].

Natural fibers and fillers have become more popular with the movement today for biodegradability for composite materials. These natural fibers and fillers are relatively cheap, low density, biodegradable, and provide relatively high mechanical strength. Although these natural fibers have some advantages, they also have some disadvantages such as compatibility issues, low moisture resistance, and inconsistency in the chemical composition. Some examples of these natural fibers are flax, hemp, and Cordenka. Flax and hemp are naturally occurring fibers which are grown rather than manufactured [35, 36].

The main properties that are affected by natural fibers depend on a number of variables such as fiber aspect ratio, orientation, volume fraction of the fibers, the

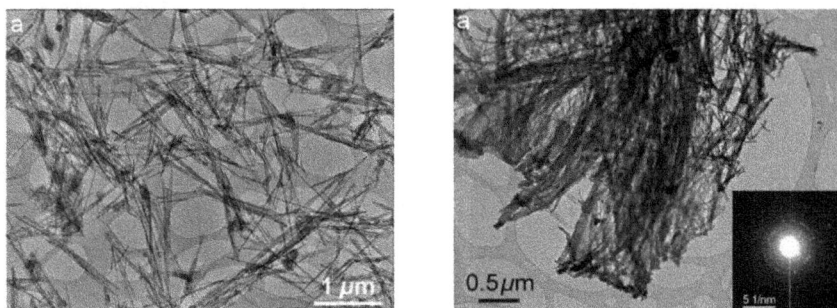

Figure 12.4: TEM images of non-treated clay powder (left) and TEM images of heat-treated clay powder with electron diffraction pattern (right). Reproduced with permission from Xu et al. [33].

adhesion between fiber-matrix sections, and the stress transfer at the interference. If good tensile strength is required in an application, fiber orientation and good adhesion properties are needed. If interfacial adhesion is weak between the polymer and fiber matrix, premature failure can occur. Interfacial adhesion is often difficult with natural fibers because of the main component being cellulose. The cellulose structure causes the fiber to display very hydrophilic behavior. This issue of hydrophilicity can cause issues in polymers if too many natural fibers are used [37, 38].

Cordenka is a man-made natural fiber that has been gaining interest in the polymer industry. The process of Cordenka production first starts with a viscous process, where the natural cellulose pulp is dissolved in a sodium hydroxide and carbon disulfide mixture. Mechanical treatment is performed, and the fiber is then extruded into its final form. Cordenka is known for having good dimensional stability, excellent adhesion, and high thermal stability. When added to PLA, otherwise known as polylactic acid, Cordenka fibers doubled the impact strength of the composite PLA material. Cordenka fibers have been researched extensively to be used in PLA because the material has relatively low impact resistance for a polymer material [35, 39].

Hemp fibers are naturally occurring fibers that are of much interest in the polymer industry today. Hemp is a plant native to the central Asia region and requires no pesticides or fertilizer to grow. The hemp fibers are extracted from the plant and are now being used in polymer composites. Hemp shives are a byproduct during the production of hemp fibers. Shives have a very similar composition to hemp fibers, although the shives contain less cellulose. Hemp shives are often used as a lime binder in construction, but these hemp byproducts are currently being researched to be used in polymer composites. Hemp fibers are naturally hydrophilic, so this factor should be considered when working with hydrophobic materials such as polyester. The difference in attractiveness can lead to the weak interaction between the polymer matrix and fiber. If too much filler is present in the polymer matrix, the weak interaction between matrices can cause stress concentrations leading to premature failure. Hemp fibers are used in polymer materials to make the material fully biodegradable [37, 40, 41].

Flax is another common natural fiber often added to biodegradable composite materials. Flax fibers are also natural fibers and contain mostly cellulose. For this reason, interfacial forces between the polymer matrix and the filler matrix need to be examined to ensure no lack of mechanical properties will occur [37, 42]. Flax material is commonly used because of the filler's good recyclability, mechanical properties, and low cost. Recently polymer composites containing flax have been covered with an outer laminate layer of basalt to help fight against aging. The resistance against aging in polymer composites containing flax is key in applications that could be exposed to water because of the cellulose in flax wanting to absorb water [29].

Another filler commonly used in composites with biodegradable properties is silica (SiO_2). Silica particles in polymers have the potential to increase the biodegradability and bioactivity of the composite material. PLA/SiO_2 composites have been found to degrade faster than pure PLA, which in turn indicates that the biodegradability of PLA material can be enhanced by silica. When introduced to PLA, silica may attack the enzyme and water molecules from the PLA chains. Because the silica material is very hydrophilic, this attack may cause the PLA material to visibly degrade faster [43, 44].

Carbon fibers have shown much improvement in bridging the performance gap of biodegradable SMPs. Carbon fibers improve the mechanical properties of SMPs and their shape recovery force. Carbon fibers are often chosen as a filler because of their high stiffness and strength properties. Aside from these properties, carbon fibers also improve the polymer's thermal and chemical stability [45–47]. Other carbon materials often used in SMPs include graphene and carbon nanotubes. These materials can be used to improve the conductive properties of polymers if needed. Multiple-walled carbon nanotubes have been used with SMPs to enhance shape memory capability, flexibility, and energy storage properties [1].

Calcium phosphate ceramic composites have recently shown to have good biocompatibility and the ability to bond with bone tissues. Calcium phosphate often makes the polymer matrix very brittle, which is why the materials have been limited in use inside the skeleton where loads could cause failure. The degradation of calcium phosphate ceramics depends on crystallinity and phase composition. Despite the challenges in loaded applications, many efforts have been made to develop calcium phosphate ceramic composite materials in the polymer industry [43, 48].

Bioactive glass, otherwise known as BG, is evolving and becoming very versatile for polymer composites. Two specific types of BGs are silicate bioactive glasses and borate-borosilicate bioactive glasses. The borate-borosilicate-based BGs have high reactivity and bioactivity. Biocompatibility of borate-borosilicate BG materials remains an issue because of toxicity from the high concentration of boron released. On the other hand, silicate bioactive glasses have been extensively researched and have been considered as fillers for polymer composites used for bone tissue engineering. BGs are made by either the melt-quench method or the sol-gel method. The BGs made from gel have improved bioactivity and cellular responses because of a larger volume of mesopores and more silanols on the surface. This allows for sol-gel-based BG materials to be modified in terms of phase composition, texture, structure, and microstructure. BGs can also be doped with trace elements to further improve biological response and surface reactivity [43, 49].

Electrospun glass nanofibers, otherwise known as EGNFs, are gaining interest in the polymer industry. EGNFs are of interest because the material can be ductile, which is desirable for tougher polymer composites. Reinforcing and toughening are two properties in which EGNFs excel. EGNFs have also been surface modified to create amine and epoxy end groups. Figure 12.5 below displays scanning electron microscope images of EGNFs. The letter A designates the fibers which underwent ultrasonication whereas the letter B displays the glass microfibers after they were cut from glass wool. The number 1 designates the untreated fibers. Epoxide-functionalized fibers are represented through the number 2 whereas the amine-functionalized fibers can be seen in the number 3 images [32, 50].

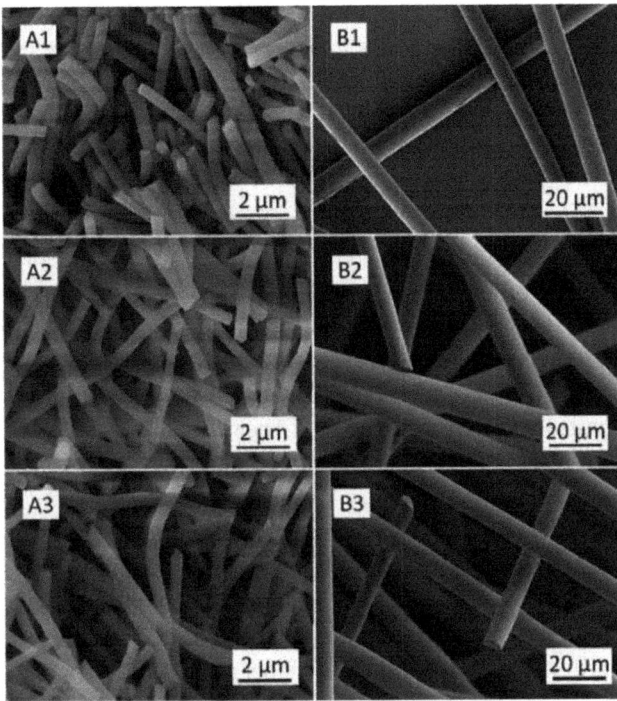

Figure 12.5: Scanning electron microscope images of EGNFs (electrospun glass nanofibers). Reproduced with permission from Wang et al. [32].

Kevlar fibers are becoming more popular because of being an organic filler. Kevlar fibers have been researched as a potential replacement for polymer composite materials in polypropylene because of glass materials having high filler content, easy to fracture fibers, and the potential to wear machinery. Kevlar fibers can be further improved by surface hydrolyzation or ball milling in phosphoric acid. Surface treatment of Kevlar fibers can be done by boiling the fibers in acetone for around 4 h. This surface treatment process is used to remove impurities and sizing agents from the Kevlar fibers. These processes can improve the interfacial interaction between the fiber and polymer matrix while also improving mechanical properties. Kevlar fibers are currently more expensive when compared to glass fibers. Although Kevlar fibers are more expensive, the fibers have good toughness and reinforcement properties [51, 52].

Another common material used as a filler in polymer composites is carbon nanotubes or CNTs. Carbon nanotubes are cylindrical in shape and formed by wrapping graphene sheets which are bonded with carbon atoms through hybridization. CNTs have been used in polymer composites because of their excellent chemical and physical properties. One downside of CNTs is price, as these materials are typically expensive. CNTs and polymer composites are typically combined by synthesized mixing of the two materials or dispersing CNTs in polymer melts. To improve the dispersion of CNTs in a polymer matrix, monomers are introduced followed by situ polymerization [53]. The homogeneous dispersion of CNTs throughout polymer composites is important because CNTs display strong van der Waals forces and have a high aspect ratio. This can cause CNTs to group together in the polymer composite, which can ultimately lead to a decrease in mechanical properties [54, 55].

Graphene is another filler commonly used in polymer composites and is the strongest material which has been discovered by mankind. Graphene fillers display good tensile strength while also being more thermally stable than CNTs. There are two different methods of synthetic routes which have been developed to synthesize polymer and graphene composites including mixing and template methods. During solution mixing, a polymer and graphene are combined in organic or aqueous solvents. Evaporation of the solvents occurs to complete the process. The melt blending method has become more attractive because this method does not involve solvents in the creation of a polymer-graphene material [53, 56, 57].

Along with improving mechanical properties such as tensile strength and Young's modulus, the use of graphene in biodegradable SMPs can increase the antibacterial nature of the composite. The antibacterial properties of PLA with the addition of graphene oxide (GO) have been studied, and it has been found that PLA has a lower cytocompatibility because of its hydrophobic nature. As a higher concentration of GO is incorporated in the composite, it begins to show an increase in cytocompatibility and antibacterial properties. After reaching 5% GO, the cytocompatibility begins to decrease again. The increase in cytocompatibility observed is accredited to the increase in surface roughness when adding graphene to the matrix, which allows the PLA to be more hydrophilic, enhancing the compatibility with important cell functions while in the human body [58].

Another common carbon-based material used in polymer composites is carbon black. Carbon black has been used to achieve Joule-based heating shape memory behaviors. In Joule-based heating, carbon black material can reduce the resistance of SMPs. This resistance can be reduced by carbon black because carbon black filler is spherical in shape. The spherical shape of the carbon black fillers can cause uniformly distributed carbon black particles to not come in contact with each other, further causing an increase in resistance. SMPs that contain carbon black fillers can also achieve fast recovery when a voltage is applied to the SMP. Carbon black polymer composites have also shown strain-induced softening effects, which should be considered if using an application containing carbon black fillers [59, 60].

12.4 Applications

One of the most important aspects of a polymer is the real-world situations the polymer can be applicable in. Certain polymers are selected for an application because they satisfy the criteria of the given application. SMPs have gained a reputation of being a versatile product in the ever-changing market. SMPs first gained recognition in the application field in the 1960s when the material was used to create PE thermal contraction tubes. The mass production of PE thermal contraction tubes by using SMPs made the material quickly gain reputation, which led to further scientific research [1].

There are many factors that need to be taken into consideration for SMPs in medical applications including biocompatibility, sterilization, biodegradability, shape transformation triggers, and mechanical properties. Any implanted material must be biocompatible to prevent adverse effects in the human body. Along with this, most implantable materials for the medical field display shape changes after implantation. These implantable devices need a triggered stimulus from the body environment or an external source. Often in medical applications, thermal stimuli are used [61]. Biodegradability is key to the medical field because biodegradable materials implanted in the body require no secondary operation of removal. Biodegradable polymers in the medical field are often used for embolization, drug delivery, tissue engineering, stents, and wound closure [10, 62].

These favorable properties of biodegradable SMPs make sterilization more difficult, as the materials react differently to sterilization methods than typical polymers. Sterilization methods typically used in the medical field are ethylene oxide, gamma or electron beam radiation, and steam. It was found that with any SMP, steam sterilization is not a viable option because the required temperature is above the transition temperature of the SMP. This can change the morphology of the polymer matrix or even eliminate the previously established shape memory effect, which is not desired [63]. Irradiation techniques use a lower temperature and are common with polymers because of their ability to penetrate packaging, but are not viable for biodegradable SMPs because of their susceptibility to chain scission and reduction in molecular weight. This in turn decreases important properties such as T_g and swelling ratio of materials such as PCL, which can allow actuation of the shape memory effect before desired [61, 64].

With this in mind newer, more unconventional sterilization methods using low temperatures were experimented with to sterilize biodegradable SMPs. Low-temperature plasma (LTP) was shown successfully in sterilizing biodegradable SMPs by using hydrogen peroxide and gas plasma to sterilize polyurethane foams. They showed some surface changes and porosity, but the sterilization method showed no effect on the shape memory abilities of the material. The disadvantage of this method proved to be the cytotoxic response after sterilization, in which sterilization caused the breakdown of the cells. Owing to this, ethylene oxide sterilization has proven to be the best method in sterilizing biodegradable SMPs. This method helps to maintain the integrity of the medical device while not negatively affecting the shape memory abilities of the polymer. Despite its toxicity, proper removal of any toxic residue makes this method the most viable for thermally induced, biodegradable SMPs. With this, it should not be used for any water-activated biodegradable SMPs, as the high humidity during sterilization can affect the integrity of the polymer [61].

Recently, endovascular treatment methods such as occlusion with platinum coils have been used to help treat embolization otherwise known as aneurysm rupture. These platinum coils today are often coated with biocompatible polymers to increase treatment success. Polyurethane-based biodegradable shape-memory foam has been a common coating used for the platinum coils in embolization applications. Although biodegradable shape-memory polyurethane foam has helped increase treatment success for aneurysms, totally eliminating the platinum coil from the assembly is being looked into. Purely polyurethane-based SMP coils are being researched as a potential for replacing the platinum material. The coil design used in embolization treatment has certain risks which could potentially be avoided with the development of a biodegradable shape-memory polyurethane foam. This compact foam would first be sent to the location of the aneurysm, then a thermal stimulus would trigger the material to expand and fill the cavity. Work for this potential fully foam solution is still in the primary stages of development but has been promising. Figure 12.6 below displays a flower-shaped medical instrument made of SMP material, aneurysm coils, and aneurysm SMP foam material [1, 65, 66].

Controlled release formulations (CRFs) have recently been researched for drug release applications. One advantage of biodegradable CRF is that the polymer matrix can be programmed to release the drugs at faster or slower rates along with pulsating movements. In biodegradable shape-memory-based CRFs, the polymer matrix has two major roles: shape recovery either reduces or enhances drug release, or shape recovery has no effect on the amount of drug released into the surrounding area. Biodegradable CRFs also degrade in the body, which can alter the amount of drug released. As the polymer matrix degrades more, drug release rates into the body will increase because the barrier is eroding. Certain drugs need to be studied before being used in controlled release systems because some drugs have been found to eliminate the shape-memory behavior of the polymer [65, 67].

Stents have been used in the medical field specifically to treat coronary arterial stenosis, otherwise known as the narrowing of arteries which provide blood to the heart. Metal stents have been used in the medical field to treat coronary arterial stenosis, but these stents have been shown to not function properly after six months. Coronary stents need to maintain a scaffolding strength for at least six months to resist stresses from vessel remodeling. Biodegradable stents are being discussed as a potential replacement for metal stents, especially because of the elimination of a secondary removal operation. Drug-eluting biodegradable SMP stents are currently being developed to help in healing. These biodegradable SMP stents are also in development with the potential to be used in other areas of the body including bile ducts, intracranial arteries, trachea, and the urethra [65].

Biodegradable SMPs for tissue engineering have recently gained much attention in the medical field. Tissue engineering refers to the ability to heal damaged tissues by

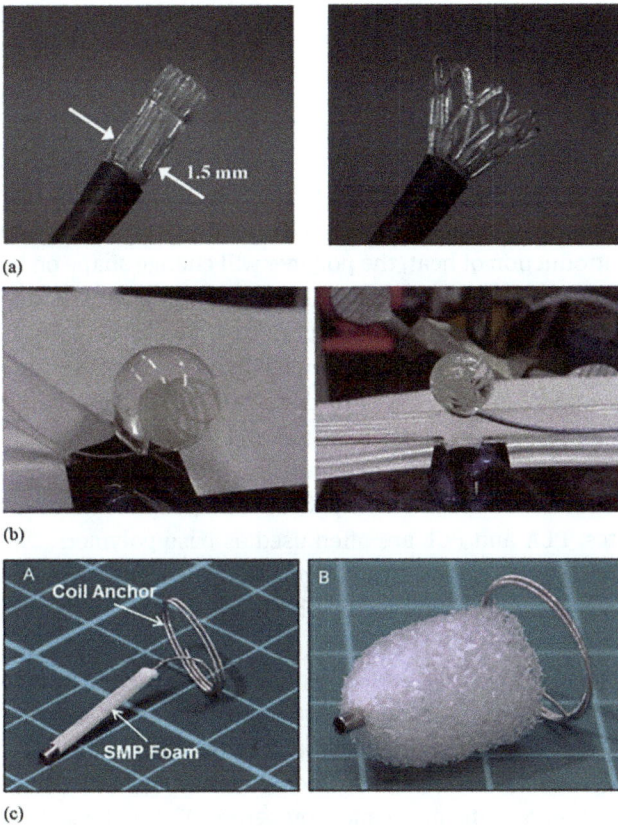

Figure 12.6: Medical devices that include a shape-memory polymer material.
(a) Surgical instruments composed of SMP material being activated by magnetic doping. (b) SMP-based aneurysm coil. (c) SMP foam-based aneurysm embolization structure. Reproduced with permission from Mu et al. [1].

including materials which serve as biological substitutes which restore, improve, and maintain tissue function. Biodegradable SMPs have more specifically been researched with regard to bone tissue engineering. One potential advantage of biodegradable SMPs in bone tissue engineering is that the material can match irregular shapes in bone defects [65, 68].

Biodegradable SMP wound closure devices have recently gained interest in the medical field. These devices include clamps, strings, and sutures. Biodegradable SMP sutures contain fibers which are expanded to 200–1000% strain. These structures are then met with heat, which causes the shape-memory effect to be triggered and further closes the wounded area. These self-tightening sutures could simplify wound closure because of the material being biodegradable and performing their function properly [69, 70].

12.5 Conclusion

SMPs have the ability to change shape as a response to an external stimulus. External stimuli that trigger the shape memory effect include heat, electricity, magnetics, light, and water. Thermally-induced SMPs are most common, and a change in shape is obtained by heating the material past the transition temperature, or T_{trans} of the material. By holding the polymer in a manipulated shape as it cools, the temporary phase is established and upon the introduction of heat, the polymer will change shape on its own. Within a biodegradable SMP composite, a natural composition gives the sustainability desired in the industry to create a greener future.

SMPs can be classified using four categories: composition and structure, net points, stimulus, and shape memory function. Classification of an SMP allows a better representation of what is taking place on a molecular level and also shows how the difference in composition affects the shape memory effects and properties of the material. An SMP composite is characterized based on the polymer and additives used. For biodegradable SMP composites, PLA and PCL are often used as base polymers. The addition of a filler or additive creates a polymer composite, which enhances the properties of the polymer and oftentimes aids in the shape memory effect by the inclusion of a thermal reactive filler.

With biodegradable polymers, natural fillers are also popular because of concerns with sustainability. By including an environmentally friendly filler, the material maintains its sustainability and biocompatibility while still enhancing necessary properties. Biodegradable SMP composites use materials such as bioactive glass, clay, and natural fibers such as hemp and flax to reinforce the mechanical properties. These fillers allow the shape memory effect to stay intact while improving recovery and fixation rate, whereas inorganic fillers will interfere with the shape memory effect of these smart materials. Overall, the combination of biodegradability and shape memory function makes SMPs a growing topic within the medical field, where minimal interaction with the body and the ability to break into naturally occurring elements is desired.

Author contributions: All the authors have accepted responsibility for the entire content of this submitted manuscript and approved submission.
Research funding: None declared.
Conflict of interest statement: The authors declare no conflicts of interest regarding this article.

References

1. Mu T, Liu L, Lan X, Liu Y, Leng J. Shape memory polymers for composites. Compos Sci Technol 2018;160:169–98.

2. Lendlein A, Gould OE. Reprogrammable recovery and actuation behaviour of shape-memory polymers. Nat Rev Mater 2019;4:116–33.
3. Madbouly SA, Lendlein A. Shape-memory polymer composites. Adv Polym Sci 2010;226:41–95.
4. Vijayan PP. Mechanical properties of shape-memory polymers, polymer blends, and composites. In: Parameswaranpillai J, Siengchin S, George J, Jose S, editors. Advanced structured materials. Singapore: Springer; 2019:199–217 pp.
5. Tsujimoto T, Takayama T, Uyama H. Biodegradable shape memory polymeric material from epoxidized soybean oil and polycaprolactone. Polymers 2015;7:2165–74.
6. Vinod A, Sanjay M, Suchart S, Jyotishkumar P. Renewable and sustainable biobased materials: an assessment on biofibers, biofilms, biopolymers and biocomposites. J Clean Prod 2020;258: 120978.
7. Lu H, Yu K, Liu Y, Leng J. Mechanical and shape-memory behavior of shape-memory polymer composites with hybrid fillers. Polym Int 2010;59:766–71.
8. Gu J, Leng J, Sun H, Zeng H, Cai Z. Thermomechanical constitutive modeling of fiber reinforced shape memory polymer composites based on thermodynamics with internal state variables. Mech Mater 2019;130:9–19.
9. Strzelec K, Sienkiewicz N, Szmechtyk T. Classification of shape-memory polymers, polymer blends and composites. In: Parameswaranpillai J, Siengchin S, George J, Jose S, editors. Advanced structured materials. Singapore: Springer; 2019:21–52 pp.
10. Qi X, Wang Y. Novel techniques for the preparation of shape-memory polymers, polymer blends and composites at mirco and nanoscales. In: Parameswaranpillai J, Siengchin S, George J, Jose S, editors. Advanced structured materials. Singapore: Springer; 2019:53–83 pp.
11. Basit A, L'hostis G, Durand B. The recovery properties under load of a shape memory polymer composite material. Mater Werkst 2019;50:1555–9.
12. Liu J, Tan CSY, Yu Z, Li N, Abell C, Scherman OA. Tough supramolecular polymer networks with extreme stretchability and fast room-temperature self-healing. Adv Healthcare Mater 2017;26:1–7.
13. Jing Z, Li J, Xiao W, Xu H, Hong P, Li Y. Crystallization, rheology and mechanical properties of the blends of poly(l-lactide) with supramolecular polymers based on poly(d-lactide)–poly(ε-caprolactone-co-δ-valerolactone)–poly(d-lactide) triblock copolymers. RSC Adv 2019;9: 26067–79.
14. Ruiz-Rubio L, Pérez-Álvarez L, Vilas-Vilela JL. Biodegradable shape-memory polymers. In: Parameswaranpillai J, Siengchin S, George J, Jose S, editors. Advanced structured materials. Singapore: Springer; 2019:216–36 pp.
15. Haskew MJ, Hardy JG. A mini-review of shape-memory polymer-based materials: stimuli-responsive shape-memory polymers. Johnson Matthey Technol Rev 2020;64:425–42.
16. Buckley CP, Prisacariu C, Caraculacu A. Novel triol-crosslinked polyurethanes and their thermorheological characterization as shape-memory materials. Polymer 2007;48:1388–96.
17. Liu T, Zhou T, Yao Y, Zhang F, Liu L, Liu Y, et al. Stimulus methods of multi-functional shape memory polymer nanocomposites: a review. Compos A Appl Sci Manuf 2017;100:20–30.
18. Dai L, Tian C, Xiao R. Modeling the thermo-mechanical behavior and constrained recovery performance of cold-programmed amorphous shape-memory polymers. Int J Plast 2020;127: 102654.
19. Kai D, Tan MJ, Prabhakaran MP, Chan QY, Liow SS, Ramakrishna S, et al. Biocompatible electrically conductive nanofibers from inorganic-organic shape memory polymers. Colloids Surf, B 2016;148: 557–65.
20. Murugan MS, Rao S, Chiranjeevi MC, Revathi A, Rao KV, Srihari S, et al. Actuation of shape memory polymer composites triggered by electrical resistive heating. J Intell Mater Syst Struct 2017;28: 2362–71.

21. Chen H, Wang L, Zhou S. Recent progress in shape memory polymers for biomedical applications. Chin J Polym Sci 2018;36:905–17.
22. Fang T, Cao L, Chen S, Fang J, Zhou J, Fang L, et al. Preparation and assembly of five photoresponsive polymers to achieve complex light-induced shape deformations. Mater Des 2018;144:129–39.
23. Liu Y, Chen H, Yang G, Zheng X, Zhou S. Water-induced shape-memory poly(d,l-lactide)/microcrystalline cellulose composites. Carbohydr Polym 2014;10:101–98.
24. Liu C, Lu H, Li G, Hui D, Fu Y-Q. A 'cross-relaxation effects' model for dynamic exchange of water in amorphous polymer with thermochemical shape memory effect. J Phys D Appl Phys 2019;52:345305.
25. Yang J, Zheng Y, Sheng L, Chen H, Zhao L, Yu W, et al. Water induced shape memory and healing effects by introducing carboxymethyl cellulose sodium into poly(vinyl alcohol). Ind Eng Chem Res 2018;57:15056–3.
26. Zare M, Prabhakaran MP, Parvin N, Ramakrishna S. Thermally-induced two-way shape memory polymers: mechanisms, structures, and applications. Chem Eng J 2019;374:706–20.
27. Wu Y, Hu J, Zhang C, Han J, Wang Y, Kumar B. A facile approach to UV/heat dual-responsive triple shape memory polymer. J Mater Chem 2014;1:1–4.
28. Tian M, Gao W, Hu J, Xu X, Ning N, Yu B, et al. Multidirectional triple-shape-memory polymer by tunable cross-linking and crystallization. ACS Appl Mater Interfac 2020;12:6426–35.
29. Živković I, Fragassa C, Pavlović A, Brugo T. Influence of moisture absorption on the impact properties of flax, basalt and hybrid flax/basalt fiber reinforced green composites. Compos B Eng 2017;111:148–64.
30. Feng X, Zhang G, Zhuo S, Jiang H, Shi J, Li F, et al. Dual responsive shape memory polymer/clay nanocomposites. Compos Sci Technol 2016;129:53–60.
31. Zhang H, Wang D, Wu N, Li C, Zhu C, Zhao N, et al. Recyclable, self-healing, thermadapt triple-shape memory polymers based on dual dynamic bonds. ACS Appl Mater Interfac 2020;12:9833–41.
32. Wang G, Yu D, Kelkar AD, Zhang L. Electrospun nanofiber: emerging reinforcing filler in polymer matrix composite materials. Prog Polym Sci 2017;75:73–107.
33. Xu B, Huang WM, Pei YT, Chen ZG, Kraft A, Reuben R, et al. Mechanical properties of attapulgite clay reinforced polyurethane shape-memory nanocomposites. Eur Polym J 2009;45:1904–11.
34. Liang W, Wang R, Wang C, Jia J, Sun H, Zhang J, et al. Facile preparation of attapulgite-based aerogels with excellent flame retardancy and better thermal insulation properties. J Appl Polym Sci 2019;136:47849.
35. Frackowiak S, Ludwiczak J, Leluk K. Man-made and natural fibres as a reinforcement in fully biodegradable polymer composites: a concise study. J Polym Environ 2018;26:4360–8.
36. Ferreira FV, Pinheiro IF, Mariano M, Cividanes LS, Costa JC, Nascimento NR, et al. Environmentally friendly polymer composites based on PBAT reinforced with natural fibers from the amazon forest. Polym Compos 2019;40:3351–60.
37. Terzopoulou ZN, Papageorgiou GZ, Papadopoulou E, Athanassiadou E, Reinders M, Bikiaris DN. Development and study of fully biodegradable composite materials based on poly(butylene succinate) and hemp fibers or hemp shives. Polym Compos 2016;37:407–21.
38. Krishnasamy S, Thiagamani SM, Kumar CM, Nagarajan R, Shahroze RM, Siengchin S, et al. Recent advances in thermal properties of hybrid cellulosic fiber reinforced polymer composites. Int J Biol Macromol 2019;141:1–3.
39. Meredith J, Coles SR, Powe R, Collings E, Cozien-Cazuc S, Weager B, et al. On the static and dynamic properties of flax and Cordenka epoxy composites. Compos Sci Technol 2013;80:31–8.
40. Murali B, Mohan DC. Chemical treatment on Hemp/polymer composites. J Chem Pharmaceut Res 2014;6:419–23.

41. Tanasă F, Zănoagă M, Teacă CA, Nechifor M, Shahzad A. Modified hemp fibers intended for fiber-reinforced polymer composites used in structural applications—a review. I. Methods of modification. Polym Compos 2020;41:5–31.

42. Abida M, Gehring F, Mars J, Vivet A, Dammak F, Haddar M. A viscoelastic–viscoplastic model with hygromechanical coupling for flax fibre reinforced polymer composites. Compos Sci Technol 2020; 189:108018.

43. Dziadek M, Stodolak-Zych E, Cholewa-Kowalska K. Biodegradable ceramic-polymer composites for biomedical applications: a review. Mater Sci Eng C 2017;71:1175–91.

44. Rizal S, Fizree HM, Hossain MS, Gopakumar DA, Ni EC, Khalil HA. The role of silica-containing agro-industrial waste as reinforcement on physicochemical and thermal properties of polymer composites. Heliyon 2020;6:1–9.

45. Li F, Scarpa F, Lan X, Liu L, Liu Y, Leng J. Bending shape recovery of unidirectional carbon fiber reinforced epoxy-based shape memory polymer composites. Compos A Appl Sci Manuf 2019;116: 169–79.

46. Li F, Leng J, Liu Y, Remillat C, Scarpa F. Temperature dependence of elastic constants in unidirectional carbon fiber reinforced shape memory polymer composites. Mech Mater 2020;148: 103518.

47. Ren Z, Liu L, Liu Y, Leng J. Damage and failure in carbon fiber-reinforced epoxy filament-wound shape memory polymer composite tubes under compression loading. Polym Test 2020;85: 106387.

48. Park J, Park SY, Lee D, Song YS. Shape memory polymer composites embedded with hybrid ceramic microparticles. Smart Mater Struct 2020;29:1–9.

49. Jäger F, Mohn D, Attin T, Tauböck TT. Polymerization and shrinkage stress formation of experimental resin composites doped with nano-vs. micron-sized bioactive glasses. Dent Mater J 2020;40:110–5.

50. Nazarnezhad S, Baino F, Kim HW, Webster TJ, Kargozar S. Electrospun nanofibers for improved angiogenesis: promises for tissue engineering applications. Nanomaterials 2020;10:1609.

51. Fu S, Yu B, Tang W, Fan M, Chen F, Fu Q. Mechanical properties of polypropylene composites reinforced by hydrolyzed and microfibrillated Kevlar fibers. Compos Sci Technol 2018;163:141–50.

52. Vasudevan A, Senthil Kumaran S, Naresh K, Velmurugan R. Layer-wise damage prediction in carbon/Kevlar/S-glass/E-glass fibre reinforced epoxy hybrid composites under low-velocity impact loading using advanced 3D computed tomography. Int J Crashworthiness 2020;25:9–23.

53. Sun X, Sun H, Li H, Peng H. Developing polymer composite materials: carbon nanotubes or graphene? Adv Mater 2013;25:5153–76.

54. Wang E, Dong Y, Islam MZ, Yu L, Liu F, Chen S, et al. Effect of graphene oxide-carbon nanotube hybrid filler on the mechanical property and thermal response speed of shape memory epoxy composites. Compos Sci Technol 2019;169:209–16.

55. Blokhin AN, Dyachkova TP, Maksimkin AV, Stolyarov RA, Suhorukov AK, Burmistrov IN, et al. Polymer composites based on epoxy resin with added carbon nanotubes. Fullerenes, Nanotub Carbon Nanostruct 2020;28:45–9.

56. Gao W, Zhao N, Yu T, Xi J, Mao A, Yuan M, et al. High-efficiency electromagnetic interference shielding realized in nacre-mimetic graphene/polymer composite with extremely low graphene loading. Carbon 2020;157:570–7.

57. da Luz FS, del-Río MT, Nascimento LF, Pinheiro WA, Monteiro SN. Graphene-incorporated natural fiber polymer composites: a first overview. Polymers 2020;12:1601.

58. Arriagada P, Palza H, Palma P, Flores M, Caviedes P. Poly(lactic acid) composites based on graphene oxide particles with antibacterial behavior enhanced by electrical stimulus and biocompatibility. J Biomed Mater Res 2017;106:1051–60.

59. Lei M, Chen Z, Lu H, Yu K. Recent progress in shape memory polymer composites: methods, properties, applications and prospects. Nanotechnol Rev 2019;8:327–51.
60. Lan X, Liu L, Liu Y, Leng J. Thermomechanical and electroactive behavior of a thermosetting styrene-based carbon black shape-memory composite. J Appl Polym Sci 2017;135:45978.
61. Wong Y, Kong J, Widjaja LK, Venkatraman SS. Biomedical applications of shape-memory polymers: how practically useful are they? Sci China Chem 2014;57:476–89.
62. Omid SO, Goudarzi Z, Kangarshahi LM, Mokhtarzade A, Bahrami F. Self-expanding stents based on shape memory alloys and shape memory polymers. J Compos Compd 2020;2:92–8.
63. Ecker M, Danda V, Shoffstall AJ, Mahmood SF, Joshi-Imre A, Frewin CL, et al. Sterilization of thiol-ene/acrylate based shape memory polymers for biomedical applications. Macromol Mater Eng 2016;302:1600331.
64. Zhao Y, Zhu B, Wang Y, Liu C, Shen C. Effect of different sterilization methods on the properties of commercial biodegradable polyesters for single-use, disposable medical devices. Mater Sci Eng C 2019;105:110041.
65. Peterson GI, Dobrynin AV, Becker ML. Biodegradable shape memory polymers in medicine. Adv Healthcare Mater 2007;6:1–16.
66. Xiao R, Huang WM. Heating/solvent responsive shape-memory polymers for implant biomedical devices in minimally invasive surgery: current status and challenge. Macromol Biosci 2020;20:2000108.
67. Wang K, Stradman S, Zhu XX. A mini review: shape memory polymers for biomedical applications. Front Chem Sci Eng 2017;11:143–53.
68. Ramaraju H, Akman RE, Safranski DL, Hollister SJ. Designing biodegradable shape memory polymers for tissue repair. Adv Funct Mater 2020;30:1–40.
69. Mao Q, Hoffmann O, Yu K, Lu F, Lan G, Dai F, et al. Self-contracting oxidized starch/gelatin hydrogel for noninvasive wound closure and wound healing. Mater Des 2020;194:108916.
70. Li M, Chen J, Shi M, Zhang H, Ma PX, Guo B. Electroactive anti-oxidant polyurethane elastomers with shape memory property as non-adherent wound dressing to enhance wound healing. Chem Eng J 2019;375:121999.

Supplementary Material: The online version of this article offers supplementary material (https://doi.org/10.1515/psr-2020-0077).

Israd H. Jaafar*, Sabrina S. Jedlicka and John P. Coulter

13 Poly(glycerol sebacate) – a revolutionary biopolymer

Abstract: Novel materials possessing physical, mechanical, and chemical properties similar to those found *in vivo* provide a potential platform for building artificial microenvironments for tissue engineering applications. Poly(glycerol sebacate) is one such material. It has tunable mechanical properties within the range of common tissue, and favorable cell response without surface modification with adhesive ligands, and biodegradability. In this chapter, an overview of the material is presented, focusing on synthesis, characterization, microfabrication, use as a substrate in *in vitro* mammalian cell culture, and degradation characteristics.

Keywords: biodegradable; biopolymer; elastomer.

13.1 Introduction

Poly(glycerol sebacate) (PGS) is a biodegradable and biocompatible elastomer that has been used in a wide range of biomedical applications, including drug delivery, microfluidic devices, and tissue engineering scaffolds. The material possesses similar mechanical properties to those of soft body tissues and is mechanically tunable by altering cure temperature, curing time, and molar ratio of the sebacic acid and glycerol constituents. As an example, the Young's Moduli for living tissue such as muscle is 0.01–0.5 MPa and 0.7–16 MPa for skin [1, 2]. An increased cure temperature in PGS preparation correlates to an increased amount of cross-linking, resulting in a greater elastic modulus. This tunability in mechanical properties, with elastic modulus compared to that of soft tissue, has resulted in its focused use as a replacement/scaffold in biomedical applications [3].

The material was first developed by Wang et al. [4]. The elastomer forms a covalently crosslinked three-dimensional network of random coils with hydroxyl groups attached to its backbone. Crosslinking and hydrogen bonding interactions between hydroxyl groups are thought to influence its properties such as tunability in mechanical properties and biodegradability. PGS is synthesized through the polycondensation of glycerol and sebacic acid. The molecular structure of glycerol $(CH_2(OH)CH(OH)CH_2OH)$ is given in Figure 13.1. Glycerol is an organic compound that is

*Corresponding author: Israd H. Jaafar, Mechanical Engineering, Utah Valley University, 800 W University Parkway, Orem, UT, USA, E-mail: Israd.Jaafar@uvu.edu
Sabrina S. Jedlicka, Materials Science and Engineering, Lehigh University, Bethlehem, PA, USA
John P. Coulter, Mechanical Engineering and Mechanics, Lehigh University, Bethlehem, PA, USA

This article has previously been published in the journal Physical Sciences Reviews. Please cite as: I. H. Jaafar, S. S. Jedlicka and J. P. Coulter "Poly(glycerol sebacate) – a revolutionary biopolymer" *Physical Sciences Reviews* [Online] 2021, 6. DOI: 10.1515/psr-2020-0071 | https://doi.org/10.1515/9781501521942-013

colorless, odorless, viscous, and used widely in medical and pharmaceutical formulations. Its molecule consists of a backbone of three carbon atoms where the carbon atoms at each end are bonded to two hydrogen atoms (C–H) and a hydroxyl group (–OH), and the central carbon atom is bonded to a single hydrogen atom (C–H) and a hydroxyl group (–OH). It is referred to as tryhydric or trifunctional alcohol because of these three hydrophilic hydroxyl/alcohol groups that are responsible for its slight solubility in water and hygroscopic nature. Its substructure is a basic building block in many lipids. By choosing specific molar ratios where glycerol exceeds that of sebacic acid in PGS synthesis, unreacted hydroxyl groups may be present on the PGS surface. This promotes the hydrophilic nature of the surface, which in turn may encourage cell attachment in culture. The use of trifunctional alcohol also results in a polymer with many branches that plays a role in crosslinking. Researchers have also studied the effect of synthesizing PGS with higher molar ratios of glycerol to sebacic acid on the physical, mechanical, and biodegradation characteristics of the resulting polymer [5, 6].

Sebacic acid is a dicarboxylic acid. Its molecular structure ($HOOC(CH_2)_8COOH$) is given in Figure 13.2. It is a derivative of castor oil. In its pure state, it is a white flake or powdered crystal that is soluble in ethanol, ether, and soluble slightly in water. Sebacic acid is also the natural metabolic intermediate in ω-oxidation of medium-to-long-chain fatty acids [4]. It also has the appropriate chain length for reaction with glycerol. Shorter chained dicaroboxylic acids are more acidic and are more likely to cyclize (reaction resulting in the formation of an aromatic or ring structure) during polymerization. Longer chained dicarboxylic acids on the other hand are more hydrophobic and would react poorly with glycerol.

The US Food and Drug Administration has approved glycerol and polymers containing sebacic acid for medical applications. PGS pre-polymer is formed by the polycondensation reaction of equimolar amounts of these components, in which a small number of crosslinks are formed. For a given stoichiometric mixture of glycerol and sebacic acid, the elasticity of the resulting polymer depends on the curing

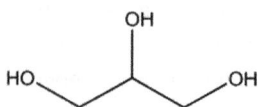

Figure 13.1: Molecular structure of glycerol.

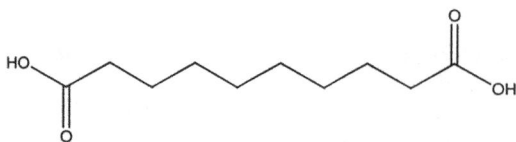

Figure 13.2: Molecular structure of sebacic acid.

temperature and time [7, 8]. Its physical and mechanical property is also tunable based on glycerol to sebacic ratios [5, 6]. The molecular structure of a PGS monomer is given in Figure 13.3. Upon further heating in a vacuum environment to draw out the moisture, crosslinks are formed between the polymeric chains, resulting in a crosslinked thermoset, as shown in Figure 13.4.

PGS has been shown to have mechanical properties comparable to that of common soft tissue [9]. It is a biodegradable, soft, and mechanically stable elastomer, analogous to vulcanized rubber [4]. PGS in the fully cured state is also both biocompatible and nontoxic [10]. *In vitro* tests have shown that PGS demonstrates a favorable response to various cells, such as NIH 3T3 fibroblasts, human aortic smooth muscle cells (SMCs), and human aortic endothelial cells that is as good as PLGA and tissue culture PS [11]. Comparative *in vivo* biocompatibility tests, between PGS and PLGA, through subcutaneous implantation in Sprague-Dawley rats, showed no granulation or scar formation in the former (at 60 days), and a slower formation of fibrous encapsulation [4]. Studies have also found that PGS is superior, in terms of mechanical properties, biodegradation characteristics, and cell response and morphology to PLGA [12, 13]. Researchers have shown that PGS is osteoconductive and contributes to bone regeneration by attracting host progenitor/stem cell populations, with mechanical properties that transmit signals favoring differentiation and matrix maturation toward bone regeneration [14, 15].

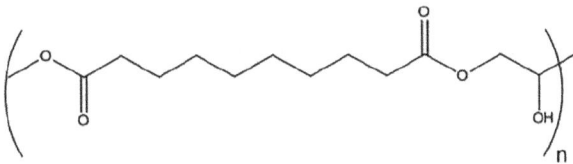

Figure 13.3: Molecular structure of a PGS monomer.

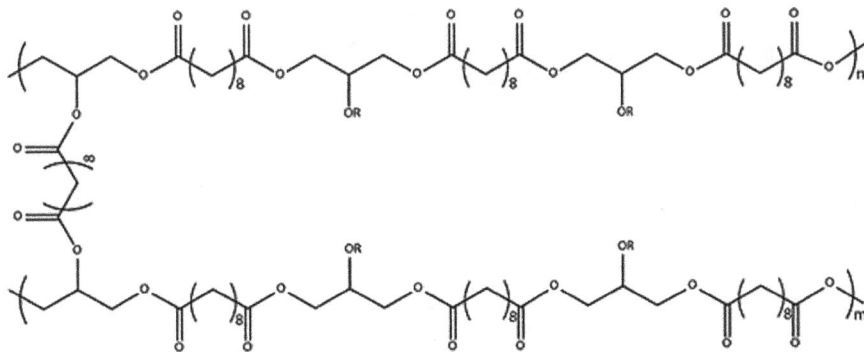

Figure 13.4: Crosslinked PGS.

Unlike PLGA, PGS primarily degrades by surface erosion, resulting in a linear degradation profile of mass, preservation of geometry and intact surface, and retention of mechanical strength. Unlike PLGA, PGS also has many carboxyl and hydroxyl groups within its chains that allow for linkages with bioactive moieties such as ester, amide, ether, and acetal bonds [16].

13.2 Synthesis and material characterization

Characterization studies on PGS have looked at its mechanical characteristics [8], thermal analysis via differential scanning calorimetry (DSC) [4, 17], molecular bonding schemes via Fourier transform infrared (FTIR) spectroscopy [4, 17], *in vitro* and *in vivo* degradation characteristics [13], biocompatibility [12, 18], swelling behavior [8, 13], and shape-memory effect [17].

PGS pre-polymer is synthesized based on a well-established method [4]. Jaafar et al. [7] reported that for the synthesis, equimolar (1:1) amounts of anhydrous glycerol (Sigma–Aldrich) and sebacic acid (*Sigma–Aldrich*) is mixed in an airtight glass jar that is partially immersed in a heated silicone bath. The mixture is gradually heated to 120 °C under nitrogen gas flow and stirred with a mechanized rotor at 50 rpm for 24 h. The gas flow is then stopped, and vacuum (at 150 mmHg) is applied for 48 h. This will result in a highly viscous PGS prepolymer. This experimental setup is illustrated in Figure 13.5. It is noted that others have investigated a method of microwave-assisted pre-polymerization which eliminates this time-consuming prepolymer preparation [19, 20].

Curing of the prepolymer at elevated temperatures (starting from 120 °C) and durations results in an elastomer with varying rigidity that depends on the number of

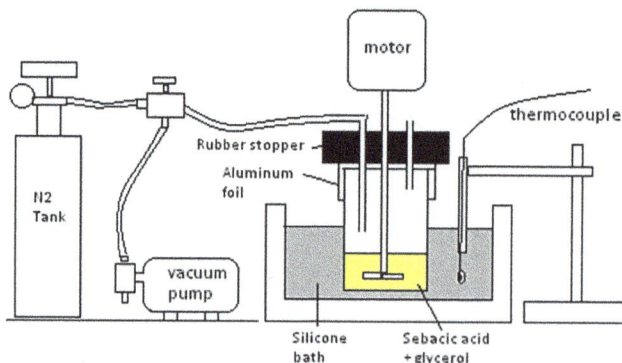

Figure 13.5: Schematic of the apparatus used for PGS prepolymer synthesis.

crosslinks that have formed. A higher temperature and duration of cure results in a stiffer and more rigid elastomer. Researchers have proposed that curing results in an esterification reaction where glycerol is removed from the polymer [21]. Covalent bonds are formed which lock the glycerol-sebacate chains into a random three-dimensional network of coils, which transforms the highly viscous pre-polymer into a solid elastomer [4].

13.2.1 Thermal analysis

The prepolymer was cured in a vacuum oven set at 150 mmHg at different temperatures (130, 140, and 165 °C) and durations as indicated in the legend of Figure 13.6. Differential Scanning Calorimetry (DSC) was used to quantify the extent of cure using heat of reaction or changes in the glass transition temperature (Tg), which would indicate a change in polymeric chain mobility. Neither was observed for PGS. However, it was useful in detecting the presence of crystallinity in the material. DSC measured both a recrystallization exotherm during cooling and a melting endotherm during heating. The Tg, which is observed as a slightly discernable step in the curves between −30 and −40 °C, does not appear to shift for all the cases. A broad melting transition is exhibited, initiating approximately at −15 °C and extending to 35 °C for the prepolymer. The melting transition ranges for the cured samples occur below 15 °C which indicates that PGS is fully amorphous at room temperature.

The trend noted for the endothermic melting transition of PGS, with increasing cure temperature and duration, is a narrowing of the transition region, reduction in peak magnitude, and a general shift of the peaks toward lower temperatures. This is an apparent indicator that the polymer chains are "locked-in" because of a greater number of crosslinks that form with higher curing temperature and duration. Hence the latent energies decrease with a higher cure temperature and duration. The relative differences in the sample's degree of crystallinity can be quantified by measuring the relative differences between areas under the melting peaks, as given in Table 13.1. A tenfold decrease in energy for the sample cured at 165 °C compared to those cured at 130 and 140 °C, indicates that the relative percentage crystallinity in the material drops significantly when this curing temperature and cure time is applied. Hence this indicates that as the material is cured and thus acquires more crosslinks, crystallinity is suppressed. In the crosslinked regions of the PGS elastomer, polymer chains are bound together and are unable to flow past one another. However, melting still does occur during DSC temperature ramping because of heterogeneity in the PGS structure, where this is attributed to the un-crosslinked regions. Analysis of PGS prepared in a porous format also exhibited the same trend of decreasing endotherms with cure of PGS, at 120, 140, and 165 [22].

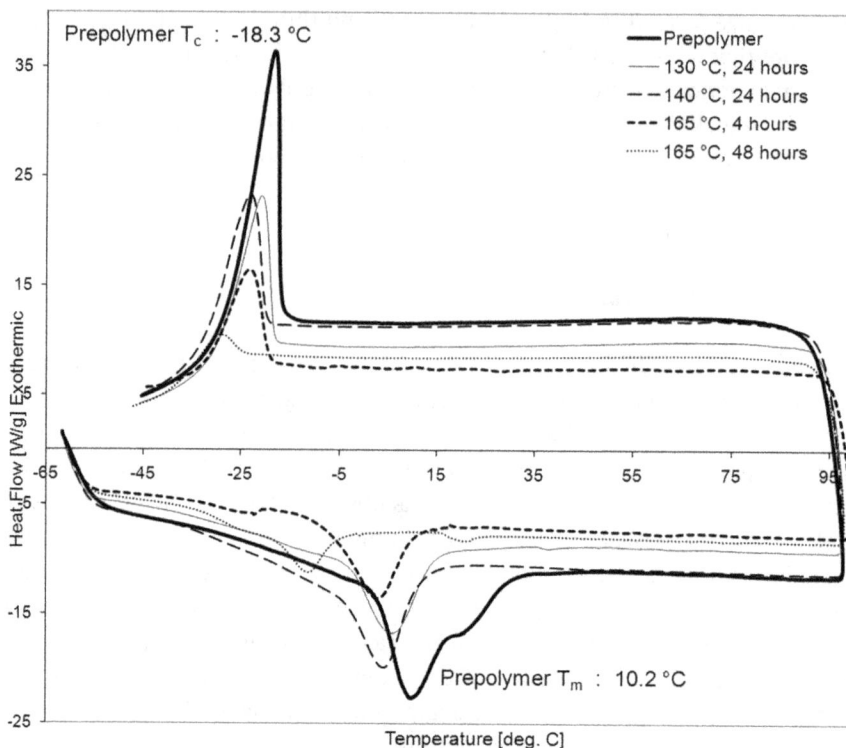

Figure 13.6: DSC curves for PGS prepolymer and samples cured at different temperatures and cure durations as indicated in the legend [7] Reproduced with permission.

Table 13.1: Relative differences in melting enthalpies between cured samples and the prepolymer.

Sample	Melting endotherm peak area [J/g]	Relative difference with pre-polymer [J/g]
Pre-polymer	49.57	0.00
130 °C	48.69	0.88
140 °C	45.26	4.31
165 °C, 4 h	28.50	21.10
165 °C, 24 h	27.94	21.60

13.2.2 ATR-FTIR

The ATR-FTIR spectra of PGS pre-polymer, glycerol, and sebacic acid are shown in Figure 13.7. For the alcohol component of PGS, i.e. glycerol, a broad and intense –OH stretch in the 3000–3600 cm^{-1} range is observed. This indicates that the hydroxyl groups are hydrogen-bonded. This stretch is also observed to a lesser extent for the PGS pre-polymer, which appears comparatively weaker for the PGS cured sample. For sebacic acid, peaks at approximately 1699, 1300, and 930 cm^{-1} are the absorption bands of the aliphatic acid/carboxylic acid (C=O) group. These separate peaks are no longer evident in both the PGS pre-polymer and cured samples. Evidently observed in all the spectra are peaks at 2924 and 2853 cm^{-1}, which are for the stretch of the alkane – CH$_2$ group. For this group, the peaks are more intense for the PGS pre-polymer and glycerol, particularly when compared to sebacic acid. The intense peaks at approximately 1172 (C–O) and 1700 cm^{-1} (C=O) are for ester bonds that form within the three-dimensional PGS network.

The ATR-FTIR spectra for PGS cured at 165 °C at different durations are plotted in Figure 13.8. The ATR-FTIR spectra for PGS cured at 165 °C at different durations are plotted in Figure 13.9. The main changes in the spectra when cure time is increased is (i) the significant reduction in the carboxylic acid –COOH bend at 1418 cm^{-1} (Figure 13.9 inset A) signifying that as the cure time proceeds acids further react with alcohols in the mixture to form the ester bonds, and (ii) the O–H stretch at approximately 3300 cm^{-1} (Figure 13.8 inset B), signifying that the hydroxyls are consumed to form the ester-based photoresist-coated. These appear to be the most prominent spectral indicators of the extent of cure in the material. Similar trends in spectral changes are observed for PGS cured at different temperatures, where an increase in cure temperature results in changes to spectral peaks that are similar to the increase in

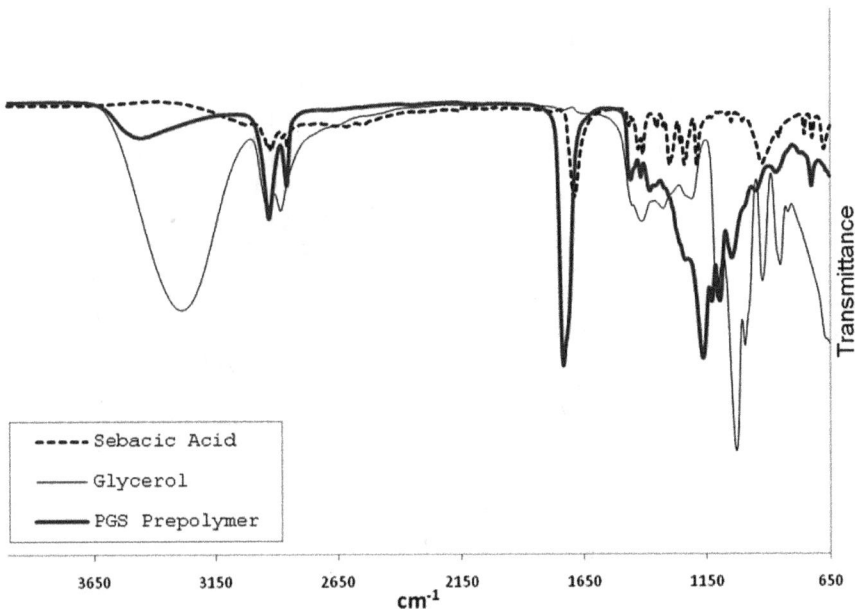

Figure 13.7: ATR-FTIR spectra of sebacic acid, glycerol, and PGS prepolymer [23].

cure duration [22]. A marked difference however is in the shift and sharpening of the C=O bond with increasing temperature, reflecting a greater degree of crosslinking.

It is noted that for PGS samples that underwent a 16-week *in vitro* degradation period, in simulated body fluid, spectral regions that exhibited notable changes include the C=O stretch with peaks that reduce in intensity, resembling that of the prepolymer [22]. Furthermore, the peak at approximately 1700 cm^{-1}, a characteristic of carboxylic acids, re-emerges. Both these observations suggest a breakdown in crosslinks.

13.2.3 Mechanical properties

The stiffness of PGS can be tuned based on its curing temperature and duration, and a molar ratio of glycerol and sebacic acid [6, 8, 24–26]. The Young's Modulus has been reported to be 0.07, 0.12, 0.43, 2.02, and 2.30 MPa for samples cured at 120, 130, 140, 150, and 165 °C cured samples, respectively [7], which is consistent with the increase in crosslinks that form in the polymer network. Chen et al. reporting Young's Modulus values of 0.06, 0.22, and 1.2 MPa for curing temperatures of 110, 120, and 130 °C. Reported results may differ amongst researchers [8, 27–29], which may be because of differences in synthesis parameters, specimen geometry, and testing method and apparatus. Compressive moduli for porous PGS samples, cured at 120, 140, and 165 °C

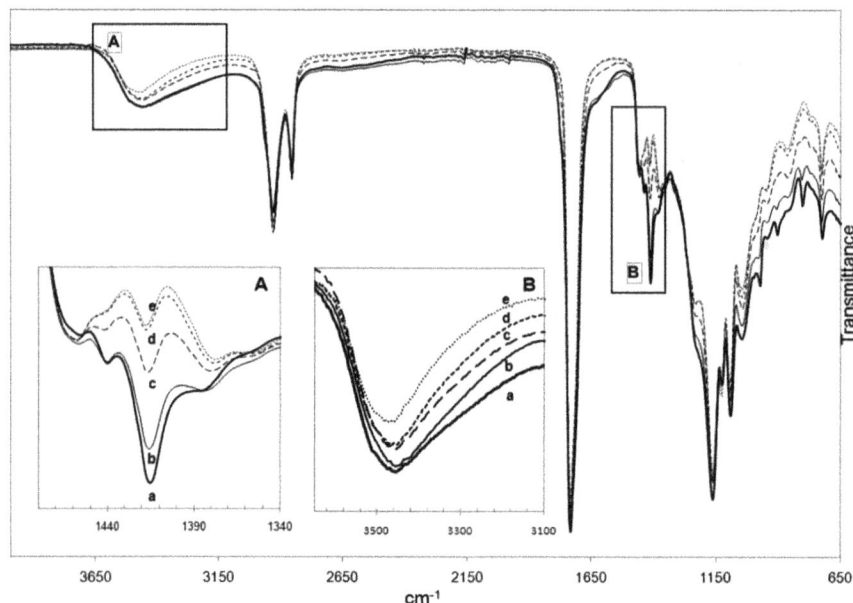

Figure 13.8: ATR-FTIR transmittance spectra for PGS (a) pre-polymer, and samples cured at 165 °C for (b) 2 h, (c) 4 h, (d) 10 h, and (e) 24 h [23].

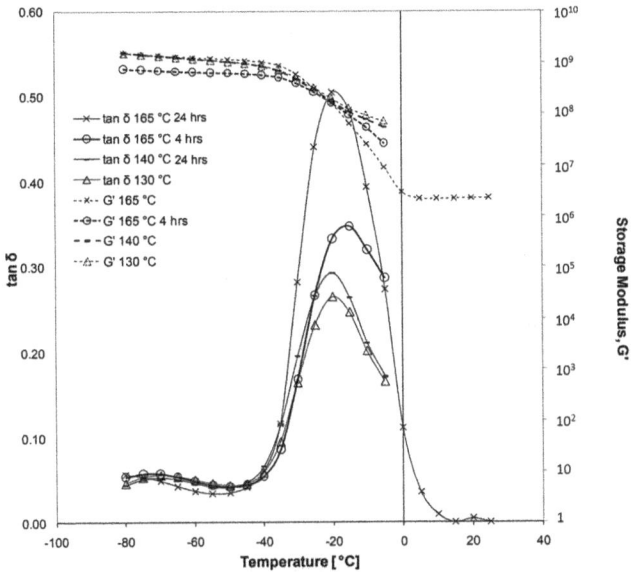

Figure 13.9: Loss factor curves for PGS cured at 130, 140, and 165 °C [7] Reproduced with permission.

has been reported to be 0.030, 0.034, and 0.115 kPA, respectively. Similar results linking an increase in mechanical properties to the degree of crosslinking have also been reported by others [30]. PGS is an elastomer that can undergo large deformations without losing its original geometry after the removal of load. For example, the material has shown to have a 330% strain to failure [4, 8, 12].

The loss factor (tan δ) curves obtained from dynamic mechanical analysis (DMA) at 1 Hz and 5 °C/min for PGS cured at 130, 140, and 165 °C for 4 and 24 h are plotted in Figure 13.9 [7]. The position of the peaks suggests, as in the DSC analysis, that the glass transition temperature remains constant regardless of the degree of cure. Interestingly, the tan δ peak heights are observed to increase with the degree of cure. This can be explained, as in the previous discussion, by noting that the degree of crystallinity of the 130 and 140 °C cured samples would be higher compared to the heavily crosslinked 165 °C cure sample.

13.3 Microfabrication

Researchers have shown that PGS can be microfabricated with microscale topography and features [16, 31–34]. Casting is one of the methods adopted for fabricating films of PGS with microfeatures and the results presented here is that which are reported by Jaafar [23]. Silicon wafers with microscale topography prepared using established methods used in the microelectronics industry are typically used as molds for this purpose. To aid in demolding, a layer of 90% w/v sucrose in dH$_2$O is spin-coated onto the wafers. This method is adapted from that used by others where the coating is used as a sacrificial layer through dissolution in dH$_2$O after the casting process [31]. To ensure fluidity and ability of PGS to enter into microcavities on the silicon mold, PGS prepolymer is mixed with tetrahydrofuran (THF) at 50% w/v. THF should be allowed to evaporate under a laboratory fume hood before subsequent curing of PGS under vacuum pressure. Cooling of the cured material is also done under vacuum pressure. Sample results of this procedure are shown in the SEM images provide in Figure 13.10–13.12. This capability for microscale feature fabrication of PGS may widen its use as a substrate that provides biomimetic surface features for enhanced *in vitro* cell–substrate interaction. In addition to providing topographical cues for cell-substrate interaction, mechanical cues may also be tuned by varying the topographical geometries. The stiffness of each micropillar, for example, may be varied by customizing both the height and diameter.

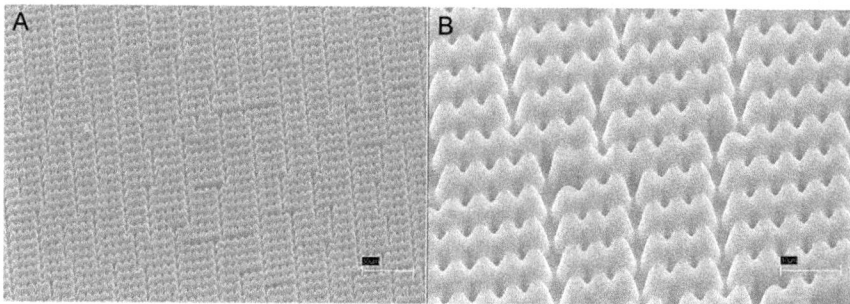

Figure 13.10: Casted PGS with microscale surface features shown in (A), and magnified in (B). The scale shown is 10 and 20 µm respectively [23].

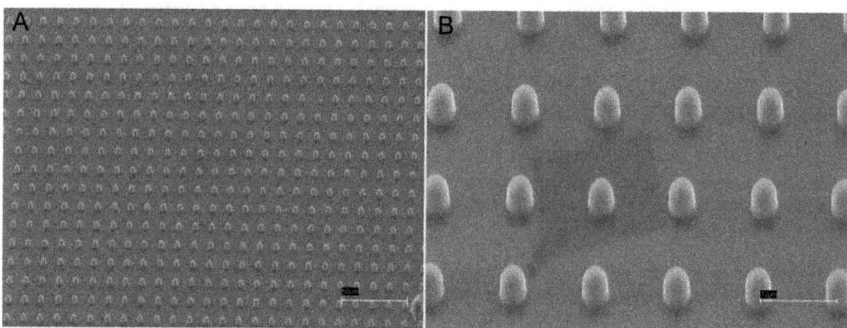

Figure 13.11: PGS with 12 µm center-center array of 3 µm pillars shown in (A) and magnified in (B). The scale shown is 50 and 12 µm respectively [23].

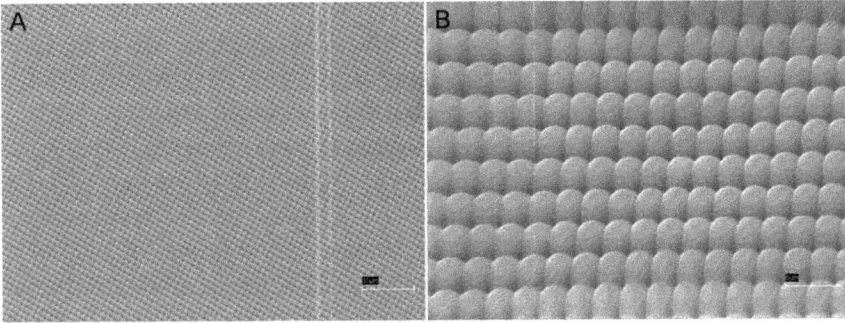

Figure 13.12: Sample caption PGS with 3 μm center-center array of 3 μm pillars shown in (A) and magnified in (B). The scale shown is 20 and 6 μm respectively [23].

Researchers have also prepared PGS in a microporous format, primarily using a salt leaching method [22, 35, 36]. In this method, NaCl particles were leached out from the bulk material matrix via immersion in distilled water. Tracking of complete salt removal is tracked by noting the pH value of the distilled water [36].

13.4 Human mesenchymal stem cell culture on PGS

PGS has tunable mechanical properties that depend on its degree and duration of cure, as well the molarity ratio between its sebacic acid and glycerol constituents. Human mesenchymal stem cells (hMSCs) were cultured *in vitro* on PGS at varying bulk modulus (0.12, 1.11, and 2.30 MPa). It is noted, however, through degradation studies, soft and lightly crosslinked PGS may degrade at a fast rate in simulated bodily fluid [22]. It has been remarked that acidic degradation products of very soft PGS are cytotoxic [37]. This limits the duration of *in vitro* cell culture on such substrates, beyond a certain time span. Other researchers have also looked at human cardiac mesenchymal stem cell culture on PGS membranes [36].

13.4.1 Cell proliferation

A study reported hMSC morphology and actin cytoskeletal organization on these PGS substrates [23]. The cells on the softest substrate exhibited a branched and filopodia-rich morphology, with a contracted, raised, and rounded cell body whereas the cells grown on the stiffest film became confluent and exhibited a striated, aligned growth. f-actin fibers were found to be more pronounced and stretched on the glass coverslips and stiffer films, as compared to the softest film which exhibited a diffuse cytoskeleton. The stem cell marker, stro-1, was expressed by the control hMSCs, but progressively disappeared on the various material substrates. These reported results demonstrate that the mechanical cues provided by PGS affect hMSC morphology and actin cytoskeletal organization. It also appears that the material property and substrate elasticity of PGS may play a part in the commitment of stem cells to specific lineages.

For cell culture, it should be noted that PGS cured at a low temperature (120 °C for 24 h), resulting in an elastic modulus of 0.07 MPa results in a sticky elastomer. This is likely because of unreacted monomers in the material, which could be addressed via washing of the elastomer with sequentially decreasing v/v% solutions of ethanol in dH_2O. This will act to leach the unreacted molecules out of the network [32]. Others have also suggested that the amount of unreacted oligomers and monomers is inversely proportional to curing time, and that ethanol can be used as a leaching agent [25]. However, such an approach leads to shrinkage and warpage of the substrates. To ensure a complete reaction, the cure duration could be extended. This may cause an unintended increase in the elastic modulus. Other methods to address this issue may be via increasing the molar ratio of glycerol to sebacic acid. This however may result in a reduction of unreacted sebacic acid recrystallization that occurs upon cool-down. For the purpose of cell culture, leaching of unreacted monomers may be achieved via pre-conditioning the material by immersing it for long durations in cell culture media, before cell seeding [38].

Figure 13.13 shows the differences in hMSC proliferation on a glass control, and PGS substrates cured at 120, 140, and 165 °C with corresponding substrate elasticities of 0.12, 1.11, and 2.30 MPa, over a 9-day period. The results indicate that the stiffer PGS substrates (relative to the softest PGS 120) result in comparable cell proliferation to that of glass. The maximum count may be because of cell confluence. By observing cell counts on Days two and six, before confluency, it does appear that the hMSCs proliferate at a faster rate on PGS 140 and PGS 165.

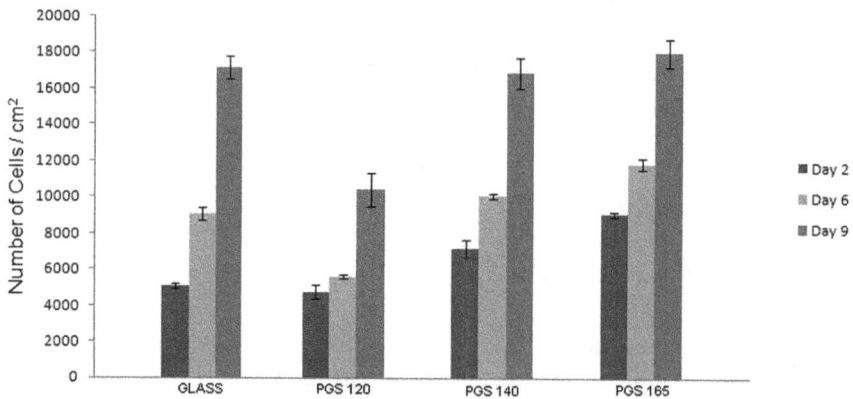

Figure 13.13: Count of hMSCs on samples at days two, six, and nine from the start of cell seeding. The results were extrapolated to per cm^2 from observed data over a 750 × 750 µm region [23].

13.4.2 Phase contrast microscopy

The phase-contrast images shown in Figure 13.14 indicate that the hMSCs proliferate and grow as expected on glass. Upon reaching confluency, the cells appear striated and aligned. Similar observations were also made by others with hMSCs on glass and basal media [39]. The same general observation is made for hMSCs growing on PGS 165. The hMSCs on PGS 140 also appear slightly striated but to a relatively reduced degree compared to those on glass and PGS 165. However, a stark difference is observed for hMSCs cultured on the softer PGS 120. The cells appear to develop long, branched, and thin filopodia. Others have made similar observations of hMSCs on very soft substrate matrices [40].

13.4.3 Scanning electron microscopy images

Figures 13.15–13.17 show representative SEM images of hMSCs cultured on PGS 120, PGS 140, and PGS 165 that were fixed after nine days of incubation. There are fewer cells cultured on the PGS 120 (Figure 13.15) as compared to PGS 140 (Figure 13.16) and PGS 165 (Figure 13.17). In Figure 13.15C, pronounced filopodia are observed (black arrowheads), which are longer than 10 µm. In contrast, the filopodia of hMSCs on PGS 140 and PGS 165 are relatively shorter. On PGS 120, these longer cell extensions appear to be probing the substrate or are being used for motion, which may be indicative of increased activity in protein families such as Cdc42. Some spreading of the lamellipodia can also be observed

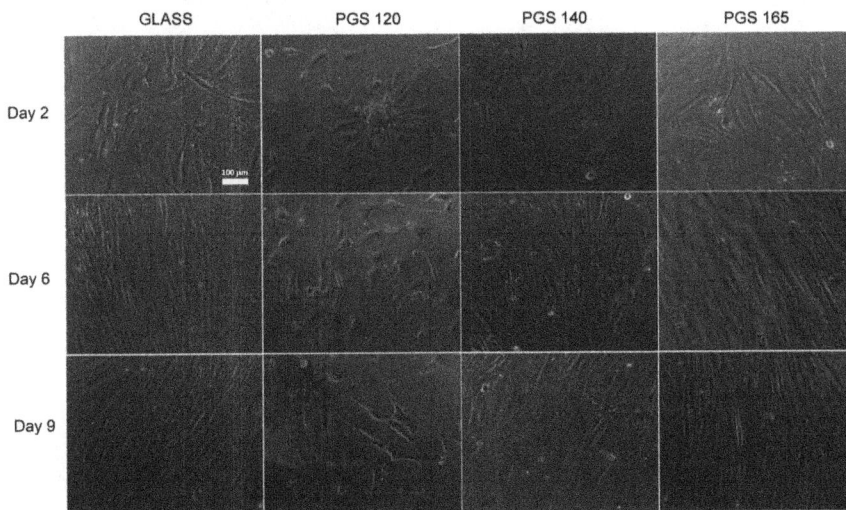

Figure 13.14: Phase-contrast images of hMSCs on glass and PGS, all images are at equal magnification. The scale bar at the upper left indicates 100 µm [23].

for hMSCs on PGS 120 (Figure 13.15C, solid white arrowheads). Close observation also reveals tiny membrane ruffles in the lamellipodia region, which may be an indicator of inefficient adhesion to the substrate. It has been proposed that lamellipodia that do not form stable attachments to the substrate may detach and retract toward the cell body, and this, in turn, results in membrane ruffles [41]. The hMSCs on PGS 120 appear raised and rounded, which may be a result of the unstable lamellipodia attachments. In contrast to this, the cells on PGS 140 in particular have a relatively more spread out appearance whereas those on PGS 165 are highly striated. Inspection of the hMSCs spreading on the stiffer substrate reveals a dense network of stretched actin stress fibers coursing within the cell body (Figure 13.18).

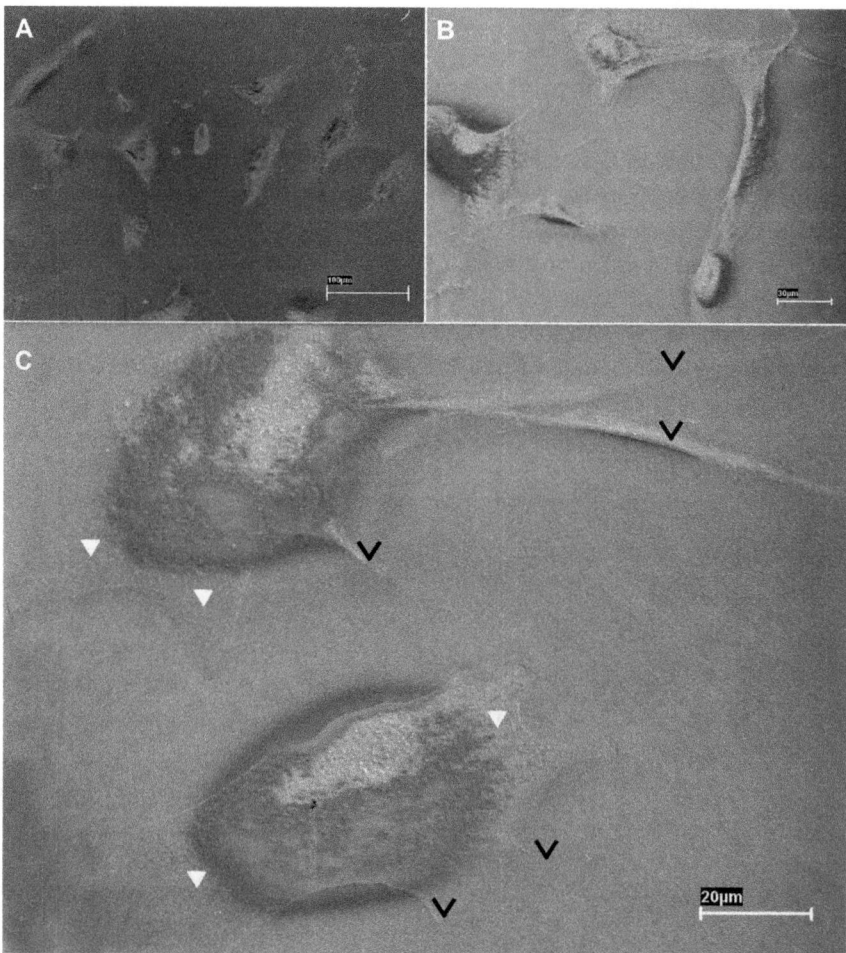

Figure 13.15: SEM images of hMSCs on PGS 120. The magnified images in (B) and (C) show the cells appear to have a raised morphology, as opposed to being flattened and spread out on the surface. In (C), the solid white arrowheads mark the cell's lamellipodia, and the black arrowhead marks the filopodia. A long filopodium (over 60 μm in length) is observed in (C), extending from the top hMSC. Both hMSCs also have shorter filopodia approximately 10 μm extending around the periphery. The scale bars in (A), (B), and (C) are 100, 30, and 20 μm respectively [23].

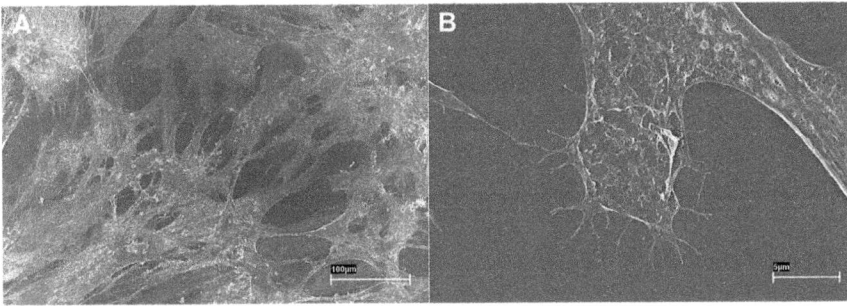

Figure 13.16: SEM images of hMSCs on PGS 140. The cells are relatively more numerous, flat, and spread out as shown in (A). The magnified image in (B) shows detail on hMSC very short (2–3 µm in length) filopodia cell extensions that appear to probe the PGS surface. The scale bars in (A) and (B) are 100 and 5 µm, respectively [23].

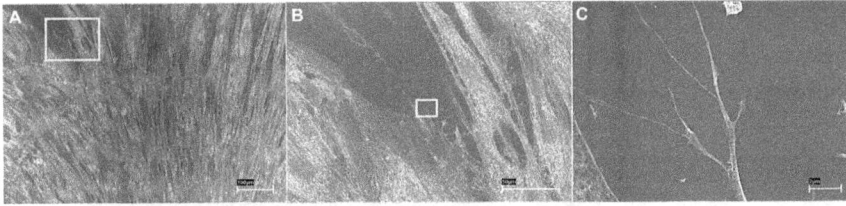

Figure 13.17: SEM images of hMSCs on PGS 165. The cells appear to be even more numerous, spread out, robust, and striated. Image (B) is magnified from the rectangular region in the image (A). Image (C) is magnified from the rectangular region in the image (B), showing thin nano-scale filopodia that appear to probe the PGS surface. The scale bars in (A), (B), and (C) are 100, 50, and 2 µm, respectively [23].

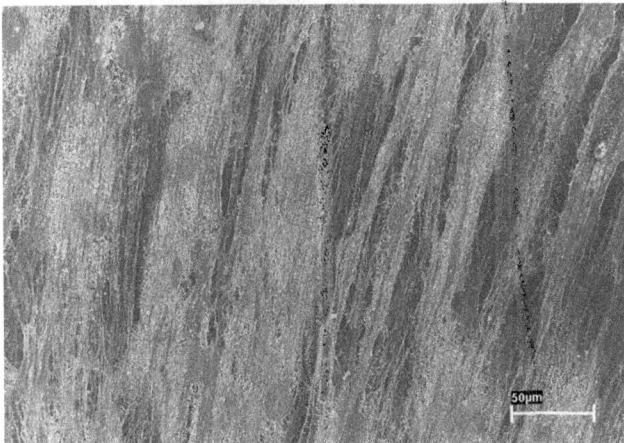

Figure 13.18: hMCs on PGS 165 show a highly striated and dense network of filaments [23].

13.4.4 Immunocytochemistry (ICC) and cytoskeletal examination

Figure 13.19 shows immunofluorescent images of *f-actin*-stained hMSCs. These images indicate that the hMSC actin cyctoskeleton responds differently to substrate stiffness. The *f-actin* filaments on glass (Figure 13.19A) and PGS 140 (Figure 13.19B) and 165 (Figure 13.19D) appear to be stretched stress bundles. Typically, this is correlated to activation of the Rho protein. The actin on PGS 120 (Figure 13.19B) appears relatively diffused, which may be an indicator that these cells have not formed strong focal adhesion sites with the substrate.

On PGS 165, the dense network of *f-actin* stress fibers appears to form a criss-cross mesh. In comparison, those on glass and PGS 140 appear long and parallel. Others have also observed this criss-cross mesh pattern exhibited by hMSCs that have undergone osteogenic differentiation [42]. A representative image that provides detail of this dense network of actin stress fibers is provided in Figure 13.20. These stress fibers are observed to be less pronounced on the softer PGS 120. From SEM observations, the cells are also observed to be raised and round, compared to being flat and spread out on the stiffer substrates. This may account for the diffused actin structure on PGS 120, and these results indicate that organization of the actin cytoskeleton may be influenced by the PGS substrate bulk modulus.

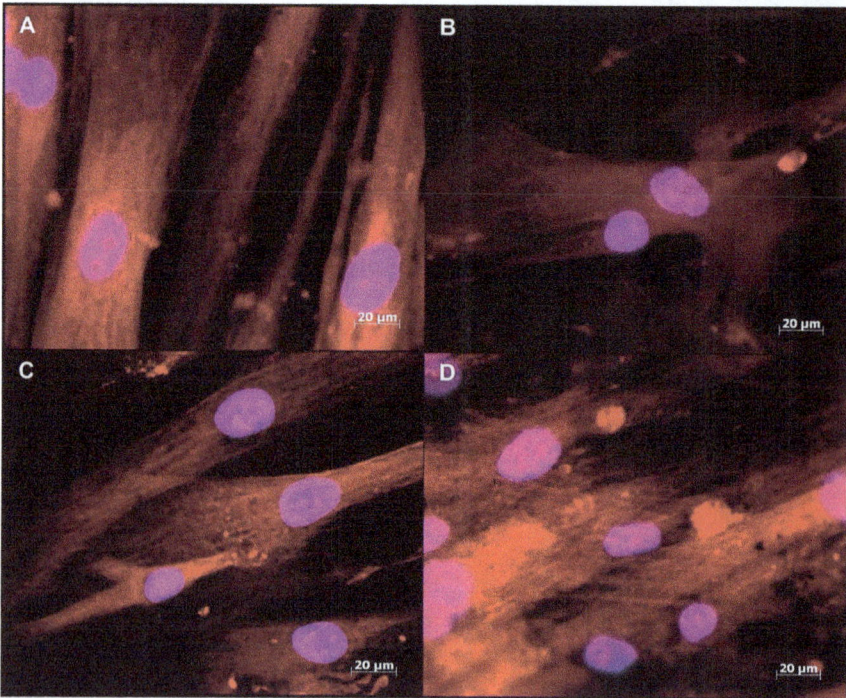

Figure 13.19: Immunofluorescent images of *f-actin*-stained (red) with nuclei (magenta) hMSCs on (A) glass, (B) PGS 120, (C) PGS 140, and (D) PGS 165 [23].

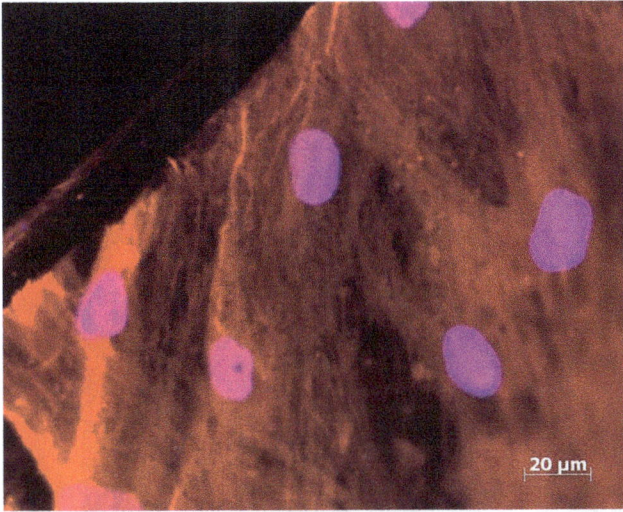

Figure 13.20: Dense network f-actin stress bundles (red) with nuclei (magenta) on PGS 165 [23].

The cells were also examined for stem cell character using the *Stro-1* hMSC surface marker, fluorescently tagged with a secondary antibody (anti-mouse IgM). The results are shown in Figure 13.21. Figures 13.21A, C and 13.21D are for the glass, PGS 140, and PGS 165 respectively. The cells on the glass control show that the hMSCs still have an abundance of hMSC surface receptors for the surface marker molecules to attach to, indicating the hMSCs were still hMSCs. However, the surface marker is progressively fainter on the relatively softer PGS 140 and PGS 165 substrates. Interestingly, it disappears on the softest PGS 120 (Figure 13.21B) substrate. These results appear to indicate that the combination of mechanical and perhaps, material extracellular signals provided by the substrates may have had an effect in initiating the hMSCs to commit to differentiation pathways. It also appears that the softest substrate had the most profound effect in this regard. Investigations in this regard should be furthered in future studies.

13.5 Degradation

Researchers have suggested that PGS scaffolds undergo hydrolytic degradation by scissions of cross-links in the elastomer; the rate of degradation can be tailored by the degree of cross-linking [3]. In fact, with regards to biodegradability, PGS was developed based on the criteria of: (i) being able to undergo hydrolytic degradation to minimize variation in degradation kinetics caused by enzymatic degradation, (ii) having hydrolyzable ester bonds in the structure, (iii) having crosslinks in the polymer chains,

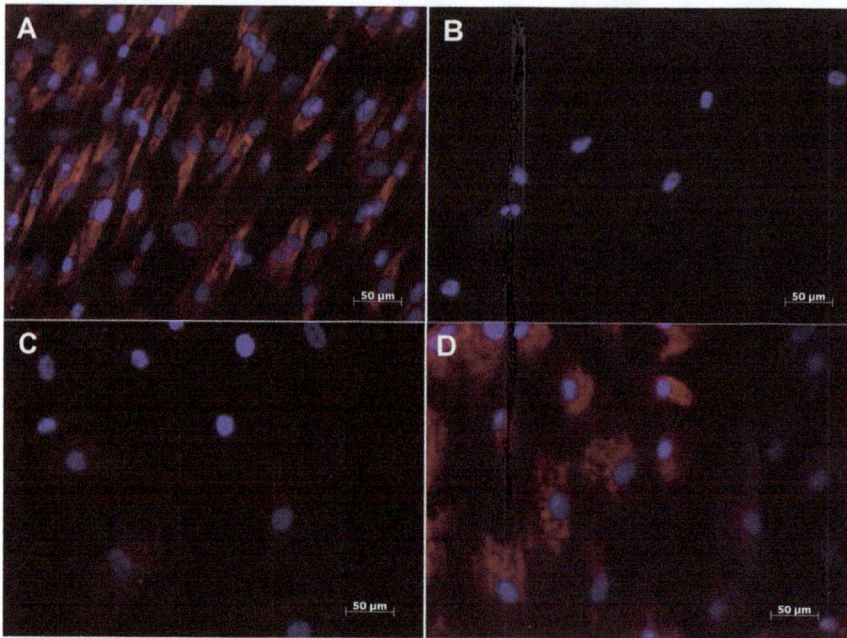

Figure 13.21: Fluorescent images of hMSCs stained with *Stro-1* on (A) glass, (B) PGS 120, (C) PGS 140, and (D) PGS 165 [23].

and that the chemical bonds of these crosslinks should be hydrolyzable and identical to those on the polymer backbone [3].

More recently, Krook et al. found that the porosity and mechanism of degradation ultimately return the PGS porous scaffold material to its pre-polymeric state [22]. This extent of degradation can be inferred from relative comparisons of DSC and ATR-FTIR data with that of the pre-polymer, as discussed and pointed out in Sections 13.2.1 and 13.2.2. Morphological changes because of degradation in simulated body fluid were also reported in that study [22]. The most significant change occurred with PGS 120 cured specimens, in which the integrity of the scaffolds was significantly diminished relative to PGS 140 and 165. Consistent with observations made via thermal (DSC) and spectroscopic (ATR-FTIR) analysis DSC, the final material appears to match that of the viscous pre-polymer.

Others have reported that *in vivo* PGS primarily degrades via surface erosion, where there is a preservation of geometry and intact surface, with retention of overall mechanical strength that does not depend on curing time [13]. However, Sun et al. reported that degradation of PGS, when used as a drug carrier, is faster *in vivo* compared to an *in vitro* comparison study [43]. Liang et al. also noted that PGS-based materials degraded *in vitro* at a faster rate when immersed in a culture medium, relative to a buffered solution at the optimum pH 8 [44]. In this regard, PGS degradation studies have demonstrated that the correlation between PGS degradation behavior *in vivo* and *in vitro* is difficult [3, 4]. It appears, however, that researchers agree that degradation kinetics of PGS cured at a lower temperature is faster compared to those cured at a higher temperature. For example, Chen et al. [8], observed that *in vitro* degradation of PGS cured at 110 °C, is faster compared to that cured at 120 °C.

13.6 Summary

PGS is a biocompatible elastomer, with an elastic modulus that can be tuned depending on cure temperature, duration, and molar ratio of the sebacic acid and glycerol used in its preparation. It undergoes hydrolytic degradation by scissions of crosslinks in the polymer. The rate at which this occurs is dependent on the degree of crosslinking. Degradation ultimately returns PGS back to its pre-polymeric state. The extent of the crosslinking and subsequent degradation can be inferred from relative comparisons of data from thermal analysis (DSC) and spectroscopy (ATR-FTIR). In addition to being biocompatible for cell culture, the ability of PGS to be mechanically tuned, controlled in its degradation makes PGS an attractive material for biomedical applications.

Author contributions: All the authors have accepted responsibility for the entire content of this submitted manuscript and approved submission.
Research funding: None declared.
Conflict of interest statement: The authors declare no conflicts of interest regarding this article.

References

1. Fung YC. Biomechanics: mechanical properties of living tissues. New York: Springer-Verlag; 1993.
2. Meyers MA, Chen P-Y, Lin AY-M, Seki Y. Biological materials: structure and mechanical properties. Prog Mater Sci 2008;53:1–206.
3. Rai R, Tallawi M, Grigore A, Boccacccini AR. Synthesis, properties and biomedical applications of poly(gycerol sebacate) (PGS): a review. Prog Polym Sci 2012;37:1051–78.
4. Wang Y, Ameer GA, Sheppard BJ, Langer R. A tough biodegradable elastomer. Nat Biotechnol 2002; 20:602–6.
5. Guo X-L, Lu X-L, Dong D-L, Sun Z-J. Characterization and optimization of glycerol/sebacate ratio in poly(glycerol-sebacate) elastomer for cell culture application. J Biomed Mater Res A 2014;102: 3903–7.
6. Hollister SJ, Kemppainen JM. Tailoring the mechanical properties of 3D-designed poly(glycerol sebacate) scaffolds for cartilage applications. J Biomed Mater Res A 2010;94:9–18.
7. Jaafar IH, Ammar MM, Jedlicka SS, Pearson RA, Coulter RA. Spectroscopic evaluation, thermal, and thermomechanical characterization of poly(glycerol-sebacate) with variations in curing temperatures and durations. J Mater Sci 2010;45:2525–9.

8. Chen Q-Z, Bismarck A, Hansen U, Junaid S, Tran MQ, Harding SE, et al. Characterisation of a soft elastomer poly(glycerol sebacate) designed to match the mechanical properties of myocardial tissue. Biomaterials 2008;29:47–57.
9. Zhang L-Q, Shi R. Novel elastomers for biomedical applications. Current Topics in elastomers research. Boca Raton: CRC Press; 2008.
10. Martina M, Hutcmacher DW. Biodegradable polymers applied in tissue engineering research: a review. Polym Int 2006;56:145–57.
11. Bhowmick AK. Current topics in elastomers research. Boca Raton: CRC Press; 2008.
12. Sundback CA, Shyu JY, Wang Y, Faquin WC, Langer RS, Vacanti JP, et al. Biocompatibility analysis of poly(glycerol sebacate) as a nerve guide material. Biomaterials 2005;26:5454–64.
13. Wang Y, Kim YM, Langer RS. In vivo degradation characteristics of poly(glycerol sebacate). J Biomed Mater Res A 2003;66A:192–7.
14. Zaky SH, Lee KW, Gao J, Jensen A, Verdelis K, Wang Y, et al. Poly (glycerol sebacate) elastomer supports bone regeneration by its mechanical properties being closer to osteoid tissue rather than to mature bone. Acta Biomater 2017;54:95–106.
15. Zaky SH, Lee K-W, Gao J, Jensen A, Close J, Wang Y, et al. Poly(glycerol sebacate) elastomer: a novel material for mechanically loaded bone regeneration. Tissue Eng 2014;20:1–2.
16. Neeley WL, Redenti S, Klassen H, Tao S, Desai T, Young MJ, et al. A microfabricated scaffold for retinal progenitor cell grafting. Biomaterials 2008;29:418–26.
17. Cai W, Liu L. Shape-memory effect of poly(glycerol-sebacate) elastomer. Matter Lett 2008;62: 2171–3.
18. Motlagh D, Yang J, Lui KY, Webb AR, Ameer GA. Hemocompatibility evaluation of poly(glycerol-sebacate) in vitro for vascular tissue engineering. Biomaterials 2006;27:4315–24.
19. Aydin HM, Salimi K, Rzayev ZM, Piskin E. Microwave-assisted rapid synthesis of poly(glycerol-sebacate) elastomers. Biomater Sci 2013;1:503–9.
20. Tang J, Knowles JC, Bayazit MK, Lau CC. Tailoring degree of esterification and branching of poly(glycerol sebacate) by energy efficient microwave irradiation. Polym Chem 2017;8:3937–47.
21. Gao J, Crapo PM, Wang Y. Macroporous elastomeric scaffolds with extensive micropores for soft tissue engineering. Tissue Eng 2006;12:917–25.
22. Krook NM, Jaafar IH, Sarkhosh T, LeBlon CE, Coulter JP, Jedlicka SS. In vitro examination of poly(glycerol sebacate) degradation kinetics: effects of porosity and cure temperature. Int J Polym Mater Po 2019;69:535–43.
23. Jaafar IH. An investigation of mechanically tunable and nanostructured polymer scaffolds for directing human mesenchymal stem cell development [PhD Dissertation]t. Ann Arbor: ProQuest LLC; 2011.
24. Liu Q, Tian M, Shi R, Zhang L, Chen D, Tian W. Structure and properties of thermoplastic poly(glycerol sebacate) elastomers originating from prepolymers with different molecular weights. J Appl Polym Sci 2007;104:1131–7.
25. Pomerantseva I, Krebs N, Hart A, Neville CM, Huang AY, Sundback CA. Degradation behavior of poly(glycerol sebacate). J Biomed Mater Res A 2009;91:1038–47.
26. Li Y, Huang W, Cook WD, Chen Q. A comparative study on poly(xylitol sebacate) and poly(glycerol sebacate): mechanical properties, biodegradation and cytocompatibility. Biomed Mater 2013;8.
27. Chung DJ, Kim J, Hwang MY, Kim MJ. Biodegradable and elastomeric poly(glycerol sebacate) as a coating material for nitinol bare stent. BioMed Res Int 2014;2014:7.
28. Mitsak AG, Dunn AM, Hollister SJ. Mechanical characterization and non-linear elastic modeling of poly(glycerol sebacate) for soft tissue engineering. J Mech Behav Biomed Mater 2012;11:3–15.
29. Ding X, Chen Y, Chao CA, Wu Y-L, Wang Y. Control the mechanical properties and degradation of poly(glycerol sebacate) by substitution of the hydroxyl groups with palmitates. Macromol Biosci 2020:20.

30. Conejero-Garcia A, Gimeno HR, Saez YM, Vilarino-Feltrer G, Ortuno-lizaran I, Valles-Lluch A. Correlating synthesis parameters with physicochemical properties of poly(glycerol sebacate). Eur Polym J 2017;87:406–19.

31. Bettinger CJ, Kulig EJ, Vacanti JP, Wang Y, Borenstein JT, Langer RS. Three-dimensional microfluidic tissue-engineering scaffolds using a flexible biodegradable polymer. Adv Mater 2005;18:165–9.

32. Bettinger CJ, Orrick B, Misra A, Langer RS, Borenstein JT. Microfabrication of poly (glycerol-sebacate) for contact guidance applications. Biomaterials 2006;27:2258–65.

33. Pashneh-Tala S, Owen R, Bahmaee H, Rekstyte S, Malinauskas M, Claeyssens F. Synthesis, characterization and 3D micro-structuring via 2-photon polymerization of poly(glycerol sebacate)-methacrylate–an elastomeric degradable polymer. Front Physiol 2018;6:41.

34. Massoumi N, Jean N, Jean A, Zugates JT, Johnson KL, Engelmayr GC Jr. Laser microfabricated poly(glycerol sebacate) scaffolds for heart valve tissue engineering. J Biomed Mater Res A 2012; 101A:104–14.

35. Jian B, Wu W, Song Y, Tan N, Ma C. Microporous elastomeric membranes fabricated with polyglycerol sebacate improved guided bone regeneration in a rabbit model. Int J Nanomed 2019; 14:2683–92.

36. Rai R, Tallawi M, Barbani N, Frati C, Madeddu D, Cavalli S, et al. Biomimetic poly(glycerol seacate) (PGS) membranes for cardiac patch application. Mater Sci Eng C 2013;33:3677–87.

37. Liang S-L, Cook WD, Thouas GA, Chen Q-Z. The mechanical characteristics and in vitro biocompatibility of poly(glycerol sebacate)-bioglass elastomeric composites. Biomaterials 2010; 31:8516–29.

38. Chen Q-Z, Ishii H, Thouas GA, Lyon AR, Wright JS, Blaker JJ, et al. An elastomeric patch derived from poly(glycerol sebacate) for delivery of embryonic stem cells to the heart. Biomaterials 2010;31: 3885–93.

39. Rodriguez JP, Gonzalez M, Rios S, Cambiazo V. Cytoskeletal organization of human mesenchymal stem cells (MSC) changes during their osteogenic differentiation. J Cell Biochem 2004;93:721–31.

40. Engler AJ, Sen S, Sweeney HL, Discher DE. Matrix elasticity directs stem cell lineage specification. Cell 2006;126:677–89.

41. Borm B, Requardt RP, Herzog V, Kirfel G. Membrane ruffles in cell migration: indicators of inefficient lamellipodia adhesion and compartments of actin filament reorganization. Exp Cell Res 2005;302:83–95.

42. Yourek G, Hussain MA, Mao JJ. Cytoskeletal changes of mesenchymal stem cells during differentiation. ASAIO J 2007;53:219–28.

43. Sun Z-J, Chen C, Sun M-Z, Ai C-H, Lu X-L, Zheng Y-F, et al. The application of poly(glycerol-sebacate) as a biodegradable drug carrier. Biomaterials 2009;30:5209–14.

44. Liang S-L, Yang X-Y, Fang X-Y, Cook WD, Thouas GA, Chen Q-Z. In vitro enzymatic degradation of poly (glycerol sebacate)-based materials. Biomaterials 2011;32:8486–96.

Index

https://doi.org/10.1515/9781501521942-014